Analytical Chemistry: Emerging Trends

Analytical Chemistry: Emerging Trends

Edited by
Taylor Spencer

WILLFORD PRESS
www.willfordpress.com

Published by Willford Press,
118-35 Queens Blvd., Suite 400,
Forest Hills, NY 11375, USA

ISBN: 978-1-68285-803-5

Cataloging-in-Publication Data

Analytical chemistry : emerging trends / edited by Taylor Spencer.
    p. cm.
Includes bibliographical references and index.
ISBN 978-1-68285-803-5
1. Chemistry, Analytic. 2. Chemistry, Analytic--Technological innovations. 3. Chemistry. I. Spencer, Taylor.
QD75.22 .A53 2020
543--dc23

For information on all Willford Press publications
visit our website at www.willfordpress.com

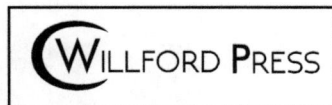

WILLFORD PRESS

Contents

Permissions

List of Contributors

Index

Preface

This book aims to highlight the current researches and provides a platform to further the scope of innovations in this area. This book is a product of the combined efforts of many researchers and scientists, after going through thorough studies and analysis from different parts of the world. The objective of this book is to provide the readers with the latest information of the field.

Analytical chemistry is a branch of chemistry which is concerned with the separation, quantification and identification of matter. Every naturally occurring or artificially produced substance contains a combination of different elements. It is possible to identify and quantify the different types of impure species in a substance with the help of analytic tests. These tests also help in determining the properties of pure materials. Such methods have been improved with the aid of computer technology and sophisticated analytic instruments. Gas chromatography is one such technique which separates the different components of a gaseous mixture. The mass spectrometer is an analytic instrument that separates substances in order of the mass of the constituent ions. Some techniques offering greater sensitivity are polarography, atomic absorption and neutron activation. This book presents the complex subject of analytical chemistry in the most comprehensible and easy to understand language. Some of the diverse topics covered herein address the varied techniques that fall under this category. It will serve as a valuable source of reference for graduate and post graduate students.

I would like to express my sincere thanks to the authors for their dedicated efforts in the completion of this book. I acknowledge the efforts of the publisher for providing constant support. Lastly, I would like to thank my family for their support in all academic endeavors.

Editor

A Highly Sensitive and Selective Colorimetric Hg^{2+} Ion Probe using Gold Nanoparticles Functionalized with Polyethyleneimine

Kyung Min Kim (ID),[1,2] Yun-Sik Nam,[3] Yeonhee Lee,[3] and Kang-Bong Lee (ID)[1]

[1]*Green City Technology Institute, Korea Institute of Science and Technology, Hwarang-ro 14 gil 5, Seoul 02792, Republic of Korea*
[2]*Department of Chemistry, Korea University, Anam-ro, Seongbuk-gu, P.O. Box 145, Seoul 136-701, Republic of Korea*
[3]*Advanced Analysis Center, Korea Institute of Science and Technology, Hwarang-ro 14 gil 5, Seoul 02792, Republic of Korea*

Correspondence should be addressed to Kang-Bong Lee; leekb@kist.re.kr

Academic Editor: Silvana Andreescu

A highly sensitive and selective colorimetric assay for the detection of Hg^{2+} ions was developed using gold nanoparticles (AuNPs) conjugated with polyethyleneimine (PEI). The Hg^{2+} ion coordinates with PEI, decreasing the interparticle distance and inducing aggregation. Time-of-flight secondary ion mass spectrometry showed that the Hg^{2+} ion was bound to the nitrogen atoms of the PEI in a bidentate manner (N–Hg^{2+}–N), which resulted in a significant color change from light red to violet due to aggregation. Using this PEI-AuNP probe, determination of Hg^{2+} ion can be achieved by the naked eye and spectrophotometric methods. Pronounced color change of the PEI-AuNPs in the presence of Hg^{2+} was optimized at pH 7.0, 50°C, and 300 mM·NaCl concentration. The absorption intensity ratio (A_{700}/A_{514}) was correlated with the Hg^{2+} concentration in the linear range of 0.003–5.0 μM. The limits of detection were measured to be 1.72, 1.80, 2.00, and 1.95 nM for tap water, pond water, tuna fish, and bovine serum, respectively. Owing to its facile and sensitive nature, this assay method for Hg^{2+} ions can be applied to the analysis of water and biological samples.

1. Introduction

Mercury ion (Hg^{2+}) is ubiquitously distributed in the environment, and it is considered to be one of the major environmental pollutants to be widely used in industry, agriculture, and medicine. It is nonessential and toxic to the human body. Hg^{2+} is considered to be one of the major environmental pollutants to be widely used in industry, agriculture, and medicine. This mercury ion exists in inorganic and organic mercury ions. Upon entering the body, inorganic mercury ions are accumulated mainly in the kidneys and give rise to vomiting and diarrhea, followed by hypovolemic shock, oliguric renal failure, and possibly death [1–4]. Organic mercury ions such as methylmercuric (MeHg$^+$), ethylmercuric (EtHg$^+$), and phenylmercuric (PhHg$^+$) ions can also cause injuries at the central nervous system and lead to paresthesias, headaches, ataxia, dysarthria, visual field constriction, blindness, and hearing impairment [5–7]. Therefore, detection of mercury ion in various sample matrices has been an urgent issue.

Several analytical techniques, such as direct mercury analyzer (DMA) [8], ion chromatography (IC) [9], and high performance liquid chromatography (HPLC) [10, 11], have been utilized to detect mercury ions. However, these ways generally require complicated sample pretreatment process, skillful technicians, and sophisticated instrumentations. Therefore, the low-cost and facile analytical method for selective detection of mercury ions remains to be a challenge for analytical chemists.

Various nanoparticle assays for a simple detection of Hg^{2+} ions have recently been investigated using gold nanorods (AuNRs) [12], carbon nanoparticles (CNPs) [13], silver nanoparticles (AgNPs) [14], silver nanoprisms (AgNPRs) [15], and AuNPs [16–18]. These colorimetric methods are especially promising in the analysis of Hg^{2+} with their naked eye or UV-Vis applications, due to their high extinction coefficients and the interparticle distance-dependent optical properties.

Polyethyleneimine (PEI) was applied to use a chemical functionalizer as a harmless gene delivery mediator, templates, stabilizers, and molecular gum to arrange metal nanoparticles

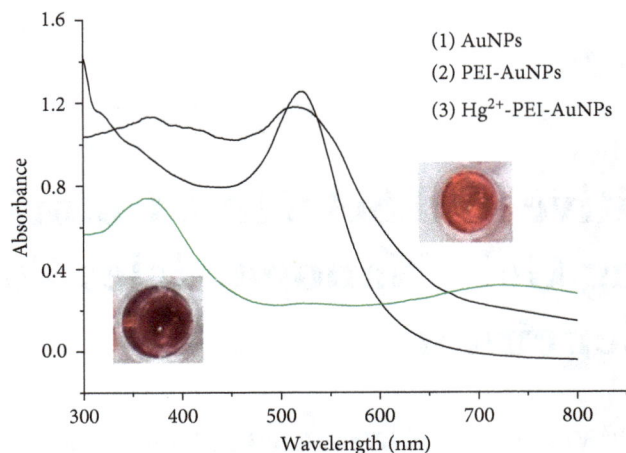

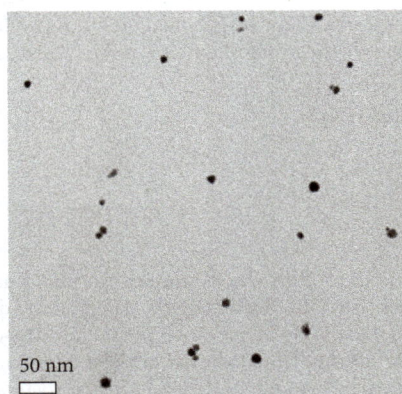 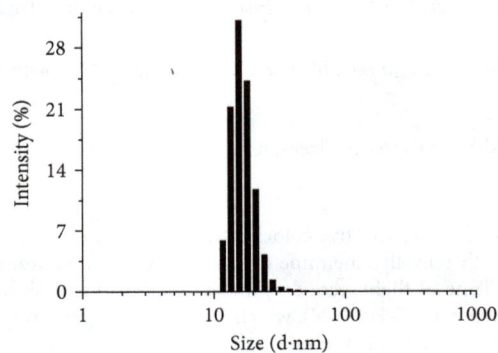

(b)

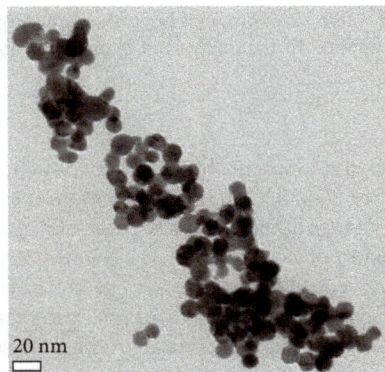 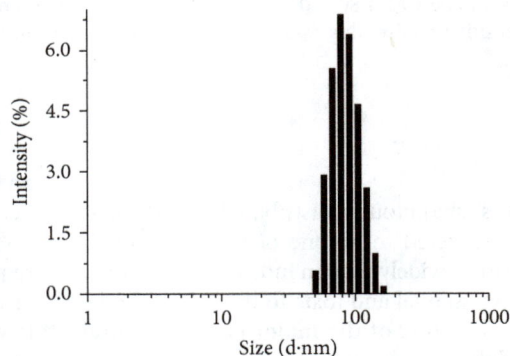

(c)

FIGURE 1: (a) UV-Vis absorption spectra of (1) AuNPs (black line), (2) PEI-AuNPs (red line), and (3) PEI-AuNPs conjugated with Hg^{2+} (green line). (b) TEM image of PEI-AuNPs (left) and the corresponding particle-size distribution histogram (right). (c) TEM image of PEI-AuNPs conjugated with Hg^{2+} (left) and particle-size distribution histogram (right).

[19–22]. Various amine groups could give sufficient active sites for strong combining capability, and these characteristics of PEI can be useful to control the selectivity of different ions. For its application, AuNPs conjugated with PEI (PEI-AuNPs) have been utilized as delivery of drug and gene in breast cancer therapy [23].

This study showed that PEI-AuNPs were aggregated with inorganic and organic mercury ions, and these ions induced the definite color change of AuNPs selectively among other diverse ions. Dispersed and aggregated AuNPs were characterized by ultraviolet-visible spectroscopy (UV-Vis), high-resolution transmission electron microscopy (HR-TEM), and dynamic light scattering (DLS) upon addition of Hg^{2+}. Hg^{2+} ion binding sites on the surface of PEI-AuNPs were elucidated by ^{13}C nuclear magnetic spectroscopy (^{13}C NMR), X-ray photoelectron spectroscopy (XPS), and time-of-flight secondary ion mass spectrometry (TOF-SIMS) [24, 25]. The

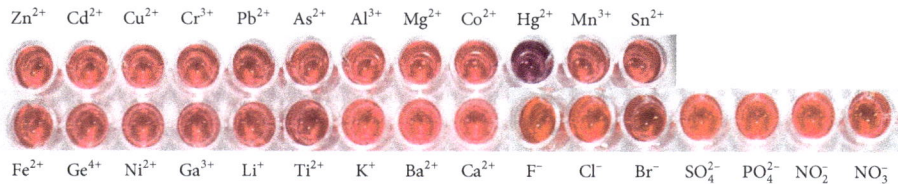

(a)

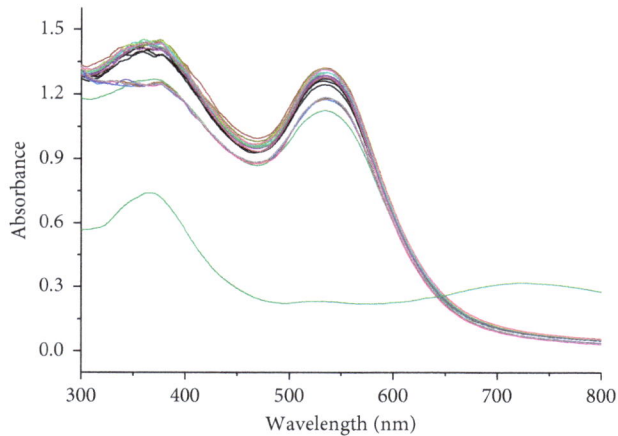

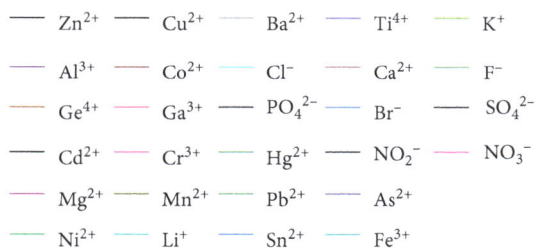

(b)

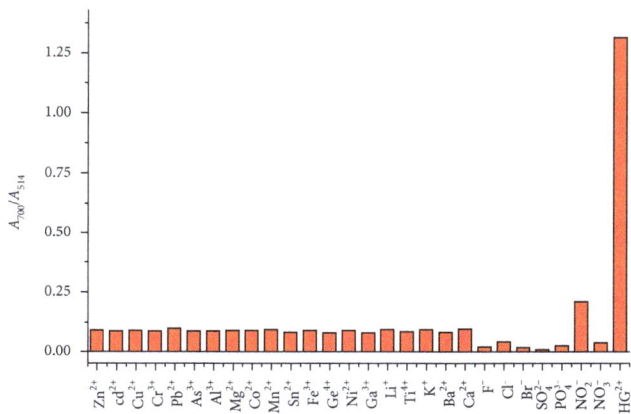

(c)

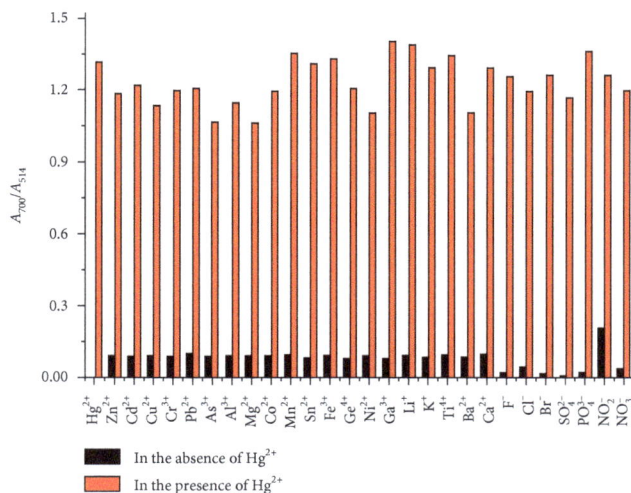

(d)

FIGURE 2: (a) Photographic images, (b) UV-Vis absorption spectra, (c) absorption ratios (A_{700}/A_{514}) of PEI-AuNPs with $5.0\,\mu M\cdot Hg^{2+}$ ion and $50\,\mu M$ various ions at pH 7, 50°C, and 500 mM·NaCl concentration, and (d) absorption ratios (A_{700}/A_{514}) of PEI-AuNPs with $50\,\mu M$ various ions in the presence and absence of $5.0\,\mu M\cdot Hg^{2+}$.

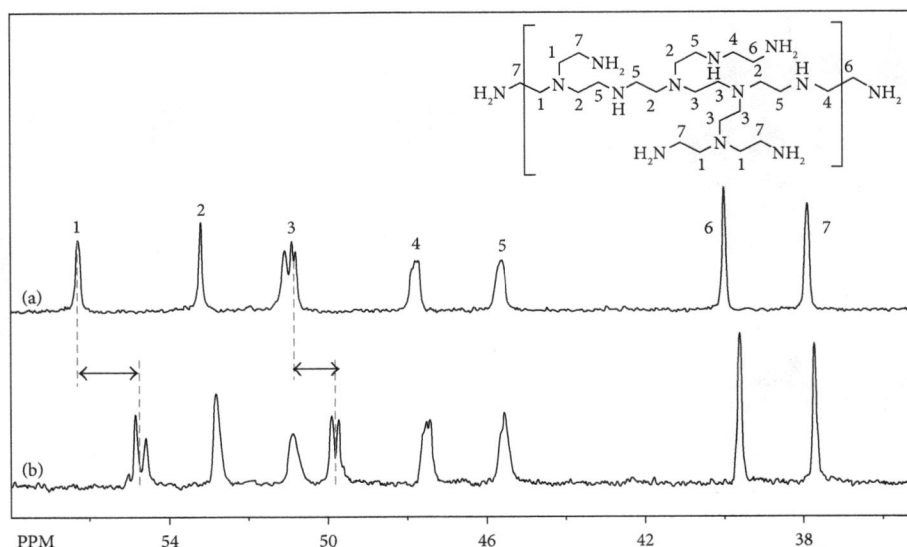

FIGURE 3: ^{13}C NMR spectrums for (a) PEI and (b) PEI-AuNPs in ^2H_2O at room temperature.

interference effects were tested in the presence of other metal ions and anions. Also, PEI-AuNP assay method for detection of Hg^{2+} ion was optimized in terms of pH, temperature, and salt condition.

This present assay using PEI-AuNP was very simple, cost-efficient, and allowed for the on-site detection of Hg^{2+} in real time. The limit of detection (LOD) was ~2 nM in various samples. Therefore, this technique could be utilized to monitor Hg^{2+} ions in a wide range of practical samples.

2. Experimental

2.1. Materials. Gold (III) chloride trihydrate ($HAuCl_4 \cdot 3H_2O$), PEI, methylmercuric chloride, and phenylmercuric chloride were sourced from Sigma-Aldrich (St. Louis, MO, USA). Ethylmercuric chloride was obtained from Chem Service (Tower Lane, West Chester, USA) Salts of NO_3^-, NO_2^-, PO_4^{3-}, SO_4^{2-}, Br^-, Cl^-, F^-, Ca^{2+}, Cd^{2+}, Fe^{3+}, Ba^{2+}, Mn^{2+}, Ga^{3+}, Ti^{4+}, Al^{3+}, Mg^{2+}, K^+, Ge^{4+}, Cr^{3+}, Cu^{2+}, Li^+, As^{3+}, Co^{2+}, Sn^{2+}, Pb^{2+}, Hg^{2+}, Ni^{2+}, and Zn^{2+} were purchased from AccuStandard (New Haven, CT, USA). NaCl, HCl, and NaOH were purchased from Samchun Chemical (Gyeonggi-Do, Korea). Tap water was acquired from our laboratory and pond water was obtained from a pond at the Korea Institute of Science and Technology (KIST). Tuna fish was sourced from Dongwon (Seoul, Republic of Korea). Bovine serum was purchased from Sigma-Aldrich (St. Louis, MO, USA). Citrus leaf sample was obtained from Swan Leaf Pty Ltd (Perth, WA, Australia). Distilled water was obtained using a Milli-Q water purification system (Millipore, Bedford, MA, USA).

2.2. Preparation of PEI-AuNPs. PEI-AuNPs were synthesized as following literature procedures by mixing aqueous solutions of $HAuCl_4 \cdot 3H_2O$ (1 mL, 0.025 M) and PEI (7.8 mL, 9.74 mM), with subsequent reduction of $HAuCl_4$ at ~pH 7.0 [26]. The mixture was kept to react for three days at ambient temperature, and ~10 nm PEI-AuNPs were produced.

2.3. Sample Preparation and Hg^{2+} Sensing Test Using PEI-AuNPs. The suspended particles in all water samples were removed by a syringe filter (0.20 μm pore size) prior to analysis, and sample aliquots (9 mL) were mixed with a 500 μM$\cdot Hg^{2+}$ solution (1 mL) to produce a 50 μM$\cdot Hg^{2+}$ stock solution.

Tuna fish was obtained from a local supermarket. Its muscle tissues were crushed and dried on petri dishes overnight at 85°C. A small portion (ca. 0.3 g) was incubated in 2 mL HNO_3 for 1 h before addition of 0.5 mL $HClO_4$ (70%). Then, the samples were irradiated under a UV lamp for 3 h to convert all possibly contained organic mercury to inorganic mercury. Finally, the acid extracts were transferred to 50 mL volumetric flasks, and the volume was adjusted to 50 mL with Milli-Q water [27]. These samples were spiked with 100 μg$\cdot mL^{-1}$ of Hg^{2+}.

Stock solution of bovine serum was made to be 0.1% concentration in water. The suspensions were stirred and centrifuged alternately for 30 min and the solutions (9 mL) were blended with 1 mL of a 100 μg$\cdot mL^{-1}$ Hg^{2+} solution.

0.9 g of citrus leaf was added to 5 mL concentrated HNO_3 and heated for 2 h in the boiling water bath. After being cooled to room temperature, the samples were added with 2 mL of 30% H_2O_2, followed by heating for 1 h in the boiling water bath. Finally, the sample volume was made to 25 mL with double-distilled water. Before conducting experiments, the pH value of samples solution was neutralized by solid NaOH [28].

To evaluate the utility of our proposed method, Hg^{2+} concentrations added in tap, pond water, tuna fish, and bovine serum were measured with 1 mL of the PEI-AuNP solution, and followed by UV-Vis spectrophotometry. Those analytical results were confirmed with DMA.

2.4. Instrumentation. The absorption spectra were recorded by UV-Vis spectrophotometer (S-3100, Sinco, Seoul, Republic of Korea). UV-Vis spectra were acquired in the range

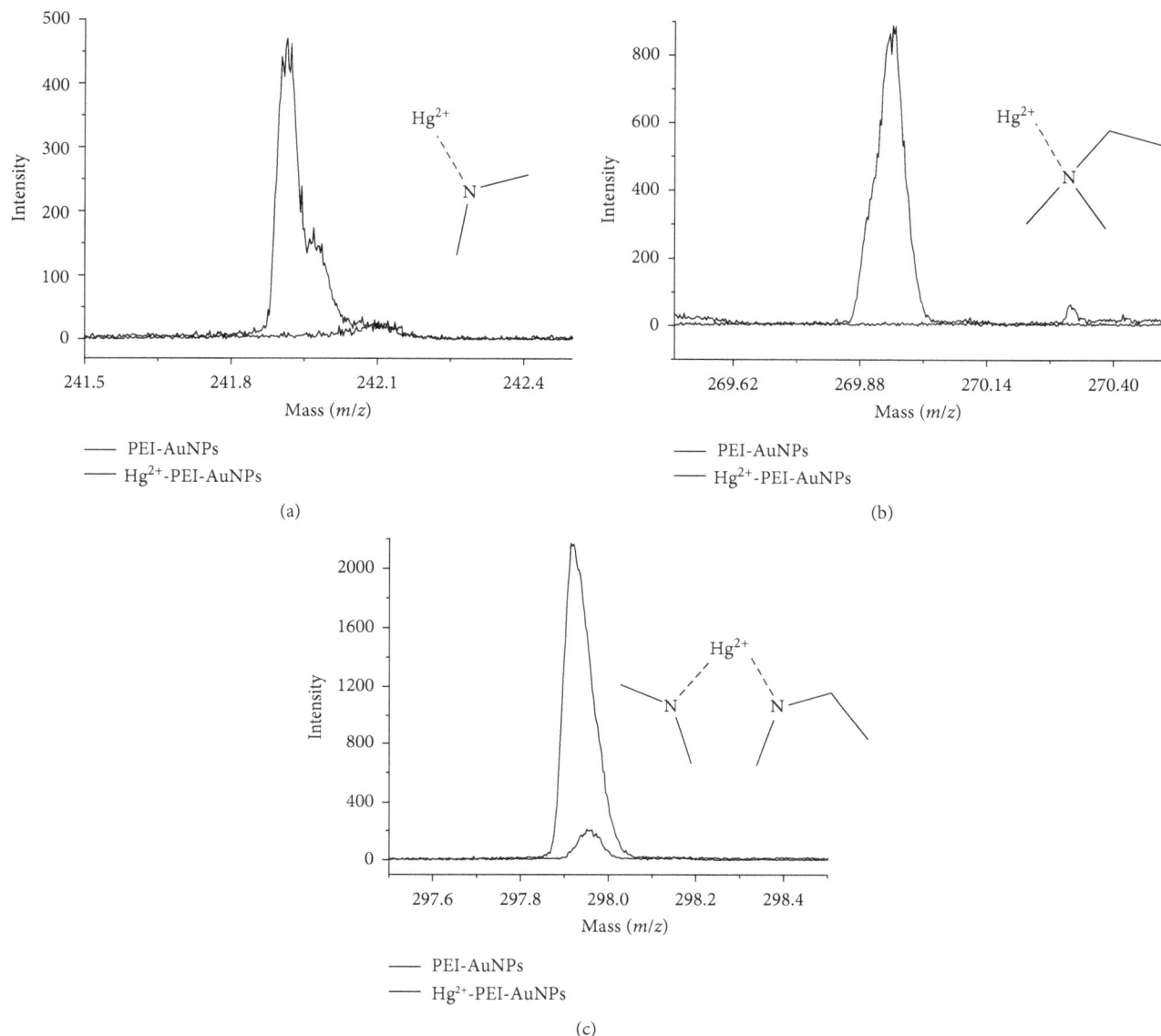

FIGURE 4: Mass peaks for (a) $(CH_2)_2NHg^{2+}$ (m/z: 242), (b) $CH_2CH_2(CH_2)_2NHg^{2+}$ (m/z: 270), and (c) $(CH_2)_2NHg^{2+}N(CH_2)CH_2CH_2$ (m/z: 298) fragments in TOF-SIMS spectra of PEI-AuNPs (black) and PEI-AuNPs-Hg^{2+} (red). These molecular fragments are expected based on PEI-AuNPs-Hg^{2+} structural elements in the zoomed circle in Scheme 1.

of 300–800 nm by using 4 mm path length quartz cells. The pH measurements were conducted with an HI 2210 pH meter (Hanna Instruments, Woonsocket, RI, USA). The concentrations of Hg^{2+} ions in various samples were measured by DMA (DMA 80, Milestone, Italy). ^{13}C NMR spectra were measured on an Avance III 400 MHz ^1H NMR spectrometer (Bruker, Billerica, MA, USA). XPS analysis was performed using a PHI 5000 VersaProbe III instrument (ULVAC-PHI, Chigasaki, Japan). Mass spectra were measured using TOF-SIMS (TOF-SIMS 5, ION-TOF, Münster, Germany). The size distributions of nanoparticles were recorded by a Zetasizer (Malvern Instruments Ltd., Worcestershire, UK). The images and sizes of PEI-AuNPs and their Hg^{2+}-induced aggregates were measured on a micrograph using transmission electron microscope (TEM; CM30, Philips, NC, USA). TEM

samples were obtained by settling the scattered AuNPs and evaporating the solvent.

3. Results and Discussion

3.1. Characterization of PEI-AuNPs and Their Complexes with Hg^{2+}. PEI-AuNPs were prepared as ~10 nm size as reducing $HAuCl_4$ with amine group of PEI. The AuNPs were usually synthesized by the citrate reduction of $HAuCl_4$, and their sizes were ~33 nm as described in earlier studies [29], but AuNPs conjugated with PEI (PEI-AuNPs) under these conditions became much smaller. As a result, the mean AuNP size depended on the quantity and type of reducing agents, pH, temperature, and reaction time [30, 31]. A strong localized surface plasmon resonance (LSPR) peak of these

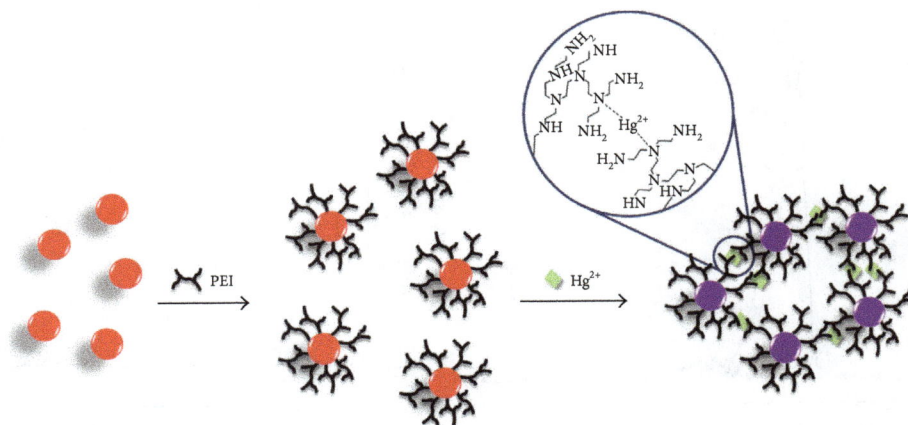

SCHEME 1: Schematic illustration of the AuNPs capped with PEI, and the aggregation of PEI-AuNPs reacted with Hg^{2+} ion, accompanied by a color change, and the predicted coordination bond between Hg^{2+} ions and PEI-AuNPs.

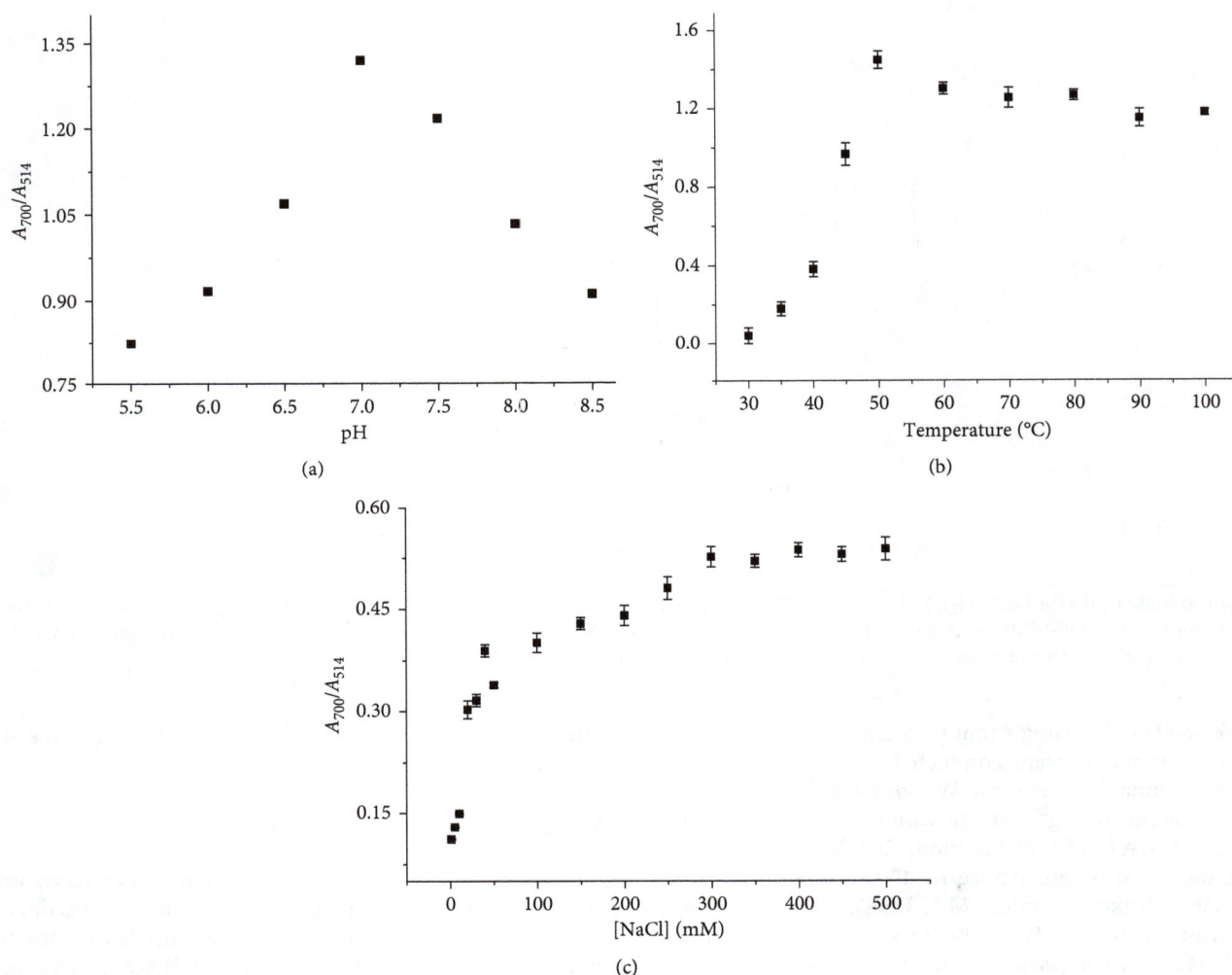

(a)

(b)

(c)

FIGURE 5: Absorption ratios (A_{700}/A_{514}) of Hg^{2+}-PEI-AuNPs as a function of (a) pH, (b) temperature, and (c) concentration of NaCl.

label-free AuNPs appeared at ca. 514 nm in their UV-Vis spectrum, resulting in the red color of the corresponding solution. The size of AuNPs influenced to change their

surface plasmon absorption maxima at 514 nm [32]. The sizes of PEI-AuNPs and Hg^{2+}-PEI-AuNPs were distributed to be ~15 and ~75 nm, respectively, according to TEM

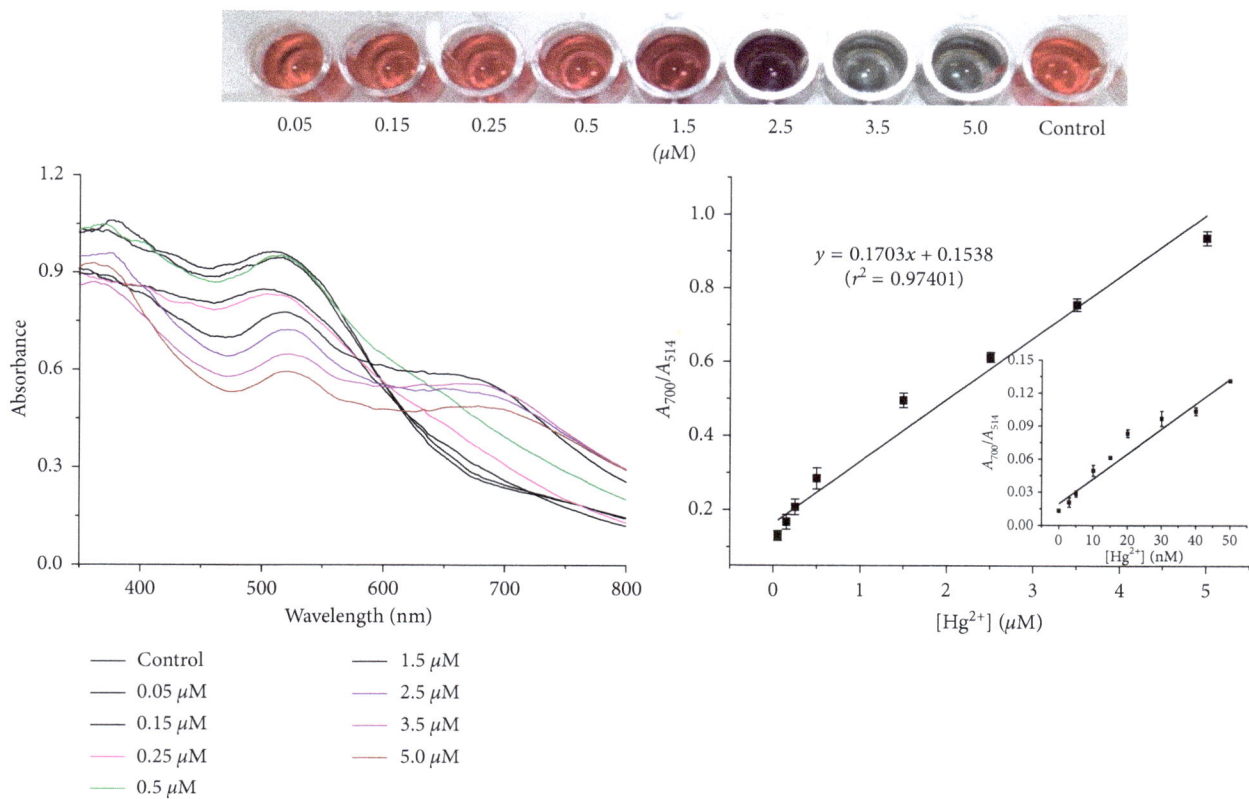

(a)

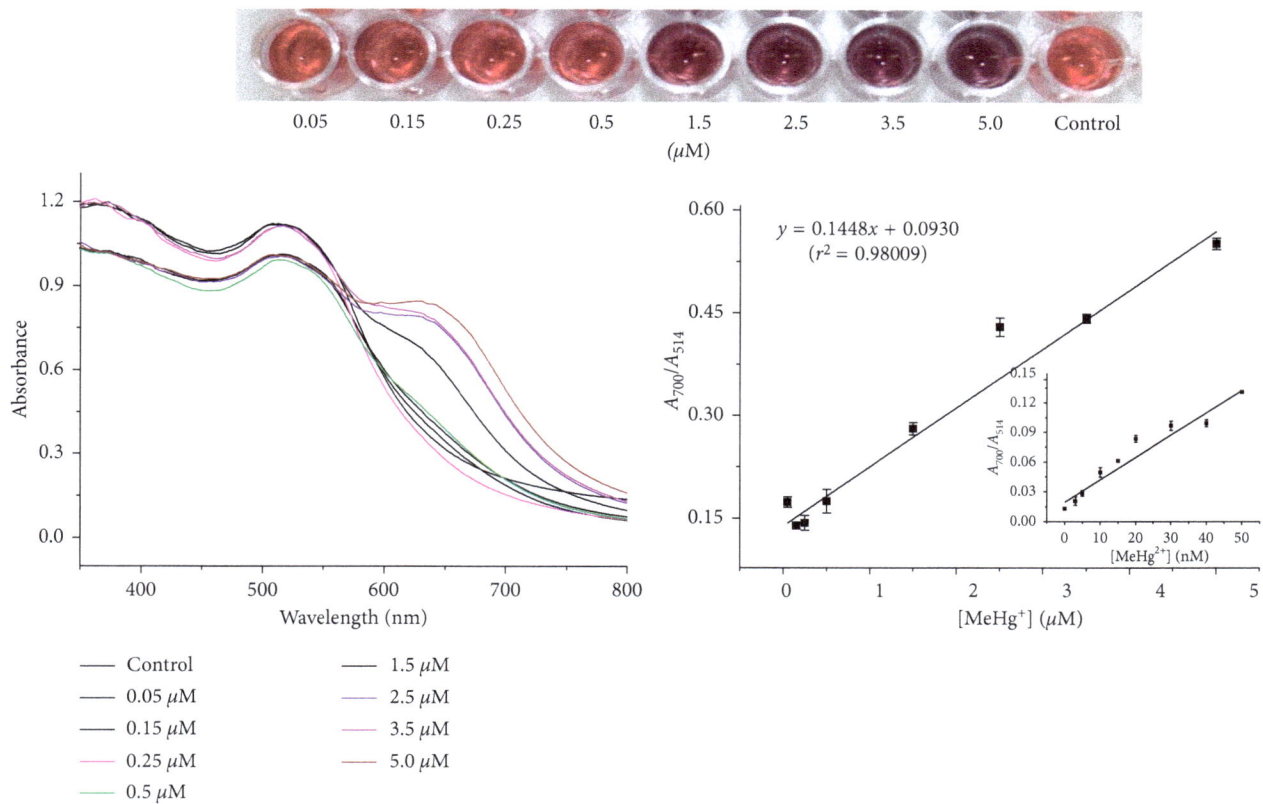

(b)

FIGURE 6: Continued.

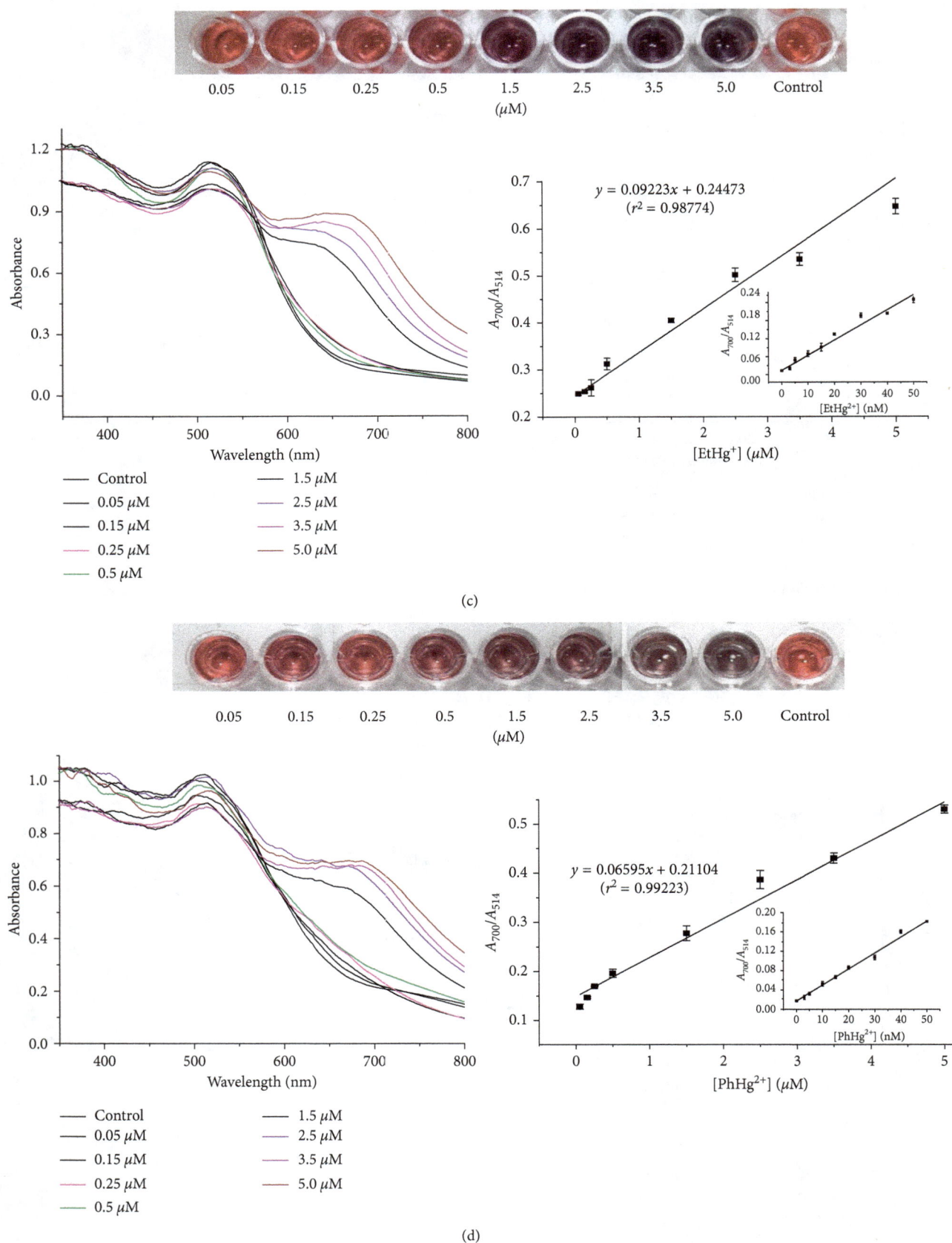

FIGURE 6: Photographic images, UV-Vis absorption spectra, and absorption ratios (A_{700}/A_{514}) of the color change of PEI-AuNPs upon addition of (a) Hg^{2+}, (b) $MeHg^+$, (c) $EtHg^+$, and (d) $PhHg^+$ with various concentrations (0.05, 0.15, 0.25, 0.5, 1.5, 2.5, 3.5, and 5.0 μM from left to right) in the presence of 300 mM·NaCl. Inset: plot of A_{700}/A_{514} versus (a) Hg^{2+}, (b) $MeHg^+$, (c) $EtHg^+$, and (d) $PhHg^+$ ion concentration (0–50 nM).

images and Zetasizer measurements (Figures 1(b) and 1(c)). The color of these PEI-AuNPs was similar to those of label-free AuNPs, but distinct color change occurred from red to dark violet upon addition of Hg^{2+} ions. UV-Vis absorption spectra for AuNP, PEI-AuNP, and Hg^{2+}-PEI-AuNP solutions are demonstrated in Figure 1(a). Upon addition of Hg^{2+} ions, the strong absorption band of PEI-AuNPs at 514 nm was gradually shifted to 700 nm, and a new absorbance band concomitantly increased in intensity upon addition of Hg^{2+} ions (Figure 1(a)). When PEI-AuNPs are aggregated, the conduction electrons near their surfaces become delocalized and are shared amongst neighboring particles. As a result, the surface plasmon resonance (SPR) shifts to lower energies, causing the shift of absorption and scattering peaks to longer wavelengths.

3.2. Selectivity of PEI-AuNPs for Hg^{2+} Ions and Related Interference Effects.

The selectivity for PEI-AuNP assay method was tested in $50\,\mu M \cdot Hg^{2+}$ and various $500\,\mu M$ metal cations (Zn^{2+}, Cd^{2+}, Cu^{2+}, Cr^{3+}, Pb^{2+}, As^{3+}, Al^{3+}, Mg^{2+}, Co^{2+}, Mn^{2+}, Sn^{2+}, Fe^{3+}, Ge^{4+}, Ni^{2+}, Ga^{3+}, Li^+, Ti^{4+}, K^+, Ba^{2+}, and Ca^{2+} ions) and anions (F^-, Cl^-, Br^-, SO_4^{2-}, PO_4^{3-}, NO_2^-, and NO_3^- ions). The interference by other $500\,\mu M$ numerous anions and cations for the selectivity of Hg^{2+} was further examined. Any metal cations and anions did not induce any color changes except Hg^{2+}, as shown in Figure 2(a). UV-Vis absorption spectra of PEI-AuNPs solutions at pH 7, 50°C, and $300\,mM \cdot NaCl$ concentration were recorded in the presence of various metal cations and anions (Figure 2(b)). The strong absorption band at 700 nm distinctly appeared for Hg^{2+}, enabling to discriminate easily from different metal cations and anions. The absorbance ratios (A_{700}/A_{514}) of the PEI-AuNPs solution upon addition of each cation and anion were measured to test the selectivity for Hg^{2+} ion (Figure 2 (c)). The absorbance ratio of PEI-AuNPs solution in the presence of Hg^{2+} was ~11 times greater than those in the presence of other ions. A high absorbance ratio of Hg^{2+}-PEI-AuNPs was attributed to the aggregation of PEI-AuNPs, whereas a low absorbance ratio of PEI-AuNPs in the presence of other ions indicated to keep well-dispersed forms of PEI-AuNPs. Therefore, Hg^{2+} ion must be selectively coordinated with a specific site of PEI-AuNPs. The interference effect by other ions in the selectivity of Hg^{2+} toward PEI-AuNPs was tested in PEI-AuNP solutions upon addition of Hg^{2+} ions mixed with other ions. Other ions did not interfere with determination of Hg^{2+}, even though their concentrations were ten times greater than that of Hg^{2+}. No metal cations and anions except Hg^{2+} perturbed absorption bands at 514 and 700 nm (Figure 2(d)).

3.3. Binding Sites of PEI to AuNPs and Hg^{2+} to PEI-AuNPs.

XPS spectra for PEI-AuNPs and Hg^{2+}-PEI-AuNPs were measured to confirm the binding site of Hg^{2+} ion to PEI-AuNPs (not shown) [33]. The high-resolution N 1s signal in Hg^{2+}-PEI-AuNPs at 406.2 eV showed the binding energy of Hg^{2+}-N bonds [34]. Thus, it was found that Hg^{2+} ion must be bound to nitrogen atom of PEI.

The binding site of Hg^{2+} to PEI was further examined with ^{13}C NMR spectra for free PEI and PEI bound to Hg^{2+}

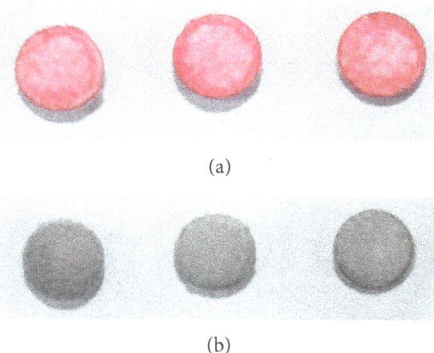

FIGURE 7: (a) Photographic image of the Whatman paper added with PEI-AuNP solution and dried. (b) Photographic image of the Whatman paper added with $0.1\,ppm \cdot Hg^{2+}$ ion solution.

(Hg^{2+}-PEI) as a model of PEI-AuNPs and Hg^{2+}-PEI-AuNPs (Figure 3). ^{13}C NMR spectra showed that two CH_2 peaks (peaks 1 and 3) resonating 56.5 ppm and 51.0 ppm in free PEI shifted significantly to 54.5 and 49.8 ppm, respectively, in comparison to those of Hg^{2+}-PEI, as shown in Figure 3. The chemical shifts of other peaks changed a little, which indicated that Hg^{2+} ions must be coordinated to nitrogen atoms of tertiary amine in PEI [35, 36].

TOF-SIMS spectra for PEI-AuNPs and Hg^{2+}-PEI-AuNPs showed an additional evidence for Hg^{2+} binding site to PEI-AuNPs (Figure 4). TOF-SIMS spectra of Hg^{2+}-PEI-AuNPs provided an additional information for Hg^{2+} binding site to PEI-AuNPs. These MS spectra showed the molecular fragments of $(CH_2)_2NHg^{2+}$ (m/z: 242), $CH_2CH_2(CH_2)_2NHg^{2+}$ (m/z: 270), and $(CH_2)_2NHg^{2+}N(CH_2)CH_2CH_2$ (m/z: 298) in Hg^{2+}-PEI-AuNPs (Figure 4). These molecular fragments did not appear for PEI-AuNPs, which indicated that Hg^{2+} must be coordinated with tertiary nitrogen atoms in PEI [37].

3.4. Optimum Conditions for PEI-AuNP Probe.

To optimize the sensitivity of the PEI-AuNP probe for Hg^{2+}, the probe was tested as functions of pH, temperature, salt concentration, PEI concentration, and reaction time. The absorbance ratios changed as a function of pH, and it was the highest at pH 7 (Figure 5(a)). This optimum pH of the PEI-AuNP probe must be something to do with pKa of tertiary amine and the conformation of PEI [38]. Thus, Hg^{2+} must be optimally coordinated to nitrogen elements of PEI in its N-tetrahedral form at pH 7, leading to the highest sensitivity of the probe [39].

The sensitivity of the PEI-AuNP probe for Hg^{2+} ions was examined as a function of temperature in the range of 30–100°C, and its sensitivity was optimized at 50°C (Figure 5(b)). Also, the sensitivity of PEI-AuNP probe was monitored as a function of NaCl concentration, and it was optimized at $300\,mM \cdot NaCl$ concentration (Figure 5(c)).

The optimum concentration of PEI conjugated to AuNPs was examined in the presence of $0.4\,\mu g \cdot mL^{-1}$ Hg^{2+} solution,

TABLE 1: Concentrations of Hg^{2+} ions measured by the PEI-AuNP colorimetric probe and DMA in tap water, pond water, tuna fish, and bovine serum samples spiked with Hg^{2+} ions.

| | Content of Hg^{2+} added to various samples ($n = 7$) | | | | | |
| | PEI-AuNP probe | | | | | DMA |
Sample	Added amount (μM)	Detected amount (μM)	Coefficient of variation (%)	Recovery (%)	LOD (nM)	Detected amount (μM)
Tap water	0.150	$0.149 \pm 1.53 \times 10^{-4}$	0.102	99.9 ± 0.102	1.72	0.153
	0.500	0.499 ± 0.0101	2.03	99.9 ± 2.03		0.505
Pond water	0.150	$0.150 \pm 6.92 \times 10^{-4}$	0.460	100.2 ± 0.461	1.80	0.156
	0.500	$0.500 \pm 3.41 \times 10^{-3}$	0.682	100.1 ± 0.683		0.504
Tuna fish	0.150	$0.149 \pm 7.09 \times 10^{-4}$	0.473	99.8 ± 0.473	2.00	0.152
	0.500	0.503 ± 0.0155	3.08	100.6 ± 3.10		0.504
Bovine serum	0.150	$0.150 \pm 6.43 \times 10^{-4}$	0.427	100.1 ± 0.428	1.95	0.155
	0.500	$0.503 \pm 7.68 \times 10^{-3}$	1.52	100.7 ± 1.53		0.507

TABLE 2: Analytical results for the detection of Hg^{2+} in real citrus leaf samples.

| | This method | | | | DMA |
Samples	Average value (nM)	Recovery (%)	RSD (%)	Accuracy (%)	Average value (nM)
1	22.8 ± 0.34	101.8	1.54	0.22	22.4
2	23.9 ± 0.42	98.7	1.72	1.29	24.3
3	23.1 ± 0.50	98.1	2.13	1.90	23.5

RSD: relative standard deviation.

TABLE 3: Comparison of previously reported instrumental methods and nanoparticle assay methods proposed for the detection of Hg^{2+}.

	Sensing principles	Matrices	LOD	Reference
Instruments				
DMA	—	River sediment, bovine liver, tomato leaves, spinach leaves, sewage sludge, mussel tissue, fish tissue, fish protein	1.04 μM	[8]
IC	IC	Tunny fish, oyster, and trumpet	0.5 μM	[9]
HPLC	SPE	Tap water, river water, sea water, and coal-washing waste water	14.9 nM	[10]
HPLC	SAX	Drinking water, lake water, river water, tap water, and sea water	0.8 nM	[11]
Functionalized nanoparticles				
AuNRs	Colorimetric	—	14.9 nM	[12]
CNPs	Fluorometric	Tap water and commercial bottled mineral water	10 nM	[13]
AgNPs	Colorimetric	Tap water	1 nM	[14]
AgNPRs	Colorimetric	Lake water and tap water	3.0 nM	[15]
Lysine–AuNPs	Colorimetric	Tap water	2.9 nM	[16]
DDTC–AuNPs	Colorimetric	Drinking water	2.9 nM	[17]
TCA–AuNCs	Colorimetric	Tap water and lake water	0.5 nM	[18]
PEI-AuNPs	Colorimetric	Tap water and pond water	1.72 nM	This study

SPE: solid phase extraction; SAX: strong anion exchange column; DDTC: diethyldithiocarbamate; TCA: thiocyanuric acid.

and the absorbance ratio of UV-Vis spectra as a function of PEI concentration revealed that optimum concentration of PEI was ~33 μM (data not shown).

3.5. Quantitation of Hg^{2+}, $EtHg^+$, $MeHg^+$, and $PhHg^+$ Using the PEI-AuNP Assay Method. The change for color, UV-Vis spectra, and TEM image of the PEI-AuNP probe upon addition of inorganic (Hg^{2+}) and organic mercury ions ($MeHg^+$, $EtHg^+$, and $PhHg^+$) were monitored. The color of PEI-AuNPs changed gradually from red to dark violet as the concentration of mercuric ions increased (Figure 6). Also, the absorbance increases at 700 nm and decreases concomitantly at 514 nm, as the concentration of mercuric ions increases (0.05, 0.15, 0.25, 0.50, 1.5, 2.5, 3.5, and 5.0 μM) in PEI-AuNPs solutions. The absorbance ratios for concentrations of each mercuric ion were measured in triplicate. Linear regression analysis of the calibration curve showed a good linearity (r^2 was 0.9817 for Hg^{2+}, 0.98009 for $MeHg^+$, 0.98774 for $EtHg^+$, and 0.99223 for $PhHg^+$) within the linear dynamic range of 0.003–3.0 μM. The limits of detection of this probe in tap, pond water, bovine serum, and tuna fish using this probe were measured as 1.72, 1.80, 2.00, and 1.95 nM, respectively, using [3σ/slope].

Paper-type sensor was fabricated, and the present probe solution was dropped to the Whatman paper, and dried it. The color of the Whatman paper turned red (Figure 7(a)), and the functionality of the sensor was tested on water sample containing Hg^{2+} ion. When water sample of 0.1 ppm Hg^{2+} was added onto the Whatman paper disc, its color turned dark purple (Figure 7(b)). This fact showed this paper disc coated with PEI-AuNP solution can be utilized as a paper-type Hg^{2+} sensor.

3.6. Application of the PEI-AuNP Probe in the Analyses of Real Samples. To validate the present assay method, the colorimetric responses in real water samples were tested. The tap water, pond water, bovine serum, and tuna fish samples spiked with 0.6 and 1.8 μM·Hg^{2+} were analyzed using the PEI-AuNP probe and DMA. As shown in Table 1, the analytical results of the proposed probe are nearly identical to those obtained using DMA. Hg^{2+} ions in real citrus leaf samples were also determined using both the colorimetric AuNP probe and DMA, as shown in Table 2, and their analytical results are almost the same. Thus, present AuNP-based probe in determination of Hg^{2+} ions seemed to be more advantageous than instrumental methods in terms of simplicity, sensitivity, cost, and time.

The previously reported instrumental methods and nanoparticle assay methods for the detection of Hg^{2+} ions are compared in Table 3, showing that the colorimetric PEI-AuNP probe offers the lowest LOD for the determination of Hg^{2+} ions in aqueous samples.

4. Conclusions

A highly sensitive and selective colorimetric probe to determine Hg^{2+} ions was developed using AuNPs conjugated with branched polyethyleneimine. The sensing mechanism of this colorimetric probe was originated from the aggregation of PEI-AuNPs in the presence of Hg^{2+}, and the Hg^{2+} ion was found to be selectively coordinated by nitrogen element of PEI conjugated with AuNPs. This method offers simple, highly sensitive, highly selective, and cost-efficient on-site monitoring of the Hg^{2+} ion, allowing the detection of concentrations as low as 1.72 nM to be visually achieved within 40 min.

Conflicts of Interest

The authors declare that there are no conflicts of interest regarding the publication of this paper.

Authors' Contributions

Kyung Min Kim and Yun-Sik Nam equally contributed to this work.

Acknowledgments

This research was financially supported by Korea Institute of Science and Technology (2E27070) and the Korea Ministry of Environment (2016000160008) as "Public Technology Program based on Environmental Policy."

References

[1] J. A. Cowan, *Inorganic Biochemistry: An Introduction*, Wiley-VCH, Weinheim, Germany, 1997.

[2] M. Gibb and K. G. O'Leary, "Mercury exposure and health impacts among individuals in the artisanal and small-scale gold mining community: a comprehensive review," *Environmental Health Perspectives*, vol. 122, no. 7, pp. 667–672, 2014.

[3] J. D. Park and W. Zheng, "Human exposure and health effects of inorganic and elemental mercury," *Journal of Preventive Medicine and Public Health*, vol. 45, no. 6, pp. 344–352, 2012.

[4] Y. Fang, X. Sun, W. Yang et al., "Concentrations and health risks of lead, cadmium, arsenic, and mercury in rice and edible mushrooms in China," *Food Chemistry*, vol. 147, pp. 147–151, 2014.

[5] Y. S. Hong, Y. M. Kim, and K. E. Lee, "Methylmercury exposure and health effects," *Journal of Preventive Medicine and Public Health*, vol. 45, no. 6, pp. 353–363, 2012.

[6] J. G. Dórea, M. Farina, and J. B. Rocha, "Toxicity of ethylmercury (and Thimerosal): a comparison with methylmercury," *Journal of Applied Toxicology*, vol. 33, no. 8, pp. 700–711, 2013.

[7] D. A. Geier, B. S. Hooker, J. K. Kern, P. G. King, L. K. Sykes, and M. R. Geier, "A dose-response relationship between organic mercury exposure from thimerosal-containing vaccines and neurodevelopmental disorders," *International Journal of Environmental Research and Public Health*, vol. 11, no. 9, pp. 9156–9170, 2014.

[8] C. C. Windmoller, N. C. Silva, P. H. M. Andrade, L. A. Mendes, and C. M. do Valle, "Use of a direct mercury analyzer® for mercury speciation in different matrices without sample preparation," *Analytical Methods*, vol. 9, pp. 2159–2167, 2017.

[9] Q. Liu, "Determination of mercury and methylmercury in seafood by ion chromatography using photo-induced chemical vapor generation atomic fluorescence spectrometric detection," *Microchemical Journal*, vol. 95, no. 2, pp. 255–258, 2010.

[10] Y. Yin, M. Chen, J. Peng, J. Liu, and G. Jiang, "Dithizone-functionalized solid phase extraction–displacement elution-high performance liquid chromatography–inductively coupled plasma mass spectrometry for mercury speciation in water samples," *Talanta*, vol. 81, no. 4-5, pp. 1788–1792, 2010.

[11] H. Cheng, C. Wu, L. Shen, J. Liu, and Z. Xu, "Online anion exchange column preconcentration and high performance liquid chromatographic separation with inductively coupled plasma mass spectrometry detection for mercury speciation analysis," *Analytica Chimica Acta*, vol. 828, pp. 9–16, 2014.

[12] T. Placido, G. Aragay, J. Pons, R. Comparelli, M. L. Curri, and A. Merkoçi, "Ion-directed assembly of gold nanorods: a strategy for mercury detection," *ACS Applied Materials & Interfaces*, vol. 5, no. 3, pp. 1084–1092, 2013.

[13] Y. Guo, Z. Wang, H. Shao, and X. Jiang, "Hydrothermal synthesis of highly fluorescent carbon nanoparticles from sodium citrate and their use for the detection of mercury ions," *Carbon*, vol. 52, pp. 583–589, 2013.

[14] L. Li, L. Gui, and W. Li, "A colorimetric silver nanoparticle-based assay for Hg(II) using lysine as a particle-linking reagent," *Microchimica Acta*, vol. 182, pp. 1977–1981, 2015.

[15] N. Chen, Y. Zhang, H. Liu et al., "High-performance colorimetric detection of Hg^{2+} based on triangular silver nanoprisms," *ACS Sensors*, vol. 1, no. 5, pp. 521–527, 2016.

[16] G. Sener, L. Uzun, and A. Denizli, "Lysine-promoted colorimetric response of gold nanoparticles: a simple assay for ultrasensitive mercury(II) detection," *Analytical Chemistry*, vol. 86, no. 1, pp. 514–520, 2014.

[17] L. Chen, J. Li, and L. Chen, "Colorimetric detection of mercury species based on functionalized gold nanoparticles," *ACS Applied Materials & Interfaces*, vol. 6, no. 18, pp. 15897–15904, 2014.

[18] Z. Chen, C. Zhang, H. Ma et al., "A non-aggregation spectrometric determination for mercury ions based on gold nanoparticles and thiocyanuric acid," *Talanta*, vol. 134, pp. 603–606, 2015.

[19] W. J. Song, J. Z. Du, T. M. Sun, P. Z. Zhang, and J. Wang, "Gold Nanoparticles capped with polyethyleneimine for enhanced siRNA delivery," *Small*, vol. 6, no. 2, pp. 239–246, 2010.

[20] Z. Yuan, N. Cai, Y. Du, Y. He, and E. S. Yeung, "Sensitive and selective detection of copper ions with highly stable polyethyleneimine-protected silver nanoclusters," *Analytical Chemistry*, vol. 86, no. 1, pp. 419–426, 2014.

[21] Y. Liu, Z. Li, J. Liu, L. Xu, and X. Liu, "An unusual red-to-brown colorimetric sensing method for ultrasensitive silver(I) ion detection based on a non-aggregation of hyperbranched polyethylenimine derivative stabilized gold nanoparticles," *Analyst*, vol. 140, no. 15, pp. 5335–5343, 2015.

[22] V. V. Kumar, M. K. Thenmozhi, A. Ganesan, S. S. Ganesan, and S. P. Anthony, "Hyperbranched polyethylenimine-based sensor of multiple metal ions (Cu^{2+}, Co^{2+} and Fe^{2+}): colorimetric sensing via coordination or AgNP formation," *RCS Advances*, vol. 5, pp. 88125–88132, 2015.

[23] M. Li, Y. Li, X. Huang, and X. Lu, "Captopril-polyethyleneimine conjugate modified gold nanoparticles for co-delivery of drug and gene in anti-angiogenesis breast cancer therapy," *Journal of Biomaterials Science, Polymer Edition*, vol. 26, no. 13, pp. 813–827, 2015.

[24] N. D. Huston, B. C. Attwood, and K. G. Scheckel, "XAS and XPS characterization of mercury binding on brominated activated carbon," *Environmental Science & Technology*, vol. 41, no. 5, pp. 1747–1752, 2007.

[25] J. B. Lindén, M. Larsson, S. Kaur et al., "Polyethyleneimine for copper absorption II: kinetics, selectivity and efficiency from seawater," *RSC Advances*, vol. 5, pp. 51883–51890, 2015.

[26] J. D. S. Newman and G. J. Blanchard, "Formation of gold nanoparticles using amine reducing agents," *Langmuir*, vol. 22, no. 13, pp. 5882–5887, 2006.

[27] J. Bell, E. Climent, M. Hecht, M. Buurman, and K. Rurack, "Combining a Droplet-based microfluidic tubing system with gated indicator releasing nanoparticles for mercury trace detection," *ACS Sensors*, vol. 1, no. 4, pp. 334–338, 2016.

[28] X. Cui, L. Zhu, Y. Hou, P. Wang, Z. Wang, and M. Yang, "A fluorescent biosensor based on carbon dots-labeled oligodeoxyribonucleotide and graphene oxide for mercury (II) detection," *Biosensors and Bioelectronics*, vol. 63, pp. 506–512, 2015.

[29] M. J. Rak, N. K. Saadé, T. Friščić, and A. Moores, "Mechanosynthesis of ultra-small monodisperse amine-stabilized gold nanoparticles with controllable size," *Green Chemistry*, vol. 16, pp. 86–89, 2014.

[30] C. Lin, K. Tao, D. Hua, Z. Ma, and S. Zhou, "Size Effect of gold nanoparticles in catalytic reduction of *p*-nitrophenol with $NaBH_4$," *Molecules*, vol. 18, no. 10, pp. 12609–12620, 2013.

[31] C. Deraedt, L. Salmon, S. Gatard et al., "Sodium borohydride stabilizes very active gold nanoparticle catalysts," *Chemical Communications*, vol. 50, pp. 14194–14196, 2014.

[32] S. K. Ghosh and T. Pal, "Interparticle coupling effect on the surface plasmon resonance of gold nanoparticles: from theory to applications," *Chemical Reviews*, vol. 107, no. 11, pp. 4797–4862, 2007.

[33] M. Min, L. Shen, G. Hong et al., "Micro-nano structure poly (ether sulfones)/poly(ethyleneimine) nanofibrous affinity membranes for adsorption of anionic dyes and heavy metal ions in aqueous solution," *Chemical Engineering Journal*, vol. 197, pp. 88–100, 2012.

[34] F. F. Tao, "Design of an *in-house* ambient pressure AP-XPS using a bench-top X-ray source and the surface chemistry of ceria under reaction conditions," *Chemical Communications*, vol. 48, pp. 3812–3814, 2012.

[35] D. R. Holycross and M. Chai, "Comprehensive NMR studies of the structures and properties of PEI polymers," *Macromolecules*, vol. 46, no. 17, pp. 6891–6897, 2013.

[36] W. Y. Seow, K. Liang, M. Kurisawa, and C. A. E. Hauser, "Oxidation as a facile strategy to reduce the surface charge and toxicity of polyethyleneimine gene carriers," *Biomacromolecules*, vol. 14, no. 7, pp. 2340–2346, 2013.

[37] D. Schaubroeck, Y. Vercammen, L. V. Vaeck, E. Vanderleyden, P. Dubruel, and J. Vanfleteren, "Surface characterization and stability of an epoxy resin surface modified with polyamines grafted on polydopamine," *Applied Surface Science*, vol. 303, pp. 465–472, 2014.

[38] K. A. Curtis, D. Miller, P. Millard, S. Basu, F. Horkay, and P. L. Chandran, "Unusual salt and pH induced changes in polyethylenimine solutions," *PLoS One*, vol. 11, no. 9, Article ID e0158147, 2016.

[39] J. Gaffney and N. Marley, "In-depth review of atmospheric mercury: sources, transformations, and potential sinks," *Energy and Emission Control Technologies*, vol. 2, pp. 1–21, 2014.

Benzo[g]coumarin-Based Fluorescent Probes for Bioimaging Applications

Yuna Jung,[1] **Junyang Jung** ⓘ**,**[1,2] **Youngbuhm Huh** ⓘ**,**[1,2] **and Dokyoung Kim** ⓘ[1,2,3]

[1]*Department of Biomedical Science, Graduate School, Kyung Hee University, 26 Kyungheedae-Ro, Dongdaemun-Gu, Seoul 02447, Republic of Korea*
[2]*Department of Anatomy and Neurobiology, College of Medicine, Kyung Hee University, 26 Kyungheedae-Ro, Dongdaemun-Gu, Seoul 02447, Republic of Korea*
[3]*Center for Converging Humanities, Kyung Hee University, 26 Kyungheedae-Ro, Dongdaemun-Gu, Seoul 02447, Republic of Korea*

Correspondence should be addressed to Dokyoung Kim; dkim@khu.ac.kr

Academic Editor: Verónica Pino

Benzo[g]coumarins, which consist of coumarins fused with other aromatic units in the linear shape, have recently emerged as an interesting fluorophore in the bioimaging research. The pi-extended skeleton with the presence of electron-donating and electron-withdrawing substituents from the parent coumarins changes the basic photophysical parameters such as absorption and fluorescence emission significantly. Most of the benzo[g]coumarin analogues show red/far-red fluorescence emission with high two-photon absorbing property that can be applicable for the two-photon microscopy (TPM) imaging. In this review, we summarized the recently developed benzo[g]coumarin analogues including photophysical properties, synthesis, and applications for molecular probes that can sense biologically important species such as metal ions, cell organs, reactive species, and disease biomarkers.

1. Introduction

Coumarin (2H-chromen-2-one) is a chemical compound in the benzopyrone chemical class that can be found in many natural species [1, 2]. Coumarins possess variety of biological activities and unique photophysical properties. Among them, the fluorescent property received much attention recently due to its high quantum yield, high stability, and biological compatibility [3, 4]. The coumarin-based fluorescent dyes and molecular probes have been applied not only for basic science such as physics, chemistry, medical science, and clinical science but also for industry and engineering [5, 6]. In progress, various kinds of the expanded or combined structure of coumarin derivatives have been discovered. Among them, linearly extended derivatives, benzo[g]coumarin, show superior photophysical properties in the bioimaging applications over the other derivatives [7]. Moreover, a large two-photon absorbing ability with longer excitation and emission wavelengths in optical window, high photostability, and high chemical stability are the key advantageous features of benzo[g]coumarin derivatives. In this review, we describe a brief explanation of benzocoumarin analogues with photophysical properties, their synthetic methods, and the recently developed benzo[g]coumarin-based one/two-photon excitable fluorescent dyes/probes that can sense biologically important species.

2. Benzo[g]coumarins

Benzocoumarin derivatives can be categorized into four types depending on the position of the fused aromatic ring in the parent coumarin backbone (Figure 1): (1) benzo[c]coumarin (3,4-benzocoumarin) fused on 3,4-position; (2) benzo[g]coumarin (6,7-benzocoumarin) fused on 6,7-position; (3) benzo[f]coumarin (5,6-benzocoumarin) fused on 5,6-position; and (4) benzo[h]coumarin (7,8-benzocoumarin) fused on 7,8-position.

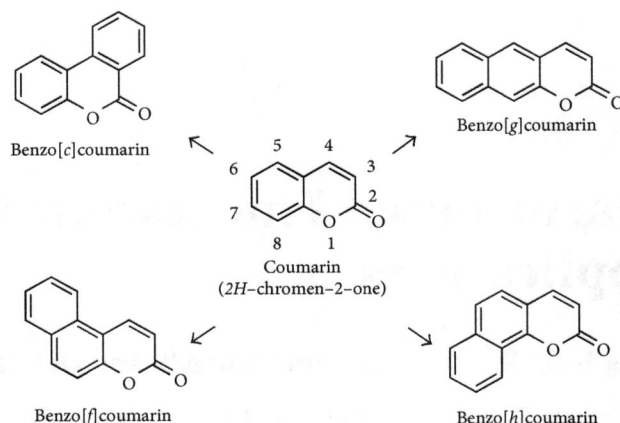

FIGURE 1: Chemical structures of coumarin and benzocoumarin derivatives.

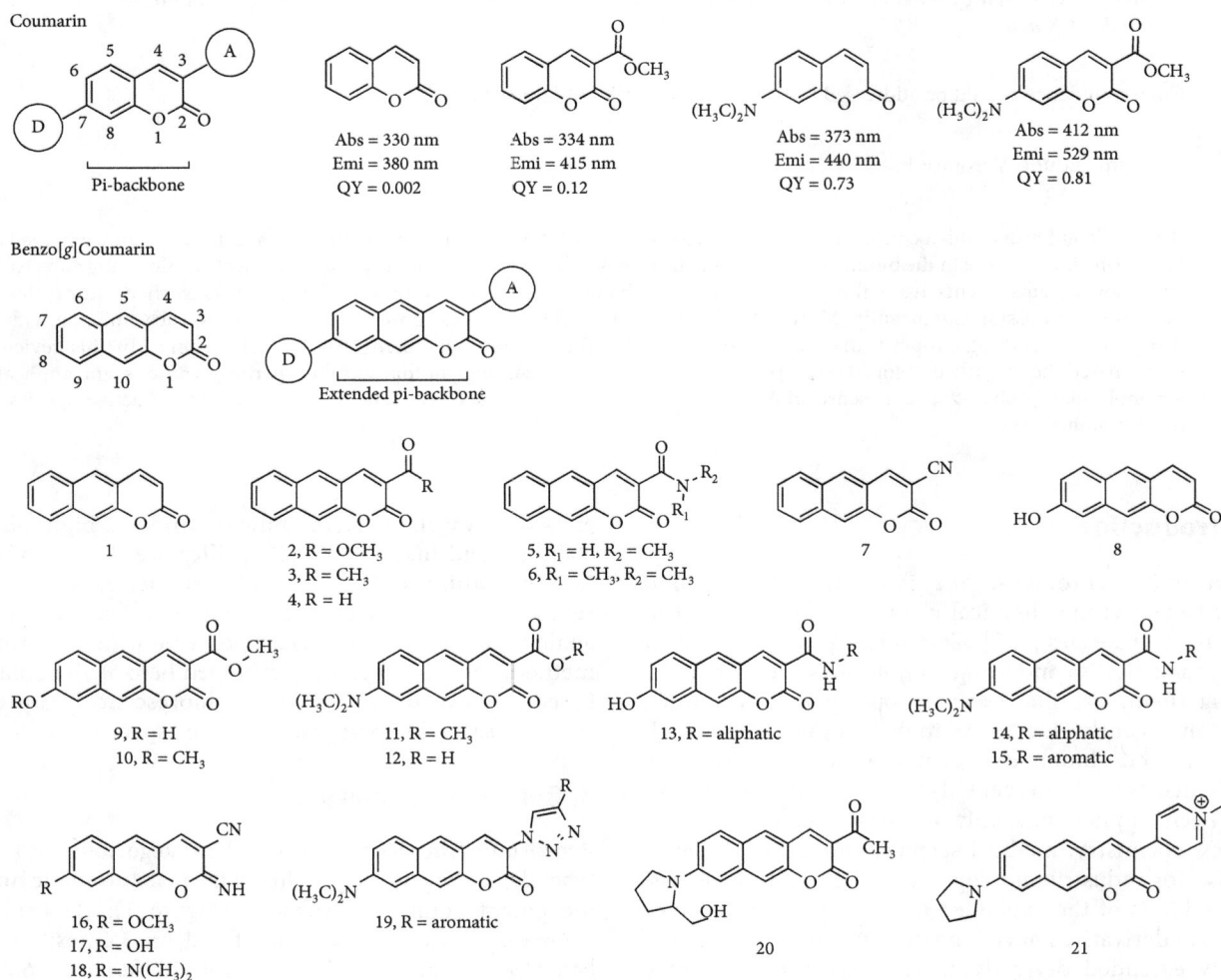

FIGURE 2: Chemical structure and basic photophysical properties of coumarin and benzo[g]coumarin derivatives. D: electron-donating group; A: electron-accepting group. The wavelengths are derived from the highest intensity values in the absorption and fluorescence emission spectra.

2.1. Photophysical Properties of Benzo[g]coumarin Analogues.

The photophysical property such as absorption and fluorescence emission of coumarin derivatives has been identified by many scientists in various fields. Among the derivatives, the functional group substitution on the 3- and 7-position gives large property changes from the original backbone. Typically, coumarin itself shows maximum absorbance and emission at a short wavelength ($\lambda_{max,abs}$ = 330 nm; $\lambda_{max,emi}$ = 380 nm) with poor fluorescence quantum yield (Φ_F) (Figure 2) [7]. However, the appropriate substitution, electron donor-acceptor- (D-A-) type structure on the pi-backbone, induces intramolecular charge transfer (ICT) accompanied with the quantum yield increment and the red shift of the fluorescence emission wavelength (see the absorption and emission wavelengths with quantum yield information in Figure 2) [8, 9].

Considering the bioimaging application of the coumarin fluorophore, the excitation and emission at a shorter wavelength give drawbacks such as (i) interference of autofluorescence from the biological substances, (ii) limited imaging depth, (iii) photodamage of the sample, and (iv) photobleaching of the fluorophore [10, 11]. In that sense, pi-extended benzo[g/f/h]coumarins are expected to provide several advantageous features over the corresponding coumarins due to their extended aromatic backbone that evokes longer wavelength absorption and fluorescence emission. Also, the conformationally restricted pi-backbone extension gives high photostability with a high quantum yield.

The photophysical properties of benzocoumarin analogues may also be predicted based on the type and substitution position. Recently, Ahn et al. reported a systematic analysis result for photophysical properties of benzocoumarin analogues [12]. They revealed that linear-type benzo[g]coumarins give similar substitution-induced property changes like parent coumarins, and the fluorescence property is suitable for bioimaging application over the others (very poor or no fluorescence from the benzo[f/h]coumarin in aqueous media).

The electron-withdrawing functional group-substituted benzo[g]coumarin at 3-position induces a red shift of fluorescence from the parent benzocoumarin (1, $\lambda_{max,emi}$ = 459 nm); ester moiety (2, 534 nm), ketone (3, 549 nm), aldehyde (4, 547 nm), amide (5, 515 nm), and nitrile (7, 533 nm) (Figure 2; Table 1). Interestingly, the combination of electron-withdrawing and electron-donating moieties at 3- and 8-position gives a significant shift of both absorption and emission spectra to the longer wavelength region (8–21; Figure 2; Table 1). The substitution of the hydroxyl group alone at 7-position (electron-donating position) shows no significant red shift ($\lambda_{max,emi}$ = 466 nm) (8), but the combination with electron-withdrawing substitution (methyl ester) at 3-position induces large changes (9, 10, and 13; $\lambda_{max,emi}$ > 500 nm). The alkylamine (-NR$_2$, R = alkyl) substitution at 8-position gives more significantly red-shifted absorption (>430 nm) and fluorescence emission (>580 nm) with a combination of ester, amide, nitrile, triazole, and pyridinium salts (11–21). The details of each compound with their photophysical properties are covered in the next chapter with the reported applications.

2.2. Synthesis of Benzo[g]coumarin Analogues. Synthesis of benzo[g]coumarin analogues follows the established synthetic routes including the Knoevenagel condensation or Wittig reaction with intramolecular cyclization reaction to

TABLE 1: Photophysical properties of the benzo[g]coumarin derivatives **11–21**. The wavelengths are derived from the highest intensity values in the absorption and fluorescence emission spectra in the described solvent.

Compound	λ_{abs} (nm)	λ_{emi} (nm)	Solvent	Reference
1	320	459	CH$_3$CN	[13]
2	332	534	CH$_3$CN	[13]
3	337	549	CH$_3$CN	[13]
4	345	547	CH$_3$CN	[13]
5	334	515	CH$_3$CN	[13]
6	326	510	CH$_3$CN	[13]
7	335	533	CH$_3$CN	[13]
8	357	466	CHCl$_3$	[14]
9	450	603	DI H$_2$O	[12]
10	355	539	DI H$_2$O	[12]
11	467	623	EtOH	[15]
12	413	599	DI H$_2$O	[12]
13	370/453	542/604	pH 4/pH 7 buffer	[16]
14	470	626	pH 7.4 buffer	[17]
15	450	600	pH 7.4 buffer	[18]
16	357	522	DI H$_2$O	[12]
17	435	582	DI H$_2$O	[12]
18	444	607	EtOH	[15, 19]
19	431	591	CH$_3$CN	[20]
20	487	633	pH 7.4 buffer	[21]
21	527	691	EtOH	[22]

o-hydroxynaphthaldehyde (Figure 3(a)) [7]. Alternatively, the direct electrophilic substitution of naphthols with β-keto ester followed by cyclization also gives benzo[g]coumarin in the presence of a catalyst. The metal-catalyzed aryl C–H functionalization of alkynoates is also feasible (Figure 3(a)).

Synthetic methods to make an electron-donating moiety on benzo[g]coumarin analogues were developed by many scientists including Ahn et al. [12, 15]. Representative derivatives which have a primary/secondary amine or a hydroxy/methoxy moiety at the 8-position were synthesized from the key intermediates A and B (Figure 3(b)). Intermediate A analogues were prepared by monoamination through the Bucherer reaction, protection of the hydroxyl group by methoxymethyl ether (MOM), and formylation through directed lithiation. On the contrary, methoxy or hydroxy group-substituted intermediate B analogues were prepared by monoprotection of the hydroxyl group by MOM first and then by methylation and formylation [12].

The MOM deprotection of the intermediates A and B in acidic condition gives an o-hydroxynaphthaldehyde intermediate, and the cascade intramolecular cyclization reaction generates benzo[g]coumarin derivatives (Figure 3(b)) [15, 19].

3. Benzo[g]coumarins for Fluorescent Probes

Benzo[g]coumarin derivatives have been used in various research areas. In particular, their unique photophysical property gives many advantages in the bioimaging applications such as fluorescent probes and tags and photolabile materials.

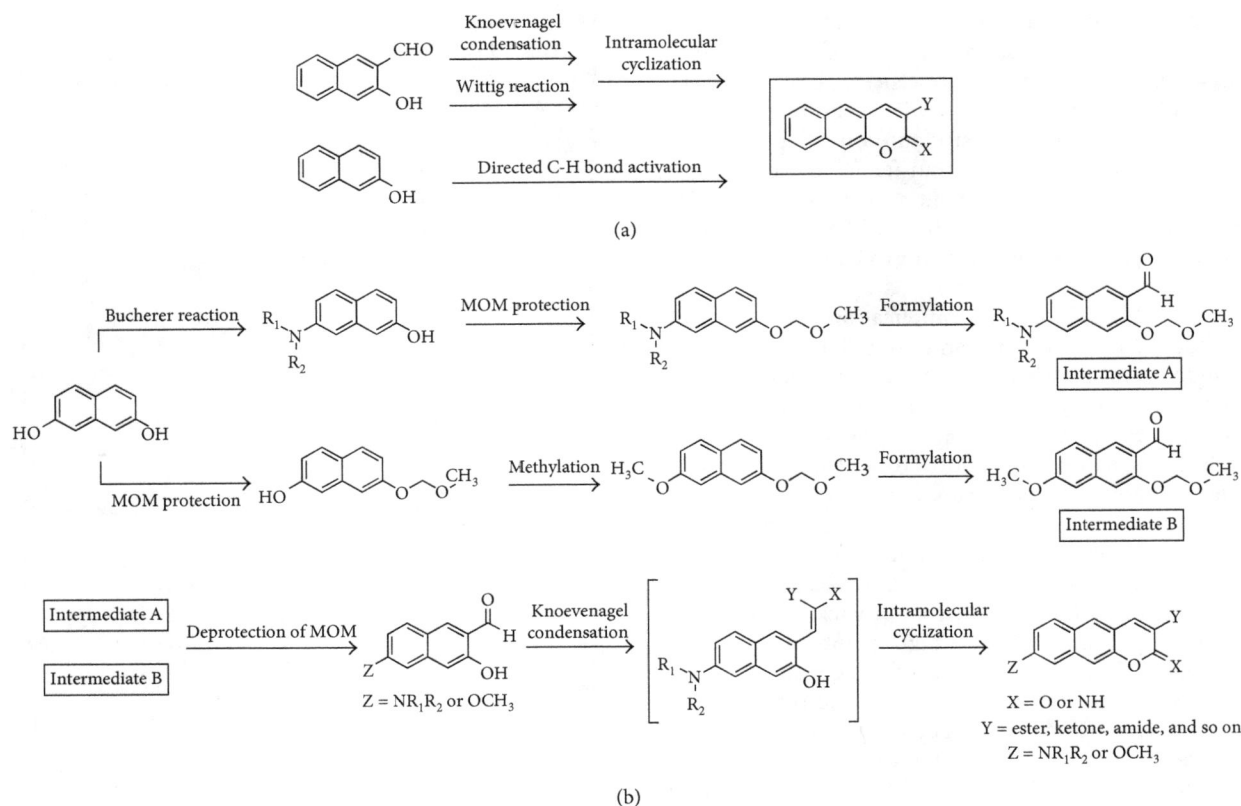

FIGURE 3: Synthetic routes for benzo[g]coumarin derivatives. (a) Routes for nonsubstitution at C-8 position. (b) Routes for substitution at C-8 position. X = O or NH; Y = ester, ketone, amide, nitrile, and so on; Z = NR₁R₂ or OR (R = alkyl).

Recently, a few examples of notable applications using benzo[g]coumarin derivatives for the fluorescent probes were reported. The fluorescent probe is undoubtedly an essential and useful tool in the biological, medical, and environmental sciences to investigate molecular interactions and biological activities, among others [23]. As we described above, benzo[g]coumarin analogues with suitable substitution show the absorption and fluorescence emission at the longer wavelength region (red and near-infrared) that gives better cellular or tissue imaging results than the shorter wavelength. Moreover, the pi-extended structure with proper substitutions gives a sufficiently large two-photon absorbing property that is applicable for the two-photon excitation microscopy [19]. Two-photon excitation microscopy is a fluorescence imaging technique that allows imaging of ex vivo and *in vivo* tissue up to millimeter depths [10, 24].

In this chapter, we summarized recently reported fluorescent probes based on the benzo[g]coumarin analogues (one/two-photon absorbing) with their interesting applications in the (i) sensing and bioimaging of biologically important species including metal ions, cell organs, reactive oxygen species (ROS), and disease biomarkers and (ii) deep tissue imaging.

3.1. Sensing of Metal Ions

3.1.1. Copper Ions (Cu²⁺). Copper ion plays crucial roles in living systems including signal transduction, oxygen

transportation via copper metalloenzymes, cellular energy generation, and cofactors of protein activity. As a result, the homeostasis of copper ions in the biological system is very important and directly related with various diseases: Alzheimer's disease (AD), Wilson's disease, Prion disease, and Menkes disease [25, 26].

Cho et al. reported a fluorescent probe for Cu²⁺ and quantitatively estimated ion concentrations in human tissues by two-photon microscopy imaging and analysis [27]. The amide- and dimethylamino-substituted benzo[g]coumarin analogue is linked with a benzo[h]coumarin analogue as an internal reference (internal reference: insensitive toward substrates or environment and maintains a steady fluorescence intensity) (22; Figure 4). The fluorescence intensity of benzo[g]coumarin at the red region (emission: 550–650 nm) was decreased through a chelation to the copper ion with a piperazine linker. Benzo[g]coumarin analogue (sensing part) and benzo[h]coumarin analogue (internal reference part) serve sufficiently high two-photon action cross section ($\Phi\delta$, GM value), 32 GM and 46 GM in ethanolic water (EtOH/HEPES 9 : 1 v/v, pH 7.0) at 750 nm two-photon excitation, respectively. Probe **22** shows high sensitivity ($0.84\,\mu$M) and selectivity toward Cu²⁺ with no perturbation due to high concentration of biological alkali and alkaline earth metal ions. They investigated the quantitative estimation of the Cu²⁺ concentration in live cells, rat brain tissue, and human colon tissue samples by using two-photon microscopy and analyzed the results. The higher

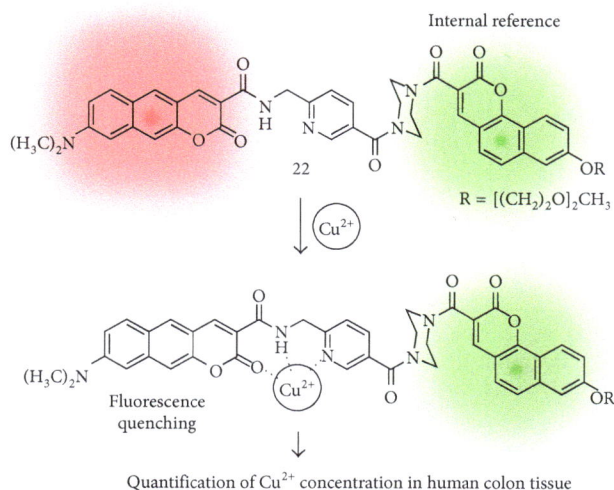

FIGURE 4: Benzo[g]coumarin- and benzo[h]coumarin-based fluorescent probe for the copper ion (22).

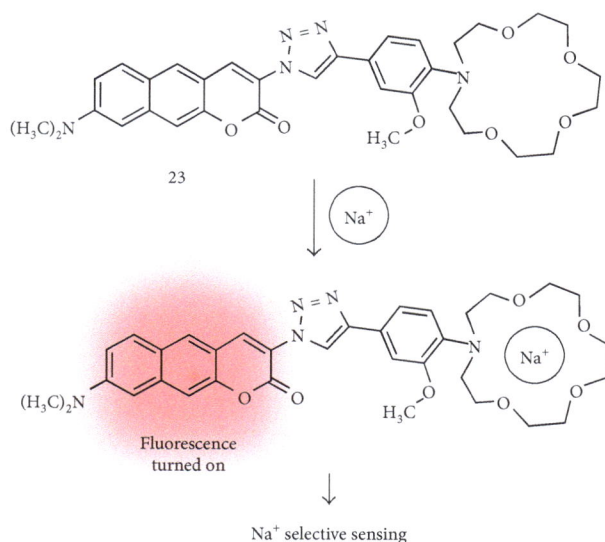

FIGURE 5: Benzo[g]coumarin-based fluorescent probe for the sodium ion (23).

concentration of Cu^{2+} in the cancer tissue ($22 \pm 3\,\mu M$) was observed than in polyp ($13 \pm 2\,\mu M$) or normal ($8.2 \pm 0.3\,\mu M$) samples, and it revealed that estimation of Cu^{2+} concentration may be useful for the diagnosis of colon cancer.

3.1.2. Sodium Ions (Na^+).

Sodium ion is one of the most important analytes in life science. It is necessary for live species for the nerve impulses, heart activity, metabolic functions, and biological balance [28, 29]. The Na^+ concentration range in the intracellular (5–30 mM) and extracellular (100–150 mM) space was observed, and competitive cation K^+ also showed similar concentrations. Accordingly, a development of tools for selective detection of Na^+ over K^+ is very important and challenging. Recently, Holdt et al. designed a Na^+ selective fluorescent probe based on the benzo[g]coumarin derivatives which have an N-(o-methoxyphenyl)aza-15-crown-5 moiety (23; Figure 5) [20]. A higher Na^+/K^+ selectivity of 23 was observed, but also, it gives a higher K_d value (223 mM) as a limitation for detection of lower concentration of Na^+. In this study, the bioimaging application was not reported, but they proposed a design strategy to develop benzocoumarin-based fluorescent probes.

3.1.3. Mercury Ions (Hg^{2+}).

Mercury ion is a chemical that is widely used in industry and basic science [30]. However, mercury is a highly poisonous element and causes damage to the central nervous system and other organs. So far, various kinds of detecting methods for mercury species (Hg^{2+}, $MeHg^+$, etc.) including fluorescent probes have been developed [30]. By using the latent probe approach with the benzo[g]coumarin platform, Ahn et al. reported new fluorescent probes for mercury ion sensing (Figure 6) [15]. The cleavage of sensing moiety on the platform by selective and sensitive chemical reaction toward the target analyte generates a chemically unstable intermediate,

which undergoes a fast cyclization reaction to afford an iminobenzo[g]coumarin derivative (24; Figure 6(a)). For the mercury ion sensing, they introduce a vinyl ether group to the Hg^{2+}-promoted hydrolysis (25; Figure 6(b)). The probe shows negligible fluorescence emission in the aqueous media due to the generation of the free rotation-induced nonradiative decay pathway from the dicyanoalkene moiety (molecular rotor moiety) and gives significant fluorescence enhancement upon adding mercury ions followed by iminobenzo[g]coumarin formation. Compound 24 shows absorption and fluorescence emission maximum at 446 nm and 585 nm, respectively, with a high quantum yield (QY = 0.67).

3.1.4. Fluoride Ions (F^-).

Fluoride ion plays an important role in chemistry, environment, medicine, and biology. Therefore, analytical methods that can selectively detect the fluoride ion have been requested in various fields [31]. In this vein, Ahn et al. developed an iminobenzo[g]coumarin precursor for fluoride ion sensing (26; Figure 6(c)) [32]. The desilylation of silyl enol ether moiety by fluoride ions followed by the intramolecular cyclization produced a compound 24. In this study, they showed the distribution of fluoride ions in cells and in a live vertebrate, zebrafish, using two-photon microscope (TPM) for the first time. The clear images at deep tissue regions, ~350 μm depth, represent the superior property of iminobenzo[g]coumarin 24 for the two-photon bioimaging.

3.2. Imaging of Cell Organs

3.2.1. Mitochondria.

Mitochondria is an organelle found in almost all eukaryotic organisms and plays important roles such as production of ATP, protein regulation, storage of calcium ions, and cellular metabolism regulation, among others [33]. Therefore, the defects of mitochondrial function

(a)

(b)   (c)

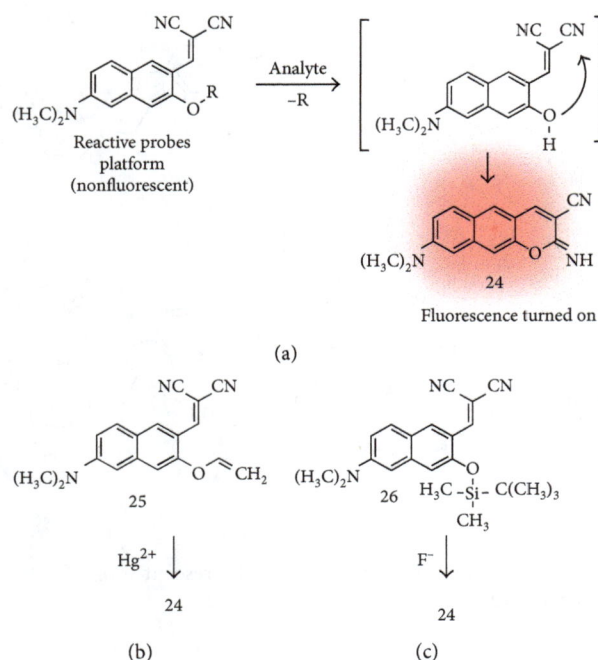

FIGURE 6: (a) Iminobenzo[*g*]coumarin analogue (**24**) and structure of its precursor (reactive probes platform) with a proposed sensing mechanism. Reaction-based fluorescent probes for mercury ion (**25**) (b) and fluoride ion (**26**) (c).

FIGURE 7: Triphenylphosphonium salt-linked benzo[*g*]coumarin probe for mitochondria tracking (**27**) and measuring mitochondrial pH values (**28**).

could be directly related to many diseases. So far, various techniques to understand the biological and pathological roles of mitochondria have been developed, and recently, fluorescence methods with imaging materials are used as a standard method to monitor mitochondria dynamics at the subcellular level. In 2014, Kim et al. reported a red-emissive two-photon probe (**27**; Figure 7) based on the benzo[*g*] coumarin for the real-time imaging of mitochondria tracking [17]. The mitochondrial-targeting moiety, triphenylphosphonium (TPP) salt, is linked on the electron-withdrawing part of the benzo[*g*]coumarin core via the amide bond, and the resulting compound **27** exhibited absorption and emission maximum at 470 nm and 626 nm, respectively, with no pH-sensitive changes in the biologically relevant pH range. The staining ability of **27** toward mitochondria was verified by the costaining experiment in the T98G cell line with MitoTracker Green (MTG) as a known mitochondrial labeling marker, and the high Pearson's colocalization coefficient value (0.96) indicates the organ specificity of **27** for mitochondria. In the TPM tissue imaging application using **27**, they observed the evenly distributed mitochondria in the CA1–CA3 region of the rat hippocampal tissue slice at a 200 μm depth.

3.2.2. Mitochondrial pH.

The monitoring of pH values and its dynamics in the cellular organs is very important to understand the pH-related biological, physiological, and pathological roles of cells and organisms [34]. In 2016, Kim et al. reported follow-up results that can monitor pH values in the mitochondria using a benzo[*g*]coumarin

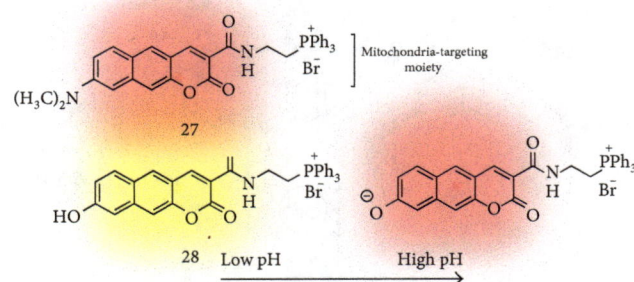

analogue (**28**; Figure 7) [16]. Probe **28** has a hydroxyl group on the electron-donating position (C-8), and it is protonated or deprotonated at the pK_a value near 8.0 which is a known pH value of mitochondria. At low pH (pH 4.0), **28** shows absorption and emission maximum at 370 nm and 542 nm, respectively, and the peak is shifted to 453 nm and 604 nm at high pH (pH 10.0) in a ratiometric manner. The high two-photon absorption cross section values (20–70 GM) of **28** at pH 4.0 and 10.0 and the high Pearson's colocalization coefficient (0.95) indicate the ability of selective imaging for mitochondria in the tissue samples. In the cellular imaging, a dense population of mitochondria around the nucleus than in the periphery was observed, and higher mitochondrial pH values in the perinuclear position than in the periphery of cells were also monitored. In a further study, they measured the mitochondrial pH values in the astrocyte from the Parkinson's disease (PD) mouse model and in the rat hippocampal tissue slice. Slightly acidic average pH values in the PD model astrocytes are observed compared with the wild-type astrocytes. The deep tissue imaging results provide average mitochondrial pH values in 7.86–7.88 at CA1, CA3, and the dentate gyrus region.

3.3. Sensing of Reactive Species

3.3.1. Nitric Oxide (NO).
Nitric oxide is a reactive nitrogen radical species, and its functions in living systems have been recognized to be related with cardiovascular, immune, and central nervous systems [35, 36]. So far, various kinds of chemical tools are developed to monitor the location, amount, and retention time in complex microenvironments such as cell and tissue, and these have been applied for the disease study and management. In 2017, Liu et al. reported a new fluorescent probe specifically for NO based on the *N*-nitrosation of the aromatic amine (**29**; Figure 8) [18]. Probe **29** has a benzo[*g*]coumarin backbone with dimethylamine and amide groups at the 3- and 8-position as an electron-donating and electron-withdrawing moiety, respectively. The original fluorescence emission of benzo[*g*]coumarin at 608 nm (at 473 nm excitation) is quenched by the photo-induced electron transfer (PET) from *p*-phenylenediamine

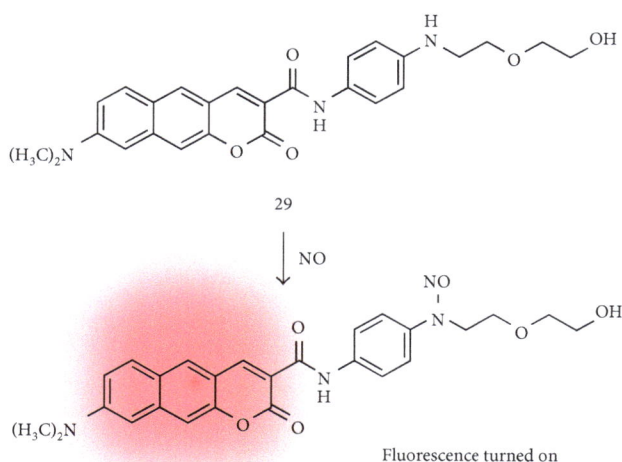

FIGURE 8: N-nitrosation-based benzo[g]coumarin probe for nitric oxide (NO) (**29**).

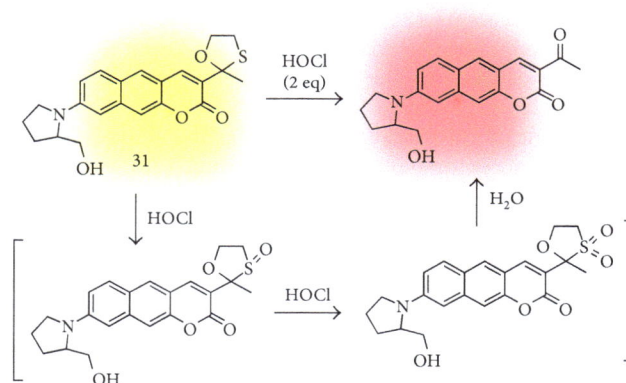

FIGURE 9: Michael addition and aldol condensation-based fluorescent probe for hydrogen sulfide (H$_2$S) (**30**).

moiety and recovered by N-nitrosation reaction of NO. The sensitivity toward NO was verified in the screening with the other reactive species such as ClO$^-$, H$_2$O$_2$, $^•$OH, O$_2^-$, NO$_2^-$, and ONOO$^-$. The two-photon action cross section values were increased from 2.4 GM to 54 GM at 830 nm under the excitation at 760–900 nm after adding the NO species. The TPM imaging studies for the exogenous and endogenous NO detection were carried out in the live cells (HepG2 cell line), mouse brain tissue, and ischemia/reperfusion injury (IRI) mouse model. Higher fluorescence signals in the TPM images of the IRI model compared with the healthy control represent that NO probe can be applied as a practical tool for studying NO-related biological processes.

3.3.2. Hydrogen Sulfide (H$_2$S).

Hydrogen sulfide is an endogenous gaseous transmitter along with carbon monoxide (CO) and nitric oxide (NO) [37]. Recent studies of H$_2$S revealed that it has a close relationship with neuronal activity, muscle relaxation, insulin management, inflammation, and aging [38]. Very recently, Ahn et al. reported a benzo[g] coumarin-based fluorescent probe for monitoring of exogenous and endogenous H$_2$S (**30**; Figure 9) [39]. The original fluorescence of the benzo[g]coumarin analogue (628 nm fluorescence emission at 485 nm excitation) was enhanced by Michael-type addition followed by aldol condensation of the α,β-unsaturated carbonyl group with H$_2$S [40]. Probe **30** shows the fast response (~8 min), high selectivity (negligible changes toward biological species), and high sensitivity (detection limit = 0.9 μM) for H$_2$S. Bioimaging accessibility for the H$_2$S was verified by TPM cellular imaging in the HeLa cell line.

3.3.3. Hypochlorous Acid (HOCl).

Hypochlorous acid is a kind of reactive oxygen species (ROS) [41], and the high level of HOCl is reported in several disorders such as cancer,

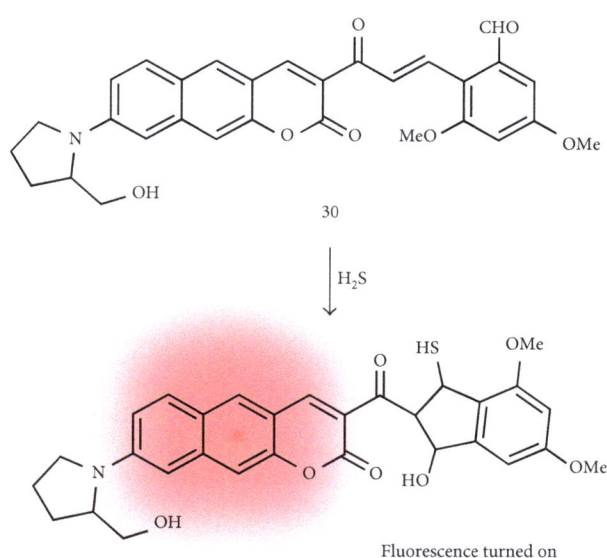

FIGURE 10: Oxathiolane deprotection-based fluorescent probe for hypochlorous acid (HOCl) (**31**).

arthritis, and neurodegenerative disease [42]. Therefore, monitoring the HOCl level and physiological distribution with a pathological mechanism is an important issue. However, the detection of endogenous HOCl is a challenging task due to the low biological concentration, a short lifetime, and a strong oxidizing property [43]. In 2017, Ahn et al. reported a benzo[g]coumarin-based ratiometric probe for endogenous HOCl imaging in live cells and tissues (**31**; Figure 10) [21]. An oxathiolane group is substituted at the electron-withdrawing position, and the deprotection into acetyl of this moiety by HOCl causes the intramolecular charge-transfer (ICT) character change of the benzo[g]coumarin dye in a ratiometric manner: emission maximum shift from 598 nm (with 424 nm absorption maximum) to 633 nm (with 598 nm absorption maximum). Probe **31** shows a low detection limit at the nanomolar level (34.8 nM) with high sensitivity toward HOCl over various reactive species including H_2O_2, $^\bullet OH$, O_2^-, $^1O_2$, and reactive nitrogen species (RNS). The probe **31** and the reaction product give good two-photon action cross section values, 142 GM and 439 GM, respectively. The level of HOCl in the hippocampal slices of the mouse was analyzed by TPM ratiometric imaging with probe **31**, and the slightly higher concentration of HOCl was observed at the dentate gyrus (DG) which is linked to the cognitive ability and memory retention.

3.4. Sensing of Disease Biomarkers

3.4.1. Amyloid-Beta Plaque (Aβ Plaque). Amyloid-beta plaque is an abnormal aggregate of the chemically sticky form of the amyloid-beta peptide (up to 42 or 43 amino acids long) that builds up between nerve cells in the AD patients [44, 45]. Therefore, the extracellular Aβ plaque deposition in the brain is considered as a hallmark of AD. So far, various kinds of contrast agents have been developed for the diagnosis of AD by direct detection of plaques [46]: (i) magnetic resonance imaging (MRI), (ii) positron emission tomography (PET), (iii) single-photon emission computed tomography (SPECT), and (iv) fluorescence imaging. By using the benzo[g]coumarin analogue, Ahn et al. found out the selective Aβ plaque staining ability of the iminobenzo[g] coumarin analogue (**24**) [47]. In this study, probe **24** selectively stains the Aβ plaques including cerebral amyloid angiopathy (CAA) in the whole brain region successfully. Probe **24** is accumulated in the Aβ plaques accompanying with significant fluorescence increments due to the nature of the donor-acceptor-type dye [48]; strong fluorescence in hydrophobic or viscous environment likes inside of the Aβ plaque. The *in vivo* TPM deep tissue imaging of Aβ plaques in the AD mouse model (5XFAD) treated with probe **24** via intraperitoneal injection shows the high blood-brain barrier (BBB) permeability of **24** and its superior deep tissue imaging ability (~600 nm depth) with high resolution.

3.4.2. Monoamine Oxidases (MAOs). Monoamine oxidases are a key enzyme responsible for the regulation of intracellular levels of biogenic amines and amine-based neurotransmitters such as dopamine, adrenaline, and serotonin [49]. A recent study revealed that suppressed or overregulated activity of MAOs is observed in several diseases including cancer and neurodegenerative diseases [50], AD, and Parkinson's disease (PD). In 2012, Ahn et al. developed a fluorescent probe (**32**) that can sense the activity of MAOs by enzymatic cleavage of the aminopropyl moiety followed by intramolecular cyclization and generation of an iminobenzo[g]coumarin (**24**) (Figure 11) [19]. In the intensive study, they applied probe **32** to find a correlation between activity of MAOs and AD progress in the animal model by using TPM [47]. Interestingly, significant background signal enhancement that correlated with MAO's activity was observed in older AD mice. The MAO's enzymatic product **24** is accumulated from the outside of Aβ plaques to the inside, and the fluorescence intensity is increased as growing older (increased numbers and size of Aβ plaques in the brain of the mouse model) (Figure 12). This is

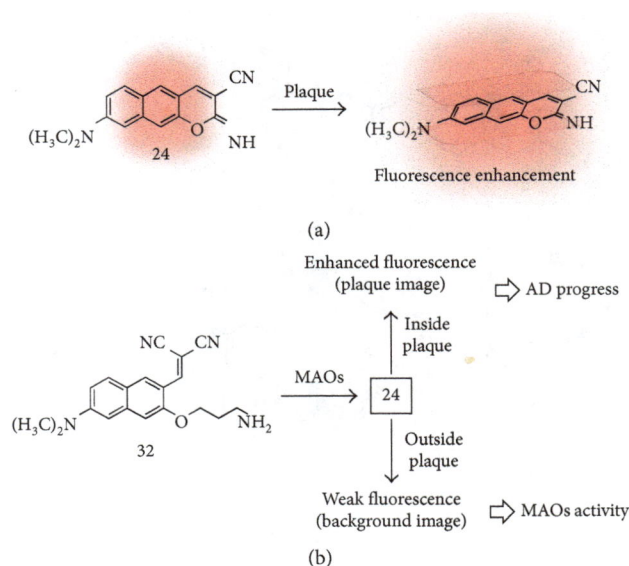

FIGURE 11: (a) Intercalation-based fluorescent probe for amyloid-beta plaque (**24**). (b) Amine oxidation-based fluorescent probe for monoamine oxidases (MAOs) activity in Alzheimer's disease (AD) (**32**).

the first demonstration for following activity of MAOs and AD progress *in vivo*.

3.5. Deep Tissue Imaging.
The fluorescence tissue imaging has emerged as the strong tool for studying biological events and clinical applications. In particular, the fluorescence deep tissue imaging with high resolution offers collective information of the cellular processes in a macroscopic view. Among the various imaging techniques, TPM has shown superior performance for deep tissue imaging. However, a key limitation for TPM-based deep tissue imaging is the autofluorescence interference from intrinsic biomolecules in the tissue such as nicotinamide adenine dinucleotide (NADH) and its phosphate analogue (NADPH), riboflavin, and flavoproteins [22]. The autofluorescence issue when using the known two-photon absorbing dyes has been solved by technical methods such as tuning the excitation wavelength, reducing the laser power, and changing the detection channel and/or sensitivity.

To overcome this issue, Ahn et al. focused on the systematic study of the new two-photon absorbing dyes based on benzo[g]coumarin analogues. In 2017, they reported pyridyl/pyridinium-benzo[g]coumarin analogues which have far-red-emitting (585–691 nm) fluorescence (**33–42**; Figure 13; Table 2) [22]. They optimized the wavelength of benzo[g]coumarin analogues that can address the autofluorescence issue. The pyridinium group at electron-withdrawing position (C-3) makes the significant wavelength shift to the far-red region (660–691 nm) from the parent benzo[g]coumarin or pyridyl-benzo[g]coumarin. In the brain tissue imaging with Py + BC690 (**42**), the clear deep tissue TPM imaging after an optical clearing process (BABB clearing) [51] was observed at the stage down to 1380 nm

(a)

A: healthy (9-month old), B: 5XFAD (9-month old) + selegiline

(b)

(c)

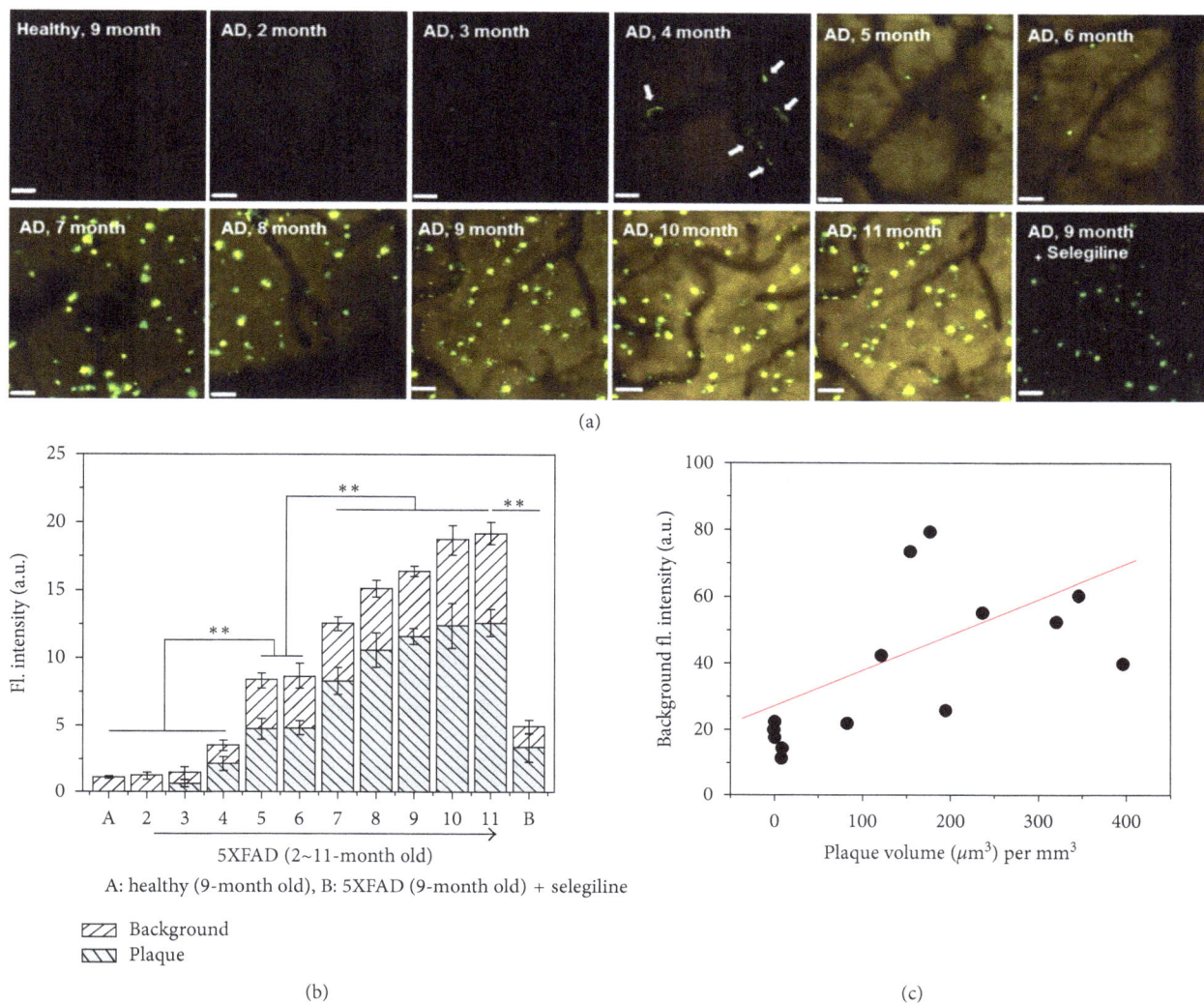

FIGURE 12: *In vivo* TPM coimaging of MAO activity and Aβ plaques using probe **32**. (a) *In vivo* fluorescence images (from (z)-stack, magnified 20x) of the frontal cortex region of transgenic and healthy mice, obtained after intraperitoneal injection of probe **32**. The scale bar is 60 μm. The images were acquired at 200–300 μm depth from the surface of the cortex. (b) Plots of the average fluorescence intensity of Aβ plaques and background images in (a), respectively. (c) A plot of the background fluorescence intensity versus the plaque volume (μm^3) per mm^3. **p-value < 0.01. Reproduced from [43] with permission from the American Chemical Society.

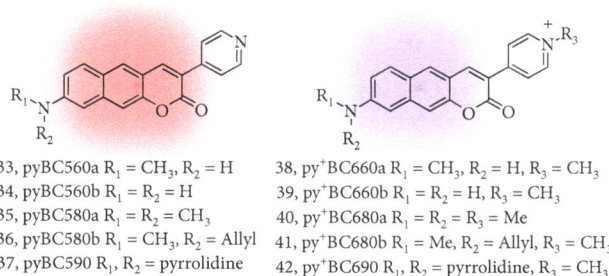

33, pyBC560a $R_1 = CH_3$, $R_2 = H$
34, pyBC560b $R_1 = R_2 = H$
35, pyBC580a $R_1 = R_2 = CH_3$
36, pyBC580b $R_1 = CH_3$, $R_2 = $ Allyl
37, pyBC590 $R_1, R_2 = $ pyrrolidine

38, py⁺BC660a $R_1 = CH_3$, $R_2 = H$, $R_3 = CH_3$
39, py⁺BC660b $R_1 = R_2 = H$, $R_3 = CH_3$
40, py⁺BC680a $R_1 = R_2 = R_3 = Me$
41, py⁺BC680b $R_1 = Me$, $R_2 = $ Allyl, $R_3 = CH_3$
42, py⁺BC690 $R_1, R_2 = $ pyrrolidine, $R_3 = CH_3$

FIGURE 13: Red/far-red-emitting pyridinium-benzo[g]coumarin derivatives (**33–42**).

TABLE 2: Photophysical properties of pyridinium-benzo[g]coumarin derivatives (**33–42**).

Compound	λ_{abs} (nm)	λ_{emi} (nm)	Brightness	GM
33	n.r.	n.r.	n.r.	n.r.
34	n.r.	n.r.	n.r.	n.r.
35	445	585	n.r.	n.r.
36	n.r.	n.r.	n.r.	n.r.
37	n.r.	n.r.	n.r.	n.r.
38	506	663	749	n.d.
39	499	660	510	n.d.
40	513	681	2089	n.d.
41	511	680	1173	150
42	527	691	799	160

The wavelengths, brightness, and GM values are derived from the highest intensity values in the absorption and fluorescence emission spectra. Brightness: molar extinction coefficient ($LMol^{-1} \cdot cm^{-1}$) × quantum yield (Φ_F); GM: two-photon absorption cross section (TPACS, Goeppert-Mayer unit); n.r.: not reported; n.d.: not determined.

depth. The imaging depth indicated the high tissue uptake of dye and penetration ability of **42** which are important features as a bioimaging agent.

4. Summary and Outlook

Since the first report about the pi-extended structure of coumarin, the tremendous knowledge and experimental results have been accumulated. In this focused review, the basic photophysical property, synthetic method, and applications of benzo[g]coumarin analogues are summarized. Molecular structures of linearly pi-extended benzo[g]coumarin analogues are expected to provide a longer excitation and emission wavelength at the red/near-infrared region with larger two-photon absorbing ability, and the experiment results have given evidences. In addition, their rigid conformation with facile function granting serves the high quantum yield, superior photostability/chemical stability, and applicability for the development of molecular probes. Some of the benzo[g]coumarin analogues showed promising two-photon absorbing properties holding great promise in the development of two-photon bioimaging probes to sense biologically important species. Most of the bioimaging applications of benzo[g]coumarin analogues are carried out very recently; therefore, we hope that this review inspires scientists to develop more advanced systems with useful practical applications such as disease biomarker sensing for prognosis and diagnosis.

Conflicts of Interest

The authors declare that there are no conflicts of interest.

Acknowledgments

This work was supported by a grant from the Kyung Hee University in 2017 (KHU-20170857).

References

[1] S. M. Sethna and N. M. Shah, "The chemistry of coumarins," *Chemical Reviews*, vol. 36, no. 1, pp. 1–62, 1945.

[2] A. Thakur, R. Singla, and V. Jaitak, "Coumarins as anticancer agents: a review on synthetic strategies, mechanism of action and SAR studies," *European Journal of Medicinal Chemistry*, vol. 101, pp. 476–495, 2015.

[3] X. Chen, F. Wang, J. Y. Hyun et al., "Recent progress in the development of fluorescent, luminescent and colorimetric probes for detection of reactive oxygen and nitrogen species," *Chemical Society Reviews*, vol. 45, no. 10, pp. 2976–3016, 2016.

[4] M. H. Lee, A. Sharma, M. J. Chang et al., "Fluorogenic reaction-based prodrug conjugates as targeted cancer theranostics," *Chemical Society Reviews*, vol. 47, no. 1, pp. 28–52, 2018.

[5] J. Wu, W. Liu, J. Ge, H. Zhang, and P. Wang, "New sensing mechanisms for design of fluorescent chemosensors emerging in recent years," *Chemical Society Reviews*, vol. 40, no. 7, pp. 3483–3495, 2011.

[6] S. R. Trenor, A. R. Shultz, B. J. Love, and T. E. Long, "Coumarins in polymers: from light harvesting to photo-cross-linkable tissue scaffolds," *Chemical Reviews*, vol. 104, no. 6, pp. 3059–3078, 2004.

[7] M. Tasior, D. Kim, S. Singha, M. Krzeszewski, K. H. Ahn, and D.T. Gryko, "π-expanded coumarins: synthesis, optical properties and applications," *Journal of Materials Chemistry C*, vol. 3, no. 7, pp. 1421–1446, 2015.

[8] Z. Yang, J. Cao, Y. He et al., "Macro-/micro-environment-sensitive chemosensing and biological imaging," *Chemical Society Reviews*, vol. 43, no. 13, pp. 4563–4601, 2014.

[9] G. Jones, S. F. Griffin, C. Y. Choi, and W. R. Bergmark, "Electron donor-acceptor quenching and photoinduced electron transfer for coumarin dyes," *Journal of Organic Chemistry*, vol. 49, no. 15, pp. 2705–2708, 1984.

[10] H. M. Kim and B. R. Cho, "Small-molecule two-photon probes for bioimaging applications," *Chemical Reviews*, vol. 115, no. 11, pp. 5014–5055, 2015.

[11] J. Chen, W. Liu, B. Zhou et al., "Coumarin- and rhodamine-fused deep red fluorescent dyes: synthesis, photophysical properties, and bioimaging in vitro," *Journal of Organic Chemistry*, vol. 78, no. 12, pp. 6121–6130, 2013.

[12] D. Kim, Q. P. Xuan, H. Moon, Y. W. Jun, and K. H. Ahn, "Synthesis of benzocoumarins and characterization of their photophysical properties," *Asian Journal of Organic Chemistry*, vol. 3, no. 10, pp. 1089–1096, 2014.

[13] C. Murata, T. Masuda, Y. Kamochi et al., "Improvement of fluorescence characteristics of coumarins: syntheses and fluorescence properties of 6-methoxycoumarin and benzo-coumarin derivatives as novel fluorophores emitting in the longer wavelength region and their application to analytical reagents," *Chemical and Pharmaceutical Bulletin*, vol. 53, no. 7, pp. 750–758, 2005.

[14] J. A. Key, S. Koh, Q. K. Timerghazin, A. Brown, and C. W. Cairo, "Photophysical characterization of triazole-substituted coumarin fluorophores," *Dyes and Pigments*, vol. 82, no. 2, pp. 196–203, 2009.

[15] I. Kim, D. Kim, S. Sambasivan, and K. H. Ahn, "Synthesis of π-extended coumarins and evaluation of their precursors as reactive fluorescent probes for mercury ions," *Asian Journal of Organic Chemistry*, vol. 1, no. 1, pp. 60–64, 2012.

[16] A. R. Sarkar, C. H. Heo, L. Xu et al., "A ratiometric two-photon probe for quantitative imaging of mitochondrial pH values," *Chemical Science*, vol. 7, no. 1, pp. 766–773, 2016.

[17] A. R. Sarkar, C. H. Heo, H. W. Lee, K. H. Park, Y. H. Suh, and H. M. Kim, "Red emissive two-photon probe for real-time imaging of mitochondria trafficking," *Analytical Chemistry*, vol. 86, no. 12, pp. 5638–5641, 2014.

[18] Z. Mao, H. Jiang, Z. Li, C. Zhong, W. Zhang, and Z. Liu, "An N-nitrosation reactivity-based two-photon fluorescent probe for the specific in situ detection of nitric oxide," *Chemical Science*, vol. 8, no. 6, pp. 4533–4538, 2017.

[19] D. Kim, S. Sambasivan, H. Nam et al., "Reaction-based two-photon probes for in vitro analysis and cellular imaging of monoamine oxidase activity," *Chemical Communications*, vol. 48, no. 54, pp. 6833–6835, 2012.

[20] T. Schwarze, H. Müller, D. Schmidt, J. Riemer, and H.-J. Holdt, "Design of Na+-selective fluorescent probes: a systematic study of the Na+-complex stability and the Na+/K+ selectivity in acetonitrile and water," *Chemistry–A European Journal*, vol. 23, no. 30, pp. 7255–7263, 2017.

[21] Y. W. Jun, S. Sarkar, S. Singha et al., "A two-photon fluorescent probe for ratiometric imaging of endogenous hypochlorous acid in live cells and tissues," *Chemical Communications*, vol. 53, no. 78, pp. 10800–10803, 2017.

[22] Y. W. Jun, H. R. Kim, Y. J. Reo, M. Dai, and K. H. Ahn, "Addressing the autofluorescence issue in deep tissue imaging by two-photon microscopy: the significance of far-red emitting dyes," *Chemical Science*, vol. 8, no. 11, pp. 7696–7704, 2017.

[23] J. Chan, S. C. Dodani, and C. J. Chang, "Reaction-based small-molecule fluorescent probes for chemoselective bioimaging," *Nature Chemistry*, vol. 4, no. 4, pp. 973–984, 2012.

[24] W. Denk, J. Strickler, and W. Webb, "Two-photon laser scanning fluorescence microscopy," *Science*, vol. 248, no. 4951, pp. 73–76, 1990.

[25] D. J. Waggoner, T. B. Bartnikas, and J. D. Gitlin, "The role of copper in neurodegenerative disease," *Neurobiology of Disease*, vol. 6, no. 4, pp. 221–230, 1999.

[26] M. G. Savelieff, S. Lee, Y. Liu, and M. H. Lim, "Untangling amyloid-β, tau, and metals in Alzheimer's disease," *ACS Chemical Biology*, vol. 8, no. 5, pp. 856–865, 2013.

[27] D. E. Kang, C. S. Lim, J. Y. Kim, E. S. Kim, H. J. Chun, and B. R. Cho, "Two-photon probe for Cu^{2+} with an internal reference: quantitative estimation of Cu^{2+} in human tissues by two-photon microscopy," *Analytical Chemistry*, vol. 86, no. 11, pp. 5353–5359, 2014.

[28] E. Murphy and D. A. Eisner, "Regulation of intracellular and mitochondrial sodium in health and disease," *Circulation Research*, vol. 104, no. 3, pp. 292–303, 2009.

[29] D. Landowne, "Movement of sodium ions associated with the nerve impulse," *Nature*, vol. 242, no. 5398, pp. 457–459, 1973.

[30] H. N. Kim, W. X. Ren, J. S. Kim, and J. Yoon, "Fluorescent and colorimetric sensors for detection of lead, cadmium, and mercury ions," *Chemical Society Reviews*, vol. 41, no. 8, pp. 3210–3244, 2012.

[31] M. Cametti and K. Rissanen, "Highlights on contemporary recognition and sensing of fluoride anion in solution and in the solid state," *Chemical Society Reviews*, vol. 42, no. 5, pp. 2016–2038, 2013.

[32] D. Kim, S. Singha, T. Wang et al., "In vivo two-photon fluorescent imaging of fluoride with a desilylation-based reactive probe," *Chemical Communications*, vol. 48, no. 82, pp. 10243–10245, 2012.

[33] C. P. Baines, "The cardiac mitochondrion: nexus of stress," *Annual Review of Physiology*, vol. 72, no. 1, pp. 61–80, 2010.

[34] J. Santo-Domingo and N. Demaurex, "The renaissance of mitochondrial pH," *Journal of General Physiology*, vol. 139, no. 6, pp. 415–423, 2012.

[35] V. Calabrese, C. Mancuso, M. Calvani, E. Rizzarelli, D. A. Butterfield, and A. M. Giuffrida Stella, "Nitric oxide in the central nervous system: neuroprotection versus neurotoxicity," *Nature Reviews Neuroscience*, vol. 8, no. 10, pp. 766–775, 2007.

[36] C. Szabo, "Gasotransmitters in cancer: from pathophysiology to experimental therapy," *Nature Reviews Drug Discovery*, vol. 15, no. 3, pp. 185–203, 2015.

[37] R. Wang, "Hydrogen sulfide: the third gasotransmitter in biology and medicine," *Antioxidants and Redox Signaling*, vol. 12, no. 9, pp. 1061–1064, 2010.

[38] L. Li, P. Rose, and P. K. Moore, "Hydrogen sulfide and cell signaling," *Annual Review of Pharmacology and Toxicology*, vol. 51, no. 1, pp. 169–187, 2011.

[39] H. G. Ryu, S. Singha, Y. W. Jun, Y. J. Reo, and K. H. Ahn, "Two-photon fluorescent probe for hydrogen sulfide based on a red-emitting benzocoumarin dye," *Tetrahedron Letters*, vol. 59, no. 1, pp. 49–53, 2018.

[40] S. Singha, D. Kim, H. Moon et al., "Toward a selective, sensitive, fast-responsive, and biocompatible two-photon probe for hydrogen sulfide in live cells," *Analytical Chemistry*, vol. 87, no. 2, pp. 1188–1195, 2015.

[41] S. J. Klebanoff, "Myeloperoxidase: friend and foe," *Journal of Leukocyte Biology*, vol. 77, no. 5, pp. 598–625, 2005.

[42] O. M. Panasenko, I. V. Gorudko, and A. V. Sokolov, "Hypochlorous acid as a precursor of free radicals in living systems," *Biochemistry*, vol. 78, no. 13, pp. 1466–1489, 2013.

[43] Y. Yue, F. Huo, C. Yin, J. O. Escobedo, and R. M. Strongin, "Recent progress in chromogenic and fluorogenic chemosensors for hypochlorous acid," *Analyst*, vol. 141, no. 6, pp. 1859–1873, 2016.

[44] D. J. Selkoe and J. Hardy, "The amyloid hypothesis of Alzheimer's disease at 25 years," *EMBO Molecular Medicine*, vol. 8, no. 6, pp. 595–608, 2016.

[45] P. Lewczuk, B. Mroczko, A. Fagan, and J. Kornhuber, "Biomarkers of Alzheimer's disease and mild cognitive impairment: a current perspective," *Advances in Medical Sciences*, vol. 60, no. 1, pp. 76–82, 2015.

[46] V. L. Villemagne, V. Doré, S. C. Burnham, C. L. Masters, and C. C. Rowe, "Imaging tau and amyloid-β proteinopathies in Alzheimer disease and other conditions," *Nature Reviews Neurology*, vol. 14, no. 4, pp. 225–236, 2018.

[47] D. Kim, S. H. Baik, S. Kang et al., "Close correlation of monoamine oxidase activity with progress of Alzheimer's disease in mice, observed by in vivo two-photon imaging," *ACS Central Science*, vol. 2, no. 12, pp. 967–975, 2016.

[48] D. Kim, H. Moon, S. H. Baik et al., "Two-photon absorbing dyes with minimal autofluorescence in tissue imaging: application to in vivo imaging of amyloid-β plaques with a negligible background signal," *Journal of the American Chemical Society*, vol. 137, no. 21, pp. 6781–6789, 2015.

[49] D. E. Edmondson, C. Binda, J. Wang, A. K. Upadhyay, and A. Mattevi, "Molecular and mechanistic properties of the membrane-bound mitochondrial monoamine oxidases," *Biochemistry*, vol. 48, no. 20, pp. 4220–4230, 2009.

[50] M. B. H. Youdim, D. Edmondson, and K. F. Tipton, "The therapeutic potential of monoamine oxidase inhibitors," *Nature Reviews Neuroscience*, vol. 7, no. 4, pp. 295–309, 2006.

[51] H.-U. Dodt, U. Leischner, A. Schierloh et al., "Ultramicroscopy: three-dimensional visualization of neuronal networks in the whole mouse brain," *Nature Methods*, vol. 4, no. 4, pp. 331–336, 2007.

3

Voltammetric Determination of Anethole on La$_2$O$_3$/CPE and BDDE

Mateusz Kowalcze ⓘ**, Jan Wyrwa** ⓘ**, Małgorzata Dziubaniuk, and Małgorzata Jakubowska** ⓘ

AGH University of Science and Technology, Faculty of Materials Science and Ceramics, Mickiewicza 30, 30-059 Kraków, Poland

Correspondence should be addressed to Małgorzata Jakubowska; jakubows@agh.edu.pl

Academic Editor: Pablo Richter

In this work, DPV determination of anethole was presented using various carbon, two-diameter (1.5 and 3 mm) electrodes, that is, BDD, GC, CP, and CP doped by La$_2$O$_3$ and CeO$_2$ nanoparticles. La$_2$O$_3$/CPE to our best knowledge was proposed first time. Cyclic voltammograms confirmed totally irreversible electrode electrooxidation process, controlled by diffusion, in which two electrons take part. The most satisfactory sensitivity $0.885 \pm 0.016\ \mu$A/mg L^{-1} in 0.1 mol L^{-1} acetate buffer was obtained for La$_2$O$_3$/CPE with the correlation coefficient r of 0.9993, while for BDDE it was $0.135 \pm 0.003\ \mu$A/mg L^{-1} with r of 0.9990. The lowest detection limit of 0.004 mg L^{-1} was reached on La$_2$O$_3$/CPE (3 mm), what may be compared with the most sensitive conjugate methods, but in the proposed approach, no sample preparation and analyte separation was needed. Anethole was successfully determined in specially prepared ethanol extracts of herbal mixtures of various compositions, which imitated real products. The proposed procedure was verified in analysis of commercial products, that is, anise essential oil, which contains a large concentration of anethole, and in alcohol drinks like Metaxa, Ouzo, and Rakija, in which the considered analyte occurs on trace levels. Structure and properties of the considered nanopowders and graphite pastes were investigated by EDX, SEM, and EIS.

1. Introduction

1.1. Anethole: Properties, Application, and Methods of Determination.

Anethole—anise camphor, *para*-methoxyphenylpropene, CAS number: 104-46-1—is a colorless, crystalline aromatic terpenoid analogue with a characteristic sweet taste and pleasant aroma, occurring in many essential oils obtained from plants belonging to the family *Apiaceae*, such as anise (*Pimpinella anisum*), star anise (*Illicium verum*), fennel (*Foeniculum vulgare*), liquorice (*Glycyrrhiza glabra*), and caraway (*Carum carvi*) [1–4] (Table 1). It occurs naturally in the form of two isomers: *trans-* (CAS number: 4180-23-8) and *cis-* (CAS number: 25679-28-1), where naturally the *trans*-isomer is much more common [4–7].

Due to its organoleptic properties, pleasant aroma and sweet taste, essential oils containing anethole have been used for centuries in the perfume, pharmaceutical, and spirit industries [1–6, 11, 12]. In pharmaceutical applications, anethole properties such as oestrogenic action, depressive action to the central nervous system, psycholeptic, insecticide, bactericidal, anticarcinogenic, anti-inflammatory, and anesthetics activity [1–6, 11, 12] are very important. The bactericidal properties of anethole are, due to lack of free phenolic group, weaker than their natural analogue—eugenol (Figure 1) [6]. In the spirit industry, anethole is present in various types of alcoholic beverages based on anise, fennel, or licorice—mainly Absinthe, Pastise, Ouzo, Rakija, and Metaxa [6, 12]. In addition, some alcohols must contain exactly the specified amount of anethole; for example, Pastise contains 1.5–2.0 g L^{-1} of this compound [12]. So, accurate determination of anethole content is one of the important stages of drinks production.

Among the quantitative methods of anethole assays in various matrices (Table 2), chromatographic techniques are dominant. Voltammetric techniques are used to evaluate the antioxidant properties of the compounds containing

Table 1: Content of anethole in various essential oils.

Type of plants	Country of origin	Part of plants	Method of obtaining oil and analysis	Content of anethole in the oil (%)	Reference
Fennel	Brazil	Seed	Supercritical fluid extraction (SFE), GC-MS	11.0–47.4-*trans*	[1]
Star anise	India	Fruit	Distillation in Clevenger-type apparatus, GC-MS	1.07-*cis* 75.62-*trans*	[7]
Anise	Germany	Seed	Distillation in Clevenger-type apparatus, GC-MS	0.14-*cis* 82.10-*trans*	[8]
Caraway	Different European countries	Fruit	Steam-distillation, GC-FID	0–2.2-*trans*	[9]
Bitter orange	Algeria	Peel	Distillation in Clevenger-type apparatus, GC-MS	2.3-*trans*	[10]

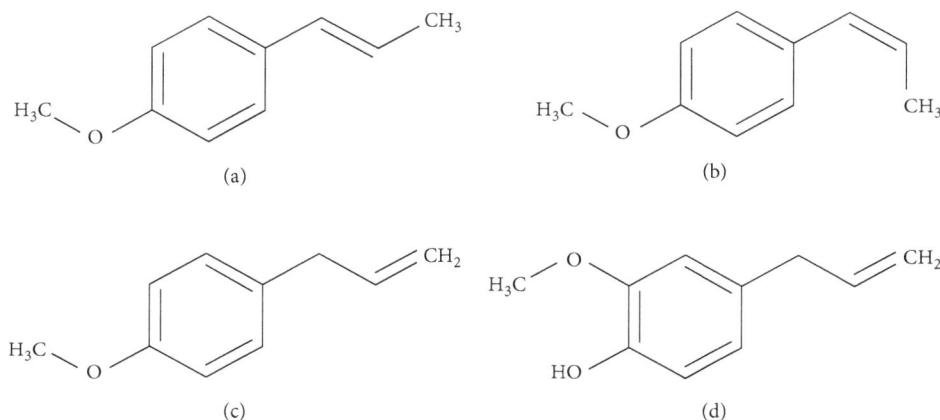

Figure 1: Structural formulas: (a) *trans*-anethole and (b) *cis*-anethole, and their analogues (c) estragole and (d) eugenol.

Table 2: Selected methods for quantification of both isomers of anethole in various matrices.

Type of matrix	Analytical technique	Linearity range (mg L^{-1})	LOD (mg L^{-1})	LOQ (mg L^{-1})	*Trans*-anethole conc. (mg L^{-1})	Reference
Fennel essential oil	HPLC	10–100	0.95	3.0	0.2–30.5	[15]
Fennel essential oil	GC-MS	10–550	0.002	0.006	0.16–40	[15]
Aniseed drinks	HPLC	2–16	0.0023	0.0077	125–4040	[16]
Human blood serum	HS-SPME-GC-MS	0.002–0.2	0.0036	0.0053	0.0054–0.0176	[17]

anethole [13] and may be useful in the classification of alcoholic beverages [14]; therefore, developing the method of anethole determination seems to be justified and interesting.

In this work, we present the possibility of *trans*-anethole determination by differential pulse voltammetry (DPV) technique. Various carbon electrodes, that is, glassy carbon electrode (GCE), boron-doped diamond electrode (BDDE), carbon paste electrode (CPE), carbon paste electrode doped by cerium(IV) oxide (CeO$_2$/CPE), and carbon paste electrode doped by lanthanum(III) oxide (La$_2$O$_3$/CPE), were used and tested. After designation of the analytical parameters of the method and optimization, quantitative and qualitative assays of anethole were applied in four specially prepared herbal matrices similar to anethole-containing

beverages and in various commercially available products of natural origin.

The obtained results are very promising and can be used in determination of anethole in a variety of matrices without analyte separation or sample preparation.

2. Experimental

2.1. Measuring Apparatus and Software. A multipurpose Electrochemical Analyzer M161 with the electrode stand M164 (both MTM-ANKO, Poland) was used for all voltammetric measurements. The classical three-electrode quartz cell of 10 mL volume was applied. Various carbon sensors were utilized as the working electrodes, that is, glassy carbon electrode (BASi, $\phi = 3$ mm and home-made, $\phi = 1.5$ mm), boron diamond-doped electrode (Windsor Scientific, $\phi = 3$ mm), carbon paste electrode, carbon paste electrode doped by cerium(IV) oxide, and carbon paste electrode doped by lanthanum(III) oxide. Carbon paste electrodes were prepared in our laboratory. Also a double-junction reference electrode Ag/AgCl/KCl (3 M) with replaceable outer junction (2.5M KNO$_3$) and a platinum wire as an auxiliary electrode were used. The ambient temperature was ca. 23°C. The MTM-Anko EAPro 1.0 software enabled electrochemical measurements, data acquisition, and processing of the results.

Electrochemical impedance spectroscopy measurements were performed using a frequency analyzer (Solartron model FRA 1260) coupled with dielectric interface (model 1296). The surface morphology of electrode material was observed using ultrahigh-resolution scanning electron microscope with field emission (FEG-Schottky emitter; Nova Nano-SEM 200, FEI Europe BV) cooperating with EDAX EDS analyzer.

2.2. Carbon Paste Electrodes Doped by La$_2$O$_3$ and CeO$_2$.

Carbon-based electrodes are useful for voltammetric determination of wide range of analytes in liquid solutions. Moreover, their applicative properties can be improved by doping with metal oxide modifiers. The usage of different doping oxides was reported for modification of carbon-based electrodes so far [18–22]. According to the literature, addition of controlled amounts of cerium(IV) oxide to glassy carbon electrodes material led to the significant enhancement of sensitivity, selectivity, reproducibility, and response time in the amperometric quantification of eugenol [22].

In this work, the experimental data obtained by use of lanthanum oxide-doped graphite paste electrodes applied in voltammetric analysis are presented for the first time. The carbon pastes were prepared by hand mixing an adequate amount of graphite powder and rare earth oxide powder with paraffin oil using a pestle and mortar for at least 30 minutes in the case of each batch. Nanopowders of lanthanum(III) oxide (99.99%) and cerium(IV) oxide (99.9%) were provided by Acros Organics. The ratio of used paraffin oil and graphite powder was determined based on literature repots as well as our experience in order to get electrodes characterized by high chemical and mechanical stability during performance in liquid solutions. After standing overnight, the resulting homogenous pastes were packed into the well of the working electrodes to depth of 2 mm with two different diameters (1.5 and 3 mm). The body of working electrode was a Teflon tube with stainless steel rode of 1.5 mm diameter serving as electric contact. To provide the required smoothness of electrodes, working surfaces the forehead of electrodes were polished on a print paper or tissue paper.

The amounts of used reagents, details of prepared electrodes, and pastes are presented in Table 3.

2.3. Chemicals and Glassware.

As a supporting electrolyte, buffers of a different pH were prepared in our laboratory (from reagents pure for analysis, POCH, Poland): acetate buffer—mix acetic acid and sodium acetate; Britton–Robinson buffer—mix boric acid, phosphoric acid, acetic acid, and sodium hydroxide; Sørensen phosphate buffer—mix sodium hydrogen phosphate and sodium dihydrogen phosphate; ammonia buffer—mix ammonia and ammonium chloride. As a standard solution, trans-anethole (analytical standard, Sigma-Aldrich) was used. 1 μL of solution contains 3.48 μg of trans-anethole. Reagents used to determine the impact of interferents are 99% eugenol (Reagent Plus, Sigma-Aldrich), 99% carvacrol (food grade, Sigma-Aldrich), ≥98.5% thymol (pure, Sigma-Aldrich), and zinc, lead, cadmium, bismuth, aluminum, thallium, chromium, and vanadium (all metals from Certipur, Merck). The other chemicals were 95% ethanol (food grade, Polmos, Poland) and 0.1 mol L^{-1} solution of sulfuric acid (pure for analysis, POCH, Poland) for activation of BDD electrode. All reagents used were prepared using quadruply distilled water (two last stages from quartz). Glassware was first immersed in 6 M nitric acid and then rinsed repeatedly with distilled water.

2.4. Samples.

To verify the possibility to determine anethole in herbal matrices A, B, and D, three solutions were prepared. Also matrix C was tested, which did not contain anethole. The composition and preparation of the matrices imitated different anethole-containing beverages. Each matrix was prepared by pouring with the ethanol (95%, food grade) the appropriate herbal composition and the five-day maceration of the mixture. After this time, each matrix was rectified once.

Anethole was also determined in commercially available products such as anise oil (KEJ, Poland), Efe Rakija (Turkey), Ouzo Typnaboy (Greece), and Metaxa*** (Greece).

Three independent samples of the same type were tested.

2.5. Standard Procedure of Voltammetric Measurements.

Measurements were performed using differential pulse voltammetry (DPV). Before each series of measurements, surface of the BDD electrode was activated 15 minutes in 0.1 mol L^{-1} sulfuric acid solution by the potential of 2400 mV. Before each calibration, the BDDE surface was additionally renewed by the potential 1500 mV and time of 30 s in supporting electrolyte. GCE was activated by polishing with polishing powder MicroPolish Alumina 0.05 μm (Buehler, USA).

The investigation of anethole was performed in different supporting electrolytes depending on the working electrode used, that is, 0.1 M acetate buffer with pH 3, 4, 5, or 6; Britton–Robinson buffer with pH 2 and 3; 0.1 M Sørensen's phosphate buffer with pH 6, 7, and 8; or 0.1 M ammonia buffer with pH 9 and distilled water, giving total volume of 5 mL filling the quartz voltammetric cell. The best results were obtained in supporting electrolyte consisting 5 mL of 0.1 M acetate buffer with pH 6. The volume of added standard solution of anethole was of 1–5 μL.

The solution in cell was stirred (ca. 500 rpm) using a magnetic stirring bar. Then, after a rest period of 5 s, differential pulse voltammograms were recorded in the

TABLE 3: Details of paste preparation and CP electrodes parameters.

Electrode	Electrode diameter (mm)	Ingredients	Paste label	Weight ratio of powder ingredients
CeO_2/CPE 20%	1.5 3	2 g graphite, 1.5 ml paraffin oil, 0.5 g CeO_2	PCe20-5	0.2 CeO_2–0.8 graphite
La_2O_3/CPE 20%	1.5 3	2 g graphite, 1.5 ml paraffin oil, 0.5 g La_2O_3	PLa20-2	0.2 La_2O_3–0.8 graphite
La_2O_3/CPE 30%	1.5 3	2 g graphite, 1.5 ml paraffin oil, 0.857 g La_2O_3	PLa30-3	0.3 La_2O_3–0.7 graphite
La_2O_3/CPE 40%	1.5 3	2 g graphite, 1 ml paraffin oil, 1.333 g La_2O_3	PLa40-4	0.4 La_2O_3–0.6 graphite

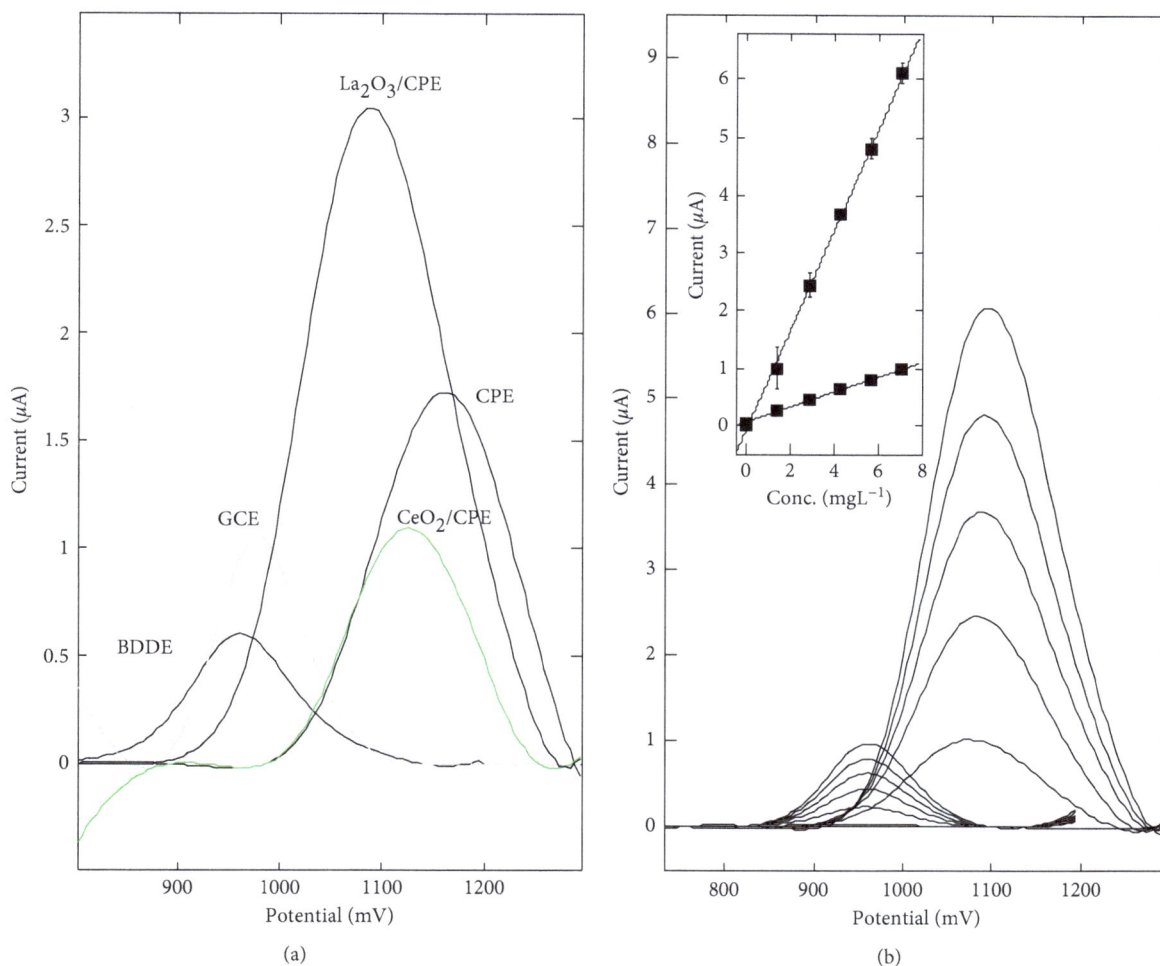

FIGURE 2: a) DP voltammograms on various carbon electrodes in 0.1 mol L^{-1} acetate buffer of 3.48 mg L^{-1} of anethole; (b) calibration in 0.1 mol L^{-1} acetate buffer of 0, 1.39, 2.78, 4.18, 5.57, and 6.96 mg L^{-1} of anethole on BDDE (black) and La_2O_3/CPE (blue).

potential window: 0–1200 mV (BDDE), 500–1300 mV (CPE, La_2O_3/CPE, CeO_2/CPE), and 600–1200 mV (GCE). The other standard experimental parameters were as follows: potential step $E_s = 5$ mV; pulse potential dE = 50 mV; and time of potential step = 40 ms (20 ms waiting time + 20 ms sampling time).

All experiments were performed at 23°C. All experiments were carried out in triplicate.

3. Results and Discussion

3.1. Carbon Electrodes in Determination of Anethole. The purpose of the study was to investigate whether carbon paste electrodes doped by two new rare earth element oxides may be useful in voltammetric determination of the anethole. Commercially available and popular sensors were used as a comparison. The well-defined DPV anethole peak

TABLE 4: DPV anethole determination in the range of 0.7–17.5 mg L^{-1} in 0.1 mol L^{-1} acetate buffer, pH 6.0 ($n = 3$).

Electrode	Anethole peak position (mV)	$a \pm SD_a$ (μA/mg L^{-1})	$b \pm SD_b$ (μA)	r	LOD (mg L^{-1})
Electrode diameter 3 mm, area 7.07 mm²					
BDDE	965	0.135 ± 0.003	0.050 ± 0.013	0.9990	0.024
GCE	990	0.306 ± 0.005	0.031 ± 0.046	0.9995	0.011
CPE	1155	0.546 ± 0.036	−0.068 ± 0.083	0.9936	0.006
CeO$_2$/CPE 20%	1125	0.341 ± 0.009	−0.060 ± 0.021	0.9989	0.010
La$_2$O$_3$/CPE 20%	1095	0.885 ± 0.016	0.076 ± 0.069	0.9993	0.004
Electrode diameter 1.5 mm, area 1.77 mm²					
GCE	970	0.022 ± 0.001	0.0035 ± 0.0051	0.9989	0.148
CPE	1110	0.111 ± 0.004	−0.005 ± 0.043	0.9983	0.030
CeO$_2$/CPE 20%	1075	0.227 ± 0.008	−0.054 ± 0.080	0.9978	0.014
La$_2$O$_3$/CPE 20%	1080	0.449 ± 0.021	−0.058 ± 0.044	0.9959	0.007
La$_2$O$_3$/CPE 30%	1095	0.350 ± 0.016	−0.039 ± 0.035	0.9958	0.009
La$_2$O$_3$/CPE 40%	1100	0.332 ± 0.010	0.029 ± 0.110	0.9980	0.010

(Figure 2) was obtained on the various carbon electrodes, that is, glassy carbon, boron-doped diamond, carbon paste, and carbon paste doped by lanthanum(III) oxide and cerium(IV) oxide, which were considered in this work. The peak position was observed between 965 and 1155 mV (Table 4, second column). Anodic shift (of 150–200 mV) of anethole oxidation potential has been obtained for carbon paste and two nanoparticles-modified electrodes compared to GCE and BDDE, confirming the lower transfer rate on CPE and nanoparticles/CPE.

Quantitative analysis was preceded by especially projected procedure of baseline modeling and subtraction (Figure 3). The first step of the proposed approach was subtraction of the experimental baseline obtained for the supporting electrolyte. Next, the typical approximation by the polynomial of the 2nd degree was utilized. These two steps were necessary, because the background shape was very different from the polynomial function.

Analytical parameters were determined and tested for two groups of electrodes (Table 4), that is, of the diameter of 3 mm (geometric area of 7.07 mm²) and of the diameter of 1.5 mm (geometric area of 1.77 mm²). After signal processing, the linear relation between peak current and concentration of the anethole in the range of 0.7–17.5 mg L^{-1} was noticed. Generally, paste electrodes were characterized by the greater sensitivity and lower detection limit in comparison to BDDE and GCE. However, the repeatability of the signal for successive analyte concentration was excellent for the latter (CV < 1%).

The highest sensitivity of 0.89 μA/mg L^{-1} among the considered electrodes was obtained on the carbon paste doped by the 20% of lanthanum(III) oxide nanoparticles, with the correlation coefficient r of 0.9993 (for averaged signals for each concentration) and the lowest detection limit of 0.004 mg L^{-1}. The sensitivity for the anethole on the electrode doped by the 20% of cerium(IV) oxide nanoparticles ($\phi = 3$ mm) was even lower (0.34 μA/mg L^{-1}) than the reference value obtained on CPE (0.55 μA/mg L^{-1}). The lowest sensitivity in the group of sensors with a diameter of 3 mm was obtained on BDDE (0.14 μA/mg L^{-1}) what was ca. 6 times less than on La$_2$O$_3$/CPE. Considering the sensors with a diameter of 1.5 mm, the highest sensitivity of 0.45 μA/mg L^{-1} was obtained on CeO$_2$/CPE. For CPE- and La$_2$O$_3$-doped CPE, the repeatability of the signal relied on the percent (w/w) of the added nanoparticles and was on the level of 5–8% (CV) when the nanopowder addition was lower than 20%. For the addition greater than 20%, the repeatability of the signal rapidly deteriorated (CV > 10%), and therefore these electrodes were not considered in further tests. CP electrodes doped by CeO$_2$ did not also show the satisfactory repeatability of the signals recorded for each concentration. It was also observed that increasing addition of the lanthanum(III) oxide nanoparticles decreased sensitivity for the anethole.

For further detailed analysis, La$_2$O$_3$/CPE ($\phi = 3$ mm) as a sensor of the greatest sensitivity for the anethole was chosen and for comparison BDDE, which is reliable after appropriate activation. Voltammograms and calibration lines for the anethole in the concentration range from 1.39 to 6.96 mg L^{-1} prepared on the mentioned two sensors are presented in Figure 2(b).

3.2. Supporting Electrolyte Effects. There are several ways in which the supporting electrolytes solvent system can influence mass transfer, the electron reaction (electron transfer), and the chemical reactions which are coupled to the electron transfer. As a supporting electrolyte, 4 different buffers (acetate, Britton–Robinson, phosphate, and ammonia) were applied in examination of the analyte behavior in pH range from 2.0 (BR buffer) to pH 9.0 (ammonia buffer). The best parameters—repeatability, sensitivity, limit of detection, and the favorable relation between signal and baseline—were obtained using acetate buffer; therefore, the pH effect was tested carefully in the pH range typical for this electrolyte, that is, from 3.0 to 6.0. In the considered

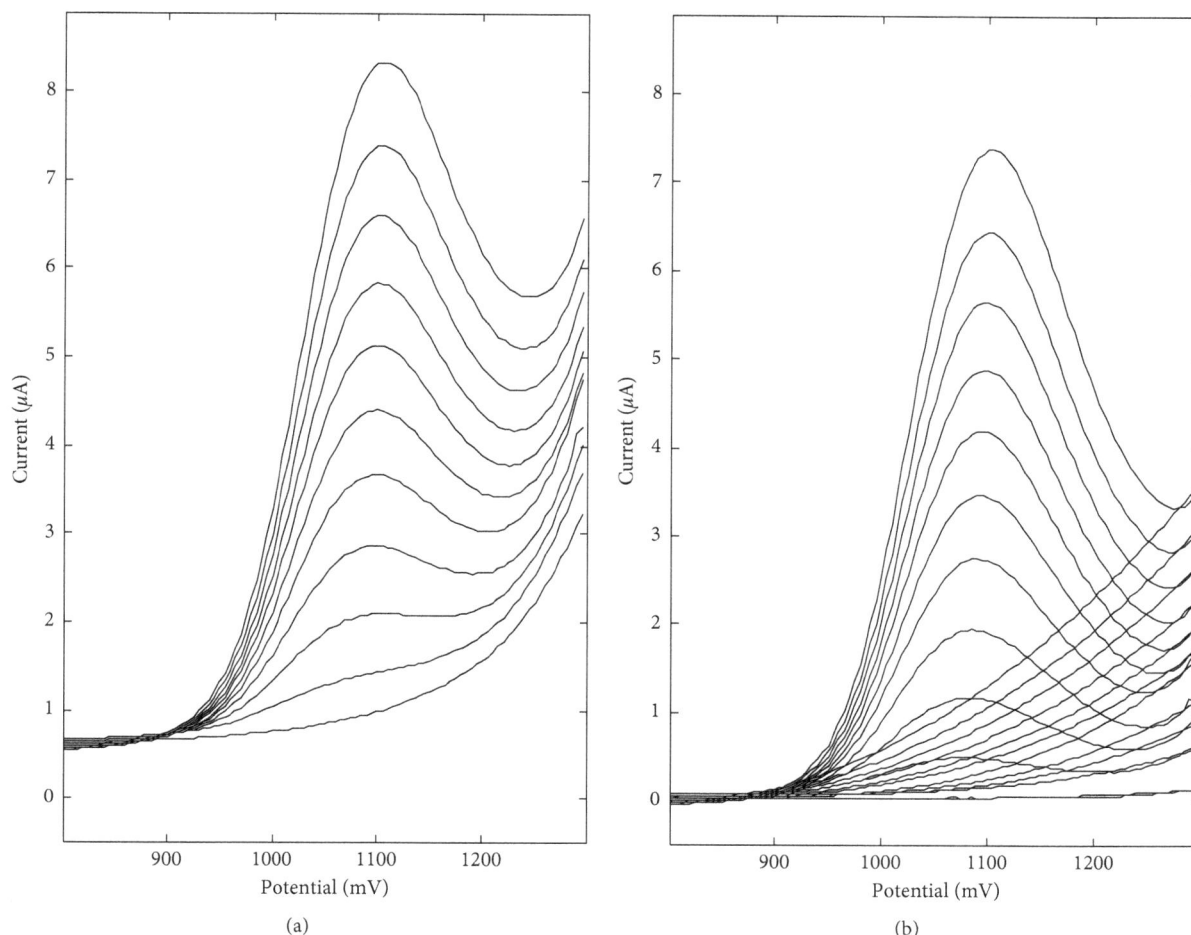

FIGURE 3: Strategy of baseline interpretation: (a) DP voltammograms in 0.1 mol L^{-1} acetate buffer of 0–6.96 mg L^{-1} of anethole on La$_2$O$_3$/CPE with experimental baseline; (b) the same voltammograms with subtracted experimental background and simulated baseline by the approximation using 2nd degree polynomials.

supporting electrolytes at strongly acidic pH (Britton–Robinson buffer, pH 2.0), neutral pH (phosphate buffer, pH 7.0), and basic pH (phosphate buffer, 8.0; ammonia buffer, pH 9.0), the investigated analyte did not show adequate analytical sensitivity and repeatability.

Figure 4 presents the influence of the acetate buffer pH on the anethole voltammetric signal. The well-defined DPV peak was observed in the whole range of the considered pH, that is, 3.0–6.0. For BDDE, the peak position changed in the range from 950 to 990 mV, without a distinct maximum current change. For La$_2$O$_3$/CPE, the oxidation peak currents decreased to pH 5.0 and then increased. Anethole oxidation potential decreased from 1180 mV to 1080 mV as pH increased. Further experiments were done by pH 6, because less positive peak position equal to 1080 mV is more suitable for oxidation. Sensitivity in this case was also ca. 44% greater in comparison with the best variant obtained for the other pH.

3.3. Parameters of Anethole Electrooxidation on BDDE and La$_2$O$_3$/CPE. The voltammetric behavior of anethole on two carbon electrodes, that is, BDD and La$_2$O$_3$-modified electrodes, in 0.1 mol L^{-1} acetate buffer of pH 6.0 has been investigated by recording cyclic voltammograms (CV) using the scan rates of 0.025, 0.05, 0.1, 0.2, 0.25, and 0.5 V s^{-1}. It was observed that anethole is irreversibly oxidized on these electrodes (Figure 5), what was confirmed by the absence of cathodic step on the backward branch of the CV. The CP electrode modification with La$_2$O$_3$ nanoparticles leads to the anodic shift of anethole oxidation potential on ca. 200 mV. The effect of potential scan rate in the range of 0.025–0.5 V s^{-1} on the voltammetric behavior of anethole is also presented in Figure 5. The anethole oxidation currents were proportional to the square root of the potential scan rate (1), confirming that the electrochemical process is diffusion controlled [23].

$$\text{BDD: } i_p\,[\mu A] = (2.59 \pm 0.12)v^{1/2}\left[\text{Vs}^{-1}\right]$$
$$+ (0.57 \pm 0.05), \quad r = 0.9953,$$

$$\text{La}_2\text{O}_3\text{/CPE: } i_p\,[\mu A] = (0.64 \pm 0.08)v^{1/2}\left[\text{Vs}^{-1}\right]$$
$$+ (1.34 \pm 0.03), \quad r = 0.9701. \quad (1)$$

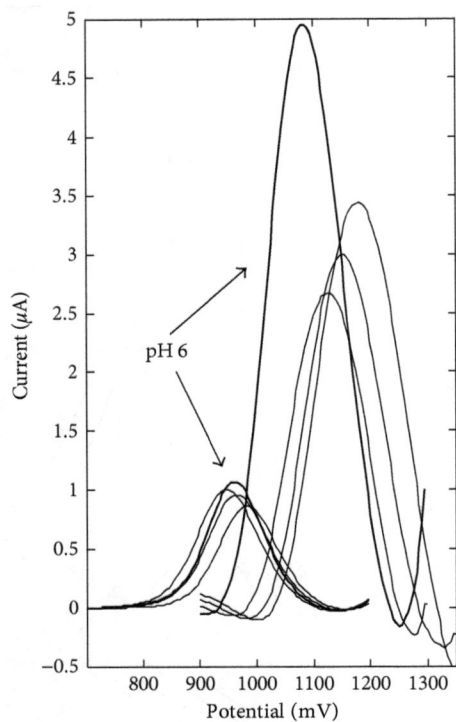

(a)

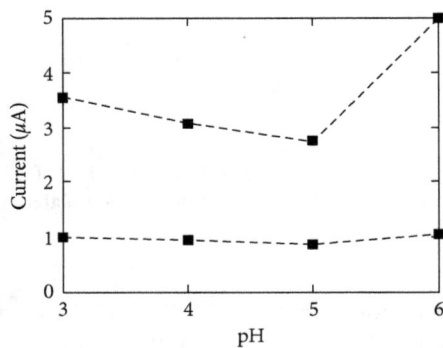

(b)

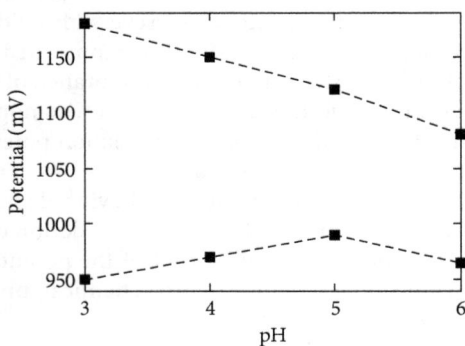

(c)

FIGURE 4: The effect of supporting electrolyte pH on the voltam-metric characteristics of anethole on BDDE (black) and La$_2$O$_3$/CPE (blue): (a) DP voltammograms obtained for 6.96 mg L^{-1} of anethole in 0.1 mol L^{-1} acetate buffer; (b) peak current in relation to pH; (c) peak potential in relation to pH.

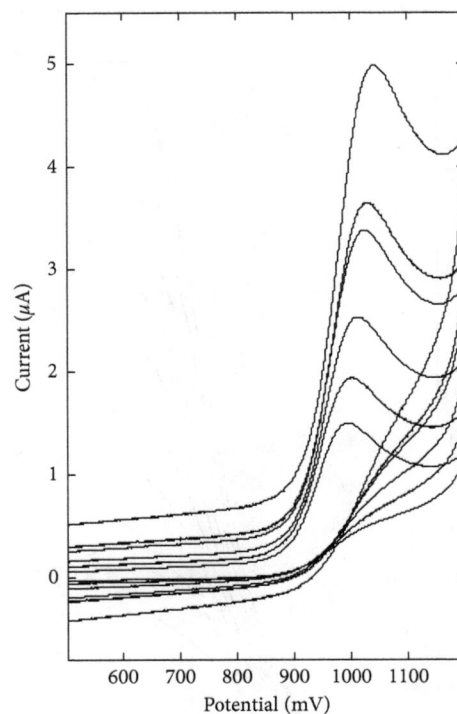

(a)

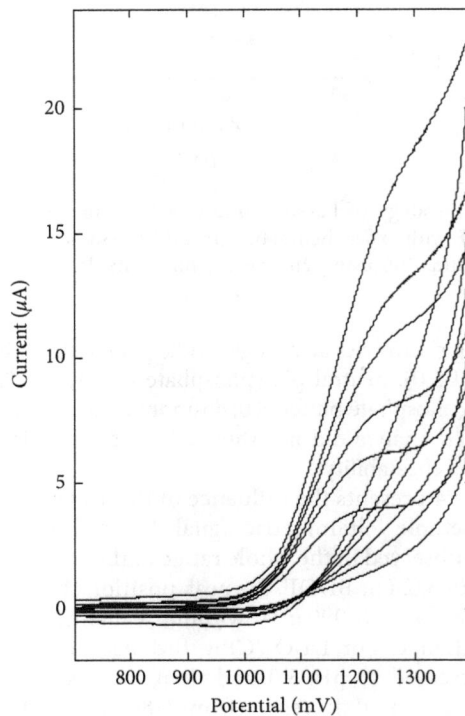

(b)

FIGURE 5: Cyclic voltammograms by the scan rate of 0.025, 0.05, 0.1, 0.2, 0.25, and 0.5 V s^{-1}, obtained for 13.92 mg L^{-1} of anethole in 0.1 mol L^{-1} acetate buffer on (a) BDDE and (b) La$_2$O$_3$/CPE.

Moreover, the natural logarithm of anethole peak cur-rent (lni_p) increases linearly with the natural logarithm of scan rate (ln$v$) in the range of potential scan rate under investigation, and the regression equation is described by

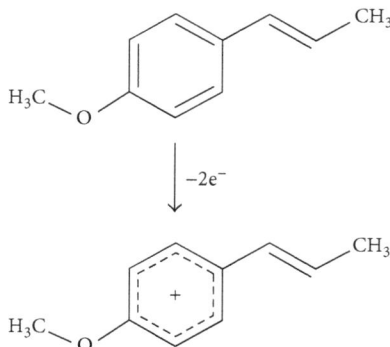

FIGURE 6: The proposed mechanism of oxidation of *trans*-anethole, on the basis of own experimental data and [25].

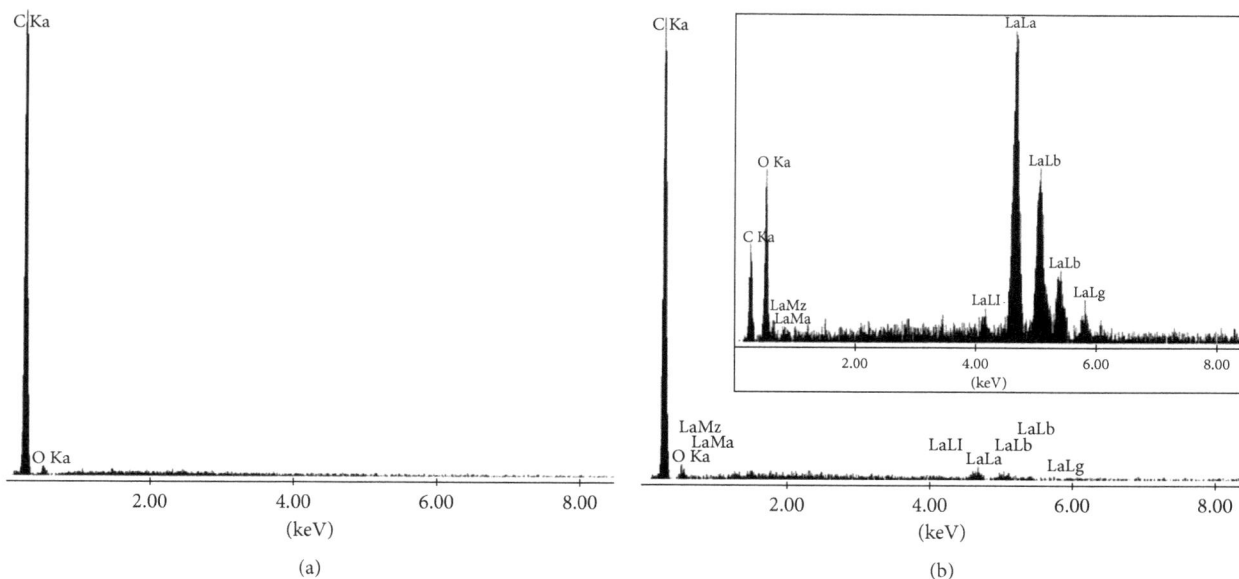

FIGURE 7: EDX spectrum for (a) nondoped carbon paste; (b) carbon paste doped by lanthanum(III) oxide. Inset: image of lanthanum(III) oxide nanopowder.

$$BDD: \ln i_p\,[\mu A] = (0.313 \pm 0.003)\ln v\left[Vs^{-1}\right]$$

$$+ (1.071 \pm 0.007), \quad r = 0.9998,$$

$$La_2O_3/CPE: \ln i_p\,[\mu A] = (0.073 \pm 0.009)\ln v\left[Vs^{-1}\right]$$

$$+ (0.617 \pm 0.021), \quad r = 0.9700. \tag{2}$$

The value of the slope is below the theoretical value of 0.5, what proves the diffusion nature of anethole oxidation peak once again [23]. A linear relationship between the oxidation potential E_p and $\ln v$ has been observed, confirming totally irreversible electrode processes:

$$BDD: E_p\,[V] = (0.017 \pm 0.001)\ln v\left[Vs^{-1}\right]$$

$$+ (1.053 \pm 0.002), \quad r = 0.9953,$$

$$La_2O_3/CPE: E_p\,[V] = (0.032 \pm 0.001)\ln v\left[Vs^{-1}\right]$$

$$+ (1.301 \pm 0.003), \quad r = 0.9972. \tag{3}$$

In this case, the number of electrons participating in the reaction can be calculated according to [24]

$$E_p - E_{p/2} = \frac{47.7}{(1-\alpha)n_\alpha}, \tag{4}$$

where α is assumed to be 0.5 for a totally irreversible electrode process. The $E_p - E_{p/2}$ is 53 mV for BDDE and 59 mV for La_2O_3/CPE. Hence, the number of electrons participating in the anethole oxidation process equals n_α to 2.22 for BDDE and 2.47 for La_2O_3/CPE, what agree well with the values reported earlier [25]. Figure 6 presents the proposition of the electrode reaction.

3.4. Investigation by Energy-Dispersive X-Ray Spectroscopy and Scanning Electron Microscopy. The chemical composition of the pastes used for construction of CPE and La_2O_3/CPE was analyzed by EDX. The EDX spectrum for the nondoped carbon paste (Figure 7(a)) confirmed the presence of carbon, as the dominant element, and a small

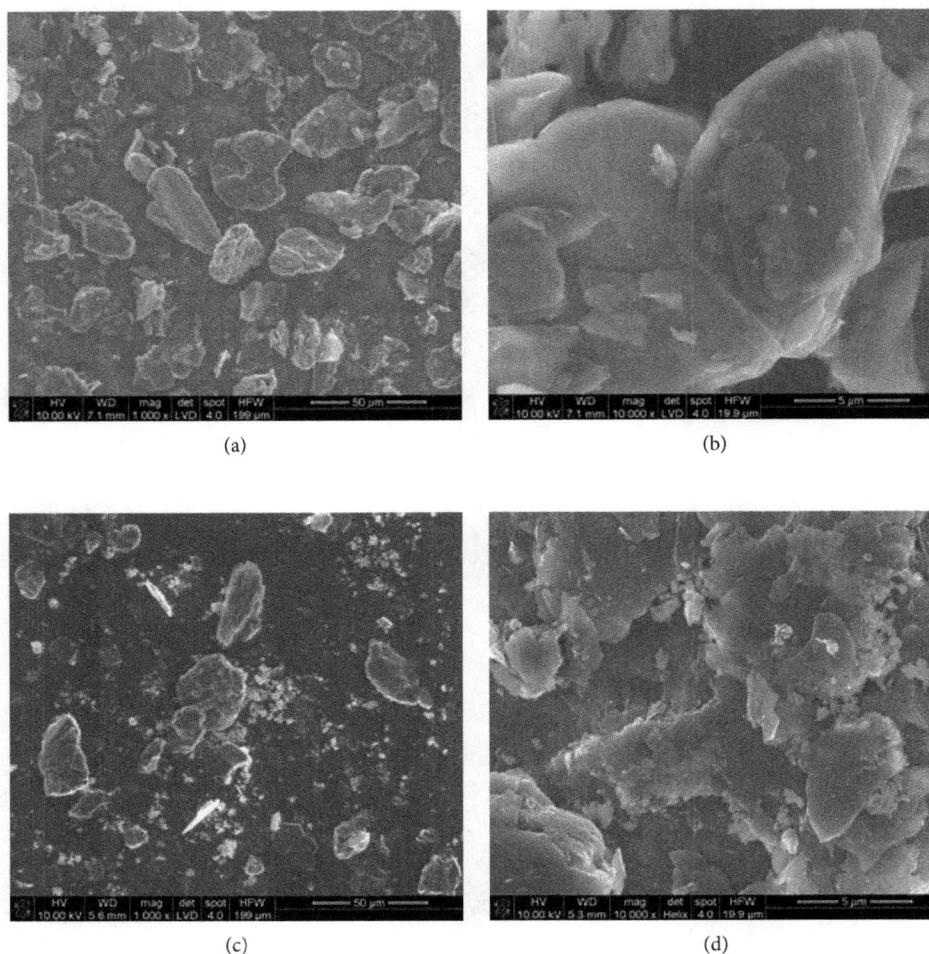

(a)   (b)

(c)   (d)

FIGURE 8: SEM image of the surface: (a) nondoped carbon paste, magnification 1000x; (b) nondoped carbon paste, 10,000x; (c) carbon paste doped by 20% of La$_2$O$_3$ (w/w), 1000x; (d) carbon paste doped by 20% of La$_2$O$_3$ (w/w), 10,000x.

amount of oxygen. The EDX spectrum for carbon paste doped by lanthanum(III) oxide (Figure 7(b)) confirmed the presence of the elements carbon, lanthanum, and oxygen.

The surface morphology of the CPE and La$_2$O$_3$/CPE was observed by SEM. The SEM test showed that the nondoped carbon paste was characterized by a surface formed by irregular graphite flakes (Figures 8(a) and 8 (b)). The surface of the La$_2$O$_3$-doped carbon paste is more porous, heterogeneous, and irregular than the surface of nondoped carbon paste (Figures 8(c) and 8(d)). This suggests that the presence of La$_2$O$_3$ molecules in carbon paste significantly increases the morphological structure of the material, which facilitates the electron transfer process in the electrode–solution interface, giving better sensitivity and higher repeatability of the voltammetric signal.

3.5. Application of Electrochemical Impedance Spectroscopy. Electrical properties of experimental set comprised of the studied carbon paste electrode and Ag/AgCl/KCl reference

electrode immersed in solution containing analyte were determined by Electrochemical Impedance Spectroscopy (EIS) method. The measurements were performed in room temperature, with the frequency range of 0.1–10 MHz and the amplitude of sinusoidal voltage signal of 20 mV. The experimental data were analyzed using the ZView software (version 2.2, Scribner Associates, Inc.), which helped in determination of equivalent circuits' optimal parameters.

The comparison of Nyquist's spectra obtained for three different carbon paste electrodes is given in Figure 9. In each case, spectrum was comprised of semicircle visible in high-frequency range and the spur in middle and low frequencies. Electrical equivalent circuits were fitted to the experimental data sets. Spectra were analyzed by connected in series two parallel equivalent circuits consisted of resistors (R) and constant phase elements (CE) and additional constant phase element indispensable to model the spur in low frequencies. The scheme of the used equivalent model is depicted in inset of Figure 9. The simulated spectra are plotted by solid black line and exhibit good agreement with experimental data presented by points. The semicircle parts of the spectra in

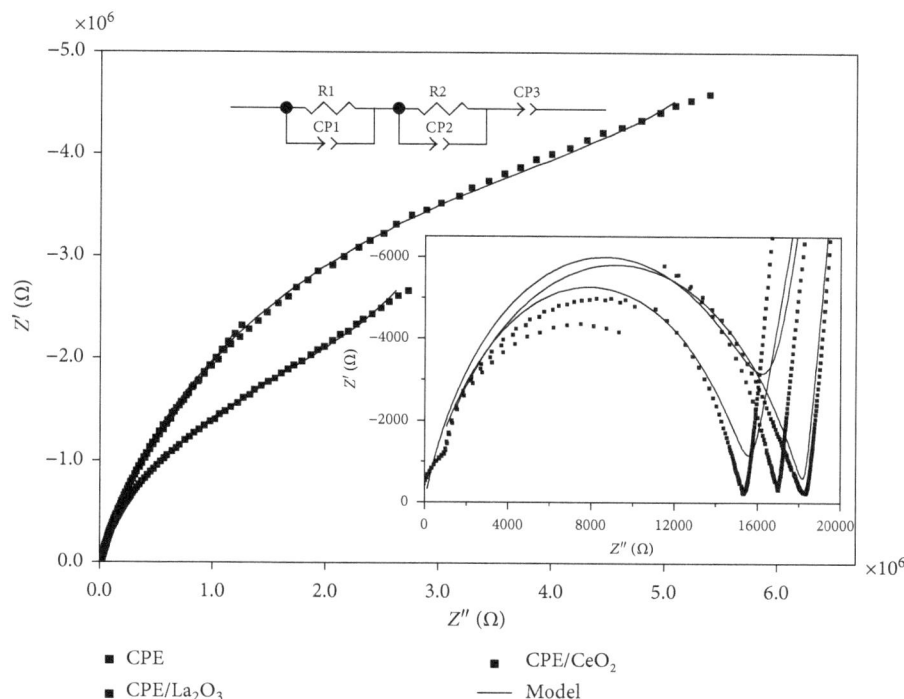

FIGURE 9: Nyquist's spectra for three different carbon paste electrodes (diameter 3 mm) with fitted equivalent circuit simulations.

TABLE 5: Parameter values of fitted equivalent circuits from Figure 9.

Frequency range	Parameter	Electrode		
		CPE	CeO$_2$/CPE	La$_2$O$_3$/CPE
High (10^6–10^7 Hz)	R1/Ω	18,248	15,680	16,751
	CE-T-1/Ss^n	7.850×10^{-10}	6.262×10^{-10}	5.009×10^{-10}
	$n1$	0.720	0.750	0.777
Medium (10^6–10^2 Hz)	R2/Ω	1.857×10^6	2.272×10^6	6.027×10^6
	CE-T-2/Ss^n	7.05×10^{-7}	2.65×10^{-7}	1.21×10^{-7}
	$n2$	0.875	0.841	0.857
Low (10^2–10^{-1} Hz)	CE-T-3/Ss^n	1.03×10^{-6}	6.66×10^{-7}	5.81×10^{-7}
	$n3$	0.940	0.765	0.881

high frequencies look similar in each case. The course of above mentioned part of the spectrum depends on the reference electrode and solution used during the measurements. Therefore, the parameters of R1 and CE1 are of similar value (Table 5). The differences in course of spectrum in the middle- and low-frequency parts indicate that it is attributable to carbon paste electrode properties. On the basis of conducted analysis, there is a strong relation between the applicable properties of carbon paste electrode and the value of resistance exhibited in middle-frequency fragments of spectra. In particular, the highest value of resistance R2 shows the electrode modified by lanthanum (III) oxide, while the electrode without rare earth oxide addition is characterized by the lowest value of this parameter. Concomitantly, the lower CE-T-2 value, the better performance of electrode. The most significant differences between behaviors of studied electrodes are visible in low

frequencies part of the spectra. It is reflected in particular in CE3 element. CE-T-3 value determined for undoped electrode is of order higher than for electrodes modified by rare earth metal oxides. Moreover, parameter $n3$ for CPE is somehow higher than for doped electrodes and close to 1, indicating stronger capacitive properties of undoped graphite than in the case of electrodes modified modified by lanthanum and cerium oxides.

3.6. Interferences. Such parameters as potential window, potential step, potential pulse, and time of potential step were tested to optimize the procedure of the anethole determination. The criteria of optimization were repeatability, sensitivity of the method, and the favorable relation between signal and baseline. It was observed that starting potential does not have influence on the anethole

TABLE 6: Influence of interferences ($n = 3$) on anethole peak of 13.92 mg L^{-1} in 0.1 mol L^{-1} acetate buffer, pH = 6.0.

Interferent	Proportion anethole: interferent	$i_{anethole+interferent}/i_{anethole}$	
		BDDE	La$_2$O$_3$/CPE 20%
Metal ions			
Zn^{2+}	20 : 0	1.00	1.00
	20 : 2	0.99	1.02
	20 : 10	0.99	1.21
	20 : 20	0.99	1.27
Cd^{2+}	20 : 0	1.00	1.00
	20 : 2	0.99	1.07
	20 : 10	0.98	1.12
	20 : 20	0.96	1.15
Pb^{2+}	20 : 0	1.00	1.00
	20 : 2	0.99	1.04
	20 : 10	0.99	1.10
	20 : 20	0.98	1.13
V^{3+}	20 : 0	1.00	1.00
	20 : 2	1.00	0.95
	20 : 10	0.99	0.88
	20 : 20	0.99	0.82
Bi^{3+}	20 : 0	1.00	1.00
	20 : 2	0.99	1.02
	20 : 10	0.98	1.04
	20 : 20	0.96	1.22
Al^{3+}	20 : 0	1.00	1.00
	20 : 2	1.00	1.02
	20 : 10	0.99	1.02
	20 : 20	0.99	1.03
Tl^{+}	20 : 0	1.00	1.00
	20 : 2	0.98	0.99
	20 : 10	0.95	0.91
	20 : 20	0.93	0.90
Cr^{3+}	20 : 0	1.00	1.00
	20 : 2	0.99	1.12
	20 : 10	0.99	1.22
	20 : 20	0.95	1.23
Organic compounds			
Eugenol	20 : 0	1.00	1.00
	20 : 1	1.19	1.14
	20 : 2	1.46	1.14
	20 : 5	1.90	0.84
Thymol	20 : 0	1.00	1.00
	20 : 1	1.11	1.11
	20 : 2	1.16	1.21
	20 : 5	1.32	1.26
Carvacrol	20 : 0	1.00	1.00
	20 : 1	1.23	1.09
	20 : 2	1.55	1.07
	20 : 5	2.00	1.04

peak. Taking into account all the criteria selected experiment conditions are potential step 5 mV, potential pulse 50 mV, and time of the potential step 40 ms (i.e., waiting time 20 ms + sampling time 20 ms).

As possible interferences, metal ions such as Zn(II), Pb (II), Cd(II), V(III), Bi(III), Al(III), Tl(I), and Cr(III) and organic compounds such as eugenol, carvacrol, and thymol were tested, which may be present in plants and products of biological origin, in which anethole also occurs. The concentration of the metal ions was in the range of 1.4–14 mg L^{-1} by the 13.92 mg L^{-1} of anethole, which was present in the measured solution. The anethole peak position was not moved, and also no additional peaks were observed. However, the impact of the analyzed ions on the height of the anethole peak after addition of metals was noticeable: change for BDDE anethole signal was in the range of 93–99%, and change for La$_2$O$_3$/CPE was in the range from 82 to 127% (Table 6). The greater sensitivity variation in the last case may be connected with the chemical reactions between metal ions and active lanthanum(III) oxide nanoparticles, what could cause the change of the number of active centres on the electrode surface.

No additional current peaks coming from eugenol, carvacrol, and thymol were observed in the considered acetate buffer (pH 6.0) and potential area where anethole peak was recorded. The concentration of the added substances was in the range of 0.7–3.5 mg L^{-1} by the 13.92 mg L^{-1} of anethole. The presence of biological compounds in the solution caused, in the experiments with BDDE, increase of the sensitivity up to 100%. This value is related to the study of the carvacrol effect at the 4 times excess of anethole. The mentioned interferents may facilitate the charge transfer between the analyte and the electrode. In the case of measurements on La$_2$O$_3$/CPE, an addition of 3 biological compounds resulted in the change of the signal amplitude in the range of 84–126%.

3.7. Determination of Anethole in Herbal Matrices and Commercial Products. Because anethole occurs in food products (beverages, herbal oils, and tinctures) and herbs such as anise, star anise, fennel, liquorice, and caraway, the problem of anethole determination of specially prepared herbal matrices was considered. The composition of these mixtures which mimics the real commercially available products is given in Table 7. It is important that some matrices contain anethole, while the other did not contain this analyte, and it was added at the stage of recovery studies. The concentration of the anethole in *matrices A, B*, and *D* was on the level 0.1–1.6 g L^{-1} (Table 7). The highest concentration was in the most complex mixture *B*, while the lowest in *D*, where only one component contained anethole. The significant decrease of the sensitivity of the method was observed in comparison to the measurements in only supporting electrolyte. The decrease was to 73% (*matrix B*) in the case of BDDE and to 66% (*matrix D*) in the case of La$_2$O$_3$/CPE. Exemplary voltammograms recorded on La$_2$O$_3$/CPE in the case of anethole determination in *matrix D* are presented in Figure 10(a).

Matrix C did not contain a detectable concentration of the anethole; therefore, this analyte was added to the herbal extract, and percent of recovery was studied (Table 8). Using BDDE, the concentration of anethole of 3.5–10.5 mg L^{-1} was successfully determined with recovery 101–108%. La$_2$O$_3$/CPE enabled determination of 1.4–4.2 mg L^{-1} of anethole with recovery 95–100%. The correlation coefficient r in each case was greater than 0.995. The presence of herbal

TABLE 7: Determination of anethole in four herbal matrices ($n = 3$).

Matrix label	Composition of herbal mixture		Amount of ethanol for maceration	In the matrix may be anethole	Anethole conc. $\pm$ SD (g L^{-1})
A	*Fruit of anise*	7.5 g	25 mL	Yes	BDDE: 0.143 ± 0.005
	Fruit of star anise	2.5 g			La$_2$O$_3$/CPE: 0.102 ± 0.024
B	Hyssop leaves	2.125 g			
	Root of the sweet flag	0.45 g			
	Lemon balm leaves	1.5 g			BDDE: 1.63 ± 0.07
	Fruit of anise	7.5 g	50 mL	Yes	La$_2$O$_3$/CPE: 1.61 ± 0.19
	Fruit of star anise	2.5 g			
	Fruit of fennel	6.25 g			
	Fruit of coriander	0.75 g			
C	Wormwood leaves	7.5 g	50 mL	No	BDDE < LOD
					La$_2$O$_3$/CPE < LOD
D	Wormwood leaves	7.5 g			
	Hyssop leaves	2.125 g			
	Root of the sweet flag	0.45 g			BDDE: 0.500 ± 0.005
	Lemon balm leaves	1.5 g	50 mL	Yes	La$_2$O$_3$/CPE: 0.475 ± 0.033
	Fruit of fennel	6.25 g			
	Fruit of coriander	0.75 g			

Italic signs herbs from which it is possible to extract anethole.

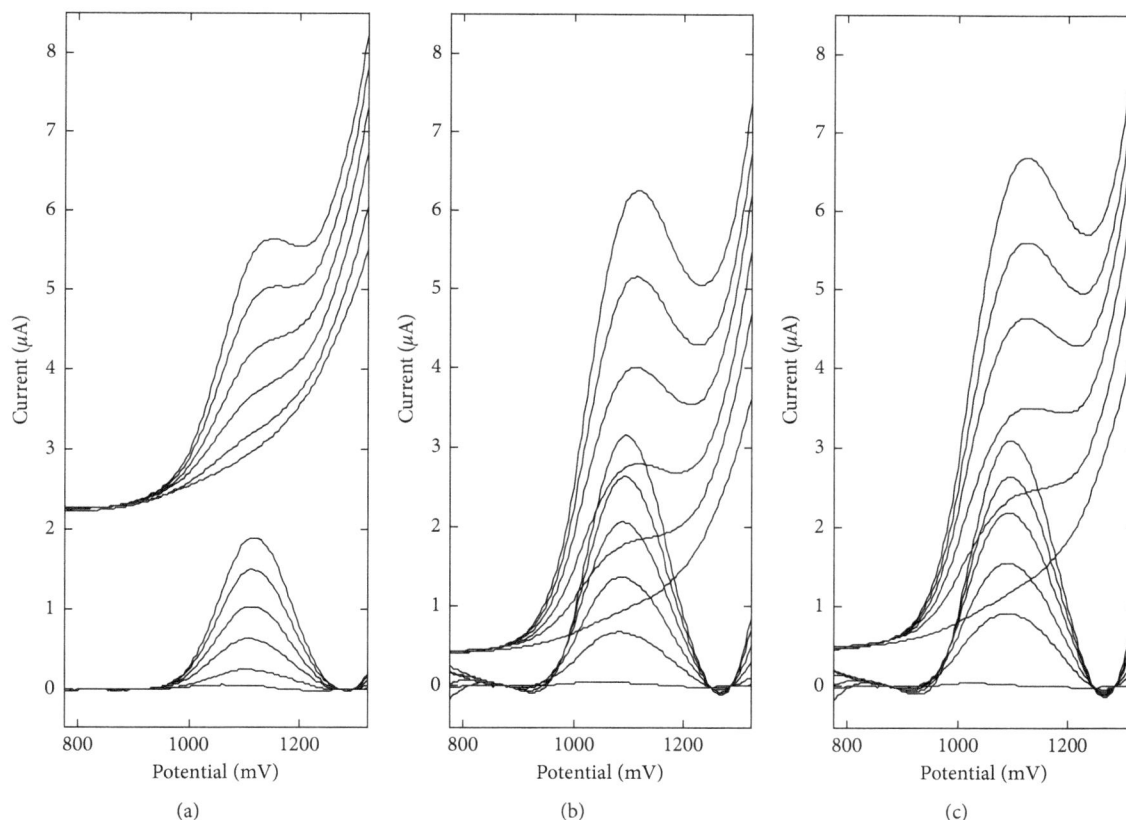

FIGURE 10: Standard addition voltammograms (black) and the same curves after subtracting experimental background, and next simulated baseline approximated using 3rd degree polynomials (blue), in determination of the anethole in 0.1 mol L^{-1} acetate buffer on La$_2$O$_3$/CPE, in (a) *matrix D*, (b) Ouzo, and (c) Raki. Each voltammogram set: background, sample, and 1.39, 2.78, 4.18, and 5.57 mg L^{-1} standard additions of anethole.

matrix C caused a significant decrease of the method sensitivity, that is, ca. 30% in the case of BDDE and even 50% in the case of La$_2$O$_3$/CPE.

Further, anethole was determined in commercially available products. Some research objects were chosen in which the mentioned analyte is present on very low and very

TABLE 8: Standard addition voltammetry of anethole in herbal matrix C (n = 3).

Electrode	Added (mg L^{-1})	$a \pm SD_a$ (μA/mg L^{-1})	$b \pm SD_b$ (μA)	Found $\pm$ SD (mg L^{-1})	Recovery (%)	r
BDDE 3 mm	3.48	0.0939 $\pm$ 0.04	0.3460 $\pm$ 0.05	3.68 $\pm$ 0.70	105.87	0.9951
	6.96	0.0973 $\pm$ 0.03	0.6859 $\pm$ 0.03	7.05 $\pm$ 0.51	101.30	0.9982
	10.44	0.0963 $\pm$ 0.04	1.0814 $\pm$ 0.02	11.22 $\pm$ 0.61	107.51	0.9987
La$_2$O$_3$/CPE 3 mm	1.39	0.4306 $\pm$ 0.02	0.5995 $\pm$ 0.11	1.39 $\pm$ 0.19	100.02	0.9951
	2.78	0.435 $\pm$ 0.003	1.186 $\pm$ 0.001	2.728 $\pm$ 0.039	95.79	0.9999
	4.18	0.4409 $\pm$ 0.01	1.8124 $\pm$ 0.01	4.11 $\pm$ 0.07	98.45	0.9999

TABLE 9: Determination of anethole in commercial products, calibration parameters ($n = 3$).

Samples	$a \pm SD_a$ (μA/mg L^{-1})	$b \pm SD_b$ (μA)	r
BDDE, electrode diameter 3 mm, area 7.07 mm^2			
Anise essential oil	0.0293 $\pm$ 0.001	0.0667 $\pm$ 0.005	0.9967
Metaxa***	0.0535 $\pm$ 0.002	0.0691 $\pm$ 0.008	0.9975
Ouzo	0.0905 $\pm$ 0.004	0.1699 $\pm$ 0.014	0.9971
Raki	0.1068 $\pm$ 0.004	0.2216 $\pm$ 0.018	0.9967
La$_2$O$_3$/CPE, electrode diameter 3 mm, area 7.07 mm^2			
Anise essential oil	0.3682 $\pm$ 0.018	0.8544 $\pm$ 0.078	0.9951
Metaxa***	0.8061 $\pm$ 0.032	1.1490 $\pm$ 0.068	0.9968
Ouzo	0.7177 $\pm$ 0.020	1.4659 $\pm$ 0.067	0.9989
Raki	0.4497 $\pm$ 0.010	0.9507 $\pm$ 0.034	0.9993

TABLE 10: Anethole concentration in commercial products ($n = 3$).

Samples	Anethole conc. $\pm$ SD/g L^{-1}		F-test	t-test
	BDDE	La$_2$O$_3$/CPE		
Anise essential oil	569.0 $\pm$ 62.5	580.00 $\pm$ 77.5	1.5376	0.1914
Metaxa***	0.129 $\pm$ 0.019	0.143 $\pm$ 0.014	1.8420	1.0275
Ouzo	0.188 $\pm$ 0.022	0.204 $\pm$ 0.014	2.4691	1.0627
Raki	0.208 $\pm$ 0.025	0.211 $\pm$ 0.012	4.3403	0.1874

high concentration level. The measurements were done without sample preparation or anethole extraction. The adequate sample volume was added directly to the electrochemical cell. It was observed (Tables 9 and 10) that, in anise essential oil, the concentration of anethole was ca. 570 g L^{-1}, while in alcohol drinks, like Metaxa, Ouzo, and Rakija, it was ca. 0.13–0.21 g L^{-1}. The results obtained using both electrodes were compatible. At 95% confidence level, the calculated Student's t-values for the replicate measurements of each sample (Table 10) using both fabricated sensors did not exceed the theoretical value (2.7765), indicating that the results obtained are not significantly different. An F-test revealed no significant difference between the standard deviations of the two sets of replicate measurements for each sample. Exemplary voltammograms recorded on La$_2$O$_3$/CPE in the case of anethole determination in Ouzo and Raki are presented in Figures 10(b) and 10(c).

4. Conclusions

In this work, a sensitive, rapid, and convenient DPV procedure of anethole determination was proposed, which does not require sample preparation and separation of the analyte, even in the case of complex matrices. Additionally, it was proved that various carbon electrodes, that is, BDD, GC, CP, and CP doped by La$_2$O$_3$, and CeO$_2$ nanoparticles, are sensitive for anethole, and the proposed analytical strategies fulfill typical validation criteria. Recording cyclic voltammograms, it was noticed that electrooxidation process has totally irreversible character, controlled by diffusion, in which two electrons take part.

The most sensitive electrode turned out to be La$_2$O$_3$/CPE with 20% of nanoparticles in graphite paste (w/w). According to our knowledge, it is the first literature report about application of such a sensor. Sensitivity obtained in DPV experiments realized by optimized parameters in 0.1 mol L^{-1} acetate buffer was for La$_2$O$_3$/CPE of 3 mm diameter equal to 0.885 $\pm$ 0.016 μA/mg L^{-1} with the correlation coefficient r of 0.9993 and the detection limit of 0.004 mg L^{-1}, while for commercially available sensor BDDE it was 0.135 $\pm$ 0.003 μA/mg L^{-1} with r of 0.9990.

Operation of the selected electrodes was verified using especially prepared herbal ethanol extracts which contained and did not contain anethole. In the last case, recovery was tested applying standard addition method. Anethole was also successfully determined in commercially available products, such as anise essential oil, which contains a large concentration of anethole, and in alcohol drinks like Metaxa, Ouzo, and Rakija, in which the considered analyte occurs on trace levels. The results obtained on La$_2$O$_3$/CPE and BDDE were statistically consistent, at 95% confidence level.

Conflicts of Interest

The authors declare that there are no conflicts of interest regarding the publication of this article.

Acknowledgments

This work was supported by the National Science Centre, Poland (Project no. 2015/19/B/ST5/01380).

References

[1] L. S. Moura, R. N. Carvalho Jr., M. B. Stefanini, L. C. Ming, and M. A. A. Meireles, "Supercritical fluid extraction from fennel (*Foeniculum vulgare*): global yield, composition and kinetic data," *Journal of Supercritical Fluids*, vol. 35, no. 3, pp. 212–219, 2005.

[2] G.-W. Wang, W.-T. Hu, B.-K. Huang, and L.-P. Qin, "*Illicium verum*: a review on its botany, traditional use, chemistry and pharmacology," *Journal of Ethnopharmacology*, vol. 136, no. 1, pp. 10–20, 2011.

[3] M. Wichtl, *Herbal Drugs and Phytopharmaceuticals*, Med-Pharm GmbH Scientific Publishers, Stuttgart, Germany, 2004.

[4] A. A. Shahat, A. Y. Ibrahim, S. F. Hendawy et al., "Chemical composition, antimicrobial and antioxidant activities of essential oils from organically cultivated fennel cultivars," *Molecules*, vol. 16, no. 12, pp. 1366–1377, 2011.

[5] R. S. Freire, S. M. Morais, F. E. A. Catunda-Juniora, and D. C. S. N. Pinheiro, "Synthesis and antioxidant, anti-inflammatory and gastroprotector activities of anethole and related compounds," *Bioorganic and Medicinal Chemistry*, vol. 13, no. 13, pp. 4353–4358, 2005.

[6] A. Kołodziejczyk, *Natural Organic Compounds*, PWN, Warsaw, Poland, 2013.

[7] S. Ariamuthu, V. Balakrishnan, and M. L. Srinivasan, "Chemical composition and antibacterial activity of essential oil from fruits of *Illicium verum* Hook. f.," *International Journal of Research in Pharmacology and Phytochemistry*, vol. 3, pp. 85–89, 2013.

[8] H. Ullah, A. Mahmood, and B. Honermeier, "Essential oil and composition of anise (*Pimpinella anisum* L.) with varying seed rates and row spacing," *Pakistan Journal of Botany*, vol. 46, no. 5, pp. 1859–1864, 2014.

[9] A. Raal, E. Arak, and A. Orav, "The content and composition of the essential oil Found in *Carum carvi* L. commercial fruits obtained from different countries," *Journal of Essential Oil Research*, vol. 24, no. 1, pp. 53–59, 2012.

[10] M. K. Abderrezak, I. Abaza, T. Aburjai, A. Kabouche, and Z. Kabouche, "Comparative compositions of essential oils of *Citrus aurantium* growing in differents oils," *ournal of Materials and Environmental Science*, vol. 5, no. 6, pp. 1913–1918, 2014.

[11] M. K. Ibrahim, Z. A. Mattar, H. H. Abdel-Khalek, and Y. M. Azzam, "Evaluation of antibacterial efficacy of anise wastes against some multidrug resistant bacterial isolates," *Journal of Radiation Research and Applied Sciences*, vol. 10, no. 1, pp. 34–43, 2017.

[12] P. Brereton, S. Hasnip, S. Bertrand, R. Wittkowski, and C. Guillou, "Analytical methods for the determination of spirit drinks," *TrAC Trends in Analytical Chemistry*, vol. 22, no. 1, pp. 19–25, 2003.

[13] M. Przygodzka, D. Zielińska, Z. Ciesarová, K. Kukurová, and H. Zieliński, "Comparison of methods for evaluation of the antioxidant capacity and phenolic compounds in common spices," *LWT–Food Science and Technology*, vol. 58, no. 2, pp. 321–326, 2014.

[14] M. Kowalcze and M. Jakubowska, "Voltammetric profiling of absinthes," *Journal of Electroanalytical Chemistry*, vol. 776, pp. 114–119, 2016.

[15] J. Fiori, M. Hudaib, L. Valgimigli, S. Gabbanini, and V. Cavrini, "Determination of *trans*-anethole in *Salvia sclarea* essential oil by liquid chromatography and GC-MS," *Journal of Separation Science*, vol. 25, no. 10-11, pp. 703–709, 2002.

[16] J. M. Murado, A. Alcazar, F. Pablos, and M. J. Martin, "LC determination of anethole in aniseed drinks," *Chromatographia*, vol. 64, no. 3-4, pp. 223–226, 2006.

[17] K. Schulz, K. Schlenzc, R. Metaschd, S. Malt, W. Romhild, and J. Dressler, "Determination of anethole in serum samples by headspace solid-phase microextraction-gas chromatography–mass spectrometry for congener analysis," *Journal of Chromatography. A.*, vol. 1200, no. 2, pp. 235–241, 2008.

[18] E. Mehmeti, D. M. Stankovic, S. Chaiyo, L. Svorc, and K. Kalcher, "Manganese dioxide-modyfied carbon paste electrode for voltammetric determination of riboflavin," *Microchimica Acta*, vol. 183, no. 5, pp. 1619–1624, 2016.

[19] N. Al-Qasmi, M. T. Soomro, M. Aslam et al., "The efficacy of the ZnO: α-Fe$_2$O$_3$ composites modified carbon paste electrode for the sensitive electrochemical detection of loperamide: a detailed investigation," *Journal of Electroanalytical Chemistry*, vol. 783, pp. 112–124, 2016.

[20] P. Khanh Quoc Nguyen and S. K. Lunsford, "Square wave anodic stripping voltammetric analysis of lead and cadmium utilizing titanium dioxide/zirconium dioxide carbon paste composite electrode," *Journal of Electroanalytical Chemistry*, vol. 711, pp. 45–52, 2013.

[21] V. Arun and K. R. Sankaran, "Nafion coated TiO$_2$ and CuO doped TiO$_2$ modified glassy carbon and platinum electrodes: preparation characterization and application of enhancement of electrochemical sensitivity of azines," *Journal of Electroanalytical Chemistry*, vol. 769, pp. 35–41, 2016.

[22] G. Ziyatdinova, E. Ziganshina, S. Romashkina, and H. Budnikov, "Highly sensitive amperometric sensor for eugenol quantification based on CeO$_2$ nanoparticles and surfactants," *Electroanalysis*, vol. 29, no. 4, pp. 1197–1204, 2017.

[23] R. S. Nicholson and I. Shain, "Theory of stationary electrode polarography. Single scan and cyclic methods applied to reversible, irreversible, and kinetic systems," *Analytical Chemistry*, vol. 36, no. 4, pp. 706–723, 1964.

[24] A. J. Bard and L. R. Faulkner, *Electrochemical Methods: Fundamentals and Applications*, John Wiley & Sons, New York, NY, USA, 2nd edition, 2001.

[25] L.-H. Wang, C.-L. Chang, and Y.-C. Hu, "Electrochemical oxidation of fragrances 4-allyl and 4-propenylbenzenes on platinum and carbon paste electrodes," *Croatica Chemica Acta*, vol. 88, no. 1, pp. 35–42, 2015.

4

Solid-Liquid Separation Properties of Thermoregulated Dicationic Ionic Liquid as Extractant of Dyes from Aqueous Solution

Rui Lv⬮, Shuya Cui⬮, Yangmei Zou, and Li Zheng

Department of Chemistry and Chemical Engineering, Mianyang Teacher's College, Mianyang, Sichuan 621000, China

Correspondence should be addressed to Shuya Cui; cuishuya@hotmail.com

Academic Editor: Ricardo Jorgensen Cassella

Two thermoregulated dicationic ionic liquids were synthesized and applied for effective extraction of the common dye malachite green oxalate (MG). The extraction parameters such as amount of ionic liquids, pH of water phase, extraction time, cooling time, and centrifugal time on the extraction efficiency were investigated systematically. It revealed that the dye has been successfully extracted into the ionic liquids, with high extraction efficiency higher than 98%, and recovery of 98.2%–100.8%, respectively. Furthermore, these ionic liquids can be recycled easily after elution. The reusable yields were 87.1% and 88.7%. The extraction of the dye into the thermoregulated ionic liquid provides a method of minimizing pollution of waste water potentially.

1. Introduction

Ionic liquid is a widely used extraction solvent, because of its high heat stability, vapor pressure, nonvolatility, and friendly environment perception [1]. In recent years, there are many researches about ionic liquid as an extractant for dye from various samples such as Congo red, Sudan dye, methyl orange, and so on [2–4], but the application of ionic liquids is restrained due to its high cost and hard to recycle [5, 6].

Thermoregulated ionic liquids are a kind of ionic liquids, with appearing liquid phase at high temperatures and solid phase at low temperatures [7]. They can be used to achieve separation and enrichment of the target compounds by changing the temperature. Further, based on the advantages, it can be easily recycled. In recent years, the reports about using temperature-sensitive ionic liquids as extractant have been mostly concentrated in the area of metal ions [8, 9], but rarely involved in dyes.

Malachite green (MG), which is one of the triphenylmethane dyes, is widely used in aquaculture to prevent fungal infections and kill parasites due to its low price. However, it was found that malachite green had potential carcinogenic, teratogenic, and mutagenic effects [10]. So, it was banned to use in aquaculture in several countries in the world [11]. Therefore, separation and determination of malachite green in water is very important to human health.

As the residues of malachite green in water are low, the efficient enrichment method is needed. Common methods of separation and enrichment are physical adsorption [12, 13], liquid-liquid extraction [14, 15], and solid-phase extraction (SPE) [16, 17]. These methods have lots of disadvantages to limit their application because they are expensive or use organic solvent which may lead to environmental pollution. Therefore, looking for a simple, efficient green method becomes necessary.

In this study, two thermoregulated ionic liquids with hydrophobic properties were synthesized and employed to separate and concentrate malachite green in water. Some parameters which would influence the separation process were investigated, and the optimum condition was obtained. Further, the feasibility of the method was validated by analyzing the real water sample and simulation water sample. The efficient, simple, and environmental friendly method was confirmed to be useful to separate and enrich the malachite green in water.

2. Materials and Methods

2.1. Chemicals. Standard malachite green was supplied by the National Research Center for Reference Material (Beijing, China). N-methylimidazole (>99%) and KPF$_6$ (>99%) were purchased from Shanghai Aladdin Reagent Co., Ltd. (Shanghai, China). Thionyl chloride, diethylene glycol, triethylene glycol, Na$_2$HPO$_4$, and NaH$_2$PO$_4$ were all of analytical grade and brought from Chengdu Kelong Reagent Co., Ltd. (Sichuan, China).

2.2. Apparatus. The melting point of ionic liquid was tested by XT-4 melting point meter (Beijing Tech Instrument Co., Ltd.) at room temperature with the thermometer uncorrected. ^{1}H NMR and ^{13}C NMR spectra were recorded on 600 MHz NMR spectrometer (Varian, USA) with the solvent of acetone-d. FTIR spectra were registered on IRAffinity-1S FTIR spectrometer (Shimadzu, Japan). UV-visible spectra were obtained from a T6 UV-visible spectrophotometer (Beijing Persee, China) at the maximum absorption wavelength of 616.9 nm. TG16-W high-speed centrifugation (Hunan Xiangyi Laboratory Instrument Development Co., Ltd., China) was employed to accelerate the phase separation process. The temperature was controlled by HH-6 thermostatic water bath (Changzhou Jintan Liangyou Instrument Co., Ltd., China). The pH of the extraction system was measured with PHSJ-3F pH meter (Shanghai Inesa Analytical Instrument Co., Ltd., China).

2.3. Water Sample Collection. Water samples were collected from fish ponds in Mianyang, China. All the water samples were filtered through 0.45 μm microporous membrane filter before extraction.

2.4. Preparation of Ionic Liquids. With diethylene glycol-bridged functionalized imidazolium dicationic ionic liquids as an example, synthesis route of polyethylene glycol-bridged functionalized imidazolium dicationic ionic liquids was shown in four steps as follows: Firstly, thionyl chloride was added to diethylene glycol in anhydrous environment and then heated to 250°C until no starting material was observed using TLC. And then, it was cooled and extracted with CHCl$_2$. The extract was washed with water, saturated NaHCO$_3$ solution, and saturated NaCl solution. The product is dried under reduced pressure. Secondly, N-methylimidazole was added to the above dichlorodiglycol in methanol at room temperature. The reaction continued for 24 h. The solvent was removed under reduced pressure to yield a viscous gum and soluble in water and in methanol. The obtained dichloride salt was used directly, without further purification. Then, the obtained dichloride salt was dissolved in methanol, and then, KPF$_6$ was added with rapid stirring. After four hours, the solvent was removed under reduced pressure. The residue was washed by water and filtrated to collect the solid. The solid was recrystallized in acetone and the colorless crystal generated, named IL-1 and 2IL-2. For IL-1, the yield was 90%. m.p.: 91–93°C. IR (νmax,

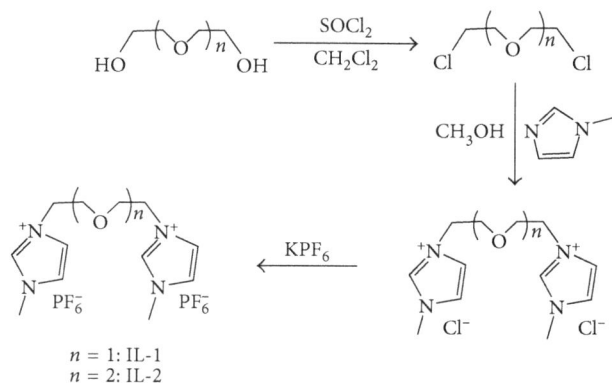

FIGURE 1: Synthesis route of two thermoregulated dicationic ionic liquids.

KBr, cm^{-1}): 3174, 2916, 1627, 1577, 1460, 1324, 1161, 850, 557. ^{1}H NMR (600 MHz, CD$_3$COCD$_3$, δ, ppm): 3.97 (s, 6H, N–CH$_3$), 4.05 (4H, s, N–CH$_2$), 4.32 (4H, s, O–CH$_2$), 7.68 (4H, C (4,5)–H), 9.03 (2H, C(2)–H); ^{13}C NMR (150 MHz, CD$_3$COCD$_3$, δ, ppm): 136.7 (C-2), 124.5, 124.0 (C-4,5), 69.5 (O–CH$_2$), 50.3 (N–CH$_2$), and 36.6 (N–CH$_3$). The structure of IL-2 was consistent with the data reported by the previous literature [18].

The synthesis route and structures of these two thermoregulated dicationic ionic liquids are shown in Figure 1.

2.5. Preparation of Standard Solution. A stock solution of malachite green (0.5 wt.%) was prepared by dissolving in distilled water. Other concentrations of the sample solution were prepared by diluting the stock solution with distilled water.

2.6. Experimental Methods. The extraction experiment was carried out in the sample solution with two ionic liquids (IL-1 and IL-2) as extractants, respectively. Firstly, mix the sample solution and ionic liquid and heat the mixture solution to make the ionic liquid melt and distribute in the sample solution for extraction (90°C for IL-1 solution and 60°C for IL-2). After that, the mixture was cooled to room temperature while the IL was separated. Finally, the mixture was centrifuged at 3000 r/min to get the supernatants, and the concentration of dye was measured using UV spectrophotometer (616.9 nm). The extraction ratio can be calculated by the following equation:

$$R\,(\%) = \frac{(C_0 - C_1)V_{\text{aq}}}{C_0 V_{\text{aq}}} \times 100\%, \qquad (1)$$

where C_0 represents the concentration of malachite green in the sample solution, C_1 refers to the concentration of malachite green in supernatants after extraction, and V_{aq} is the volume of the extraction phase. C_1 was obtained by standard curve calculation.

3. Results and Discussion

3.1. Effect of the Amount of Ionic Liquids. The amount of extractive agent was an important parameter for sample

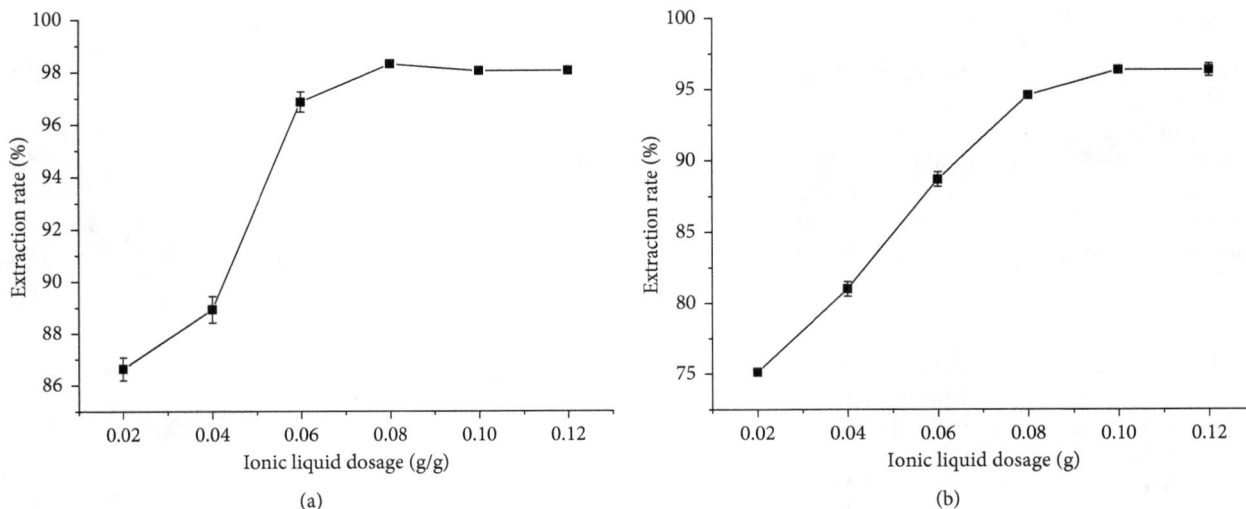

FIGURE 2: Effect of the amount of ionic liquids on extraction rate. (a) IL-1; (b) IL-2. The error bars represent the standard deviation of measurements in three parallel experiments ($n = 3$).

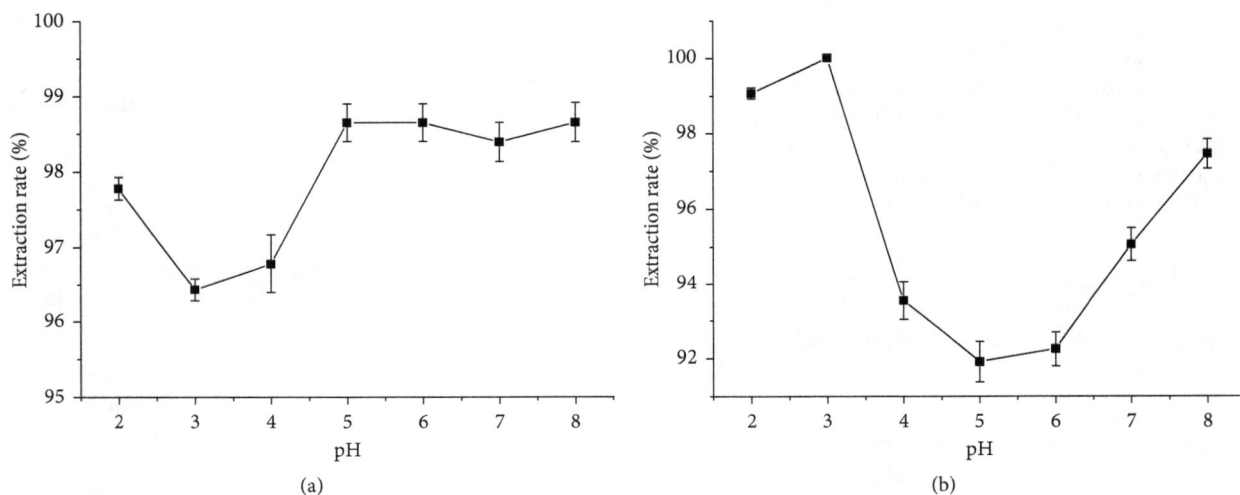

FIGURE 3: Effect of the pH on extraction rate. (a) IL-1; (b) IL-2. The error bars represent the standard deviation of measurements in three parallel experiments ($n = 3$).

extraction efficiency. The effect of the amount of ionic liquid on the extraction rate was examined and the result was shown in Figure 2. It was found that the extraction rate reached the maximum of 98.39% when the weight of IL-1 increased to 0.0800 g, while the maximum extraction rate was 96.35% when the IL-2 was 0.1000 g. Therefore, 0.0800 g IL-1 and 0.1000 g IL-2 were chosen, respectively, for the following experiments.

3.2. Effect of pH. The structure of malachite green varied with pH in solution, which affected the extraction rate of ionic liquids. Phosphate buffer was applied to control pH values. The effect of pH on the extraction rate was determined (Figure 3). Results showed that the extraction rate was varying slightly with pH when IL-1 was employed as the extractant; by comparison, when IL-2 was employed as the

extractant, the extraction rate first increased and then decreased with the increase in pH. But both of them can obtain the optimum extraction efficiency in the acidic solution. It may be because the pKa of MG is 6.9; when the pH was lower than it, the MG presents in its molecular state, which may help in extraction. But owing to the different molecular space conformation in the two ILs, they may have different effects depending on the pH value. Therefore, pH 5 was chosen as the optimal pH for IL-1, and pH 3 was chosen as the optimal pH for IL-2.

3.3. Effect of Extraction Time. Ionic liquids show the liquid state when the extraction temperature was set above their melting point. Under this temperature, the target compounds in water can be extracted. Therefore, in this study, the extraction temperature was set at 90°C for IL-1, while it

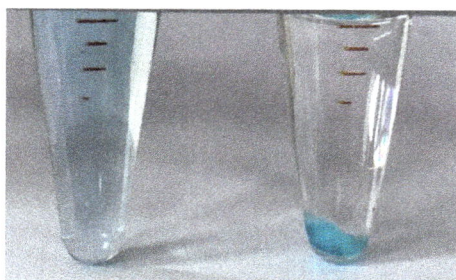

FIGURE 4: The ionic liquid at extraction stage and separation stage.

was set at 60°C for IL-2, respectively, to investigate the effects of extraction time (2–12 min with 2 min interval) on the extraction rate. From the results, it was found that when IL-1 was used as an extraction agent, the extraction rate reached the maximum when the extraction time was 4 min. At the same time, when IL-2 was used as an extraction agent, the extraction rate can reach 100% when the time was 1 min.

3.4. Effect of Cooling Time. For thermoregulated ionic liquid, when the temperature is below its melting point, it will be a hydrophobic solid precipitation, and cooling time determines the amount of precipitation. So, the effect of cooling time (1–11 min with 2 min interval) on extraction was studied. The experiment results showed that when the cooling time was 3 min for IL-1 and 7 min for IL-2, it reached the maximum extraction rate.

3.5. Effect of Centrifugal Time. The centrifugal process is to separate the solid extraction reagent and sample solution. The effect of centrifugal time (5–15 min with 5 min interval) on the extraction rate was investigated. It was found that when centrifuged at 3000 r/min for 5 min, the extraction rate reached the maximum.

In summary, the optimum condition for IL-1 was as follows: 0.0800 g IL, buffer pH 5, extraction time 4 min, cooling time 3 min, and centrifugation at 3000 r/min for 5 min, and the optimum condition for IL-2 was 0.1000 g, pH 3, extraction time 1 min, cooling time 7 min, and centrifugation time 5 min at 3000 r/min. Under the optimum conditions, the extraction rate of malachite green was 98.90% and 100% by using IL-1 and IL-2, respectively. Figure 4 shows the ionic liquid at the extraction stage and separation stage under the optimum conditions.

3.6. Influence of Interference Pigments. Effects of presence of methylene blue on extraction of malachite green by two ionic liquids were observed under the optimized condition. Results showed that the extraction rates of malachite green were more than 95% in presence of interference pigments, which indicated that these ionic liquids have good selectivity for extracting malachite green.

3.7. Recycling of Ionic Liquid. These ionic liquids can be recycled easily after elution. On addition of 1 mol/L NaOH to the ionic liquids after extraction, malachite green was

TABLE 1: The recovery test results of extraction by IL-1.

Sample	Origin (mg/L)	Added (mg/L)	Found (mg/L)	Recovery (%)	RSD
Water sample	0	5	4.91	98.20	0.00
		10	9.88	98.80	0.09
		20	19.83	99.15	0.00
Simulation water sample	9.93	5	14.92	99.74	0.11
		10	19.91	99.74	0.00
		20	29.88	99.74	0.11

TABLE 2: The recovery test results of extraction by IL-2.

Sample	Origin (mg/L)	Added (mg/L)	Found (mg/L)	Recovery (%)	RSD
Water sample	0	5	4.91	98.20	0.00
		10	9.88	98.80	0.09
		20	19.92	99.60	0.00
Simulation water sample	9.60	5	14.64	100.80	0.11
		10	19.59	99.90	0.00
		20	29.50	99.50	0.11

removed from the mixture. The recycle yields were 87.1% and 88.7% for IL-1 and IL-2, respectively.

3.8. Linearity, Precision, and Accuracy. Under the optimized conditions, the absorbance of malachite green increased linearly in the range of 0–2.5 mg·L^{-1}. The linear regression equation was $A = 0.1563C–0.0023$, and the correlation coefficient (R^2) was 0.9999. Moreover, the limit of detection was calculated to be 0.044 mg·L^{-1}.

Under the optimal experimental conditions, the relative standard deviation (RSD) of the method was determined by using 5 mg/L malachite green standard solution for 6 times, with 0.23% for IL-1 and 0.37% for IL-2, respectively. The results showed this method has better accuracy.

Experimental results with the recoveries obtained by the standard addition method are shown in Tables 1 and 2. The recovery results were in the range from 98.20% to 100.80%. From these results, it was found that the method has good accuracy and it can be used for extraction of malachite green accurately.

4. Conclusions

In this study, two thermoregulated ionic liquids were firstly employed as extractants to separate and concentrate malachite green in water. The optimum extraction conditions were achieved with the extraction rate of approximately 100%. Two ionic liquids can be recycled after the elution process, with the recovery higher than 87%. The method provided in this study is not only simple and efficient, selective for extraction of malachite green in the water, but also can avoid the environmental pollution by the extraction agent.

Conflicts of Interest

The authors declare that there are no conflicts of interest regarding the publication of this paper.

Acknowledgments

This research was supported by the Administration Department of Education, Sichuan Province (18ZB0298 and 18ZA0257) and the Scientific Research Fund of Mianyang Teacher's College (MYSY2017QN04).

References

[1] R. D. Rogers and K. R. Seddon, "Ionic liquids: solvents of the future?," *Science*, vol. 302, no. 5646, pp. 792-793, 2003.

[2] M. Gharehbaghi and F. Shemirani, "A novel method for dye removal: ionic liquid-based dispersive liquid–liquid extraction (IL-DLLE)," *Acta Hydrochimica et Hydrobiologica*, vol. 40, no. 3, pp. 290–297, 2012.

[3] Y. C. Fan, M. L. Chen, C. Shentu et al., "Ionic liquids extraction of Para Red and Sudan dyes from chilli powder, chilli oil and food additive combined with high performance liquid chromatography," *Analytica Chimica Acta*, vol. 650, no. 1, pp. 65–69, 2009.

[4] X. Chen, F. Li, C. Asumana et al., "Extraction of soluble dyes from aqueous solutions with quaternary ammonium-based ionic liquids," *Separation and Purification Technology*, vol. 106, no. 106, pp. 105–109, 2013.

[5] C. F. Poole and S. K. Poole, "Extraction of organic compounds with room temperature ionic liquids," *Journal of Chromatography A*, vol. 1217, no. 16, pp. 2268–2286, 2010.

[6] H. Passos, M. G. Freire, and J. A. Coutinho, "Ionic liquid solutions as extractive solvents for value-added compounds from biomass," *Green Chemistry*, vol. 16, no. 12, p. 4786, 2014.

[7] Q. Zhou, H. Bai, G. Xie et al., "Temperature-controlled ionic liquid dispersive liquid phase micro-extraction," *Journal of Chromatography A*, vol. 1177, no. 1, pp. 43–49, 2008.

[8] R. Rahnama, N. Mansoursamaei, and M. R. Jamali, "Preconcentration and trace determination of cadmium in spinach and various water samples by temperature-controlled ionic liquid dispersive liquid phase microextraction," *Acta Chimica Slovenica*, vol. 61, no. 1, pp. 191–196, 2014.

[9] F. Shah, T. G. Kazi, Naeemullah et al., "Temperature controlled ionic liquid-dispersive liquid phase microextraction for determination of trace lead level in blood samples prior to analysis by flame atomic absorption spectrometry with multivariate optimization," *Microchemical Journal*, vol. 101, no. 3, pp. 5–10, 2012.

[10] K. V. Rao, "Inhibition of DNA synthesis in primary rat hepatocyte cultures by malachite green: a new liver tumor promoter," *Toxicology Letters*, vol. 81, no. 2-3, pp. 107–113, 1995.

[11] S. Srivastava, R. Sinha, and D. Roy, "Toxicological effects of malachite green," *Aquatic Toxicology*, vol. 66, no. 3, pp. 319–329, 2004.

[12] I. D. Mall, V. C. Srivastava, N. K. Agarwal et al., "Adsorptive removal of malachite green dye from aqueous solution by bagasse fly ash and activated carbon-kinetic study and equilibrium isotherm analyses," *Colloids and Surfaces A Physicochemical and Engineering Aspects*, vol. 264, no. 1-3, pp. 17–28, 2005.

[13] S. Chowdhury, R. Mishra, P. Saha et al., "Adsorption thermodynamics, kinetics and isosteric heat of adsorption of malachite green onto chemically modified rice husk," *Desalination*, vol. 265, no. 1-3, pp. 159–168, 2011.

[14] Z. Zhang, K. Zhou, Y. Bu et al., "Determination of malachite green and crystal violet in environmental water using temperature-controlled ionic liquid dispersive liquid–liquid microextraction coupled with high performance liquid chromatography," *Analytical Methods*, vol. 4, no. 2, pp. 429–433, 2012.

[15] M. Ramin, F. Khalil, and N. Yousef, "Trace determination of malachite green in water samples using dispersive liquid–liquid microextraction coupled with high-performance liquid chromatography-diode array detection," *International Journal of Environmental Analytical Chemistry*, vol. 92, no. 9, pp. 1026–1035, 2012.

[16] L. Guo, J. Zhang, H. Wei et al., "Nanobeads-based rapid magnetic solid phase extraction of trace amounts of leuco-malachite green in Chinese major carps," *Talanta*, vol. 97, no. 16, p. 336, 2012.

[17] M. J. M. Bueno, S. Herrera, A. Uclés et al., "Determination of malachite green residues in fish using molecularly imprinted solid-phase extraction followed by liquid chromatography-linear ion trap mass spectrometry," *Analytica Chimica Acta*, vol. 665, no. 1, pp. 47–54, 2010.

[18] J. D. Holbrey, A. E. Visser, S. K. Spear et al., "Mercury (ii) partitioning from aqueous solutions with a new, hydrophobic ethylene-glycol functionalized bis-imidazolium ionic liquid," *Green Chemistry*, vol. 5, no. 2, pp. 129–135, 2003.

5

DNA Nanotweezers with Hydrolytic Activity for Enzyme-Free and Sensitive Detection of Fusion Gene via Logic Operation

Yongjie Xu [ID],[1] Xiangrong Luo,[1] Nana Geng,[2] Mingsong Wu,[2] and Zhishun Lu [ID][1]

[1]*Department of Laboratory Medicine, Guizhou Provincial People's Hospital, College of Basic Medicine, Guizhou University, Guiyang 550002, Guizhou, China*
[2]*Special Key Laboratory of Oral Diseases Research, Higher Education Institutions of Guizhou Province, Zunyi Medical University, Zunyi 563099, Guizhou, China*

Correspondence should be addressed to Yongjie Xu; 504267611@qq.com and Zhishun Lu; 94291411@qq.com

Academic Editor: Chih-Ching Huang

Gene fusion is a molecular event occurring in cellular proliferation and differentiation, and the occurrence of irregular fusion gene results in various malignant diseases. So, sensing fusion gene with high performance is an important task for integrating individual disease information. Here, we proposed a nonenzymatic and high-throughput fluorescent assay system for the detection of fusion gene by employing DNA nanotweezers with hydrolytic activity. This tweezer was assembled by three single-stranded DNAs and engineered with sensing elements and reporting subunits. In the absence of the fusion gene, the engineered tweezer remained opened and inactive which led to no signal output. However, the addition of fusion genes would cause structure alterations of the tweezer from open to close and further DNAzyme activation with the assembly of two reporting subunits. Then, the activated DNAzyme catalyzed fluorescence substrates for signal conversion. Taking BCR/ABL fusion gene as an example, the tweezer-based assay system showed not only excellent distinguishing capability towards different input targets but also high sensitivity with a detection limit of 5.29 pM. In addition to good detection performance, this system was simple and enzyme-free, offering a powerful nanometer tool as a smart nanodevice for sensing fusion detection.

1. Introduction

Gene fusions are a molecular event in cancer [1–3]. Many fusions resulting from chromosomal rearrangements are driver mutations in tumors and are currently used as biomarkers or drug targets. Examples include BCR/ABL, a target for Gleevec in chronic myeloid leukemia [4]; EML4-ALK, a target for crizotinib in lung cancer [5]; and PAX3-FOXO1, a biomarker for alveolar rhabdomyosarcoma [6]. Meanwhile, gene fusions are also a necessary molecular event in the defense of cancer and other diseases. Such T-cell receptor excision circles (TRECs) and K-deleting recombination excision circles (KRECs), as circularized DNA elements, are formed during the fusion process that creates T- and B-cell receptors [7, 8]. Their quantity in peripheral blood can be considered as an estimation of thymic and bone marrow output, which reflects individual immunity as hallmarks. Therefore, detecting fusion gene with high sensitivity and specificity is an urgent need for clinical diagnosis.

Conventional methods for detecting fusion gene include real-time quantitative reverse transcription PCR [9], flow cytometry [4], chromosome analysis [5], fluorescence in situ hybridization [10], and more. Such methods are still time-consuming and complicated in operation to some extent. To overcome these limitations, biosensing methods have attracted substantial research efforts, and numerous electrochemical, chemiluminescent, electrochemiluminescent, fluorescent, surface plasmon and resonance biosensing systems have been developed. These methods facilitate fusion gene analysis and improve analytical performance to some extent by adopting enzyme-assisted isothermal amplification and nanomaterials. However, native enzymes and artificial nanomaterials usually suffer from instability and high cost, which put constraints on their further application. In addition, fusion event occurred in cellular development, and proliferation is similar to the "AND" logic gate event in computer science which can be harnessed for intelligent and versatile detection. Unfortunately, this uniform trait has not been well taken into consideration in these biosensing strategies. Therefore, the exploration of a smart method that meets these challenges simultaneously remains a challenge.

DNA molecules are of great utility for this purpose because the combinatorial sequence space allows for an enormous diversity of signal carriers [11], and the predictability and specificity of Watson–Crick base pairing facilitate the design of gate architectures [12]. As a versatile construction material, DNA molecules indeed have been used for engineering molecular structures, engineering biological nanodevices [13, 14], and engineering various nanodevices, including "tweezers" [15, 16], "walkers" [17, 18], "stepper" [19], and engineering more [20–22] mechanical functions through encoding information in the base sequence of DNA. These assemblies also have the ability to attain cascade amplification and logic gate operation upon including catalytic [23–25] and logical control elements [26–28] and circuits [29–32]. Besides, owing to their properties of high biocompatibility, outstanding stability, low cost, and easily custom synthesis, DNA-based assemblies have the potential to be powerful tools for biosensing and bioanalysis. It is noted that DNA tweezers are molecular devices that can sense, hold, and release target DNA upon specific interaction. Since the first demonstration of a DNA-fueled molecular tweezer by Yurke et al [33] based on the strand displacement mechanism, several DNA molecular tweezers operating on similar fashion have been reported. These DNA tweezers include the adenosine monophosphate and adenosine deaminase-triggered aptamer tweezers [34], the pH-programmable tweezers reversibly switched by pH stimuli [35], and the photoresponsive DNA tweezers operated by invertible photoswitching [36]. The operations of these tweezers, however, require either the involvement of enzymes which may be subject to thermodynamic limitations, or rigid pH control of the system which suffers from tedious preparation processes, or the use of toxic azobenzene moieties. The development of simple and cost-effective DNA tweezers with new functionality will therefore facilitate the construction of different molecular machines for various applications.

In the present study, we report a new type of smart DNA nanotweezer with catalytic function for specific recognition of BCR/ABL fusion gene and then outputting an amplified signal. The DNA nanotweezer, self-assembled from three single-stranded DNAs, is tailored with recognition elements and catalytic subunits which show promising switches for molecular computation and signal amplification [37]. Recognition elements respond to fusion gene logically, whereas the responses regulate the "open" and "closed" states of the tweeters and further regulate the "inactive" and "active" states of the DNAzyme. The activated DNAzymes successively cleave the fluorescence substrates, thus enabling the intelligent and sensitive detection of fusion events without the attending of any native enzyme.

2. Experimental Section

2.1. Reagents and Materials.
The oligonucleotides used in the experiments were supplied by Shanghai Sangon Co., Ltd. (Shanghai, China). Fluorescence substrates of MNAzyme were purified with high-performance liquid chromatography (HPLC), while other strands were purified by polyacrylamide gel electrophoresis (PAGE). The base sequences of these oligonucleotides are given in Table S1. All oligonucleotides were prepared by resuspending the lyophilized oligonucleotides in DEPC-treated water at a nominal 10 μM concentration and stored at −20°C until use. All other reagents were of analytical grade, and ultrapure water (≥18 MΩcm, Milli-Q, Millipore) was used in all experiments.

2.2. Apparatus.
The gel electrophoresis was carried out in a flat-bed electrophoresis system on the PowerPac™ Basic electrophoresis analyzer (Bio-Rad, USA). The gel was imaged on the Bio-Rad ChemiDoc™ MP imaging system under ultraviolet light (Bio-Rad, USA), and florescence images were collected by the Bio-Rad ChemiDoc™ MP imaging system at an excitation wavelength of Cy5 (650 nm).

2.3. Agarose Gel Electrophoresis.
The assembly process of the DNA nanotweezer and its response to target input were tested by agarose gel electrophoresis which was conducted in 1 × TBE buffer (90 mM Tris-HCl, 90 mM boric acid, 2 mM EDTA, and pH 7.9) at 150 V for 30 min. Then the gel was imaged with the ChemiDoc™ MP imaging system.

2.4. Nanotweezer Preparation.
Prior to detection, the open nanotweezer was prepared by mixing stoichiometric quantities of stock solutions of strands A, B, and C to a final concentration of 1 μM in a microtube. After that, the mixture was left for 30 min at room temperature to allow the reaction to reach near-completion. The assembled nanotweezer was then used for the sequent experiment.

2.5. Determination of Target Fusion Gene.
For target fusion gene measurement, 5 μL of 1 μM DNA nanotweezer and 2 μL of 10 μM fluorescence substrates (Strand F, Cy5-GTTTCCTCguCCCTGG-BHQ1) were firstly transferred into wells of the microplate followed by adding different concentrations of BCR-ABL fusion gene. The mixed solution was carried out in 50 μL of 1 × GeneAmp® PCR Buffer II with Mg^{2+} ion (10 mM Tris-HCl, 50 mM KCl, and 50 mM $MgCl_2$; pH 8.3) and incubated at 40°C for 50 min. Then the resultant solutions were subjected to the Bio-Rad ChemiDoc™ MP imaging system at the excitation wavelength of 650 nm, and the fluorescence images at the emission wavelength of 670 nm were collected from the integration of fluorescence intensities for 10 s. Finally, the captured images were analyzed with the software of quantity one (Bio-Rad) to get the quantitative intensities. All the measurements were run in triplicates, and the assay system without target fusion gene was used as the blank control. All measurements were performed at room temperature.

3. Results and Discussion

3.1. Design of the Smart Nanodevice. The design of the smart nanotweezer for nonenzymatic and sensitive detection of fusion gene is illustrated in Scheme 1. The nanotweezer is assembled from three single-stranded strains, termed as strands A, B, and C, respectively. Strand C consists of two 18-base sequences which hybridize with complementary sequences at the ends of strands A and B to form two stiff arms; the hinge is formed from a four-base single-stranded region of C between the regions hybridized to strands A and B. The free ends of strand C are loaded with sequences complementary to BCR and ABL sequences, respectively, served as reorganization elements for fusion gene. Meanwhile, strands A and B were designed with catalytic subunits which are at the distal terminus of the tweezers. In the absence of BCR gene and ABL gene (0, 0), the tweezers remain in the open state which is a catalytically inactive state so no signal output (0). Meanwhile, in the presence of BCR gene or ABL gene (0, 1 or 1, 0) separately, the tweezers remain in a catalytically inactive state and produce zero signal outputs (0). Rather, once BCR gene and ABL gene are linked together and form BCR/ABL fusion gene due to chromosomes rearrangement events (1, 1), the tweezers are closed since the binding between recognition elements and BCR/ABL fusion gene pulls the ends of the tweezers together. Thereafter, the stiff arms of the tweezers are held together by the closing strand, and catalytic activity is activated due to the proximity of the two DNAzymes subunits attached to a single strand of DNA, forming active DNAzyme structures. The DNAzymes cleave the fluorescent substrates and generate signal output (1). The alterations between active DNAzyme structures and a catalytically inactive nanostructure corresponding to molecular events easily attain logical operation, and the nanodevice smartly reflects the gene expression behavior of each individual cell.

3.2. Confirmation of Nanotweezers Abilities. To confirm the construction of the functional nanotweezers, we used agarose gel electrophoresis to compare closed tweezers' (A, B, C) + target fusion gene with dimmers of (A, B, C), (A, C), and (B, C). Figure 1 shows an image of electrophoresis bands. The bands in lanes 1, 2, 3, and 4 represent strand A, strand B, strand C, and BCR/ABL fusion gene, respectively, which are experimental controls. Lanes 5 and 6 correspond to incomplete structures of strand A + C intermediate and strand B + C intermediate. Lanes 7 contains open tweezers assembled by strands A, B, and C, while lane 8 contains closed nanotweezers corresponding to additions of the target fusion gene. Lane 9 stands for a 500 bp DNA ladder. In comparison with control bands, these bands from lane 5 to lane 8 show decreasing electrophoresis mobility with successive addition of different strands as nanotweezer components. The reason for the results is due to the stepwise assembly of nanotweezer components to form high-molecular-weight DNA hybrids. These results indicate the successful construction of the nanotweezers as sensing

nanodevices and proper responses of the nanotweezers to target gene according to Scheme 1.

To validate the ability of nanotweezers acting as an "AND" logic operation device, BCR/ABL fusion gene as inputs were incubated with the DNA tweezers, and the fluorescence responses were recorded. The inset in Figure 1 depicts the signal responses. The control sample without the analyte or its analogues used for measuring the background fluorescence shows negligible signal (image a). However, the smart nanodevices gave a largely increased signal upon the addition of BCR/ABL fusion gene (image b). This is because the fusion process put the separated BCR gene and ABL gene near each other and further put the opened stiff tweezers with arms closed after interaction with the recognition elements. These results confirm that the nanotweezers pose the ability to sense fusion gene via switchable activity of different nanostructure states.

3.3. Optimization of Experimental Conditions. To gain optimum assay performance, several important experimental parameters were optimized including incubation time, reaction temperature, and the number of bases on the hinge region. The time-dependent fluorescence changes were firstly measured through fusion gene detection experiments at 30, 40, 50, 60, and 70 min. As shown in Figure 2(a), along the prolonged time, the signal showed a gradual increase and tended to plateau at 50 min. So, the time period of 50 min was chosen for subsequent experiments. Since reaction temperature affects DNAzyme cleavage activity [38], we then tested the temperature effect on DNAzymes activity over a broad temperature range from 25, 30, 35, 40, to 45°C. Figure 2(b) shows that the highest fluorescence intensity was obtained at 40°C which can be explained by the high DNAzyme activity when the temperature is most suitable for associating and disassociating between the DNAzyme substrate strands. Consequently, to avoid the unspecific close of the nanotweezers which would lead to a rise in the background signal, we designed a complementary sequence on the hinge which kept the tweezer fixed and opened by strands binding with each other. The number of sequence bases was optimized by measuring the signal responses of signal and noise when the bases are at 4, 5, 6, 7, and 8, respectively. As presented in Figure 2(c), the number of bases in the hinge showed an optimum at around 5 bases for the following assay.

3.4. Analytical Performance. Under the optimized experimental conditions, the analytical performance of the catalytic sensing system was investigated. Both the concentrations of the nanotweezers and fluorescence substrates were fixed at 100 nM upon incubation for 50 min at 40°C. Then the fluorescence intensities were also measured in response to different concentrations of fusion gene from 0 to 10 nM. As shown in Figure 3(a), it can be observed clearly that the fluorescence signals rose with the increase of fusion gene mimics concentration. The fluorescence intensity was linearly dependent on the logistical value of fusion gene concentrations in the range of 0.01–10 nM, and there is

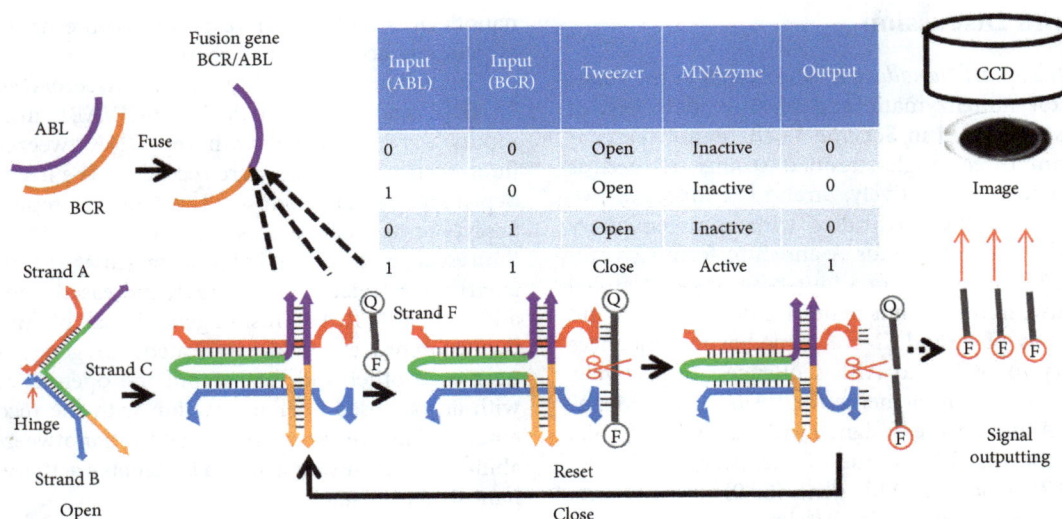

SCHEME 1: Schematic illustration of the fluorescence imaging method based on functional DNA nanotweezer regulated by fusion gene.

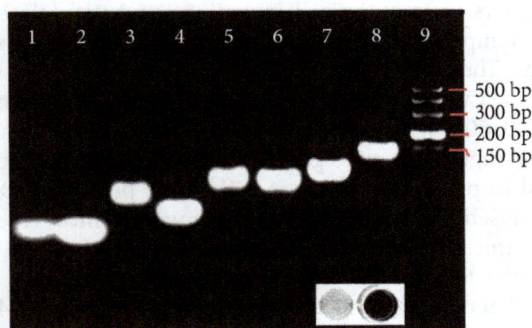

FIGURE 1: Characterization of the assembly process of DNA nanotweezer using agarose gel electrophoresis. Lanes 1, 2, 3, and 4 represents strand A, strand B, strand C, and BCR/ABL fusion gene, respectively, and lanes 5 and 6 are strand A + C and strand B + C, lane 7 represents opened nanotweezer assembled from strand A + B + C, and lane 8 represented closed nanotweezer after adding BCR/ABL fusion gene to the system in lane 7. Inset: the images a and b in lanes 7 and 8 represents the nanotweezer-based assay system without and with BCR/ABL fusion gene, respectively.

a linear regression equation of $F = 12141.35 + 1049.68 \log_{10}C$, where F is the fluorescence signal and X is the concentration of fusion gene ($R^2 = 0.9971$). Based on the signal to noise ratio repeated three times, the limit of detection was estimated to be 5.29 pM. Then a comparison was conducted in PCR [39, 40] and PCR-free format [41, 42] that employed DNAzyme as molecular switches. Although PCR-based methods were more sensitive, this proposed strategy showed a low detection limit and a wide linear range which can be attributed to the logical signal response and catalytic MNAzyme-assisted signal amplification strategy.

In addition to sensitivity, specificity of the assay system was also evaluated. Because in one individual cell, both BCR gene and ABL gene separately coexist even if no molecular event of chromosomes rearrangement occurs, the detection specificity was challenged by measuring the signal towards different targets as logic inputs, including a random DNA sequence, BCR gene, ABL gene, BCR/ABL fusion gene as well as the mixture of BCR gene, ABL gene and BCR/ABL fusion gene. As depicted in Figure 4, the smart nanodevice gives a different action towards different targets. The presence of a random DNA sequence and the absence of any target which were taken as inputs (0, 0) caused no signal output (images a and b). However, in the presence of BCR gene (1, 0) or ABL gene (0, 1), negligible signal changes in fluorescence were observed (0) as shown in images c and d, respectively. In these situations, binding of each existing gene with tweezers was taken as one separate logic input and thus lose the ability to pull the opened tweezers to be closed active DNAzymes structures. In contrast, the presence of BCR/ABL fusion gene (1, 1) that resulted from molecular fusion events led to significantly increased output signal (1) as shown in image e. These results validated the distinctive discrimination ability of nanotweezers acting as an "AND" logic operation device, demonstrating the good specificity of the smart nanosystem. This result is ascribed to the formation mechanism of the closed tweezers which rely on simultaneous recognition and binding with the fusion gene. Such an operating mechanism for the close of tweezers needs the fusion of two target inputs, which avoids the unspecial close induced by two separate signal inputs coexisting in one circumstance.

3.5. Recovery Test. To evaluate the accuracy and application potential of the proposed DNA nanodevice-based imaging system for fusion gene detection, recovery experiments were performed via spiking BCR/ABL fusion gene into 100-fold diluted human serum samples. The assay solutions were incubated in a dark circumstance for 50 min at 40°C and then immediately imaged with the fluorescence imaging system with a cooled CCD camera. As shown in Table 1, fluorescence intensities resulted from six parallel experiments generated an almost identical value to the spiked standard concentrations at 100 pM and 1 nM, with recoveries of 95.32% and 101% and coefficients of variation of

(a)

(b)

Signal
Noise

(c)

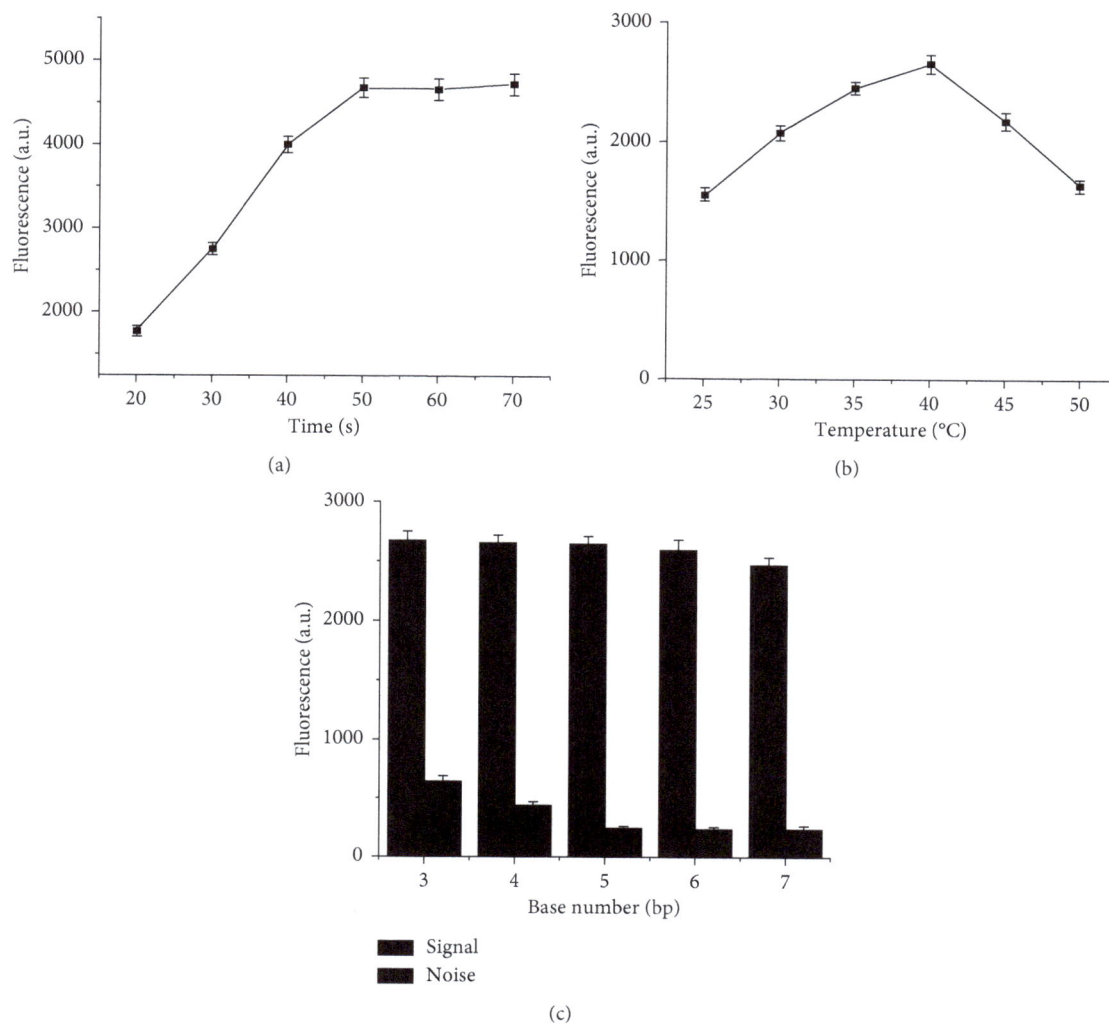

FIGURE 2: Optimization of experimental conditions. (a) Time-dependent fluorescence changes from 20, 30, 40, 50, 60, to 70 min. (b) DNAzyme cleavage activity on varied reaction temperature at 25, 30 35, 40, 45, and 50°C. (c) The effect of bases number on hinge region on signal to noise by setting at 3, 4, 5, 6, and 7 nucleotides.

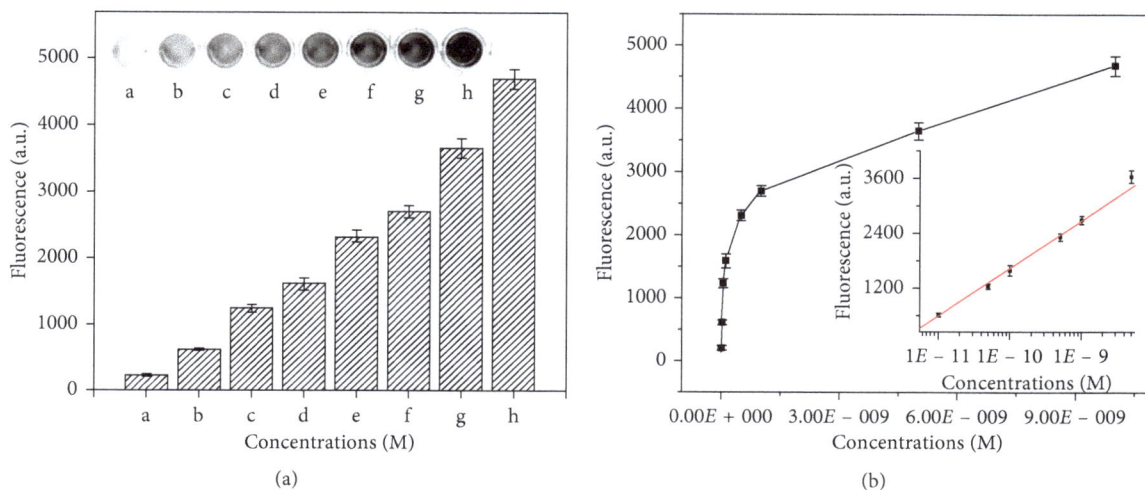

(a)

(b)

FIGURE 3: (a) Fluorescence images of BCR/ABL fusion gene at 0, 1×10^{-11}, 5×10^{-11}, 1×10^{-10}, 5×10^{-10}, 1×10^{-9}, 5×10^{-9}, and 1×10^{-8} M (image and bar a to h). (b) Dose-response curves for BCR/ABL fusion gene from 1×10^{-11} to 1×10^{-8} M. Inset: the corresponding calibration curve.

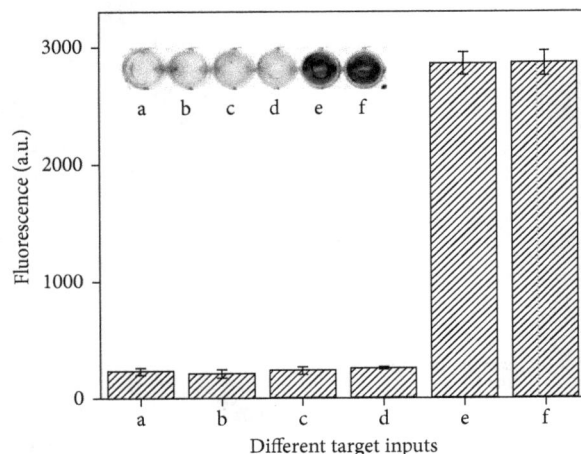

FIGURE 4: Fluorescence responses to different target inputs: random sequence (image and bar a), blank control (image and bar b), ABL gene (image and bar c), BCR gene (image and bar d), BCR/ABL fusion gene (image and bar e), and the mixture of ABL gene, BCR gene, and BCR/ABL fusion gene (image and bar f).

TABLE 1: Interference investigation via spiking BCR/ABL fusion gene into 100-fold diluted healthy human serum with six parallel experiments.

Spiked	Found	Recoveries	Coefficients of variation
50 pM	48.16 pM	96.32%	5.0%
100 pM	95.32 pM	95.32%	3.9%
1 nM	1.01 nM	101.00%	4.1%
5 nM	4.92 nM	98.40%	3.6%

3.9% and 4.1%. These results suggest that possible interferences from matrix on fusion gene detection are negligible assuming to the fact the complex components in serum did not interfere with the assembly of nanotweezers and subsequent activation of MNAzyme for catalytic cleavage. Thus, the assay system has high potential for detecting fusion gene in complex biological samples.

4. Conclusions

In summary, a nonenzymatic fluorescent method has been fabricated for sensitive and specific detection of fusion gene based on a novel functional nanotweezer. Various molecular events correspond to different signal inputs of the "AND" logic gate, which determine the structure state of the engineered tweeter and further activity of DNAzyme. These concessive responses result in a signal switching effect, attaining a logic gate operation. In this strategy, complicated molecular events occurred in the cell can be readably interpreted with the different logic outputs based on the engineered nanodevice. The sensitivity and applicability of the nanosystem were tested with the BCR/ABL fusion fragment as a target model, and results demonstrated its high

potential for fusion gene measurement with a detection limit of 5.29 pM. Furthermore, to realize real biological detection, we first need to get DNA or RNA transcript followed by incubating the generated transcript with the nanotweezer system for signal conversion and amplified detection. Therefore, the smart nanosystem is expected to have high potential in analyzing fusion gene.

Conflicts of Interest

The authors declare that there are no conflicts of interest.

Authors' Contributions

Yongjie Xu and Xiangrong Luo contributed equally to this work.

Acknowledgments

This work was supported by the Science and Technology Foundation of Guizhou Provincial Health and Family Planning Commission (gzwjkj2018-1-013 and gzwjkj2015-1-007) and Startup Foundation for Doctors of Guizhou Provincial People's Hospital (GZSYBS[2017] 05).

References

[1] T. H. Rabbitts, "Chromosomal translocations in human cancer," Nature, vol. 372, no. 6502, pp. 143–149, 1994.
[2] J. D. Rowley, "A new consistent chromosomal abnormality in chronic myelogenous leukaemia identified by quinacrine fluorescence and Giemsa staining," Nature, vol. 243, no. 5405, pp. 290–293, 1973.
[3] S. Heim and F. Mitelman, "Molecular screening for new fusion genes in cancer," Nature Genetics, vol. 40, no. 6, pp. 685-686, 2008.
[4] F. D'Alessio, P. Mirabelli, E. Mariotti et al., "Miniaturized flow cytometry-based BCR-ABL immunoassay in detecting leptomeningeal disease," Leukemia Research, vol. 35, no. 10, pp. 1290–1293, 2011.
[5] S. Soverini, A. Hochhaus, F. E. Nicolini et al., "BCR-ABL kinase domain mutation analysis in chronic myeloid leukemia patients treated with tyrosine kinase inhibitors: recommendations from an expert panel on behalf of European LeukemiaNet," Blood, vol. 118, no. 5, pp. 1208–1215, 2011.
[6] J. R. Downing, A. Khandekar, S. A. Shurtleff et al., "Multiplex RT-PCR assay for the differential diagnosis of alveolar rhabdomyosarcoma and Ewing's sarcoma," American Journal of Pathology, vol. 146, no. 3, pp. 626–634, 1995.
[7] A. Sottini, F. Serana, D. Bertoli et al., "Simultaneous quantification of T-cell receptor excision circles (TRECs) and K-deleting recombination excision circles (KRECs) by real-time PCR," Journal of Visualized Experiments, vol. 30, no. 94, pp. 207–223, 2014.
[8] A. A. Mauracher, F. Pagliarulo, L. Faes et al., "Causes of low neonatal T-cell receptor excision circles: a systematic review," Journal of Allergy and Clinical Immunology: In Practice, vol. 5, no. 5, pp. 1457–1460, 2017.
[9] A. Bennour, I. Ouahchi, M. Moez et al., "Comprehensive analysis of BCR/ABL variants in chronic myeloid leukemia

patients using multiplex RT-PCR," *Clinical Laboratory*, vol. 58, no. 5-6, pp. 433–439, 2012.

[10] S. Semrau, N. Crosetto, M. Bienko et al., "FuseFISH: robust detection of transcribed gene fusions in single cells," *Cell Reports*, vol. 6, no. 1, pp. 18–23, 2014.

[11] O. I. Wilner and I. Willner, "Functionalized DNA nano-structures," *Chemical Reviews*, vol. 112, no. 4, pp. 2528–2556, 2012.

[12] D. Y. Zhang and E. Winfree, "Engineering entropy-driven reactions and networks catalyzed by DNA," *Science*, vol. 318, no. 5853, pp. 1121–1125, 2007.

[13] N. C. Seema, "From genes to machines: DNA nanomechanical devices trends," *Trends in Biochemical Sciences*, vol. 30, no. 3, pp. 119–125, 2005.

[14] Y. Krishnan and F. C. Simmel, "Nucleic acid based molecular devices," *Angewandte Chemie International Edition*, vol. 50, no. 14, pp. 3124–3156, 2011.

[15] B. Jonathan and A. J. Turberfield, "DNA nanomachines," *Nature Nanotechnology*, vol. 2, no. 5, pp. 275–284, 2007.

[16] J. Elbaz, Z. G. Wang, R. Orbach, and I. Willner, "pH-stimulated concurrent mechanical activation of two DNA "tweezers". A "SET-RESET" logic gate system," *Nano Letters*, vol. 9, no. 12, pp. 4510–4514, 2009.

[17] T. Omabegho, R. Sha, and N. C. Seeman, "A bipedal DNA Brownian motor with coordinated legs," *Science*, vol. 324, no. 5923, pp. 67–71, 2009.

[18] J. S. Shin and N. A. Pierce, "A synthetic DNA walker for molecular transport," *Journal of the American Chemical Society*, vol. 126, no. 35, pp. 10834–10835, 2004.

[19] J. Elbaz, T. V. Ran, R. Freeman, H. B. Yildiz, and I. Willner, "Switchable motion of DNA on solid supports," *Angewandte Chemie International Edition*, vol. 48, no. 1, pp. 133–137, 2009.

[20] Z. G. Wang, J. Elbaz, and I. Willner, "DNA machines: bipedal walker and stepper," *Nano Letters*, vol. 11, no. 1, pp. 304–309, 2011.

[21] C. Buranachai, S. A. McKinney, and T. Ha, "Single molecule nanometronome," *Nano Letters*, vol. 6, no. 3, pp. 496–500, 2006.

[22] Y. Tian and C. Mao, "Molecular gears: a pair of DNA circles continuously rollsagainst each other," *Journal of the American Chemical Society*, vol. 126, no. 37, pp. 11410-11411, 2004.

[23] G. F. Joyce, "Directed evolution of nucleic acid enzymes," *Annual Review of Biochemistry*, vol. 73, no. 1, pp. 791–836, 2004.

[24] G. Seelig, B. Yurke, and E. Winfree, "Catalyzed relaxation of a metastable DNA fuel," *Journal of the American Chemical Society*, vol. 128, no. 37, pp. 12211–12220, 2006.

[25] S. J. Green, D. Lubrich, and A. J. Turberfield, "DNA hairpins: fuel for autonomous DNA devices," *Biophysical Journal*, vol. 91, no. 8, pp. 2966-2975, 2006.

[26] M. N. Stojanovic, T. E. Mitchell, and D. Stefanovic, "Deoxyribozyme-based logic gates," *Journal of the American Chemical Society*, vol. 124, no. 14, pp. 3555–3561, 2002.

[27] J. Macdonald, Y. Li, M. Sutovic et al., "Medium scale integration of molecular logic gates in an automaton," *Nano Letters*, vol. 6, no. 11, pp. 2598–2603, 2006.

[28] H. Lederman, J. Macdonald, D. Stefanovic, and M. N. Stojanovic, "Deoxyribozyme-based three-input logic gates and construction of a molecular full adder," *Biochemistry*, vol. 45, no. 4, pp. 1194–1199, 2006.

[29] M. Levy and A. D. Ellington, "Exponential growth bycross-catalytic cleavage of deoxyribozymogens," in *Proceedings of the National Academy of Sciences*, vol. 100, no. 11, pp. 6416–6421, 2003.

[30] M. N. Stojanovic, S. Semova, D. Kolpashchikov, J. Macdonald, C. Morgan, and D. Stefanovic, "Deoxyribozyme-based ligase logic gates and their initial circuits," *Journal of the American Chemical Society*, vol. 127, no. 19, pp. 6914-6915, 2005.

[31] R. Penchovsky and R. R. Breaker, "Computational design and experimental validation of oligonucleotide-sensing allosteric ribozymes," *Nature Biotechnology*, vol. 23, no. 11, pp. 1424–1433, 2005.

[32] G. Seelig, D. Soloveichik, D. Y. Zhang, and E. Winfree, "Enzyme-free nucleic acid logic circuits," *Science*, vol. 314, no. 5805, pp. 1585–1588, 2006.

[33] B. Yurke, A. J. Turberfield, A. P. Mills, F. C. Simmel, and J. L. Neumann, "A DNA-fuelled molecular machine made of DNA," *Nature*, vol. 406, no. 6796, pp. 605–608, 2000.

[34] J. Elbaz, M. Moshe, and I. Willner, "Coherent activation of DNA tweezers: a "SET-RESET" logic system," *Angewandte Chemie International Edition*, vol. 48, no. 21, pp. 3834–3837, 2009.

[35] S. Shimron, N. Magen, and J. Elbaz, "pH-programmable DNAzyme nanostructures," *Chemical Communications*, vol. 47, no. 31, pp. 8787–8789, 2011.

[36] X. Liang, H. Nishioka, N. Takenaka, and H. Asanuma, "A DNA nanomachine powered by light irradiation," *ChemBioChem*, vol. 9, no. 5, pp. 702–705, 2008.

[37] S. M. Bone, N. E. Lima, and A. V. Todd, "DNAzyme switches for molecular computation and signal amplification," *Biosensors and Bioelectronics*, vol. 70, pp. 330–337, 2015.

[38] E. Mokany, S. M. Bone, P. E. Young, T. B. Doan, and A. V. Todd, "MNAzymes, a versatile new class of nucleic acid enzymes that can function as biosensors and molecular switches," *Journal of the American Chemical Society*, vol. 132, no. 3, pp. 1051–1059, 2010.

[39] M. S. Payne, L. L. Furfaro, R. Tucker, L. Y. Tan, and E. Mokany, "One-step simultaneous detection of Ureaplasma parvum and genotypes SV1, SV3 and SV6 from clinical samples using PlexPCR technology," *Letters in Applied Microbiology*, vol. 65, no. 2, pp. 153–158, 2017.

[40] S. N. Tabrizi, L. Y. Tan, S. Walker et al., "Multiplex assay for simultaneous detection of mycoplasma genitalium and macrolide resistance using PlexZyme and PlexPrime technology," *PLoS One*, vol. 11, no. 6, Article ID e0156740, 2016.

[41] Y. V. Gerasimova and D. M. Kolpashchikov, "Folding of 16S rRNA in a signal-producing structure for the detection of bacteria," *Angewandte Chemie International Edition*, vol. 52, no. 40, pp. 10586–10588, 2013.

[42] Y. V. Gerasimova and D. M. Kolpashchikov, "Nucleic acid detection using MNAzymes," *Chemistry and Biology*, vol. 17, no. 2, pp. 104–106, 2010.

6

Design of Amperometric Biosensors for the Detection of Glucose Prepared by Immobilization of Glucose Oxidase on Conducting (Poly)Thiophene Films

Maria Pilo [ID],[1] Roberta Farre,[1] Joanna Izabela Lachowicz,[2] Elisabetta Masolo,[1] Angelo Panzanelli,[1] Gavino Sanna [ID],[1] Nina Senes,[1] Ana Sobral,[1] and Nadia Spano[1]

[1]Department of Chemistry and Pharmacy, University of Sassari, 07100 Sassari, Italy
[2]Department of Chemical and Geological Sciences, University of Cagliari, Monserrato, 09042 Cagliari, Italy

Correspondence should be addressed to Maria Pilo; mpilo@uniss.it

Academic Editor: Marek Trojanowicz

Enzyme-based sensors have emerged as important analytical tools with application in diverse fields, and biosensors for the detection of glucose using the enzyme glucose oxidase have been widely investigated. In this work, the preparation of biosensors by electrochemical polymerization of (poly)thiophenes, namely 2,2′-bithiophene (2,2′-BT) and 4,4′-bis(2-methyl-3-butyn-2-ol)-2,2′-bithiophene (4,4′-bBT), followed by immobilization of glucose oxidase on the films, is described. N-cyclohexyl-N′-(2-morpholinoethyl)carbodiimide metho-p-toluenesulfonate (CMC) was used as a condensing agent, and p-benzoquinone (BQ) was used as a redox mediator in solution. The glucose oxidase electrodes with films of 2,2′-BT and 4,4′-bBT were then tested for their ability in detecting glucose from synthetic and real samples (pear, apricot, and peach fruit juices).

1. Introduction

Qualitative and quantitative analyses in the food and beverage industry are extremely important from quality, storage, nutrition, and safety standpoints. The levels of certain sugars, like glucose, fructose, galactose, lactose, sucrose, and starch, affect intolerance conditions, diabetes, and obesity.

The determination of glucose concentration is a widespread analytical routine measurement carried out in the food industry. Glucose is often one of the most abundant monosaccharides in foods, but it is responsible for browning phenomena during dehydration and long-term storage, mainly due to the Maillard reaction. Furthermore, the accurate evaluation of the glucose amount in foods is of the utmost importance in the maintenance of its physiological level in blood of diabetics.

There are different analytical assays apt to determine glucose, colorimetric methods being the most commonly used. Today, a variety of unambiguous methods to detect and quantify glucose in assorted food matrices are available. These methods may be broadly grouped into two main categories: enzymatic approaches—including both spectrophotometric assays and glucose meters—and nonenzymatic techniques such as HPLC methods and their associated detection systems. While the former group is glucose specific, the latter is broadly adaptable and may be used to detect not only glucose but also an assortment of other carbohydrates.

In solution, glucose may exist as one of two possible anomers, termed α and β, or as the open-chain glucose aldehyde. When allowed to reach equilibrium at pH 7 and 25°C, approximately 63% of the glucose will adopt the β-glucopyranose conformation, with 37% existing as the α-glucopyranose, and less than 1% existing as either the aldehyde or as glucofuranose. Interconversion between the anomers may occur freely (albeit slowly) or via the catalytic activity of glucose mutarotases. Most enzymes preferentially bind either α or β form, but a small subset is able to utilize both anomers as the situation demands [1]. Glucose oxidase (GOx; EC 1.1.3.4) is a dimeric enzyme which catalyzes the conversion of β-D-glucose to D-glucono-1,5-lactone as part of the pentose phosphate pathway. The immobilization of the enzyme can be achieved using physical adsorption, cross-linking, and covalent bonding or entrapment in gels or membranes, but the main strategies employed require to entrap the enzymes into a conducting polymer film during its electrochemical growth or to attach them onto the surface of the functionalized film [2–4]. The process of entrapment involves the electrochemical polymerization of a monomer from a solution containing also the dissolved enzyme [3, 5–7]. This one-step procedure is very simple and rapid, but it has several possible drawbacks: (i) it requires that the concentration of the enzyme is fairly high, making it hardly applicable to expensive enzymes, (ii) the conditions required for the electropolymerization of the film have to be compatible with those necessary to allow the enzyme to retain its activity, and (iii) the physical entrapment within the polymer can induce a loss of the enzyme recognition or of the catalytic activity [2, 3, 7, 8]. An attractive alternative consists of a two-step approach involving, firstly, the electropolymerization of the conducting polymer on the electrode and, secondly, the chemical attachment of the biomolecules to the polymer surface [5, 9–12]. In contrast with the first procedure, this one has the advantages of allowing the use of optimal reaction conditions for each individual step, then facilitating macromolecular interactions and conserving a better access of the substrate to the immobilized biomolecule. To help to overcome the accessibility and proximity limitations and to reduce the susceptibility to interfering substances by lowering the electrode potentials, redox mediators have been used in biosensors. Mediators are small and soluble artificial electron transferring agents that can participate in the redox reaction by shuttling electrons from the redox center of the enzyme to the surface of the working electrode. Examples of commonly used mediators include quinones, organic conducting salts, dyes, ruthenium complexes, ferrocene, and ferricyanide derivatives.

Electrochemical biosensors are then a subclass of chemical sensors that use a biological recognition element (an enzyme, a protein, an antibody, etc.), which reacts selectively with the target analyte producing an electrical signal related to its concentration.

Our research group is involved for several years in synthesis and characterization of thiophene-based electrochemically generated conducting polymers (CPs). In particular, our interests are concerned in terthiophenes bearing free and complexed nitrogen ligands (terpyridine or phenanthroline) [13–16]. More recently, we started to study new thiophene-based structures and their application in amperometric sensors field. In this context, we want to investigate the possibility to use electrogenerated polythiophene films in the construction of GOx-based biosensors. In particular, in the present work, we compare the behavior of two biosensors based on two different films: the first one was obtained from the commercially available 2,2′-bithiophene (2,2′-BT) and the second one from a bithiophene derivative designed and synthetized in our laboratory (4,4′-bBT), both reported in Scheme 1.

Glucose oxidase was used as a (enzymatic) recognition element, and p-benzoquinone as a mediator in solution.

In this paper, GOx-based biosensors obtained by immobilization of the enzyme on a poly(2,2′-BT) or a poly

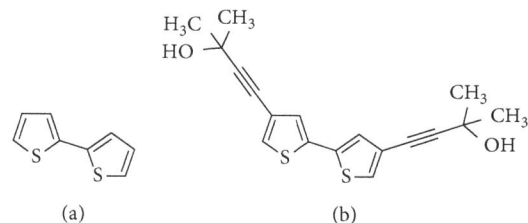

SCHEME 1: 2,2′-Bithiophene (2,2′-BT) (a) and 4,4′-bis(2-methyl-3-butyn-2-ol)-2,2′-bithiophene (4,4′-bBT) (b).

(4,4′-bBT) film will be named poly(2,2′-BT)/GOx and poly(4,4′-bBT)/GOx, respectively.

2. Experimental

2.1. Chemicals. All chemicals are commercially available as follows: tetraethylammonium hexafluorophosphate (TEAFP$_6$) from Fluka; acetonitrile, p-benzoquinone (BQ), N-cyclohexyl-N′-(2-morpholinoethyl)carbodiimide metho-p-toluenesulfonate (CMC), and glucose oxidase (GOx) from Sigma-Aldrich; 2,2′-bithiophene (2,2′-BT) from Lancaster; glucose from Carlo Erba. The synthesis of 4,4′-bBT was performed as reported in Section 2.2; all reagents used in the synthesis were from Sigma-Aldrich. Fruit juices (pear, brand "Puertosol," peach and apricot, brand "Valfrutta") were used in tests with real samples in 1/100 dilution.

All electrochemical tests were performed using a CHI-650 or an Autolab PGSTAT12 potentiostat interfaced with a PC using a specific software (CHI-650 and NOVA, resp.) in a three-electrode cell equipped with the following: a platinum disk working electrode (2 mm diameter), a graphite bar counterelectrode, and an SCE reference electrode, under argon atmosphere. Before use, platinum disk was polished with alumina powder (1 and 0.3 μm), placed in an ultrasonic bath and then rinsed with water and acetone.

2.2. Synthesis of 4,4′-bBT. A two-neck 25 mL flask equipped with a condenser, a magnetic stirrer, and an argon inlet was charged under inert atmosphere with 0.2000 g of 4,4′-dibromo-2,2′-bithiophene, 0.0053 g of [1,1′-bis(diphenylphosphino)ferrocene]dichloropalladium (Pd(dppf)Cl$_2$, 6.12·10^{-6} mol), and 3% of CuI in 5 mL of diisopropylamine. Then, 3-methyl-3-butyn-2-ol (1.24·10^{-3} mol) was added, and the mixture was left under stirring at reflux and monitored by TLC on alumina (eluent petroleum ether/ethyl acetate 3/2). The reaction was left to cool to room temperature and added to 50 mL of CH$_2$Cl$_2$ and washed with 50 mL of saturated NaHCO$_3$ solution and then with 3 × 50 mL of water. The organic phase was then treated with MgSO$_4$, filtered, and the solvent was evaporated under vacuum. The residue was purified through column chromatography on alumina using petroleum ether/ethyl acetate 3/2 as eluent. Yield 75%. ^{1}H NMR (CD$_2$Cl$_2$, ppm): δH = 7.32 (s, 2H); 7.16 (s, 2H); 2.04 (s, 2H, OH); 1.57 (s, 12H, CH$_3$). Elemental analysis: theoretical for C$_{18}$H$_{18}$O$_2$S$_2$: C, 65.40; H, 5.49; experimental: C, 65.28; H, 5.63.

2.3. Biosensor Preparation. Electrochemical polymerization of 2,2′-BT and of 4,4′-bBT was carried out in a three-electrode cell containing 0.01 M monomer solution and 0.1 M TEAFP$_6$ as supporting electrolyte in 5 mL of acetonitrile, purging argon gas in the cell for 20 minutes before each experiment. Polymerization of the monomers was performed by applying a potential value equal to +1.33 V for 2,2′-BT and to +1.45 V for 4,4′-bBT until a charge of 100 mC and 12 mC was passed through the polymerization system, respectively, in order to obtain an adequate film thickness [4]. The film thickness (*d*) was evaluated according to the following equation:

$$d = \alpha \, Q_{dep}, \tag{1}$$

where *d* = thickness in nm, α = experimental value in nm cm^2/mC, and Q_{dep} = charge density in mC/cm^2.

According to literature [17], an α value equal to 2.5 was used.

The obtained films were then neutralized by keeping them at 0 V versus SCE in a 5 mL acetonitrile solution of 0.1 M TEAFP$_6$ for 60 s. Finally, the films were characterized by cyclic voltammetry in the same solution used for the neutralization, washed with distilled water, and used for enzyme immobilization. The films were immersed in 2.5 mL of distilled water containing 7.5 mg of CMC, used as a condensing agent, and 30 mg of GOx during the night in the fridge at about 4°C [4, 5]. The GOx electrodes obtained were rinsed with ultrapure water and used in the glucose-sensing experiments. After use, all biosensors have been stored in the fridge at the temperature of 4°C.

2.4. Glucose Sensing. The glucose sensing tests were carried out at room temperature in a cell containing 20 mL of phosphate buffer 0.1 M (pH 7) and BQ 1 mM. A constant potential of +0.40 V was applied while the solution was stirred. When the background current reached a constant value, incremental amounts of a 0.2 M β-D-glucose aqueous solution were injected into the cell with a time interval of about 200 s, and the current/time response was recorded.

The biosensors were then tested in the same conditions described above, adding 200 μL of pure juice to the phosphate buffer-BQ aqueous solution. The glucose concentration was then estimated by the standard addition method.

3. Results and Discussion

3.1. Synthesis and Characterization of Poly(2,2′-BT) and of Poly(4,4′-bBT). Calibration of Poly(2,2′-BT)/GOx and of Poly (4,4′-bBT)/GOx Biosensors. The first step in the preparation of biosensors was the synthesis of a polythiophene film on an electrode surface.

A film of poly(2,2′-BT) was obtained by electrochemical polymerization of a 0.01 M solution of monomer. Cycling the potential between 0 and 1.40 V evidenced an increasing of the current peak, and a dark film on the working electrode surface was clearly seen, proving that the polymerization occurred.

Then, the electrode was polished with alumina, and the film used for the functionalization with the GOx was obtained applying a constant potential of +1.33 V, for the

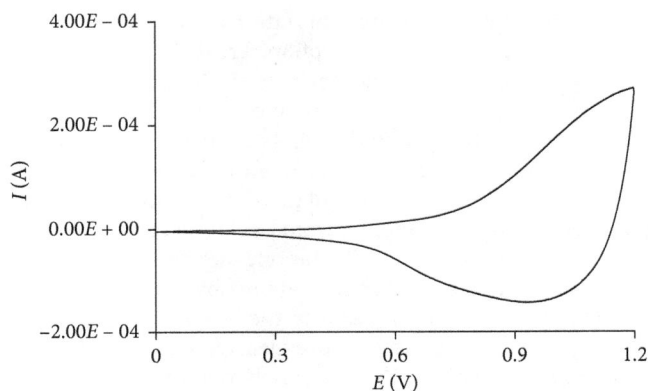

FIGURE 1: Cyclic voltammetry characterization of the poly(2,2′-BT) film in 0.1 M TEAFP$_6$/acetonitrile solvent system (Q = 100 mC; potential scan rate = 100 mV s^{-1}).

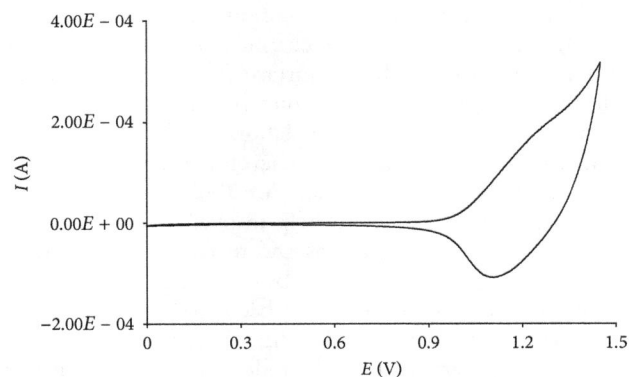

FIGURE 2: Voltammetric characterization of a poly(4,4′-bBT) film in 0.1 M TEAFP$_6$/acetonitrile solvent system (Q = 12 mC; potential scan rate = 100 mV s^{-1}).

necessary time to achieve a charge of 100 mC (corresponding to a thickness of the film of about 8 μm) [4, 18].

The dark red film obtained was characterized by cyclic voltammetry, evidencing a doping/dedoping response at 1.2/1.0 V, respectively (Figure 1).

The film of poly(4,4′-bBT) was deposited on the electrode surface in the same conditions as the unsubstituted polymer film. The voltammetric response in a potential range between 0 and 1.5 V evidences an increase in the peak current value at increasing scans, corresponding to the growth of a dark red film on the electrode surface.

The polymerization in potentiostatic conditions was performed applying a potential value of +1.45 V for a fixed time, corresponding to a final charge of 12 mC, and a thickness of 1 μm. The characterization of the obtained film by cyclic voltammetry (Figure 2) shows a doping/dedoping process at 1.2/1.1 V, respectively.

Both the films from the unsubstituted and substituted bithiophene were modified with GOx as described in the Experimental, and the biosensors obtained were tested with regard to a current increasing corresponding to the glucose concentration increasing in the phosphate buffer solution at pH 7.0. Both devices produce, at every addition of a glucose

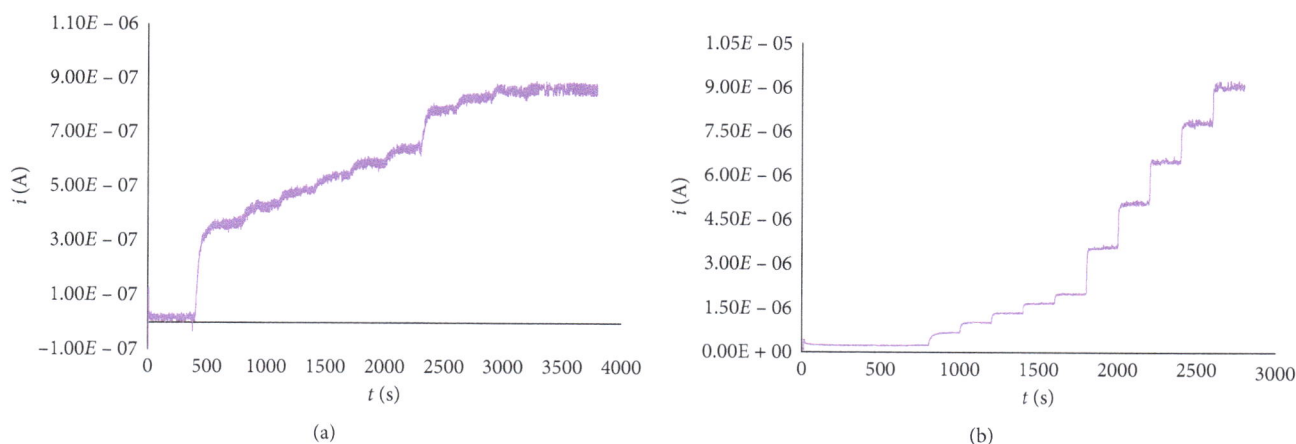

FIGURE 3: Current/time response of the biosensor poly(2,2′-BT)/GOx (a) and poly(4,4′-bBT)/GOx (b) in 0.1 M phosphate buffer (pH 7.0) and 10^{-3} M BQ with glucose concentration between 0.2 and 8.8 mM (0.2 M β-D-glucose aqueous standard solution).

solution, a jump-like increase of current, these being more intense at higher additions (Figure 3).

The two biosensors show a different behavior as a function of the standard additions of glucose solution: in particular, after each addition of analyte, the steady-state current has been reached much more later for the biosensor based on poly (2,2′-BT) rather than that based on poly(4,4′-bBT). This behavior likely depends also on the quite different thicknesses of the relevant polymeric films (the film of poly(2,2′-BT) is eight times more thick than that of the poly(4,4′-bBT) one).

The comparison between voltammetric characterization of poly(2,2′-BT) and poly(4,4′-bBT) shows that the two films, despite having different thicknesses, display similar current values. Such behavior suggests that the structure of the film and its conductivity depend not only on its thickness but even on the structure of the monomer. Furthermore, it can explain why two films originating from two different thiophene derivatives require a different thickness to obtain similar sensing performances of the GOx biosensors. On the other hand, the morphology of the film depends on the nucleation-growth mechanism, and in turn this mechanism is influenced by the structure of the starting species that allows obtaining a more or less dense structure [4, 13, 19]. The influence of the thickness of the films needs without doubt a more detailed study. The goal of the present paper was to investigate the possibility to use unsubstituted bithiophene and a simple, but not commercially available, bithiophene derivative in the construction of GOx biosensors. A more detailed study will be performed in the future, concerning the influence of different parameters (thickness of the film, monomer nature and concentration, and supporting electrolyte) on the behavior of the final biosensor. At the same time, a morphological characterization of different films will be carried out in order to investigate the nucleation-growth mechanism as well as to verify the porosity of the film that can tune the permeability of the film to the electrolyte, then influencing its conductivity.

In a biosensor arranged in the way described here, the conducting polymer film is at the same time the support to immobilize the enzyme and the conducting medium.

The electron transfer formally occurs in the polymer film, including also the process taking place at the film-solution interface [4].

3.2. Validation. Validation has been accomplished for both biosensors in terms of limit of detection (LoD), limit of quantification (LoQ), linearity, response time, intermediate precision, stability, and trueness (recovery tests). LoD has been calculated by the ratio between 3.3 times the standard deviation of the blank ($n = 11$) and the slope of the calibration curve obtained from a number of synthetic samples showing concentrations of the analyte very close to the expected LoD value, whereas the LoQ is measured as three times the LoD value. The intermediate precision was evaluated from data obtained from analyses of different aliquots of the same real sample (apricot juice) performed over five different analytical sessions held along two weeks and expressed as variation coefficient (CV). Stability of each biosensor was measured by performing, every two days, five consecutive measurements of concentration on a solution 1 mM in glucose. The biosensor was considered "stable" if the standard deviation of each cycle of measurements does not exceed three times the relevant intermediate precision value. Finally, trueness was estimated through recovery tests (multiple known additions of glucose on all real samples). Table 1 gives account for all validation parameters considered here.

Validation data reveal that the poly(2,2′-BT)/GOx biosensor is characterized by a lower LoD in comparison to the poly(4,4′-bBT)/GOx one, but the remaining parameters accounted for best performances of the biosensor bearing a substituted polythiophene. In particular, the most striking differences are relative to a sharp reduction of the response time, a clear improvement of the intermediate precision and of the stability of the biosensor, and also to a quantitative recovery in the trueness evaluation.

As evidenced in Table 2, the performances of the biosensors reported in the present work are in most cases comparable to (or better than) literature data as far as linearity range, response time, and stability is concerned, in

TABLE 1: Selected validation data for the determination of glucose with (a) poly(2,2′-BT)/GOx and (b) poly(4,4′-bBT)/GOx-based biosensors.

Biosensor	LoD (μM)	LoQ (μM)	Linearity range (mM), $Y = aX + b$; R^2	Response time (s; C glucose, mM)	Intermediate precision (CV; C glucose, mM)	Stability (days)	Trueness, recovery (% ± SD)
Poly(2,2′-BT)/GOx	30	90	0.09–5.20 $a = 1.5 \times 10^{-3} \pm 0.2 \times 10^{-3}$ $b = 5 \times 10^{-7} \pm 6 \times 10^{-7}$ $R^2 = 0.999$	180; 0.2 120; 1	7.0; 1	>15	93 ± 6
Poly(4,4′-bBT)/GOx	50	150	0.15–5.20 $a = 3.4 \times 10^{-4} \pm 0.1 \times 10^{-4}$ $b = 4 \times 10^{-8} \pm 7 \times 10^{-8}$ $R^2 = 0.997$	50; 0.2 20; 1	2.8; 1	>30	101 ± 4

TABLE 2: Comparison of the performances of biosensors proposed in the present work with literature data.

Type	Biosensor	LoD (μM)	Linearity range (mM)	Response time (s; C glucose mM)	Stability (days)	Reference
Polymer	Poly(2,2′-BT)/GOx	30	0.09–5.20	120; 1	>15	*Present work*
Polymer	Poly(4,4′-bBT)/GOx	50	0.15–5.20	20; 1	>30	*Present work*
Polymer	Poly(3,4-ethylenedioxythiophene)–poly(styrene-sulfonate)/GOx		1.1–16.5	20; 1		[20]
Polymer	o-Aminophenol/GOx	0.5	0.001–1	4; 1	>300	[21]
Polymer	Poly(ethacridine)/GOx		0.01–18	2; 5	>10	[22]
Polymer	Polypyrrole/GOx		2.5–30	30; 8	>10	[23]
NM	PtNPs-MWCNTs-PANI/GOx	1.0	0.003–8.2	3; 0.6	>48	[24]
NM	AuNPs-MWCNT/GOx	2.3	0.02–10	3; 1	>7	[25]
NM	AgNPs/PANINFs/GOx	0.25	1.0–12.0	3; 1	>7	[26]
NM	PtPd-MWCNTs/GOx	0.031	0.062–14	5; 1	>7	[27]
NM	Polyelectrolyte-SWCNTs/GOx	5.0	0.5–5.0	5; 1		[28]
NM	PdNPs/CS-graphene/GOx	0.20	0.001–1.0	10; 0.08	>7	[29]
NM	Graphene nanosheet/GOx	3.0	2.0–40.0		>21	[30]
NM	Copper nanocluster/MWCNTs	0.2	0.7–3.5	5; 0.05	>35	[31]
NM	Ni nanoparticle-modified carbon paste electrode	0.3	0.001–1.0	12; 0.05	>7	[32]
NM	CuO nanowire-modified electrode	0.049	0.0004–2	1; 1	>50	[33]

NM, nanomaterials; GOx, glucose oxidase; NPs, nanoparticles; MWCNTs, multiwalled nanotubes; PANI, polyaniline; PANINFs, polyaniline nanofibers; SWCNTs, single-walled carbon nanotubes.

particular, in the case of polymer-type biosensors. As regards LoD, biosensors here described show values higher than literature cases. Anyway, the LoD values of poly(2,2′-BT)/GOx and poly(4,4′-bBT)/GOx are adequate for the determination of glucose in beverages like fruit juices, being a typical concentration of glucose in such matrixes between 100 and 200 mM [34, 35].

Preliminary tests on possible interfering species (ascorbic acid, uric acid, lactose, and sucrose) indicate that the combined choice of a specific enzyme as GOx and of a proper working potential (+0.40 V versus SCE) allows to make negligible the possible interferences, being the possible interfering species oxidizable at higher potential values, confirming suggestions from literature data [36].

3.3. Tests in Real Sample. Glucose concentration was measured in three real samples by standard addition method using both biosensors. Each juice was tested three times.

The results obtained for the mean concentration values of glucose in the three different juices using the two biosensors are reported in Table 3.

Data obtained were statistically compared by two-tailed Student's test. Difference observed between the results obtained with the two different GOx biosensors is statistically not significant for $p < 0.05$.

4. Conclusion

In the present paper, we report the use of polythiophene-based amperometric biosensors for the determination of

TABLE 3: Mean values of glucose concentration and standard deviation in three different fruit juices (pear, peach, and apricot) with two different GOx biosensors ($n = 3$ replicates for each sample).

Juice	Glucose concentration (mM)		$Y = aX + b$; R^2	
	Poly(2,2′-BT)/GOx	Poly(4,4′-bBT)/GOx	Poly(2,2′-BT)/GOx	Poly(4,4′-bBT)/GOx
Pear	180 ± 20	170 ± 10	$a = 1.4 \times 10^{-3}$ $b = 3 \times 10^{-6}$ $R^2 = 0.994$	$a = 4.1 \times 10^{-4}$ $b = 6 \times 10^{-7}$ $R^2 = 0.996$
Peach	190 ± 10	220 ± 30	$a = 1.1 \times 10^{-3}$ $b = 2 \times 10^{-6}$ $R^2 = 0.999$	$a = 3.2 \times 10^{-4}$ $b = 5 \times 10^{-7}$ $R^2 = 0.999$
Apricot	240 ± 20	280 ± 50	$a = 3.0 \times 10^{-4}$ $b = 3 \times 10^{-6}$ $R^2 = 0.999$	$a = 4.3 \times 10^{-4}$ $b = 7 \times 10^{-7}$ $R^2 = 0.999$

glucose in aqueous solution and in real samples, namely, fruit juices. In particular, the results are focused on the comparison between a polymer film obtained from a simple bithiophene and, on the other hand, a film from a bithiophene structure leading to two unsaturated substituents on the 4 position of each heterocyclic ring. Two biosensors, poly(2,2′-BT)/GOx and poly(4,4′-bBT)/GOx, were obtained by anchoring glucose oxidase to the polymer-modified electrodes, and their behavior was tested and compared depending on the different structures of the films. According to literature references [4, 11, 12], the enzyme is reasonably assumed to be located on the surface of the polythiophene films. Being the results in aqueous solution and in real samples comparable for the poly (2,2′-BT)/GOx and poly(4,4′-bBT)/GOx, the main difference is in the different thicknesses of the film required for the optimal behavior of the biosensor. A more detailed study than that reported in this paper is required to better understand the spatial distribution of reactants and the role of polymer film thickness. Anyway, the preliminary study here reported allows doing some useful considerations. In particular, results here reported suggest that the substituted polythiophene film needs a lower thickness to obtain a biosensor performing comparably to that based on the unsubstituted film, allowing to obtain an higher stability of the film on the electrode surface and, virtually, a more effective anchoring of the enzyme and a longer life of the biosensor.

The comparison between the performances of the two biosensors allows evidencing the importance of the polymer film thickness in the behavior of the biosensor. The thickness of the film depends on the charge passed during the polymerization process. Furthermore, the thickness of the film is related to the morphology of the sensor surface that can change from a compact bilayer structure to a less dense tridimensional structure [4, 13]. The different morphologies of the film (bi- or three-dimensional) depend also on the structure of the monomer species. In this case, different tests were performed to check the more adequate thickness for both films, and the reported films are those showing the more satisfactory response concerning sensitivity, reproducibility, and stability. An adequate thickness of the film optimizes the catalytic activity of the enzyme, thus making possible a fast response of the biosensor, as well as better intermediate precision and trueness. In this context, a very important point that can influence the possibility to obtain well-performing sensors is the nature of the polymer film and its ability to give a stable adhesion to the metallic electrode surface and at the same time an efficient immobilization of the enzyme on the biosensor surface. Then, difference in the monomer units of the film can play an essential role in determining the response of the biosensor. In this paper, we explored the results deriving from an apparently simple variation on a bithiophene structure, evidencing that the proposed modification on the thiophene rings allows obtaining a more stable and more efficient sensing device. Starting from this encouraging result, we will investigate further design able to optimize the performances of new biosensors.

Conflicts of Interest

The authors declare that there are no conflicts of interest regarding the publication of this paper.

References

[1] A. L. Galant, R. C. Kaufman, and J. D. Wilson, "Glucose: detection and analysis," Food Chemistry, vol. 188, pp. 149–160, 2015.

[2] S. Cosnier, "Biosensors based on electropolymerized films: new trends," Analytical and Bioanalytical Chemistry, vol. 377, no. 3, pp. 507–520, 2003.

[3] S. Lupu, C. Lete, P. C. Balaure et al., "Development of amperometric biosensors based on nanostructured tyrosinase-conducting polymer composite electrode," Sensors, vol. 13, no. 5, pp. 6759–6774, 2013.

[4] C. Liu, T. Kuwahara, R. Yamazaki, and M. Shimomura, "Covalent immobilization of glucose oxidase on films prepared by electrochemical copolymerization of 3-methylthiophene and thiophene-3-acetic acid for amperometric sensing of glucose: effects of polymerization conditions on sensing properties," European Polymer Journal, vol. 43, no. 8, pp. 3264–3276, 2007.

[5] S. Cosnier and C. Gondran, "Fabrication of biosensors by attachment of biological macromolecules to electropolymerized conducting films," Analusis, vol. 27, no. 7, pp. 558–564, 1999.

[6] S. Cosnier, "Biomolecule immobilization on electrode surfaces by entrapment or attachment to electrochemically polymerized films. A review," Biosensors & Bioelectronics, vol. 14, no. 5, pp. 443–456, 1999.

[7] P. N. Bartlett and J. M. Cooper, "A review of the immobilization of enzymes on electropolymerized films," Journal of Electroanalytical Chemistry, vol. 362, no. 1-2, pp. 1–12, 1993.

[8] M. Hiller, C. Kranz, J. Huber, P. Bäuerle, and W. Schuhmann, "Amperometic biosensors produced by immobilization of redox enzymes at polythiophene-modified electrode surfaces," *Advanced Materials*, vol. 8, no. 3, pp. 219–222, 1996.

[9] W. Schuhmann, R. Lammert, B. Uhe, and H. L. Schmidt, "Polypyrrole, a new possibility for covalent binding of oxidoreductases to electrode surfaces as a base for stable biosensors," *Sensors and Actuators B: Chemical*, vol. 1, no. 1-6, pp. 537–541, 1990.

[10] A. Kros, R. J. M. Nolte, and N. A. J. M. Sommerdijk, "Poly(3,4-ethylenedioxythiophene)-based copolymers for biosensor applications," *Journal of Polymer Science Part A*, vol. 40, no. 6, pp. 738–747, 2002.

[11] M. Shimomura, N. Kojima, K. Oshima, T. Yamauchi, and S. Miyauchi, "Covalent immobilization of glucose oxidase on film prepared by electrochemical copolymerization of thiophene-3-acetic acid and 3-methylthiophene for glucose sensing," *Polymer Journal*, vol. 33, no. 8, pp. 629–631, 2001.

[12] T. Kuwahara, K. Oshima, M. Shimomura, and S. Miyauchi, "Immobilization of glucose oxidase and electron-mediating groups on the film of 3-methylthiophene/thiophene-3-acetic acid copolymer and its application to reagentless sensing of glucose," *Polymer*, vol. 46, no. 19, pp. 8091–8097, 2005.

[13] P. Manca, M. I. Pilo, G. Casu et al., "A new terpyridine tethered polythiophene: electrosynthesis and characterization," *Journal of Polymer Science Part A: Polymer Chemistry*, vol. 49, no. 16, pp. 3513–3523, 2011.

[14] P. Manca, S. Gladiali, D. Cozzula et al., "Oligo-thiophene tethered 1,10-phenanthroline as N-chelating moiety. Electrochemical and optical π-conjugated molecule and of the relevant conducting polymer and metallopolymers," *Polymer*, vol. 56, pp. 123–130, 2015.

[15] R. Scanu, P. Manca, A. Zucca et al., "Homoleptic Ru(II) complex with terpyridine ligands appended with terthiophene moieties. Synthesis, characterization and electropolymerization," *Polyhedron*, vol. 49, no. 1, pp. 24–28, 2013.

[16] P. Manca, R. Scanu, A. Zucca, G. Sanna, N. Spano, and M. I. Pilo, "Electropolymerization of a Ru(II)-terpyridine complex ethynyl-terthiophene functionalized originating different metallopolymers," *Polymer*, vol. 54, no. 14, pp. 3504–3509, 2013.

[17] O. A. Semenikhin, L. Jiang, T. Iyoda, K. Hashimoto, and A. Fujishima, "In situ AFM study of the electrochemical deposition of polybithiophene from propylene carbonate solution," *Synthetic Metals*, vol. 110, no. 3, pp. 195–201, 2000.

[18] J. Roncali, "Conjugated poly(thiophenes): synthesis, functionalization, and applications," *Chemical Reviews*, vol. 92, no. 4, pp. 711–738, 1992.

[19] L. A. Hernández, G. Riverso, F. Martin, D. M. González, M. C. Lopez, and M. León, "Enhanced morphology, crystallinity and conductivity of poly (3,4- ethyldioxythiophene)/ErGO composite films by in situ reduction of TrGO partially reduced on PEDOT modified electrode," *Electrochimica Acta*, vol. 240, pp. 155–162, 2017.

[20] J. Liu, M. Agarwal, and K. Varahramyan, "Glucose sensor based on organic thin film transistor using glucose oxidase and conducting polymer," *Sensors and Actuators B: Chemical*, vol. 135, no. 1, pp. 195–199, 2008.

[21] Z. Zhang, H. Liu, and J. Deng, "A glucose biosensor based on immobilization of glucose oxidase in electropolymerized o-aminophenol film on platinized glassy carbon electrode," *Analytical Chemistry*, vol. 68, no. 9, pp. 1632–1638, 1996.

[22] J.-J. Xu and H.-Y. Chen, "Amperometric glucose sensor based on glucose oxidase immobilized in electrochemically generated poly (ethacridine)," *Analytica Chimica Acta*, vol. 423, no. 1, pp. 101–106, 2000.

[23] E. Tamiya, I. Karube, S. Hattori, M. Suzuki, and K. Yokoyama, "Micro glucose using electron mediators immobilized on a polypyrrole-modified electrode," *Sensors and Actuators*, vol. 18, no. 3-4, pp. 297–307, 1989.

[24] H. Zhong, R. Yuan, Y. Chai, W. Li, X. Zhong, and Y. Zhang, "In situ chemo-synthesized multi-wall carbon nanotube-conductive polyaniline nanocomposites: characterization and application for a glucose amperometric biosensor," *Talanta*, vol. 85, no. 1, pp. 104–111, 2011.

[25] P. Si, P. Kannan, L. Guo, H. Son, and D.-H. Kim, "Highly stable and sensitive glucose biosensor based on covalently assembled high density Au nanostructures," *Biosensors and Bioelectronics*, vol. 26, no. 9, pp. 3845–3851, 2011.

[26] G. Chang, Y. Luo, W. Lu et al., "Ag nanoparticles decorated polyaniline nanofibers: synthesis, characterization, and applications toward catalytic reduction of 4-nitrophenol and electrochemical detection of H_2O_2 and glucose," *Catalysis Science & Technology*, vol. 2, no. 4, pp. 800–806, 2012.

[27] K.-J. Chen, C.-F. Lee, J. Rick, S.-H. Wang, C.-C. Liu, and B.-J. Hwang, "Fabrication and application of amperometric glucose biosensor based on a novel PtPd bimetallic nanoparticle decorated multi-walled carbon nanotube catalyst," *Biosensors and Bioelectronics*, vol. 33, no. 1, pp. 75–81, 2012.

[28] X. Pang, P. Imin, I. Zhitomirsky, and A. Adronov, "Conjugated polyelectrolyte complexes with single-walled carbon nanotubes for amperometric detection of glucose with inherent anti-interference properties," *Journal of Materials Chemistry*, vol. 22, no. 18, pp. 9147–9154, 2012.

[29] Q. Zeng, J.-S. Cheng, X.-F. Liu, H.-T. Bai, and J.-H. Jiang, "Palladium nanoparticle/chitosan-grafted graphene nanocomposites for construction of a glucose biosensor," *Biosensors and Bioelectronics*, vol. 26, no. 8, pp. 3456–3463, 2011.

[30] S. Alwarappan, C. Liu, A. Kumar, and C.-Z. Li, "Enzyme-doped graphene nanosheets for enhanced glucose biosensing," *Journal of Physical Chemistry C*, vol. 114, no. 30, pp. 12920–12924, 2010.

[31] X. Kang, Z. Mai, X. Zou, P. Cai, and J. Mo, "A sensitive non-enzymatic glucose sensor in alkaline media with a copper nanocluster/multiwall carbon nanotube-modified glassy carbon electrode," *Analytical Biochemistry*, vol. 363, no. 1, pp. 143–150, 2007.

[32] X. Cheng, S. Zhang, H. Zhang, Q. Wang, P. He, and Y. Fang, "Determination of carbohydrates by capillary zone electrophoresis with amperometric detection at a nano-nickel oxide modified carbon paste electrode," *Food Chemistry*, vol. 106, pp. 830–835, 2008.

[33] Z. Zhuang, X. Su, H. Yuan, Q. Sun, D. Xiao, and M. M. Choi, "An improved sensitivity non-enzymatic glucose sensor based on a CuO nanowire modified Cu electrode," *Analyst*, vol. 133, no. 1, pp. 126–132, 2008.

[34] K. Matsumoto, H. Kamikado, H. Matsubara, and Y. Osajima, "Simultaneous determination of glucose, fructose, and sucrose in mixtures by amperometric flow injection analysis with immobilized enzyme reactors," *Analytical Chemistry*, vol. 60, no. 2, pp. 147–151, 1988.

[35] F. Mizutani, S. Yabuki, and T. Katsura, "Amperometric enzyme electrode with fast response to glucose using a layer of lipid-modified glucose oxidase and Nafion anionic polymer," *Analytica Chimica Acta*, vol. 274, no. 2, pp. 201–207, 1993.

[36] X. Su, J. Ren, X. Meng, X. Ren, and F. Tang, "A novel platform for enhanced biosensing based on the synergy effects of electrospun polymer nanofibers and graphene oxides," *Analyst*, vol. 138, no. 5, pp. 1459–1466, 2013.

7

Solid-Phase Extraction and Large-Volume Sample Stacking-Capillary Electrophoresis for Determination of Tetracycline Residues in Milk

Gabriela Islas,[1] **Jose A. Rodriguez,**[1] **Irma Perez-Silva,**[1] **Jose M. Miranda** ⓘ**,**[2] **and Israel S. Ibarra** ⓘ[1]

[1]*Área Académica de Química, Universidad Autónoma del Estado de Hidalgo, Carretera Pachuca-Tulancingo Km. 4.5, 42076 Pachuca, Hidalgo, Mexico*
[2]*Departamento Química Analítica, Nutrición y Bromatología, Facultad de Veterinaria, Universidad de Santiago de Compostela, Pabellón 4 planta bajo, Campus Universitario s/n, 27002 Lugo, Spain*

Correspondence should be addressed to Israel S. Ibarra; isio.uaeh@gmail.com

Academic Editor: Serban C. Moldoveanu

Solid-phase extraction in combination with large-volume sample stacking-capillary electrophoresis (SPE-LVSS-CE) was applied to measure chlortetracycline, doxycycline, oxytetracycline, and tetracycline in milk samples. Under optimal conditions, the proposed method had a linear range of 29 to $200 \, \mu g \cdot L^{-1}$, with limits of detection ranging from 18.6 to $23.8 \, \mu g \cdot L^{-1}$ with inter- and intraday repeatabilities < 10% (as a relative standard deviation) in all cases. The enrichment factors obtained were from 50.33 to 70.85 for all the TCs compared with a conventional capillary zone electrophoresis (CZE). This method is adequate to analyze tetracyclines below the most restrictive established maximum residue limits. The proposed method was employed in the analysis of 15 milk samples from different brands. Two of the tested samples were positive for the presence of oxytetracycline with concentrations of 95 and $126 \, \mu g \cdot L^{-1}$. SPE-LVSS-CE is a robust, easy, and efficient strategy for online preconcentration of tetracycline residues in complex matrices.

1. Introduction

Preconcentration methods are an important tool for sample preparation because they enrich analytes in a liquid or solid sample. This improves analytical sensitivity, with the additional advantage of removing interferences [1]. Commonly employed preconcentration techniques include liquid-liquid extraction (LLE) [2], solid-phase extraction (SPE) [3], dispersive solid-phase extraction (DSPE) [4], magnetic solid-phase extraction (MSPE) [5], and quick, easy, cheap, effective, rugged, and safe (QuEChERS) [6, 7]. These techniques are termed *off-line*.

On the other hand, *online* techniques use automated systems that minimize sample manipulation. Flow techniques are commonly coupled to SPE [8, 9] using solid phases composed of molecularly imprinted polymers (MIPs) [10–12], monolithic columns [13, 14], and carbonaceous materials [15, 16].

Recently, capillary electrophoresis (CE) has received considerable attention in the development of *online* preconcentration systems such as transient isotachophoresis (tITP) [17], dynamic pH junction [18], sweeping [19, 20], and field-amplified stacking. The main advantages of these methods compared to *off-line* techniques include higher efficiency, shorter analysis time, and lower reagent and sample consumption [21–23]. *Online* preconcentration in CE is based on injection of a larger-than-normal sample volume into the capillary via hydrodynamic or electrokinetic methods [24].

Field-amplified stacking was developed for preconcentration of several analytes based on the charges of the analytes. Figure 1 shows a large-volume sample stacking

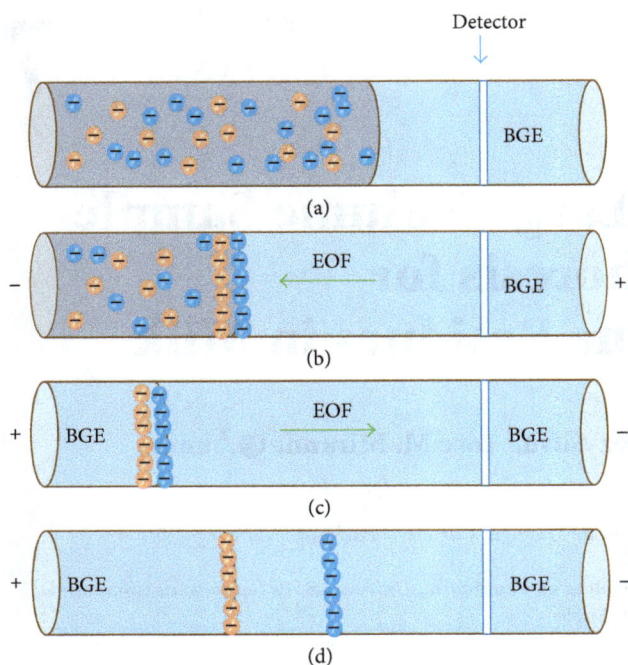

FIGURE 1: Schematic diagram of a preconcentration LVSS system. (a) Sample injection, (b) application of preconcentration potential (reverse polarity), (c) normal polarity, and (d) separation by capillary electrophoresis.

(LVSS) system, which involves a series of polarity switches in CE. The first step (Figure 1(a)) is hydrodynamic injection of a large sample volume into the capillary. Subsequently (Figure 2(b)), a voltage is applied (reverse polarity) promoting concentration of the analytes and removal of the cationic and nonionic compounds contained in the sample matrix. Finally (Figures 1(c) and 1(d)), analytes are separated in normal polarity in the background electrolyte (BGE) [25, 26].

Tetracyclines (TCs) are broad-spectrum antibiotics frequently employed in veterinary medicine for therapeutic purposes [5, 27] or incorporated into livestock feed at subtherapeutic doses as growth promoters. However, their indiscriminate use can produce enhanced bacterial resistance, allergic reactions, liver damage, and gastrointestinal issues [28, 29].

In order to protect human health from exposure of TC residues in milk, the European Union has established a maximum residue limit (MRL) of $100\,\mu g\cdot kg^{-1}$ for chlortetracycline (CT), oxytetracycline (OT), and tetracycline (TC) [30]; the Food and Drug Administration (FDA) has established a MRL of $300\,\mu g\cdot kg^{-1}$ for the combined residues CT, OT, and TC [31]; the Codex Alimentarius recommends a limit of $200\,\mu g\cdot kg^{-1}$ in milk for the combined residues CT, OT, and TC [32].

In recent years, due to the concerns caused by veterinary drugs contained in food samples, there were developed a large variety of analytical methodologies for the determination of TC residues at $\mu g\cdot kg^{-1}$ or $\mu g\cdot L^{-1}$ levels in different matrices. These methods included chemiluminescence [33], microbiological assays [34], high-performance liquid chromatography (HPLC) [35, 36], or capillary electrophoresis (CE) [37].

Taking into account the MRLs and the complexity of milk, this work develops a CE method using SPE and LVSS-CE for determination of TCs in milk that was demonstrated to be rapid, simple, and efficient. Additionally, the developed method showed higher sensitivity and accuracy than those reported by conventional methods using CZE aimed at the detection and quantification of TC residues in milk.

2. Experimental

2.1. Reagents and Chemicals. All solutions were prepared by dissolving the respective analytical grade reagent in deionized water with a resistivity not less than $18.0\,M\Omega\cdot cm$, which was provided by a Milli-Q system (Millipore, Bedford, MA, USA). Sodium phosphate was obtained from Sigma (St. Louis, MO, USA). EDTA sodium salt, sodium hydroxide, and hydrochloric acid were obtained from J.T. Baker (Phillipsburg, NJ, USA). Methanol was obtained from Mallinckrodt Baker (Xalostoc, Mexico), and 2-propanol was obtained from Fluka (St. Gallen, Switzerland).

Single stock standards of $100\,mg\cdot L^{-1}$ were prepared in methanol. The stock solutions were stored at $-4°C$. Mixed standard working solutions were prepared by diluting the standard stock solution immediately before use. The BGE solution consisted of 30 mM sodium phosphate, 2 mM EDTA disodium salt, and 2% 2-propanol. The solution pH was adjusted to 12.0 with $0.1\,M\cdot NaOH$.

2.2. Apparatus. Electrophoresis was performed using a Beckman Coulter P/ACE 5500 (Fullerton, CA, USA) with a photodiode array detector. Data were collected and analyzed with a Beckman P/ACE system with MDQ version 2.3 software. TC separations were performed in a fused silica capillary ($41.7\,cm \times 75\,\mu m$ ID). A pH/ion analyzer (model 450; Corning Science Products, NY, USA) was used to accurately adjust the pH of the electrolyte solution to within 0.01 pH units.

At the beginning of each working day, the capillary was activated with 1.0 M NaOH at 35°C for 15 min, followed by 0.1 M NaOH for 10 min, deionized water at 25°C for 10 min, and then electrolyte solution at 25°C for 10 min. The capillary was washed out between successive analyses using 1.0 M NaOH for 4 min, 0.1 M NaOH for 2 min, deionized water for 2 min, and electrolyte solution for 4 min. The detector wavelength (λ) was set at 360 nm, and the capillary was kept at 25°C. Peaks were identified by migration times and coinjection of standard solutions [5].

2.3. Sample Treatment and Analysis. A 1.0 mL milk sample was fortified with an internal standard ($50\,\mu g\cdot L^{-1}$) in polypropylene tubes. Proteins were precipitated by adding 0.2 mL of 2% acetic acid (v/v), followed by heating for 5 min (65°C) in a water bath and centrifuging at 3200 rpm for 15 min. Once completed, the protein-free liquid phase was diluted to 10 mL with deionized water. The solution was then passed through a cartridge (Sep-Pak Vac C_{18} cartridges, 1 g, 6 cc, Waters) previously activated with 5 mL of methanol,

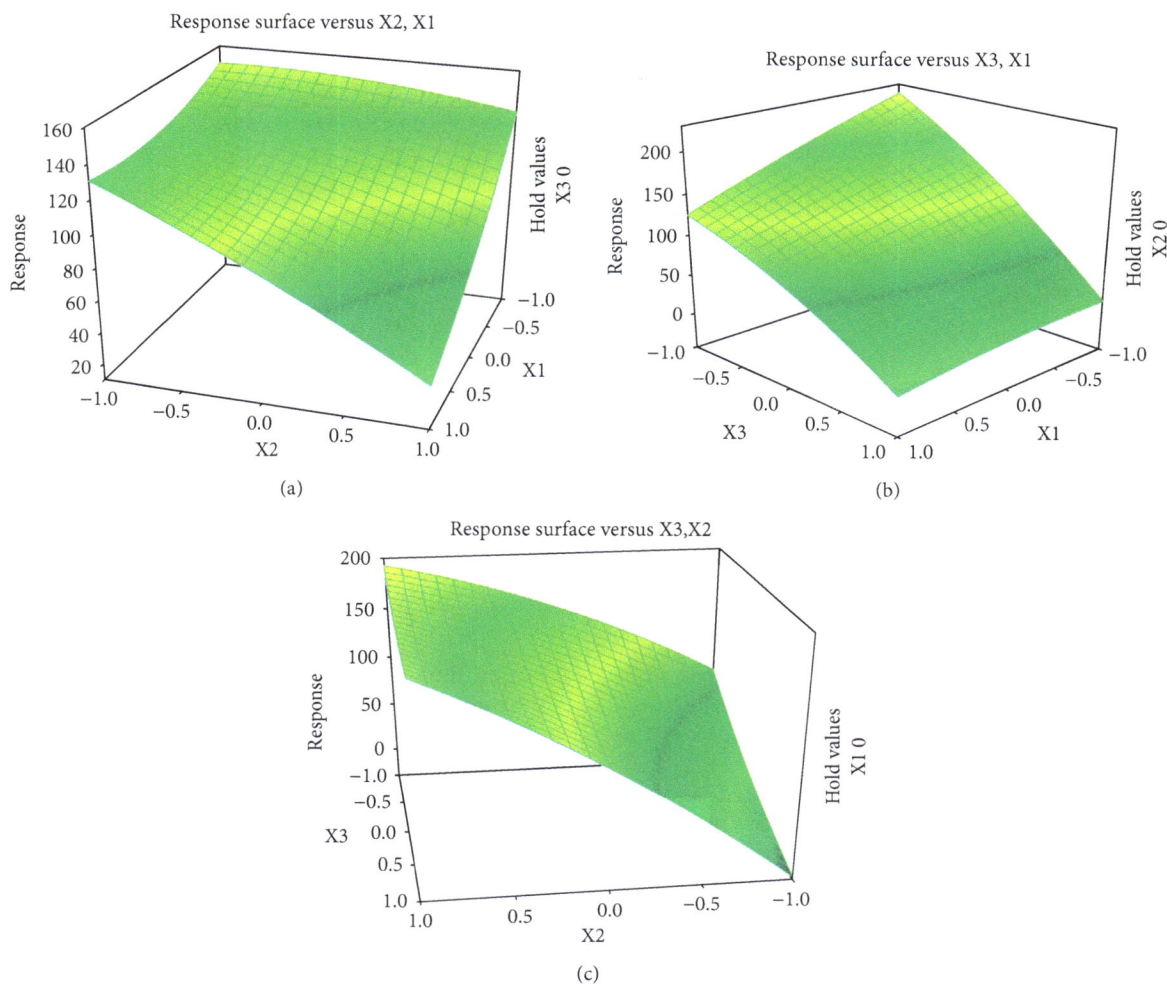

FIGURE 2: Contour and response surface plots of interactions modes for output variables (sum): (a) injection time (min) : reverse potential (kV); (b) injection time (min) : reverse polarity time (min); and (c) reverse polarity (kV) : reverse polarity time (min).

followed by 5 mL of methanol and 5 mL of deionized water at a maximum flow rate of 1 mL·min^{-1}. Analytes retained on the SPE cartridge were washed with 2.0 mL of 5.0% methanol. Retained TCs were eluted with 3.0 mL of methanol. The eluted solution was evaporated to dryness, and the residue was dissolved in 1 mL of 0.01 M NaOH containing 50 μg·L^{-1} picric acid as an internal standard.

Samples treated by SPE were introduced by hydrodynamic injection at 5 psi for 180 s (around 98% of capillary capacity). The capillary was then set in BGE vials, and a potential of 12 kV was applied for 120 s (reverse polarity) to preconcentrate TCs at the inlet, while water and other ions were removed from the capillary. Finally, polarity was returned to normal (14 kV), and CE separation was carried out.

3. Results and Discussion

3.1. LVSS Optimization. Development of an LVSS preconcentration technique for capillary electrophoresis requires optimization of control variables. Optimization

involves selection of factors that influence the analytical signal and enrichment factor. Box–Behnken design (BBD) was selected for optimization because it allows evaluation of control factors using an adjusted surface response.

The experimental design matrix describes the combination of factors in each experiment and allows simultaneous evaluation of several variables. Optimization of the system with BBD involves four steps: (i) identifying the output variable to optimize, (ii) identifying and selecting factors and levels that affect the LVSS system, (iii) data analysis and fitting of the surface response model, and (iv) confirmation under the optimal conditions obtained.

For LVSS, the output variable selected is the sum of the enrichment factors of the four TCs. The variables optimized in the procedure were the injection time (min) in the hydrodynamic mode using a pressure of 5 psi, applied potential (kV), and time (min) of reverse polarity. Injection time was varied between 2.0 and 3.0 min to evaluate the time required to fill the capillary. The reverse potential was evaluated between 8.0 and 12 kV. These values were selected to ensure

TABLE 1: Optimal conditions determined with Box–Behnken design.

Exp.	Control factors			Enrichment factors				Output variable
	Injection time (min)	Reverse potential (kV)	Applied time (min)	TC	CT	OT	DT	Sum
1	−1	−1	0	25.4	49.0	38.9	42.2	155.45
2	1	0	1	0.0	0.0	0.0	0.0	0.0
3	0	1	−1	41.2	58.0	46.1	49.1	194.34
4	0	0	0	20.4	43.7	55.7	1.2	121.02
5	1	0	−1	44.8	71.1	55.5	47.2	218.76
6	1	1	0	2.5	17.1	29.1	61.5	110.20
7	−1	0	1	0.0	0.0	0.0	0.0	0.0
8	0	−1	1	10.5	9.0	13.4	21.4	54.27
9	0	0	0	26.8	61.3	47.8	41.0	176.84
10	0	−1	−1	9.6	51.3	32.7	60.5	154.04
11	0	1	1	0.0	0.0	0.0	26.8	10.83
12	−1	0	−1	11.8	29.0	42.2	59.0	142.00
13	−1	1	0	0.0	0.0	0.0	0.0	0.0
14	0	0	0	0.0	0.0	10.8	0.0	26.75
15	1	−1	0	31.1	54.0	52.4	48.4	185.94

sufficient stacking time to remove the sample matrix from the capillary without losing analytes. Additionally, time during preconcentration (2.0–3.0 min) must be sufficient to increase analyte enrichment.

Table 1 shows the design matrix produced and the output variable in function of the sum of each enrichment factor obtained in each condition. All experiments were performed using 1 mL of a standard solution of TCs at a concentration of 1.0 mg·L^{-1}. Enrichment factors were estimated as the area ratio of the signals obtained with and without application of *online* LVSS.

Results were analyzed using MINITAB® version 17 software. Data were adjusted to the quadratic model according to the analysis of variance (ANOVA). The coefficient of determination (r^2) for the adjustment was 0.785, and the equation for the surface response was

$$Y1 = 108.2 + 27.2X1 - 29.3X2 - 80.5X3 - 4.2X1^2 + 8.9X2^2$$
$$- 13.8X3^2 + 19.9X1*X2 - 19.2X1*X3 - 20.9X2*X3, \quad (1)$$

where $Y1$ is the sum of the enrichment factor, $X1$ is the injection time (min), $X2$ is the inversion electric current (kV), and $X3$ is the applied time in the inversion electric current (min). The critical variables during LVSS are the reverse potential and applied time ($p > 0.05$). The lack-of-fit test is designed to determine if the proposed model is adequate for the observed data. The test is performed by comparing the variability of residuals from observations at replicate settings of the factors. Since the p value for lack of fit in the ANOVA table (0.744) is greater than 0.05, the model is adequate for the observed data at the 95.0% confidence level.

Based on the response surfaces (Figure 2), a clear interaction between the variables is observed, which is commonly observed for preconcentration systems employing LVSS-CE. Optimal conditions determined by BBD were $X1$:

injection time (3.0 min), $X2$: reverse potential (12 kV), and $X3$: preconcentration time (2.0 min).

The proposed methodology (LVSS-CE) was applied for the determination of TCs in commercial milk samples using a modification of the method proposed by Islas et al. [4]. However, different electrophoretic mobilities were obtained for the internal standard, which can be attributed to the ionic strength of the sample. Ionic strength significantly increases the electrophoretic mobility of analytes, thereby affecting LVSS preconcentration and causing loss of analyte if care is not taken when applying the negative polarity [38].

For these reasons and given the complexity of the sample, one of the most important steps in LVSS-CE analysis is sample cleanup. However, this may be difficult for analysis of antibiotics. For these reasons, an extraction and cleanup step was used previous to preconcentration and analysis by LVSS-CE. SPE was used for extraction and cleanup of TCs in milk samples. This technique decreases ionic strength effects, making samples suitable for analysis by LVSS-CE. For sample pretreatment, following protein removal from the milk sample, the liquid phase is diluted to 10.0 mL with deionized water and then passed through an activated C$_{18}$ SPE cartridge. Analytes retained on the SPE cartridge were washed with 2.0 mL of 5.0% methanol. Retained TCs were eluted with 3.0 mL of methanol. The eluted solution was evaporated to dryness and redissolved in 1.0 mL of 0.01 M NaOH containing 50 μg·L^{-1} picric acid [39].

3.2. *Analytical Parameters.* Under optimal conditions, analytical parameters of the LVSS-CE method were evaluated at concentrations of 0–200 μg·L^{-1} for each TC. Each standard was prepared and analyzed in triplicate using the proposed methodology. Peak areas were measured, and calibration curves were constructed from the peak area ratios (analyte : internal standard). Calibration curves showed

TABLE 2: Regression parameters of calibration: absorbance (mUA) versus TC concentration ($\mu g \cdot L^{-1}$).

Analyte	Regression parameters				
	Intercept: b0 ± ts (b0)	Slope: b1 + ts (b1)	Correlation coefficient, r	Limit of detection ($\mu g \cdot L^{-1}$)	Linear range ($\mu g \cdot L^{-1}$)
TC	−0.023 ± 0.026	0.337 + 0.013	0.994	19.93	59.79–200
CT	−0.0122 ± 003	0.030 + 0.001	0.991	23.83	71.49–200
OT	0.006 ± 0.022	0.314 + 0.011	0.995	18.60	55.8–200
DT	−0.029 ± 0.033	0.440 + 0.169	0.994	19.45	58.35–200

Analyte	Repeatability, interday (%RSD, $n = 3$)		Repeatability, intraday (%RSD, $n = 3$)	
	75 $\mu g \cdot L^{-1}$	150 $\mu g \cdot L^{-1}$	75 $\mu g \cdot L^{-1}$	150 $\mu g \cdot L^{-1}$
TC	6.60	4.72	8.64	6.01
CT	9.11	8.61	9.71	9.19
OT	7.02	1.71	9.19	6.22
DT	5.60	3.94	9.35	5.70

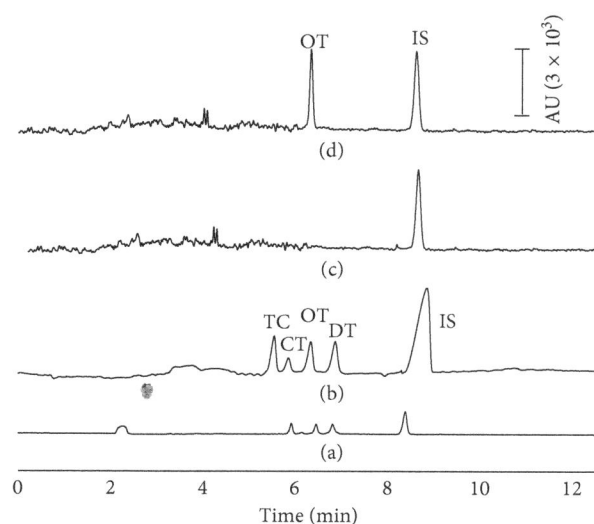

FIGURE 3: Electropherograms. (a) Standard sample of 10 mg·L^{-1} TCs and 50 mg·L^{-1} IS by CE; (b) standard sample of 1 mg·L^{-1} TCs and 5 mg·L^{-1} IS by LVSS-CE; (c) blank milk sample by SPE-LVSS-CE; and (d) real milk sample by SPE-LVSS-CE.

a linear dependence on TC concentration. Calibration regression parameters are shown in Table 2. LODs were calculated for a signal-to-noise ratio of 3.29 according to IUPAC recommendations [40].

The accuracy and precision of the method proposed was measured in terms of intra- and interday repeatabilities for migration times and peak areas. Results were determined as the relative standard deviation (%RSD) obtained in the analysis of TCs at two concentrations (75 and 150 $\mu g \cdot L^{-1}$). Based on these results and using the most restrictive MRLs established by EU regulations, the LVSS is adequate for analysis of TCs in milk samples.

3.3. *Application.* The proposed SPE-LVSS-CE method was applied for the determination of TCs in 15 commercial milk samples from different brands. Three replicate determinations of each analyte in the selected samples were performed. Two samples were determined to be positive for the presence of OT with concentrations of 95 and 126 $\mu g \cdot L^{-1}$, respectively, which

was identified by their migration times. In order to confirm the presence of the analyte, a standard addition was made to the sample extract. An increase in the peak area confirmed the presence of the antibiotic residue. Samples with TC concentrations outside the linear response range were diluted tenfold with deionized water. Confirmation using mass spectrometry is also required. The electropherograms obtained are shown in Figure 3.

4. Conclusions

The proposed SPE-LVSS-CE technique provided sensitive, rapid, simple, and efficient online preconcentration of TC residues in complex matrices such as milk. This methodology only required 1.0 mL of milk, whereas traditional methods require about 100.0 mL to reach the MRLs established by international regulations.

Additionally, this technique provides good sensitivity and accuracy compared to CZE and has a much higher stacking efficiency for the four analytes with LODs of $18.60–23.83\,\mu g \cdot L^{-1}$. The developed method allowed achieve enrichment factors from 50.33 to 70.85 compared to conventional injection mode. The SPE-LVSS-CE method achieves appropriate LODs for identification and quantification of TCs according to MRLs established by the EU, FDA, and Codex Alimentarius. The developed method was applied to preconcentrate, identify, and quantify TCs in real milk samples with satisfactory outcomes.

Conflicts of Interest

The authors declare that they have no conflicts of interest.

Acknowledgments

The authors wish to thank Programa para el Desarrollo Profesional Docente, para el Tipo Superior (PRODEP) for the approved project in the incorporation of new PTC-Profesores de Tiempo Completo; Consejo Nacional de Ciencia y Tecnología (CONACyT) (Project INFR-2014-227999 and Retention Grant no. 251112) and Xunta de Galicia and Fondo Europeo de Desarrollo Regional (FEDER) (Project GRC 2014/004), for financial support.

References

[1] L. Pillonel, J. O. Bosset, and R. Tabacchi, "Rapid preconcentration and enrichment techniques for the analysis of food volatile. A review," *LWT-Food Science and Technology*, vol. 35, no. 1, pp. 1–14, 2002.

[2] C. Wang, L. Qu, X. Liu et al., "Determination of a metabolite of nifursol in foodstuffs of animal origin by liquid–liquid extraction and liquid chromatography with tandem mass spectrometry," *Journal of Separation Science*, vol. 40, no. 3, pp. 671–676, 2017.

[3] R. Gogoi, D. C. Roy, and S. Sinha, "Determination of chlortetracycline residues in swine tissues using high performance liquid chromatography," *Pharma Innovation Journal*, vol. 6, pp. 34–36, 2017.

[4] G. Islas, J. A. Rodríguez, M. E. Páez-Hernández, S. Corona-Avendaño, A. Rojas-Hernández, and E. Barrado, "Dispersive solid phase extraction based on butylamide silica for determination of sulfamethoxazole in milk samples by capillary electrophoresis," *Journal of Liquid Chromatography & Related Technologies*, vol. 39, no. 14, pp. 658–665, 2016.

[5] I. S. Ibarra, J. A. Rodriguez, J. M. Miranda, M. Vega, and E. Barrado, "Magnetic solid phase extraction based on phenyl silica adsorbent for the determination of tetracyclines in milk samples by capillary electrophoresis," *Journal of Chromatography A*, vol. 1218, no. 16, pp. 2196–2202, 2011.

[6] C. Guo, M. Wang, H. Xiao et al., "Development of a modified QuEChERS method for the determination of veterinary antibiotics in swine manure by liquid chromatography tandem mass spectrometry," *Journal of Chromatography B*, vol. 1027, pp. 110–118, 2016.

[7] W. H. Tsai, T. C. Huang, J. J. Huang, Y. H. Hsue, and H. Y. Chuang, "Dispersive solid-phase microextraction method for sample extraction in the analysis of four tetracyclines in water and milk samples by high-performance liquid chromatography with diode-array detection," *Journal of Chromatography A*, vol. 1216, no. 12, pp. 2263–2269, 2009.

[8] W. Liu, Z. Zhang, and Z. Liu, "Determination of β-lactam antibiotics in milk using micro-flow chemiluminescence system with on-line solid phase extraction," *Analytica Chimica Acta*, vol. 592, no. 2, pp. 187–192, 2007.

[9] M. F. El-Shahat, N. Burham, and S. M. A. Azeem, "Flow injection analysis–solid phase extraction (FIA–SPE) method for preconcentration and determination of trace amounts of penicillins using methylene blue grafted polyurethane foam," *Journal of Hazardous Materials*, vol. 177, no. 1–3, pp. 1054–1060, 2010.

[10] E. Caro, R. M. Marcé, P. A. G. Cormack, D. C. Sherrington, and F. Borrull, "Synthesis and application of an oxytetracycline imprinted polymer for the solid-phase extraction of tetracycline antibiotics," *Analytica Chimica Acta*, vol. 552, no. 1-2, pp. 81–86, 2005.

[11] T. Jing, X. D. Gao, P. Wang et al., "Determination of trace tetracycline antibiotics in foodstuffs by liquid chromatography-tandem mass spectrometry coupled with selective molecular-imprinted solid-phase extraction," *Analytical and Bioanalytical Chemistry*, vol. 393, no. 8, pp. 2009–2018, 2009.

[12] F. Tan, D. Sun, J. Gao et al., "Preparation of molecularly imprinted polymer nanoparticles for selective removal of fluoroquinolone antibiotics in aqueous solution," *Journal of Hazardous Materials*, vol. 244-245, pp. 750–757, 2013.

[13] Y. K. Lv, C. L. Jia, J. Q. Zhang, P. Li, and H. W. Sun, "Preparation and evaluation of a novel molecularly imprinted hybrid composite monolithic column for on-line solid-phase extraction coupled with HPLC to detect trace fluoroquinolone residues in milk," *Analytical Methods*, vol. 5, no. 7, pp. 1848–1855, 2013.

[14] M. Seifrtová, A. Pena, C. M. Lino, and P. Solich, "Determination of fluoroquinolone antibiotics in hospital and municipal wastewaters in Coimbra by liquid chromatography with a monolithic column and fluorescence detection," *Analytical and Bioanalytical Chemistry*, vol. 391, no. 3, pp. 799–805, 2008.

[15] H. Niu, Y. Cai, Y. Shi et al., "Evaluation of carbon nanotubes as a solid-phase extraction adsorbent for the extraction of cephalosporins antibiotics, sulfonamides and phenolic compounds from aqueous solution," *Analytica Chimica Acta*, vol. 594, no. 1, pp. 81–92, 2007.

[16] S. Álvarez-Torrellas, R. S. Ribeiro, H. T. Gomes, G. Ovejero, and J. García, "Removal of antibiotic compounds by adsorption using glycerol-based carbon materials," *Chemical Engineering Journal*, vol. 296, pp. 277–288, 2016.

[17] X. Wang and Y. Chen, "Determination of aromatic amines in food products and composite food packaging bags by capillary electrophoresis coupled with transient isotachophoretic stacking," *Journal of Chromatography A*, vol. 1216, no. 43, pp. 7324–7328, 2009.

[18] L. Wang, D. MacDonald, X. Huang, and D. D. Y. Chen, "Capture efficiency of dynamic pH junction focusing in capillary electrophoresis," *Electrophoresis*, vol. 37, no. 9, pp. 1143–1150, 2016.

[19] R. Fang, G. Chen, L. Yi et al., "Determination of eight triazine herbicide residues in cereal and vegetable by micellar electrokinetic capillary chromatography with on-line sweeping," *Food Chemistry*, vol. 145, pp. 41–48, 2014.

[20] A. Šlampová, Z. Malá, P. Gebauer, and P. Boček, "Recent progress of sample stacking in capillary electrophoresis (2014–2016)," *Electrophoresis*, vol. 38, no. 1, pp. 20–32, 2017.

[21] C. Kukusamude, S. Srijaranai, M. Kato, and J. P. Quirino, "Cloud point sample clean-up and capillary zone electrophoresis with field enhanced sample injection and micelle to solvent stacking for the analysis of herbicides in milk," *Journal of Chromatography A*, vol. 1351, pp. 110–114, 2014.

[22] L. Y. Thang, H. H. See, and J. P. Quirino, "Field-enhanced sample injection-micelle to solvent stacking in non aqueous capillary electrophoresis," *Talanta*, vol. 161, pp. 165–169, 2016.

[23] L. Liu, Q. Wan, X. Xu, S. Duan, and C. Yang, "Combination of micelle collapse and field-amplified sample stacking in capillary electrophoresis for determination of trimethoprim and sulfamethoxazole in animal-originated foodstuffs," *Food Chemistry*, vol. 219, pp. 7–12, 2017.

[24] F. Kitagawa and K. Otsuka, "Recent applications of on-line sample preconcentration techniques in capillary electrophoresis," *Journal of Chromatography A*, vol. 1335, pp. 43–60, 2014.

[25] N. Wang, M. Su, S. Liang, and H. Sun, "Sensitive residue analysis of quinolones and sulfonamides in aquatic product by capillary zone electrophoresis using large-volume sample stacking with polarity switching combined with accelerated solvent extraction," *Food Analytical Methods*, vol. 9, no. 4, pp. 1020–1028, 2016.

[26] A. V. Herrera-Herrera, L. M. Ravelo-Pérez, J. Hernández-Borges, M. M. Afonso, J. A. Palenzuela, and M. A. Rodríguez-Delgado, "Oxidized multi-walled carbon nanotubes for the dispersive solid-phase extraction of quinolone antibiotics from water samples using capillary electrophoresis and large volume sample stacking with polarity switching," *Journal of Chromatography A*, vol. 1218, no. 31, pp. 5352–5361, 2011.

[27] M. E. Hume and C. J. Donskey, "Effect of vancomycin, tylosin, and chlortetracycline on vancomycin-resistant *Enterococcus faecium* colonization of broiler chickens during grow-out," *Foodborne Pathogens and Disease*, vol. 14, no. 4, pp. 231–237, 2017.

[28] J. Cornejo, E. Pokrant, D. Araya et al., "Residue depletion of oxytetracycline (OTC) and 4-epi-oxytetracycline (4-epi-OTC) in broiler chicken's claws, by liquid chromatography tandem mass spectrometry (LC-MS/MS)," *Food Additives & Contaminants: Part A*, vol. 34, no. 4, pp. 469–476, 2017.

[29] F. Granados-Chinchilla and C. Rodríguez, "Tetracyclines in food and feeding stuffs: from regulation to analytical methods, bacterial resistance, and environmental and health implications," *Journal of Analytical Methods in Chemistry*, vol. 2017, Article ID 1315497, 24 pages, 2017.

[30] Commission Regulation (EC) No 508/1999 of 4 March 1999 amending Annexes I to IV to Council Regulation (EEC) No 2377/90 laying down a community procedure for the establishment of maximum residue limits of veterinary medical products in foodstuffs of animal origin," *Official Journal of the European Community*, vol. L60, pp. 16–52, 1999.

[31] US Code of Federal Regulations, *Food and Drugs, Chapter I; Food and Drugs Administration, Department of Health and Human Services, Subchapter E; Animal Drugs, Feed, and Related Products, Part 556; Tolerances for residues of New Animal Drugs*, US Government Printing Office, Washington, DC, USA, vol. 6. 2017, http://www.fda.gov. Electronic Code of Federal Regulations (e-CFR).

[32] Codex Alimentarius, *Maximum residue limits (MRLs) and risk management recommendations (RMRs) for residues of veterinary drugs in foods*, CAC/MRL 2-2017, Codex Alimentarius Commision, pp. 1–42, 2017.

[33] C. Long, B. Deng, S. Sun, and S. Meng, "Simultaneous determination of chlortetracycline, ampicillin and sarafloxacin in milk using capillary electrophoresis with electrochemiluminescence detection," *Food Additives & Contaminants: Part A*, vol. 34, no. 1, pp. 1–8, 2016.

[34] M. Tumini, O. G. Nagel, and R. L. Althaus, "Microbiological bioassay using *Bacillus pumilus* to detect tetracycline in milk," *International Dairy Journal*, vol. 82, no. 2, pp. 248–255, 2015.

[35] E. Patyra and K. Kwiatek, "Development and validation of multi-residue analysis for tetracycline antibiotics in feed by

high performance liquid chromatography coupled to mass spectrometry," *Food Additives & Contaminants: Part A*, vol. 34, no. 9, pp. 1553–1561, 2016.

[36] D. Wei, S. Wu, and Y. Zhu, "Magnetic solid phase extraction based on graphene oxide/nanoscale zero-valent iron for the determination of tetracyclines in water and milk by using HPLC-MS/MS," *RSC Advances*, vol. 7, no. 70, pp. 44578–44586, 2017.

[37] C. Zhou, J. Deng, G. Shi, and T. Zhou, "β-cyclodextrin-ionic liquid polymer based dynamically coating for simultaneous determination of tetracyclines by capillary electrophoresis," *Electrophoresis*, vol. 38, no. 7, pp. 1060–1067, 2017.

[38] J. A. Hunt and D. G. Dalgleish, "Heat stability of oil-in-water emulsions containing milk proteins: Effect of ionic strength and pH," *Journal of Food Science*, vol. 60, no. 5, pp. 1120–1123, 1995.

[39] J. A. Rodriguez, J. Espinosa, K. Aguilar-Arteaga, I. S. Ibarra, and J. M. Miranda, "Determination of tetracyclines in milk samples by magnetic solid phase extraction flow injection analysis," *Microchimica Acta*, vol. 171, no. 3-4, pp. 407–413, 2010.

[40] K. Danzer and L. A. Currie, "Guidelines for calibration in analytical chemistry. Part I. Fundamentals and single component calibration (IUPAC Recommendations 1998)," *Pure and Applied Chemistry*, vol. 70, no. 4, pp. 993–1014, 1998.

Kinetic Studies on the Removal of some Lanthanide Ions from Aqueous Solutions using Amidoxime-Hydroxamic Acid Polymer

Fadi Alakhras (iD)

Department of Chemistry, College of Science, Imam Abdulrahman Bin Faisal University, P.O. Box 1982, Dammam 31441, Saudi Arabia

Correspondence should be addressed to Fadi Alakhras; falakhras@iau.edu.sa

Academic Editor: Manzar Sohail

Lanthanide metal ions make distinctive and essential contributions to recent global proficiency. Extraction and reuse of these ions is of immense significance especially when the supply is restricted. In light of sorption technology, poly(amidoxime-hydroxamic) acid sorbents are synthesized and utilized for the removal of various lanthanide ions (La^{3+}, Nd^{3+}, Sm^{3+}, Gd^{3+}, and Tb^{3+}) from aqueous solutions. The sorption speed of trivalent lanthanides (Ln^{3+}) depending on the contact period is studied by a batch equilibrium method. The results reveal fast rates of metal ion uptake with highest percentage being achieved after 15–30 min. The interaction of poly(amidoxime-hydroxamic) acid sorbent with Ln^{3+} ions follows the pseudo-second-order kinetic model with a correlation coefficient R^2 extremely high and close to unity. Intraparticle diffusion data provide three linear plots indicating that the sorption process is affected by two or more steps, and the intraparticle diffusion rate constants are raised among reduction of ionic radius of the studied lanthanides.

1. Introduction

Fast industrial progress throughout preceding years involved improvement and wide handling of special materials in diverse viable products. The handling of heavy metals in manufacturing introduces a huge amount of toxic metals into the atmosphere in addition to the aquatic and global surroundings [1]. Due to the certain physical and chemical features of rare earth elements (REEs), they have been used for particular applications in advanced technologies, such as special ceramics, organic synthesis, and their usage in equipment such as batteries, sensors, energy efficient lighting, nuclear technologies, and telecommunication [2–5]. REEs can be exploited in the production of alloys, magnets, catalysts, electric machines, security systems, medical applications, and in fertilizing perspectives [6–10].

In spite of the increasing concern raised by REEs, healthiness advantages versus poisonous impacts of these matters should be taken seriously. There are several environmental concerns allied with the production, processing, and utilization of REEs [11]. A few reports have accounted that the chemicals used in the refining process of REEs are involved in disease and occupational poisoning of local residents, water pollution, and farmland destruction. It was confirmed that REE's bioaccumulation through the food chain can cause diseases because of the contact with human beings even at trace concentration [12–14]. Therefore, rare earth metal ions including lanthanides in wastewaters are of main environmental interest and necessitate to be treated prior to their removal into environment.

Disposal of rare earth metal ions from large volumes of wastewaters needs a cost-effectual remediation expertise. Using novel extraction techniques having great effectiveness and biodegradability will certainly decrease expenses and environmental effects, permitting the rare earth elements to be handled very extensively as well as escalating the quantities which are regained via recovering [15]. Several techniques have been used for rare earth elements removal such as solvent extraction, organic and inorganic ion exchange, micellar ultrafiltration, chemical precipitation, solid-phase extraction, or chromatography. Chemical precipitation treatment methods need large storage facilities for the precipitated sludge and sometimes require additional treatment options. Ion exchange, which is expensive and sophisticated, would allow metallic ions recovery. Additionally, these methods are not amenable to large-scale treatment of contaminated ground water or drinking water. Purification of water contaminated with lanthanides requires the ability to selectively remove the lanthanides in the presence of other ions like Ca and Mg [16]. Therefore, it may be of value to design a chemically selective sorbent material (amidoxime-hydroxamic acid as in this study), capable of selectively sequestering the lanthanides from waste streams, ground water, and drinking water, without the need for any additional solvents or treatment processes. Adsorption process which is used in this study is highly economical and capable of removing contaminants even at trace level. The use of this technique in water and wastewater treatment due to its simplicity and low cost of operation, and its wide end-use is highly utilized [17–19].

Nevertheless, the performance of any adsorption process is exceedingly reliant on the selection of suitable medium. For the extraction of lanthanides from aqueous systems, different prospective sorbents are studied by several research groups [20–24]. It is confirmed that the sorption of lanthanides can be affected by the variation of medium's acidity, contact period, temperature, and sorbent amount. Additionally, polymers with certain chelating ligands have been used to remove valuable ionic species found in liquid systems. These chelating polymers have attractive properties including powerful capability, great selectivity, and quick removal rates of ionic species with high physical potency and stability [25–28].

In connection with our previous work [17], poly (amidoxime-hydroxamic) acid resins were used for the sorption of various lanthanide ions (La^{3+}, Nd^{3+}, Sm^{3+}, Gd^{3+}, and Tb^{3+}) from liquid systems. The impacts of pH, exposure time, counter ion, and cross-linker were investigated. The results showed that 6 hours of agitation was sufficient to accomplish utmost metal-ion removal with maximum at pH 7. The data also specified that the percentage removal declined as the quantity of cross-linker enhanced.

To understand the influence of exposure time and the kinetics of adsorption of lanthanide ions by amidoxime-hydroxamic acid polymers, more investigation were done and presented in this article. Pseudo-first-order and pseudo-second-order kinetics were utilized for data analysis. Furthermore, the intraparticle diffusion model which is developed to inspect the mechanism of sorption for a solid-liquid removal process was applied as well.

2. Materials and Methods

2.1. Materials. Unless otherwise stated all chemical materials were attained from trade sources and were utilized as obtained. Acrylonitrile, 2,2,4-trimethyl pentane, and gelatine were acquired from BDH Chemicals, Ltd. (Poole, England). Benzoyl peroxide and ethylacrylate were bought from Fluka (Buchs, Switzerland). Hydroxylamine hydrochloride was purchased from Aldrich (Milwaukee, WI). The following lanthanide salts were also used as received with no

more refinement: $LaCl_3·6H_2O$, $NdCl_3·6H_2O$, $SmCl_3·6H_2O$, and $GdCl_3·6H_2O$ from Aldrich and $TbCl_3·6H_2O$ from K&K Laboratories, Inc. (Jamaica, NY).

2.2. Preparation and Characterization of Adsorbent Material. Amidoxime-hydroxamic acid polymer was synthesized and characterized depending on a process reported by Alakhras et al. [17]. The polymer (Scheme 1) was prepared through addition of acrylonitrile and ethylacrylate by suspension polymerization. Afterwards, hydroxylamine hydrochloride was added to convert the obtained copolymer into poly (amidoxime-hydroxamic) acid resin. A detailed analysis of IR spectra, decomposition point, and water regain parameters of the obtained chelating material was reported in our previous work [17].

2.3. Sorption Kinetics of Lanthanide Ions. A batch equilibrium process was employed to investigate the removal kinetics for each metal ion (La^{3+}, Nd^{3+}, Sm^{3+}, Gd^{3+}, and Tb^{3+}). Perfectly weighted dosage of adsorbent (0.10 g) was constantly shaken at 400 rpm with 15 mL of acetic acid/acetate solution with pH 7.0 for two hours to reach equilibrium. After that, 10.0 mL of metal-ion solution with $15.0\,mg·L^{-1}$ was added, and the mixture was shaken for a definite period of time. The agitation time was changed from 0.25 to 24 h. At the end, the contents of the flasks were determined for residual adsorbates by Complexometric titration with standard ethylene diamine tetra-acetic acid (EDTA) aqueous solution using xylenol orange as the indicator [29]. The flasks were shaken with a GFL-1083 thermostated shaker kept at 25°C.

The quantity of metal ions removed at equilibrium, Q_e ($mg·g^{-1}$) was evaluated [30] according to the following equation:

$$Q_e = \frac{(C_0 - C_e)V}{W}, \quad (1)$$

where C_0 represents the initial metal amount ($mg·L^{-1}$) and C_e is the final concentration after 24 h. The volume of solution (L) is represented by V and W symbolizes the amount of dry sorbent (g). The removal amount of metal ions chelated at several intervals [31] was computed by means of the following equation:

$$Q_t = \frac{(C_0 - C_t)V}{W}, \quad (2)$$

where C_t is the metal ion concentration ($mg·L^{-1}$) in solution at different period t.

3. Results and Discussion

3.1. Sorption Kinetics. The removal rate of trivalent lanthanides (Ln^{3+}) depending on the contact time was studied by a batch equilibrium method. In this procedure, 35–60 mesh size trials of the dry adsorbent was equilibrated for two hours with acetate solution (pH = 7.0). Figure 1 demonstrates a representative reliance of lanthanides removal on exposure period. The percentage removal increased with time

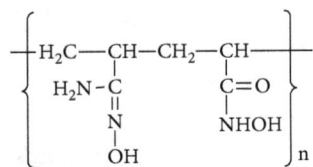

SCHEME 1: Structure of poly(amidoxime-hydroxamic) acid sorbent.

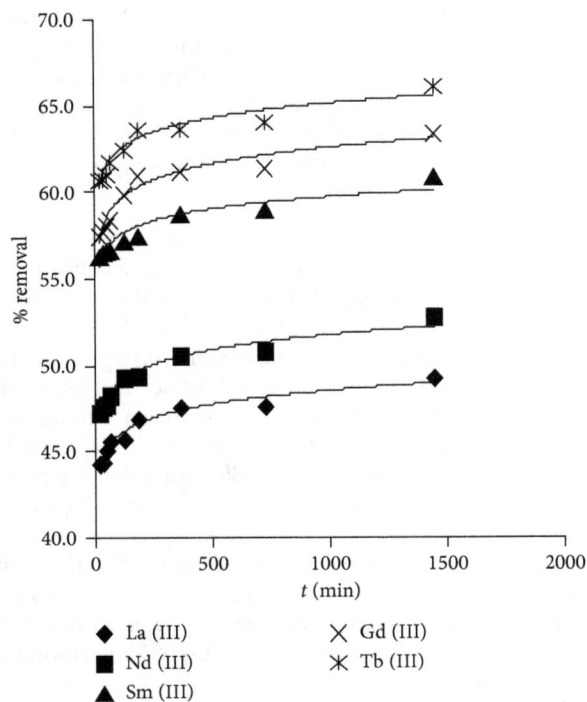

SCHEME 2: Lanthanide metal ions adsorption modes.

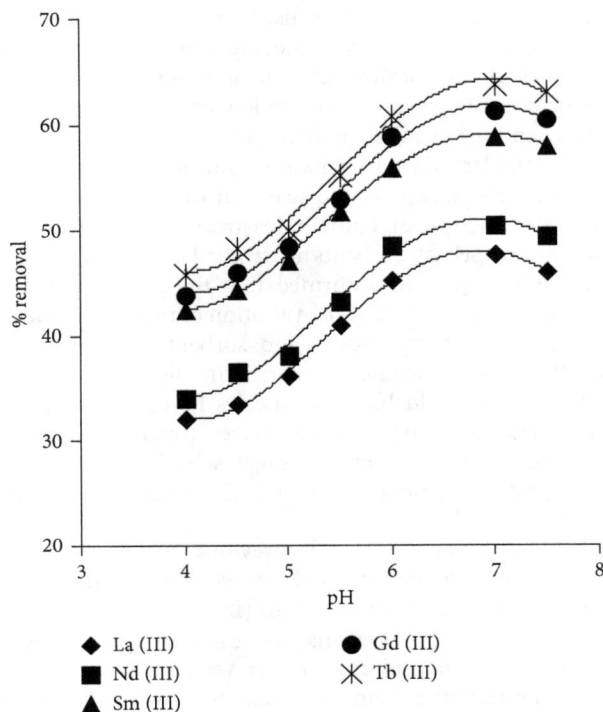

FIGURE 1: Effect of contact time on Ln^{3+} removal ($C_0 = 15$ mg·L^{-1}, $T = 25°C$, $W = 0.10$ g, and $V = 0.025$ L).

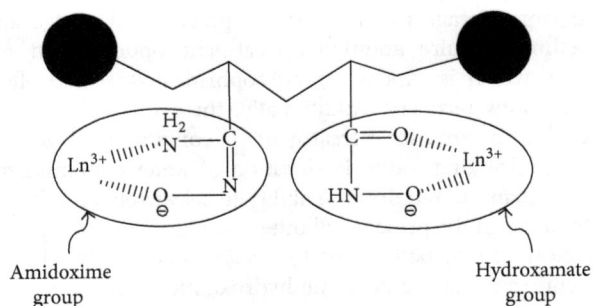

FIGURE 2: Effect of pH on Ln^{3+} removal ($C_0 = 15$ mg·L^{-1}, $T = 25°C$, $W = 0.10$ g, and $V = 0.025$ L).

until it attained a steady state after 5-6 h. In addition, the outcome pointed out fast rates of ion sorption with 60% maximum percentage achieved after 15–30 min. This behaviour can be referred to the vacant active sites on the adsorbent surface and to the high electrostatic attraction between metal ions and chelating groups [32].

The above results can also be explained depending on metal ion adsorption modes (Scheme 2). Lanthanide metal ions can strongly attach via two sites (N and O) at the same time with different modes which certainly affects on the obtained complex stability.

Furthermore, results disclosed that the lanthanide ions removal follows the following arrangement: $Tb^{3+} > Gd^{3+} > Sm^{3+} > Nd^{3+} > La^{3+}$; the polymer shows maximum removal ability toward Tb^{3+} and minimum for La^{3+}. This divergence in capabilities depicted with the studied ions by the used polymer can be interpreted by the negative steric effect on coordination with active chelating groups [33]. The ionic radius for Tb^{3+} is 106.3 pm while 115 pm for La^{3+}. The stability of the forming complex is anticipated to be less convenient for species of bigger extent; this is reliable with previous research [17, 34–36].

The removal capacity of the investigated sorbent toward Ln^{3+} was investigated in the pH range 4.0–7.5 under continuous shaking for 24 h at 25°C. The effect of pH is depicted in Figure 2, and the results revealed that metal-ion uptake increased with increase of pH and a steady state was attained at about pH 7.0. This behaviour can be attributed to the acid dissociation nature of the hydroxamate and amidoxime groups of the sorbent material. At higher pH values, Ln^{3+} ions competed favourably toward donor sites compared with hydrogen ions. Consequently, more metal ions removal was achieved. However, at pH values higher than 7, the metal ion solution can be converted to hydroxide form that has no charge, and its interaction can be decreased clearly by the sorbent. The buffer capacity of acetic acid/acetate solution at pH 7.0 is still preserved and accepted under this condition in order to accomplish the highest percent removal of lanthanide metal ions using amidoxime-hydroxamic acid resin.

TABLE 1: Linearized equations for Ln^{3+} sorption kinetics.

Kinetic model	Linear equation	Plot
Pseudo-first-order	$\ln(Q_e - Q_t) = \ln Q_e - k_1 t$	$\ln(Q_e - Q_t)$ versus t
Pseudo-second-order	$(t/Q_t) = (1/k_2 Q_e^2) + (t/Q_e)$	t/Q_e versus t
Intraparticle diffusion	$Q_t = k_{id} t^{1/2} + C$	Q_t versus $t^{1/2}$

TABLE 2: Kinetic constants for pseudo-first-order and pseudo-second-order models for Ln^{3+} sorption.

	$(Q_e)_{exp}$ (mg·g^{-1})	Pseudo-first-order			Pseudo-second-order		
		K_1 (min^{-1})	$(Q_e)_{cal}$ (mg·g^{-1})	R_1^2	K_2 (g·mg^{-1}·min^{-1})	$(Q_e)_{cal}$ (mg·g^{-1})	R_2^2
La^{3+}	1.85	1.61×10^{-3}	0.1593	0.7754	0.1831	1.80	0.9999
Nd^{3+}	1.98	1.38×10^{-3}	0.1842	0.8457	0.1548	1.91	0.9998
Sm^{3+}	2.28	1.38×10^{-3}	0.1647	0.8964	0.1700	2.22	0.9999
Gd^{3+}	2.38	1.61×10^{-3}	0.1854	0.7062	0.1735	2.31	0.9997
Tb^{3+}	2.48	1.38×10^{-3}	0.1759	0.7304	0.1887	2.41	0.9999

3.2. Kinetic Analysis of Lanthanide Ions Removal Method.

To verify the kinetic phenomena of the removal process of lanthanide ions, three kinetic models were used to evaluate the experimental data at pH 7.0, involving pseudo-first-order, pseudo-second-order, and intraparticle diffusion models. The first two models described the whole sorption process without taking into account the diffusion steps, whereas the third model is utilized to investigate the mechanism of sorption for a solid-liquid removal process that can normally be depicted by three stages [15, 18, 32]. The representative equations for the kinetic models are offered in Table 1, where Q_t and Q_e are the quantity of ions removed at time t and at equilibrium in mg·g^{-1}, correspondingly, k is the rate constant, and C is the boundary thickness layer in mg·g^{-1}.

The $(Q_e)_{cal}$ and k_1 data are presented in Table 2. Correlation coefficient R^2 was varied between 0.7062 and 0.8964. The experimental Q_e values were considerably varied from the associated evaluated Q_e ones, demonstrating that the sorption process was not following the pseudo-first-order model.

Figure 3 displays plotting of t/Q_t against t for lanthanide metal ions under investigation with obtained straight lines having $R^2 = 0.9999$. The $(Q_e)_{cal}$ data were in tremendous agreement with the investigational statistics (Table 2). Consequently, the Ln^{3+} removal process could be deduced more constructively by pseudo-second-order kinetic analysis. The outcome may propose that the removal of Ln^{3+} ions using amidoxime-hydroxamic acid polymer is based on a chemical adsorption process by means of participation or exchange of electrons between adsorbent and adsorbate [15, 37].

Intraparticle diffusion mechanism is described as the transfer of solute from liquid to solid phase during the removal process on a porous adsorbent. In this form, k_{id} (mg·g^{-1}·min$^{-1/2}$) is representing intraparticle diffusion rate constant while C (mg·g^{-1}) is demonstrating the thickness of the boundary layer. Figure 4 shows plotting of Q_t versus $t^{1/2}$ for the sorption of Tb(III) metal ion. The data provided three linear plots indicating that the sorption process is affected by two or more steps. The first region characterized macropore diffusion, and the second one represented the slow removal

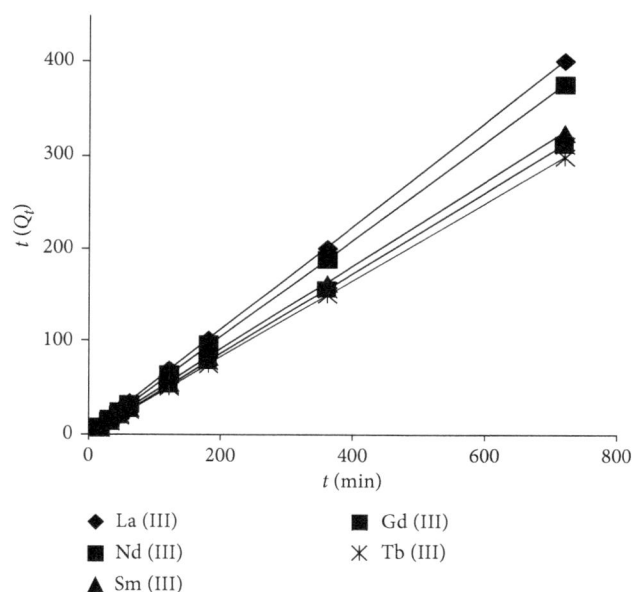

FIGURE 3: Pseudo-second-order kinetic plot for Ln^{3+} sorption.

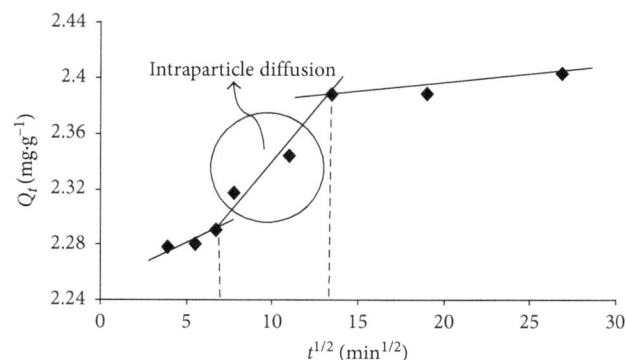

FIGURE 4: Intraparticle diffusion plot for the sorption of Tb(III) metal ion.

phase, where intraparticle diffusion is the rate-limiting step [38, 39]. The third region could be considered as the diffusion via smaller pores, which is followed by the

TABLE 3: Intraparticle diffusion kinetic parameters for Ln^{3+} sorption.

Ln^{3+}	k_{id} (mg·g^{-1}·min$^{-1/2}$)	C (mg·g^{-1})	R^2
La^{3+}	0.0047	1.6228	0.9195
Nd^{3+}	0.0056	1.7380	0.9536
Sm^{3+}	0.0098	2.0938	0.9836
Gd^{3+}	0.0102	2.1326	0.9603
Tb^{3+}	0.0134	2.2314	0.9265

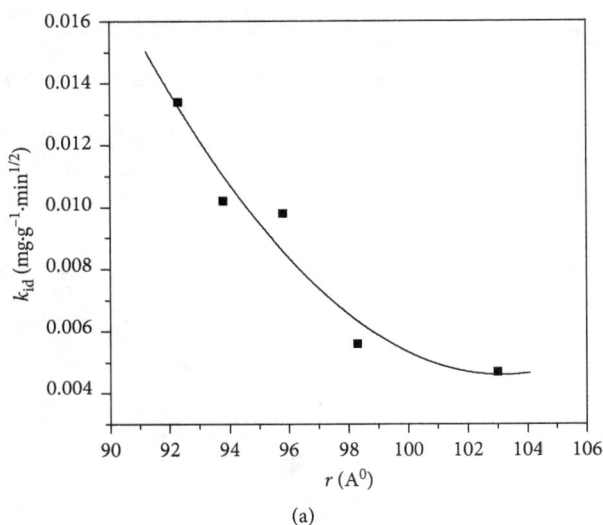

(a)

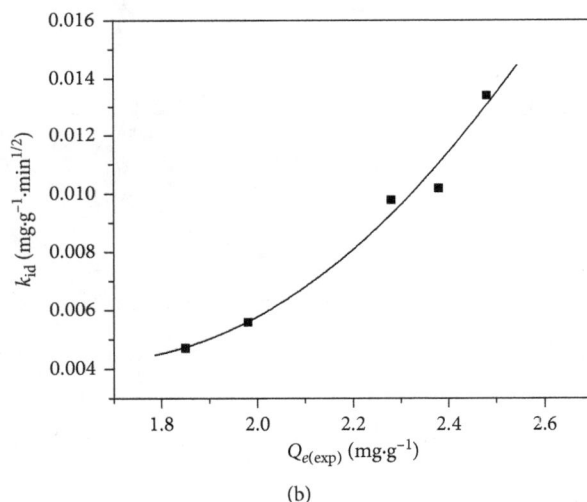

(b)

FIGURE 5: (a) Plot of ionic radius of Ln^{3+} versus k_{id} values; (b) plot of $Q_{e\,(exp)}$ of Ln^{3+} versus k_{id} values.

establishment of equilibrium. Additionally, the plot did not pass through the origin, demonstrating that pore diffusion could not be the only rate-determining step in verifying the kinetics of Ln^{3+} sorption process, and this process might be organized by a chemical reaction as well.

The values of the different parameters calculated from the plots of Q_t versus $t^{1/2}$ for the intraparticle diffusion region are presented in Table 3.

It is clearly indicated that the intraparticle diffusion rate constants are increased with decrease of ionic radius of the

studied lanthanides (Figure 5(a)). This behaviour suggested that ionic radius declined with the atomic number, and also smaller ions became more convenient to diffuse [15]. Furthermore, positive relationship was obtained by plotting k_{id} values versus experimental Q_e ones (Figure 5(b)). This performance designated that smaller metal ions can be diffused through the outer surface into the pores of the sorbent and removed more easily than bigger ones.

4. Conclusions

Amidoxime-hydroxamic acid polymer materials have been synthesized and exploited for the chelation of different lanthanides (La^{3+}, Nd^{3+}, Sm^{3+}, Gd^{3+}, and Tb^{3+}) from aqueous solutions. The sorption rate of the investigated ions relying on agitation period was analyzed by the batch equilibrium method. The removal process was carried out with fast kinetic rates, and it achieved maximum percentage after 15–30 min. The sorption of ions followed the following order: $Tb^{3+} > Gd^{3+} > Sm^{3+} > Nd^{3+} > La^{3+}$. Kinetic studies revealed that chemisorption was the rate-determining step with correlation coefficient remarkably high and close to unity. Intraparticle diffusion study afforded three regions demonstrating that the sorption course is affected by more than one step and intraparticle diffusion might be involving in controlling the diffusion of investigated lanthanides.

Conflicts of Interest

The author declares that there are no conflicts of interest regarding the publication of this paper.

References

[1] R. Saravanan and L. Ravikumar, "The use of new chemically modified cellulose for heavy metal ion adsorption and antimicrobial activities," *Journal of Water Resource and Protection*, vol. 7, no. 6, pp. 530–545, 2015.

[2] A. Fuqiang, B. Gao, X. Huang et al., "Selectively removal of Al (III) from Pr(III) and Nd(III) rare earth solution using surface imprinted polymer," *Reactive and Functional Polymers*, vol. 73, no. 1, pp. 60–65, 2013.

[3] I. Celik, D. Kara, C. Karadas, A. Fisher, and S. J. Hill, "A novel ligandless-dispersive liquid–liquid microextraction method for matrix elimination and the preconcentration of rare earth elements from natural waters," *Talanta*, vol. 134, pp. 476–481, 2015.

[4] S. Yesiller, A. E. Eroglu, and T. Shahwan, "Removal of aqueous rare earth elements (REEs) using nano-iron based materials," *Journal of Industrial and Engineering Chemistry*, vol. 19, no. 3, pp. 898–907, 2013.

[5] T. Jun, Y. Jingqun, C. Kaihong, R. Guohua, J. Mintao, and C. Ruan, "Extraction of rare earths from the leach liquor of the weathered crust elution-deposited rare earth ore with nonprecipitation," *International Journal of Mineral Processing*, vol. 98, no. 3-4, pp. 125–131, 2011.

[6] H. S. Yoon, C. J. Kim, K. W. Chung, S. D. Kim, J. Y. Lee, and J. R. Kumar, "Solvent extraction, separation and recovery of dysprosium (Dy) and neodymium (Nd) from aqueous solutions: waste recycling strategies for permanent magnet processing," *Hydrometallurgy*, vol. 165, pp. 27–43, 2016.

[7] P. K. Parhi, K. H. Park, C. W. Nam, and J. T. Park, "Liquid-liquid extraction and separation of total rare earth (RE) metals from polymetallic manganese nodule leaching solution," *Journal of Rare Earths*, vol. 33, no. 2, pp. 207–213, 2015.

[8] P. Liang, Y. Liu, and L. Guo, "Determination of trace rare earth elements by inductively coupled plasma atomic emission spectrometry after preconcentration with multiwalled carbon nanotubes," *Spectrochimica Acta Part B: Atomic Spectroscopy*, vol. 60, no. 1, pp. 125–129, 2005.

[9] R. M. Ashour, R. El-sayed, A. F. Abdel-Magied et al., "Selective separation of rare earth ions from aqueous solution using functionalized magnetite nanoparticles: kinetic and thermodynamic studies," *Chemical Engineering Journal*, vol. 327, pp. 286–296, 2017.

[10] Y. F. Xiao, X. S. Liu, Z. Y. Feng et al., "Role of minerals properties on leaching process of weathered crust elution-deposited rare earth ore," *Journal of Rare Earths*, vol. 33, no. 5, pp. 545–552, 2015.

[11] K. T. Rim, "Effects of rare earth elements on the environment and human health: a literature review," *Toxicology and Environmental Health Sciences*, vol. 8, no. 3, pp. 189–200, 2016.

[12] K. T. Rim, K. H. Koo, and J. S. Park, "Toxicological evaluations of rare earths and their health impacts to workers: a literature review," *Safety and Health at Work*, vol. 4, no. 1, pp. 12–26, 2013.

[13] L. Wang, W. Wang, Q. Zhou, and X. Huang, "Combined effects of lanthanum (III) chloride and acid rain on photosynthetic parameters in rice," *Chemosphere*, vol. 112, pp. 355–361, 2014.

[14] H. Herrmann, J. Nolde, S. Berger, and S. Heise, "Aquatic ecotoxicity of lanthanum—a review and an attempt to derive water and sediment quality criteria," *Ecotoxicology and Environmental Safety*, vol. 124, pp. 213–218, 2016.

[15] H. Zhang, R. G. McDowell, L. R. Martin, and Y. Qiang, "Selective extraction of heavy and light lanthanides from aqueous solution by advanced magnetic nanosorbents," *Applied Materials and Interfaces*, vol. 8, no. 14, pp. 9523–9531, 2016.

[16] W. Yantasee, G. E. Fryxell, R. S. Addleman et al., "Selective removal of lanthanides from natural waters, acidic streams and dialysate," *Journal of Hazardous Materials*, vol. 168, no. 2-3, pp. 1233–1238, 2009.

[17] F. A. Alakhras, K. Abu Dari, and M. S. Mubarak, "Synthesis and chelating properties of some poly(amidoxime-hydroxamic acid) resins toward some trivalent lanthanide metal ions," *Journal of Applied Polymer Science*, vol. 97, no. 2, pp. 691–696, 2005.

[18] H. M. Marwani, H. M. Albishri, T. A. Jalal, and E. M. Soliman, "Study of isotherm and kinetic models of lanthanum adsorption on activated carbon loaded with recently synthesized Schiff's base," *Arabian Journal of Chemistry*, vol. 10, pp. S1032–S1040, 2017.

[19] E. A. Abdel-Galil, A. B. Ibrahim, and M. M. Abou-Mesalam, "Sorption behavior of some lanthanides on polyacrylamide stannic molybdophosphate as organic–inorganic composite," *International Journal of Industrial Chemistry*, vol. 7, no. 3, pp. 231–240, 2016.

[20] F. Xie, T. A. Zhang, D. Dreisinger, and F. Doyle, "A critical review on solvent extraction of rare earths from aqueous solutions," *Minerals Engineering*, vol. 56, pp. 10–28, 2014.

[21] A. N. Turanov, V. K. Karandashev, N. S. Sukhinina, V. M. Masalov, A. A. Zhokhov, and G. A. Emelchenko, "A novel sorbent for lanthanide adsorption based on tetraoctyldiglycolamide, modified carbon inverse opals," *RSC Advances*, vol. 5, no. 1, pp. 529–535, 2015.

[22] Q. Chen, "Rare earth application for heat-resisting alloys," *Journal of Rare Earths*, vol. 28, p. 125, 2010.

[23] A. N. Turanov and V. K. Karandashevb, "Adsorption of lanthanides(III) from aqueous solutions by fullerene black modified with di(2-ethylhexyl)phosphoric acid," *Central European Journal of Chemistry*, vol. 7, no. 1, p. 54, 2009.

[24] E. Farahmand, "Adsorption of cerium (IV) from aqueous solutions using activated carbon developed from rice straw," *Open Journal of Geology*, vol. 6, no. 3, pp. 189–200, 2016.

[25] N. Kabay and H. Egawa, "Chelating polymers for recovery of uranium from Seawater," *Separation Science and Technology*, vol. 29, no. 1, pp. 135–150, 1994.

[26] K. A. K. Ebraheem and S. T. Hamdi, "Synthesis and properties of a copper selective chelating resin containing a salicylaldoxime group," *Reactive and Functional Polymers*, vol. 34, no. 1, pp. 5–10, 1997.

[27] K. A. K. Ebraheem, S. T. Hamdi, and J. A. Al-Duhan, "Synthesis and properties of a chelating resin containing a salicylaldimine group," *Journal of Macromolecular Science, Part A*, vol. 34, no. 9, pp. 1691–1699, 1997.

[28] F. Vernon, "Some aspects of ion exchange in copper hydrometallurgy," *Hydrometallurgy*, vol. 4, no. 2, pp. 147–157, 1979.

[29] A. I. Ismail, K. A. K. Ebraheem, M. S. Mubarak, and F. I. Khalili, "Chelation properties of some mannich-type polymers toward lanthanum(III), neodymium(III), samarium(III), and gadolinium(III)," *Solvent Extraction and Ion Exchange*, vol. 21, no. 1, pp. 125–137, 2003.

[30] F. Alakhras, N. Ouerfelli, G. M. Al-Mazaideh, T. S. Ababneh, and F. M. Abouzeid, "Optimal pseudo-average order kinetic model for correlating the removal of nickel ions by adsorption on nanobentonite," *Arabian Journal for Science and Engineering*, 2018.

[31] F. Alakhras, E. Al-Abbad, N. O. Alzamel, F. M. Abouzeid, and N. Ouerfelli, "Contribution to modelling the effect of temperature on removal of nickel ions by adsorption on nanobentonite," *Asian Journal of Chemistry*, vol. 30, no. 5, pp. 1147–1156, 2018.

[32] A. Maleki, E. Pajootan, and B. J. Hayati, "Ethyl acrylate grafted chitosan for heavy metal removal from wastewater: equilibrium, kinetic and thermodynamic studies," *Journal of the Taiwan Institute of Chemical Engineers*, vol. 51, pp. 127–134, 2015.

[33] Y. K. Agrawal and H. L. Kapoor, "Stability constants of rare earths with hydroxamic acids," *Journal of Inorganic and Nuclear Chemistry*, vol. 39, no. 3, pp. 479–482, 1977.

[34] F. Al-Rimawi, A. Ahmad, F. I. Khalili, and M. S. Mubarak, "Chelation properties of some phenolic-formaldehyde polymers toward some trivalent lanthanide ions," *Solvent Extraction and Ion Exchange*, vol. 22, no. 4, pp. 721–735, 2004.

[35] K. A. K. Ebraheem, M. S. Mubarak, Z. J. Yassien, and F. Khalili, "Chelation properties of poly(8–hydroxyquinoline 5,7–diylmethylene) crosslinked with bisphenol–a toward lanthanum(III), cerium(III), neodimium(III), samarium(III), and gadolinium(III) ions," *Separation Science and Technology*, vol. 35, no. 13, pp. 2115–2125, 2000.

[36] C. Noureddine, A. Lekhmici, and M. S. Mubarak, "Sorption properties of the iminodiacetate ion exchange resin, amberlite IRC-718, toward divalent metal ions," *Journal of Applied Polymer Science*, vol. 107, no. 2, pp. 1316–1319, 2008.

[37] Y. S. Ho and G. McKay, "Pseudo-second order model for sorption processes," *Process Biochemistry*, vol. 34, no. 5, pp. 451–465, 1999.

[38] R. R. Sheha and A. A. El-Zahhar, "Synthesis of some ferro-magnetic composite resins and their metal removal characteristics in aqueous solutions," *Journal of Hazardous Materials*, vol. 150, no. 3, pp. 795–803, 2008.

[39] N. T. Abdel-Ghani, G. A. El-Chaghaby, and E. M. Zahran, "Pentachlorophenol (PCP) adsorption from aqueous solution by activated carbons prepared from corn wastes," *International Journal of Environmental Science and Technology*, vol. 12, no. 1, pp. 211–222, 2015.

9

Development of a Flow Injection System for Differential Pulse Amperometry and its Application for Diazepam Determination

Vesna Antunović,[1] Slavna Tešanović,[2] Danica Perušković,[2] Nikola Stevanović,[2] Rada Baošić,[2] Snežana Mandić,[2] and Aleksandar Lolić ⓘ[2]

[1]Faculty of Medicine, University of Banja Luka, 78000 Banja Luka, Bosnia and Herzegovina
[2]Faculty of Chemistry, University of Belgrade, 11000 Belgrade, Serbia

Correspondence should be addressed to Aleksandar Lolić; lolix@chem.bg.ac.rs

Academic Editor: Josep Esteve-Romero

This work presents the development of a flow injection system for differential pulse amperometry (DPA) for diazepam determination in the presence of oxygen. The thin flow cell consisted of the bare glassy carbon electrode, reference silver/silver chloride, and stainless steel as the auxiliary electrode. Electrochemical reduction of diazepam (DZP) was characterised by cyclic voltammetry. Azomethine reduction peak was used for DZP quantification. The detector response was linear in the range 20–250 μmol/dm^3 of diazepam, with a calculated detection limit of 3.83 μg/cm^3. Intraday and interday precision were 1.53 and 10.8%, respectively. The method was applied on three beverage samples, energetic drink, and two different beer samples, and obtained recoveries were from 93.65 up to 104.96%. The throughoutput of the method was up to 90 analyses per hour.

1. Introduction

Amperometric detection can operate in three different modes: constant potential, pulse mode, and differential pulse (or multi pulse mode). Detection at constant potential is usually applied with flow injection analysis systems because simple and cheap potentiostats can be used. When an analyte passes through the detector, it generates a current proportional to the analyte concentration [1].

Differential pulse amperometry (DPA) is a technique described and first applied by Marcenac and Gonon in 1985 [2]. In DPA, a clean potential is applied for electrode cleaning without current sampling. Then two potential pulses are applied after the cleaning step, and the current at the end of each pulse is recorded as the function of time. During the experiment, only the difference of two current samples is displayed. Obtained dependence of ΔI (the difference of the currents recorded at the end of each potential pulse) with time is named differential pulse amperogram [3]. The main advantage of DPA is that its signal is less affected by interferences than with the other amperometric modes.

Throughout the years, DPA has been used for determination of various substances in a flow injection (FI) system. Quadrupole-pulsed amperometric detection was applied for simultaneous determination of glucose and fructose in a FI system [4]. The first pulse was used for the oxidation of glucose and the second for both analytes. The subtraction of obtained currents was used for quantification of fructose. Three-pulse FI-DPA was used for simultaneous determination of phenolic antioxidants (butylated hydroxyanisole and butylated hydroxytoluene) at a boron-doped diamond electrode [5]. Multiple-pulse amperometric flow injection system was described for simultaneous determination of pharmaceutical active substances, paracetamol and ascorbic acid [6], caffeine, ibuprofen, and paracetamol [7], and dypirone and paracetamol in pharmaceutical formulations on a bare glassy carbon electrode (GCE) [8]. A simple, fast and low-cost multiple-pulse amperometric FI system was applied for determination of sildenafil citrate (SC) in Viagra [9]. Three sequential potential pulses were applied as a function of time. SC is detected at two different irreversible oxidation processes (at 1.6 and 1.9 V), and the third pulse (1.0 V) was used for regeneration of the surface of the boron-doped diamond electrode.

Diazepam (DZP) (7-chloro-1,3-dihydro-1-methyl-5-phenyl-2H-1,4-benzodiapin-2-one) belongs to the group of 1,4-benzodiazepines. It is one of the most prescribed 1,4-benzodiazepines for a number of conditions such as anxiety, insomnia, epilepsy, and muscular spasms [10]. Its low price and availability on the black market increase the risk of its abuse. Since it is highly liposolubile, it distributes rapidly inside tissues and has the high absorption rate. The absorption rate is increased when it is mixed with alcohol which enhances its sedative effect [11]. That is why diazepam is well known as "date rape" drug, and it is classified as a drug-facilitated crimes (DFC) drug [12]. A fast and simple system for diazepam determination would be of great importance for forensic science and resolving criminal activities which are related with benzodiazepine drug abuse [13].

Determination of benzodiazepines is well documented in the literature, and various spectrophotometric [14] and electrochemical detectors [15–20] were employed. Mass spectrometers were also applied to benzodiazepine determination [21]; their wide linear ranges and low detection limits are advantageous but are still expensive compared to simple flow injection systems.

Although there are many references describing determination of benzodiazepine by electrochemical detectors, there is no paper of DPA application on diazepam determination.

Electrochemical determination of diazepam is based on reduction of 4,5-azomethine group. This simple reduction yielding dihydro specie is used for quantification of diazepam [22]. Applying more negative potentials increases interferences from oxygen, hydrogen ions, and other metal ions present in the sample [23]. Due to high background currents, sensitive determination is almost impossible. Concentration of hydrogen and metal ions is a matter of optimization process and can be controlled. However, the presence of oxygen is more problematic and can be decreased either by removing with nitrogen or by applying modifications of electrode surfaces. Often purging of nitrogen or argon prior the analyses for a few minutes or even overnight is usually enough to remove dissolved oxygen from working solutions [23]. When flow injection systems are applied, another problem arises: the choice of tubing material. Usually, standard Teflon tubings should be avoided due to absorption of oxygen through the polymer. Other polymer materials are more or less chemically inert and also affect the price of the flow system or even using glass or metallic materials but they are too rigid for handling the simple flow system. Lozano-Chaves [15] applied modified carbon-paste electrodes as sensors for the determination of diazepam and its metabolites temazepam and oxazepam in biological fluids. As a modifier, they used 5% bentonite and 5% zeolite. Bentonite-modified electrodes showed better sensitivities. Diazepam was determined at pH 10 in Britton–Robinson buffer, with the concentration range of 0.025–3.0 mg/dm^3 and the detection limit of 0.021 mg/dm^3; all experiments were performed under nitrogen atmosphere. Multiwall carbon nanotube-ionic liquid modified paste electrode was used for diazepam determination in real samples [24]. Analytical parameters for diazepam determination were

linearity in the range of 0.02–0.76 mg/dm^3 and the detection limit of 4.1 μg/dm^3. Experiments were performed under nitrogen atmosphere. As mentioned earlier, reduction of 4,5-azomethine group is often used for quantification of diazepam. Group of authors [20] noticed that there is an oxidation peak in a reverse scan in cyclic voltammetry and they determined the oxidation potential (+1.0 V versus Ag/AgCl) for diazepam in drinks on a screen-printed electrode by adsorptive stripping voltammetry. The detector response was linear in the range of 7.1–285 mg/dm^3 and a detection limit of 1.8 mg/dm^3.

This work presents the development of a flow injection for diazepam determination with differential pulse amperometry and its application in spiked beverage samples without removal of dissolved oxygen.

2. Materials and Methods

2.1. Reagents and Chemicals. All reagents used were of analytical grade quality, and all solutions were prepared in degassed and filtered deionised water. Buffer solution of HCl/KCl pH 1 was prepared by mixing appropriate volumes of the diluted solution of the hydrochloric acid (0.2 mol/dm^3, Carlo Erba, Val de Reuil, France) and potassium chloride (0.2 mol/dm^3, Betahem, Belgrade, Serbia). Britton-Robinson buffer was prepared by mixing 0.04 mol/dm^3 CH$_3$COOH (Carlo Erba, Val de Reuil, France), 0.04 mol/dm^3 H$_3$PO$_4$ (Carlo Erba, Val de Reuil, France), and 0.04 mol/dm^3 H$_3$BO$_3$ (Betahem, Belgrade, Serbia), and the pH value is adjusted with 0.02 mol/dm^3 NaOH (Betahem, Belgrade, Serbia). The stock solution of diazepam (Hemofarm, Vršac, Serbia) was prepared by dissolving the required mass in methanol to give a concentration of 10 mmol/dm^3. Stock solution was maintained under refrigerated conditions in the absence of light, and it was prepared once a week. Working standards were prepared daily by dilution of this solution with appropriate buffers.

2.2. Apparatus. Cyclic voltammetry was used for electrochemical study of diazepam. For the cyclic voltammetry, a CHI 800C potentiostat was used. A three-electrode system consisted of the glassy carbon working electrode (CH Instruments, USA, model CHI104, 3 mm in diameter), Ag/AgCl reference electrode (CH Instruments, USA; CHI111), and platinum wire as auxiliary electrode. pH measurements were performed using WTW720 pH-meter equipped with SenTix 81 pH electrode. Prior the cyclic voltammetry experiments, all solutions were purged with oxygen free nitrogen for 15 minutes.

Cleaning of glassy carbon electrodes was performed mechanically on a polishing pad using aluminium paste of different grain sizes (1, 0.3, and 0.05 μm, Buehler, USA). It was washed with distilled water between each paste, and after the finest paste, it was washed with methanol, distilled water, and then air-dried. This process was repeated each day before the start of the recording and in the case if detector response was not reproductive.

2.3. Flow Injection Analysis System. Peristaltic pump model Mini S 840 (Ismatec, Switzerland) was used. Two-position injection valve model 5020 (Rheodyne, USA) equipped with a sample loop was used. Amperometric flow cell (thin layer, BASi, USA) consists of working, reference, and auxiliary electrodes [25]. The working electrode was glassy carbon electrode (BAS Instruments, model MF-1008, USA) consisting of two circular, dual series electrodes which are embedded in a polymer based on fluorocarbons. The reference Ag/AgCl (BAS Instruments, model MF-2021, USA) electrode is filled with $3\,mol/dm^3$ NaCl solution, and the auxiliary electrode is made of stainless steel. Teflon gasket was placed between the auxiliary and working electrodes; its thickness regulates the volume of the working solution. Electrochemical measurements were performed on model CHI760b. The CHI software was used for data acquisition.

2.4. Electrochemical Procedure. Each DPA cycle applied to the working electrode consisted of three steps: first step, cleaning potential of $+0.7\,V$ for 50 ms, second $-0.75\,V$ for 50 ms (first potential pulse), and then $-0.95\,V$ for 50 ms (second potential pulse). The analytical signal was calculated as the difference between the average signal at last 25 ms of the second and the first potential pulse. DPA measurements were performed on CHI760b potentiostat. For this set of experiments, there was no need to expel the oxygen from solutions.

3. Results and Discussion

3.1. Cyclic Voltammetry of Diazepam and pH Effect. Cyclic voltammograms of a $0.5\,mol/dm^3$ diazepam in HCl/KCl buffer (pH 1) were recorded in the potential range of -0.2 to $-0.9\,V$. Voltammograms were recorded for different scan rates from 20 to $150\,mV\,s^{-1}$. In the negative scan, one reduction peak can be observed at around $-0.76\,V$ (Figure 1). The value of the peak potential became more cathodic with increasing scan rate and varied linearly with square root of scan rate, indicating irreversibility of the reduction and diffusion-controlled process on the electrode surface.

The effect of pH on reduction of diazepam was investigated when $100\,\mu mol/dm^3$ diazepam was recorded in HCl/KCl buffer (pH 1) and Britton–Robinson buffer of different pH (3–8) values in the potential range -0.2 to $-1.2\,V$. With increase of pH, the cathodic peak shifts to more negative values indicating the dependence of the reduction potential on the pH value of the medium (Figure 2). As the media becomes less acidic, the FIA signal becomes less stable and reproducible. Hence, the further experiments were performed in pH 1 solutions.

The cathodic peak is a result of a two-electron change, and the reduced product can be characterised as 4,5-dihydro-diazepam (Reaction (1)), as reported elsewhere [22, 26]:

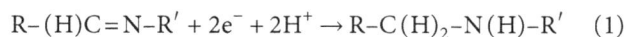

$$R-(H)C=N-R' + 2e^- + 2H^+ \rightarrow R-C(H)_2-N(H)-R' \quad (1)$$

3.2. Optimization of Flow Injection Parameters. The following parameters were investigated for the sensitivity of the thin-layer flow cell: the volume of the sample loop and the

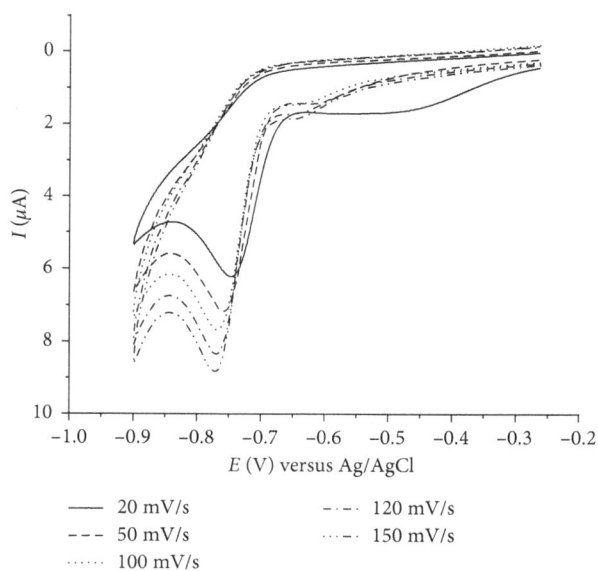

FIGURE 1: Cyclic voltammograms of $0.5\,mol/dm^3$ diazepam on glassy carbon electrode as a function of a scan rate ($20–150\,mV\,s^{-1}$).

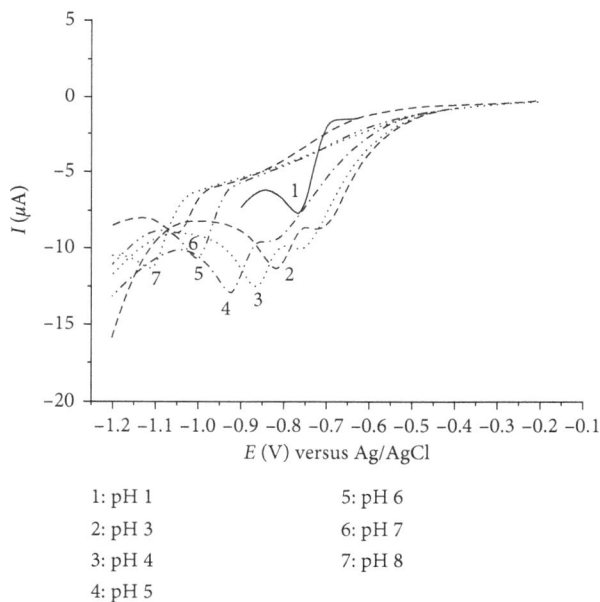

—	20 mV/s	– · –	120 mV/s
– – –	50 mV/s	· · · –	150 mV/s
· · · · ·	100 mV/s		

1: pH 1	5: pH 6
2: pH 3	6: pH 7
3: pH 4	7: pH 8
4: pH 5	

FIGURE 2: Cyclic voltammograms of $100\,\mu mol/dm^3$ diazepam as function of pH in HCl/KCl buffer (pH 1) and Britton–Robinson buffer (pH 3-8) at $100\,mV\,s^{-1}$ scan rate.

gasket thickness. The optimal conditions in FIA system were investigated with $100\,\mu mol/dm^3$ diazepam solution. Sample loops of 0.050, 0.075, and $0.120\,cm^3$ were tested. The detector response was the most sensitive with $0.050\,cm^3$ of the injected sample. The gasket determines the working volume of the flow cell. All four commercially available gaskets from BASi were tested (13, 51, 127, and $381\,\mu m$ thickness). For this type of flow cell, the best sensitivity was obtained for $51\,\mu m$ thick gasket, which was already confirmed in our previous work [25].

3.3. Optimization of Differential Pulse Amperometry.

The potentials used for the two DPA pulses were chosen according to the results obtained by cyclic voltammetry. Hence, the DPA pulses should be a little lower and a little higher than the peak potential of diazepam reduction. We investigated the following values of potential pulses, E_{clean} 0 to +1 V, E_1 −0.6 to −0.75 V, and E_2 −0.8 to −0.95 V; each potential was held for 50 ms, and the current was collected at last 25 ms for E_2 and E_1. All experiments were performed with 0.1 mmol/dm^3 diazepam solution in HCl/KCl buffer pH 1. The 0.7 V was applied for cleaning potential due to the signal shape and the system reproducibility. The sensitivity of the detector for different values of E_1 and E_2 is presented in Table 1, and it shows that the best response was obtained for $E_1 = -0.75$ V and $E_2 = -0.95$ V. Figure 3 presents the DPA waveform applied for diazepam.

Under these conditions, there was no need for purging the solutions prior the experiments or working in the inert atmosphere.

3.4. Analytical Parameters.

The calibration curve obtained from the DPA peak current for diazepam reduction at different diazepam concentration shows linear behaviour between 20 and 250 μmol/dm^3 with the regression equation ΔI (μA) $= 0.063c$ (μM) $+ 0.430$ (Figure 4). The presented calibration curve is the average obtained from the three curves recorded for the same set of standards; error bars in each point of the curve are standard deviations. The detection limit was calculated to be 13.4 μmol/dm^3, but measured as 3 s/m, where s is the standard deviation and m is the slope, and the limit of quantification was calculated to be 44.8 μmol/dm^3. Obtained limit of the detection is not comparable to LODs obtained by hyphenated techniques (chromatography/mass spectrometry) which often require complicated sample preparation [13, 27–30]. Electrochemical detection is seldom described for determination of diazepam in beverage samples. Honeychurch et al. [20] applied adsorptive stripping voltammetry with medium exchange limit, and the limit of detection was 1.8 μg cm^{-3}.

The absolute limit of detection for the sample loop volume of 0.050 cm^3 was 0.19 μg of diazepam. Reproducibility of the system was checked by six consecutive injections of 100 μmol/dm^3 DZP on the same day for investigation of intraday precision (insert picture on Figure 4), and the relative standard deviation for intraday precision was 1.53%. Interday precision was obtained by recording peak current of 100 μmol/dm^3 DZP during three consecutive days; the results showed decrease of precision, presented by the relative standard deviation of 10.8% for interday measurements. Since simple flow injection system was used and the signal frequency was dependent on tubing length, only the throughoutput was up to 90 analyses per hour (Table 2).

3.5. Interference Study.

Ascorbic acid, lactose, glucose, and citric acid are the common compounds present in the beverage samples. The effect of interferants on diazepam determination was investigated by injecting solutions containing 100 μmol dm^{-3} of each interferant and mixtures

TABLE 1: Sensitivities at different E_1 and E_2 values for FIA-DPA.

E_1 (V)	E_2 (V)	Slope (nA dm^3 mol^{-1})
−0.6	−0.8	35
−0.65	−0.8	42
−0.7	−0.8	31.3
−0.75	−0.8	14.2
−0.75	−0.9	38.8
−0.75	−0.95	92.7

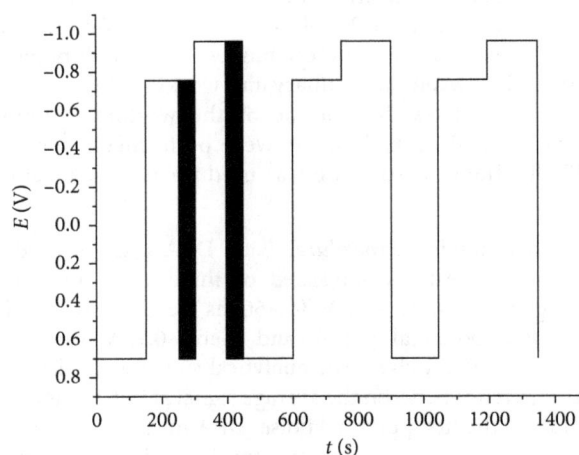

FIGURE 3: Schematic presentation of DPA waveform applied for diazepam determination. The squares are used for current measurement.

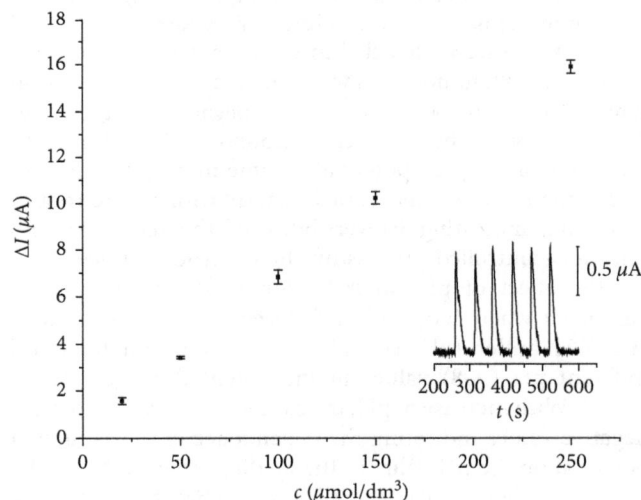

FIGURE 4: Linear response for FIA-DPA system for 20–250 μmol/dm^3 diazepam. Inset: six consecutive injections of 100 μmol/dm^3 of diazepam.

with diazepam of the same concentration. Results showed that interferants decrease diazepam signal by 4–10% (Table 3).

3.6. Determination of Diazepam in Beverage Samples.

The purpose of this method was to investigate concentrations relevant for forensic studies. On Serbian market, diazepam

TABLE 2: Analytical characteristics of the proposed methods.

Characteristics	FIA-DPA
LR, $\mu mol\, dm^{-3}$	20–250
Sensitivity, $nA\, dm^3\, \mu mol^{-1}$	63
Correlation coefficient, r	0.997
LOD, $\mu mol\, dm^{-3}$	13.4
Intraday RSD ($n = 6$)	1.53%
Interday RSD ($n = 3$)	10.8%
Throughoutput, $1\, h^{-1}$	90

TABLE 3: Results of interference study.

Interferent	Relative signal decrease (%)
Citric acid	4
Lactose	5
Ascorbic acid	8
Glucose	10

TABLE 4: Analysis results of the recovery for spiked samples ($n = 3$).

Sample	Spiked concentration ($\mu mol/dm^3$)	Results (mean $\pm$ SD) ($\mu mol/dm^3$)	Recovery ratio (%)
Guarana	50	52.48 ± 0.80	104.96
	100	101.25 ± 1.55	101.25
	150	152.05 ± 2.33	101.37
Beer I	50	47.51 ± 0.73	95.02
	100	94.31 ± 1.44	94.31
	150	139.48 ± 2.13	92.99
Beer II	50	49.12 ± 0.75	98.24
	100	96.24 ± 1.47	96.24
	150	141.66 ± 2.17	94.44

tablets containing 2, 5, or 10 mg of active substance are available; when a tablet is dissolved in 200 cm^3 of a beverage, obtained concentrations are from 35–175 $\mu mol/dm^3$ of diazepam, which fits in the linear range of the detector.

Beverage samples (Guarana and two beer samples) were purchased at a local store. 12.5 cm^3 of a sample was diluted with buffer and appropriate volume of diazepam standard solution, ultrasonicated for 30 minutes, and made up with distilled water to 25.00 cm^3. The samples were spiked with 50 $\mu mol/dm^3$ of diazepam. Solutions were injected in triplicate, and the results of recoveries for spiked samples of different concentrations are presented in Table 4.

4. Conclusions

Flow injection method with differential pulse amperometry was developed and applied for quantification of diazepam in investigated set of beverage samples. Differential pulse amperometry enables the determinations in the presence of oxygen without surface modification. Under the optimal conditions, the linear range of the system was from 20 to 250 $\mu mol/dm^3$ of diazepam, with the detection limit of 3.83 $\mu g/cm^3$, the intraday precision of 1.53%, and a throughoutput of 90 analyses per hour. The method was applied on diazepam determination in beverage samples; 0.050 cm^3 of

sample was enough for fast, precise, and sensitive quantification. With the flow rate of 1 cm^3/min, the method is economical, producing small volumes of waste. The simple sample preparation enables injection of samples without extraction or filtration.

Conflicts of Interest

The authors declare that there are no conflicts of interest regarding the publication of this paper.

Acknowledgments

This work was done within the framework of the research Project no. 172051 supported by the Ministry of Education, Science and Technological Development of Serbia.

References

[1] N. R. Stradiotto, H. Yamanaka, and M. V. B. Zanoni, "Electrochemical sensors: a powerful tool in analytical chemistry," *Journal of the Brazilian Chemical Society*, vol. 14, no. 2, pp. 159–173, 2003.

[2] F. Marcenac and F. Gonon, "Fast in vivo monitoring of dopamine release in the rat brain with differential pulse amperometry," *Analytical Chemistry*, vol. 57, no. 8, pp. 1778-1779, 1985.

[3] F. J. Heredia-López, J. L. Góngora-Alfaro, F. J. Alvarez-Cervera, and J. L. Bata-Garcia, "A PC-controlled voltage pulse generator for electroanalytical applications," *Review of Scientific Instruments*, vol. 68, no. 4, pp. 1879–1885, 1997.

[4] W. Surareungchai, W. Deepunya, and P. Tasakorn, "Quadruple-pulsed amperometric detection for simultaneous flow injection determination of glucose and fructose," *Analytica Chimica Acta*, vol. 448, no. 1-2, pp. 215–220, 2001.

[5] R. A. Medeiros, B. C. Lourenção, R. C. Rocha-Filho, and O. Fatibello-Filho, "Simple flow injection analysis system for simultaneous determination of phenolic antioxidants with multiple pulse amperometric detection at a boron-doped diamond electrode," *Analytical Chemistry*, vol. 82, no. 20, pp. 8658–8663, 2010.

[6] W. T. P. Dos Santos, E. G. N. De Almeida, H. E. A. Ferreira, D. T. Gimenes, and E. M. Richter, "Simultaneous flow injection analysis of paracetamol and ascorbic acid with multiple pulse amperometric detection," *Electroanalysis*, vol. 20, no. 17, pp. 1878–1883, 2008.

[7] S. C. Chaves, P. N. C. Aguiar, L. M. F. C. Torres et al., "Simultaneous determination of caffeine, ibuprofen, and paracetamol by flow-injection analysis with multiple-pulse amperometric detection on boron-doped diamond electrode," *Electroanalysis*, vol. 27, no. 12, pp. 2785–2791, 2015.

[8] W. T. P. dos Santos, D. T. Gimenes, E. G. N. de Almeida, S. P. Eiras, Y. D. T. Albuquerque, and E. M. Richter, "Simple flow injection amperometric system for simultaneous determination of dipyrone and paracetamol in pharmaceutical formulations," *Journal of the Brazilian Chemical Society*, vol. 20, no. 7, pp. 1249–1255, 2009.

[9] A. Carlos, V. Lopes Júnior, R. De Cássia et al., "Determination of sildenafil citrate (Viagra®) in various pharmaceutical formulations by flow injection analysis with multiple pulse amperometric detection," *Journal of the Brazilian Chemical Society*, vol. 23, no. 10, pp. 1800–1806, 2012.

[10] J. Riss, J. Cloyd, J. Gates, and S. Collins, "Benzodiazepines in epilepsy: pharmacology and pharmacokinetics," *Acta Neurologica Scandinavica*, vol. 118, no. 2, pp. 69–86, 2008.

[11] F. Charlson, L. Degenhardt, J. McLaren, W. Hall, and M. Lynskey, "A systematic review of research examining benzodiazepine-related mortality," *Pharmacoepidemiology and Drug Safety*, vol. 18, no. 2, pp. 93–103, 2009.

[12] D. S. M. Ribeiro, J. A. V. Prior, J. L. M. Santos, and J. L. F. C. Lima, "Automated determination of diazepam in spiked alcoholic beverages associated with drug-facilitated crimes," *Analytica Chimica Acta*, vol. 668, no. 1, pp. 67–73, 2010.

[13] M. Acikkol, S. Mercan, and S. Karadayi, "Simultaneous determination of benzodiazepines and ketamine from alcoholic and nonalcoholic beverages by GC-MS in drug facilitated crimes," *Chromatographia*, vol. 70, no. 7-8, pp. 1295–1298, 2009.

[14] A. M. Gil Tejedor, P. Fernández Hernando, and J. S. Durand Alegría, "A rapid fluorimetric screening method for the 1,4-benzodiazepines: determination of their metabolite oxazepam in urine," *Analytica Chimica Acta*, vol. 591, no. 1, pp. 112–115, 2007.

[15] M. E. Lozano-Chaves, J. M. Palacios-Santander, L. M. Cubillana-Aguilera, I. Naranjo-Rodríguez, and J. L. Hidalgo-Hidalgo-de-Cisneros, "Modified carbon-paste electrodes as sensors for the determination of 1,4-benzodiazepines: application to the determination of diazepam and oxazepam in biological fluids," *Sensors and Actuators B: Chemical*, vol. 115, no. 2, pp. 575–583, 2006.

[16] M. M. Correia dos Santos, V. Famila, and M. L. Simões Gonçalves, "Copper–psychoactive drug complexes: a voltammetric approach to complexation by 1,4-benzodiazepines," *Analytical Biochemistry*, vol. 303, no. 2, pp. 111–119, 2002.

[17] G. B. El-Hefnawey, I. S. El-Hallag, E. M. Ghoneim, and M. M. Ghoneim, "Voltammetric behavior and quantification of the sedative-hypnotic drug chlordiazepoxide in bulk form, pharmaceutical formulation and human serum at a mercury electrode," *Journal of Pharmaceutical and Biomedical Analysis*, vol. 34, no. 1, pp. 75–86, 2004.

[18] V. K. Gupta, R. Jain, K. Radhapyari, N. Jadon, and S. Agarwal, "Voltammetric techniques for the assay of pharmaceuticals—a review," *Analytical Biochemistry*, vol. 408, no. 2, pp. 179–196, 2011.

[19] C. G. Amorim, A. N. Araújo, M. C. B. S. M. Montenegro, and V. L. Silva, "Cyclodextrin-based potentiometric sensors for midazolam and diazepam," *Journal of Pharmaceutical and Biomedical Analysis*, vol. 48, no. 4, pp. 1064–1069, 2008.

[20] K. C. Honeychurch, A. Crew, H. Northall et al., "The redox behaviour of diazepam (Valium®) using a disposable screen-printed sensor and its determination in drinks using a novel adsorptive stripping voltammetric assay," *Talanta*, vol. 116, pp. 300–307, 2013.

[21] R. Wang, X. Wang, C. Liang et al., "Direct determination of diazepam and its glucuronide metabolites in human whole blood by μElution solid-phase extraction and liquid chromatography–tandem mass spectrometry," *Forensic Science International*, vol. 233, no. 1–3, pp. 304–311, 2013.

[22] N. Nagappa, T. Mimani, B. Sheshadri, and S. M. G. Mayanna, "Cyclic voltammetric studies of diazepam using glassy carbon electrode-estimation of diazepam in pharmaceutical samples," *Chemical & Pharmaceutical Bulletin*, vol. 46, no. 4, pp. 715–717, 1998.

[23] W. Lund, M. Hannisdal, and T. Greibrokk, "Evaluation of amperometric detectors for high-performance liquid chromatography: analysis of benzodiazepines," *Journal of Chromatography A*, vol. 173, no. 2, pp. 249–261, 1979.

[24] M. A. Zare, M. S. Tehrani, S. W. Husain, and P. A. Azar, "Multiwall carbon nanotube-ionic liquid modified paste electrode as an efficient sensor for the determination of diazepam and oxazepam in real samples," *Electroanalysis*, vol. 26, no. 12, pp. 2599–2606, 2014.

[25] A. Lolić, S. Nikolić, and P. Polić, "Optimization and application of the gas-diffusion flow injection method for the determination of chloride," *Journal of the Serbian Chemical Society*, vol. 66, no. 9, pp. 637–646, 2001.

[26] L. Thomas, J. L. Vilchez, G. Crovetto, and J. Thomas, "Electrochemical reduction of diazepam," *Journal of Chemical Sciences*, vol. 98, no. 3, pp. 221–228, 1987.

[27] M. Gros, S. Rodríguez-Mozaz, and D. Barceló, "Fast and comprehensive multi-residue analysis of a broad range of human and veterinary pharmaceuticals and some of their metabolites in surface and treated waters by ultra-high-performance liquid chromatography coupled to quadrupole-linear ion trap tandem mass spectrometry," *Journal of Chromatography A*, vol. 1248, pp. 104–121, 2012.

[28] K. Soltaninejad, M. Karimi, A. Nateghi, and B. Daraei, "Simultaneous determination of six benzodiazepines in spiked soft drinks by high performance liquid chromatography with ultra violet detection (HPLC-UV)," *Iranian Journal of Pharmaceutical Research*, vol. 15, no. 2, pp. 457–463, 2016.

[29] G. Famigliani, V. Termopoli, P. Palma, and A. Cappielo, "Liquid chromatography-electron ionization tandem mass spectrometry with the Direct-EI interface in the fast determination of diazepam and flunitrazepam in alcoholic beverages," *Electrophoresis*, vol. 37, no. 7-8, pp. 1048–1054, 2016.

[30] L. Gautam, S. Sharratt, and M. Cole, "Drug facilitated sexual assault: detection and stability of benzodiazepines in spiked drinks using gas chromatography-mass spectrometry," *PLoS One*, vol. 9, no. 2, article e89031, 2014.

10

Development of an Enzyme-Linked Immunosorbent Assay and Gold-Labelled Immunochromatographic Strip Assay for the Detection of Ancient Wool

Bing Wang [iD],[1] Jincui Gu,[1] Boyi Chen,[1] Chengfeng Xu,[1] Hailing Zheng,[2] Zhiqin Peng,[3] Yang Zhou [iD],[2] and Zhiwen Hu[3]

[1]Key Laboratory of Advanced Textile Materials and Manufacturing Technology, Ministry of Education, Zhejiang Sci-Tech University, Hangzhou 310018, China
[2]Key Scientific Research Base of Textile Conservation, State Administration for Cultural Heritage, China National Silk Museum, Hangzhou 310002, China
[3]Institute of Textile Conservation, Zhejiang Sci-Tech University, Hangzhou 310018, China

Correspondence should be addressed to Bing Wang; wbing388@163.com and Yang Zhou; cnsmzhouyang@126.com

Academic Editor: Bernd Hitzmann

The identification of ancient wool is of great importance in archaeology. Despite lots of meaningful information can be achieved by conventional detection methods, that is, light and electron microscopy, spectroscopy, and chromatography, the efficacy is likely to be limited in the detection of ancient samples with contamination or severe degradation. In this work, an immunoassay was proposed and performed for the identification of ancient wool. First, a specific antibody, which has the benefits of low cost, easy operation, and extensive applicability, was developed directly through immunizing rabbits with complete antigen (keratin). Then, an enzyme-linked immunosorbent assay (ELISA) and a colloidal gold-labelled immunochromatographic strip (ICS) were developed to qualitatively identify the corresponding protein in ancient wool samples unearthed from Kazakhstan and China. The anti-keratin antibody exhibited high sensitivity and specificity for the identification of modern and ancient wool. The limit of detection (LOD) of the ELISA method was 10 ng/mL, and no cross-reactions with other interfering antigens have been noted. It is concluded that the immunoassays are reliable methods for the identification of ancient wool.

1. Introduction

Ancient textiles are an important component of human civilization heritage. Since the Stone Age, in order to adapt to climate change, the ancients had begun to use natural resources as textile materials. Ancient textiles mainly included cotton fibre, hemp fibre, silk, and wool. The earliest evidence for the wool usage in the eastern Iran and in the northern Caucasus dates back to the 4th millennium BC [1]. The importance of wool as a major source of textile and economic trade in Eurasia has long been established [2]. Ancient wool was unearthed from time to time in the graves, tombs, or other places in many countries along the Silk Road [2–4]. The state of preservation mainly depends on the burial time and burial environment of the ancient wool. Under normal circumstances, ancient wool samples recovered from excavations, although fragmentary and fragile, are valuable finds for the study of historical textile production, its trade, and the development of sheep breeding [5].

Wool is a natural composite material consisting of keratin and keratin-associated proteins as key molecular components [6, 7]. The keratin macromolecular structure is the alignment of amino acid residues along its chain. The sequence of amino acids defines the possibility of intermolecular links and the access of amino acids to the chemical reaction [8–10]. However, ancient wool buried in different soil contexts has an intrinsic chemical complexity and the tendency to easily degrade over long periods of time

[3, 11]. This is because wool is easily affected by many factors, such as high temperature, oxidation, soil microbes, and radiation, resulting in the degradation of macromolecular chains [12]. Therefore, well-preserved woollen fabrics are rarely found in archaeological contexts, except for environments with special conditions, such as low temperature, anoxia, or extreme dryness. Until recently, species identification and detailed characterization of poorly preserved ancient wool have remained challenging issues in archaeology.

During the past several decades, there were a number of methods available for the identification of archaeological wools, such as scanning electron microscopy [1, 13], Fourier transform infrared spectroscopy [8, 14–16], nuclear magnetic resonance [17], tandem mass spectrometry [18], and gas chromatography-mass spectrometry [19]. However, archaeological wools have usually degraded into short fibres or even peptides, leaving microtraces in the soil. In addition, external contamination and compositional complexity make the identification of ancient samples highly challenging and sometimes highly uncertain. Thus, it is still difficult for archaeologists to extract enough useful information from ancient wool samples.

Immunological techniques have the potential to become a powerful analytical tool in archaeology [20, 21]. These methods offer several advantages over traditional methods used for protein analysis, including low costs, speed, increased sensitivity, and increased specificity [22, 23]. In recent years, immunological techniques have attracted increased attention from professionals involved in the research of cultural heritage [24–27]. All of these studies have demonstrated that immunoassays have the potential to identify and localize the proteins in archaeological materials rapidly and effectively. However, most of the antibodies used in these immunoassays are commercially available, and the need for tailored antibodies for targeting a specific protein with high sensitivity and specificity is highly desirable. Thus, the preparation of a tailored antibody for the detection of ancient wool is compelling, yet challenging.

In our previous research, several tailored specific antibodies were designed for the immunodetection of ancient silks [28, 29], leathers [30], and proteinaceous binders [31]. Moreover, gold-based and lanthanide-labelled immunochromatographic strip assays were developed for the on-site identification of ancient silks [32, 33]. Both the immunosensors showed high sensitivity and specificity, providing a new protocol for identifying archeological proteinaceous materials.

Herein, an immunoassay is proposed for the microtrace detection of ancient wool. First, an anti-keratin antibody was prepared by immunizing rabbits with complete antigen, and the sensitivity and specificity of the antibody were determined to verify its validity for the identification of wool samples. Then, the antibody-based immunoassays, an enzyme-linked immunosorbent assay (ELISA), and a colloidal gold-labelled immunochromatographic strip (ICS) were established and used to qualitatively identify samples from different time periods. ELISA is a powerful analytical laboratory tool for archaeology and conservation, while ICS

is especially suitable for the identification of poorly preserved ancient samples in archaeological sites. Consequently, the two methods are particularly useful when used in tandem.

2. Materials and Methods

2.1. Archaeological Samples. Several precious archaeological samples excavated from Kazakhstan and China were selected for immunological identification. Samples were provided by China National Silk Museum. The original condition of samples is shown in Figure 1. Sample I (Figure 1(a)), which was unearthed from the western Kazakhstan in the 1970s, was camel hair. Sample II (Figure 1(b)) and sample III (Figure 1(c)) were unearthed from the Shymkent region in the South Kazakhstan during the 2000s and 2010s. Sample IV (Figure 1(d)) was unearthed from the Almaty region in Kazakhstan during the 2000s and 2010s. Sample V (Figure 1(e)) and sample VI (Figure 1(f)) were unearthed from the Small River Cemetery (2000~1450 BC), located in the southwest desert of the Lop Nor Area in Xinjiang, China.

2.2. Reagents and Chemicals. Goat anti-rabbit IgG $(H + L)$ HRP-conjugated secondary antibody (100 μg at 1 mg/mL) and a TMB colour system were purchased from Hangzhou Hua'an Bio-Technology Co., Ltd. Bovine serum albumin (BSA), human serum albumin (HSA), chicken ovalbumin, and collagen were purchased from Sigma-Aldrich. Natural wool was obtained from Hangzhou Fusigongmao Co., Ltd. Natural silk was obtained from Zhejiang Misai Silk Co., Ltd. Hemp fibre and cotton fibre were obtained from Hangzhou Fusi Industry and Trade Co., Ltd. NaOH, H_2O_2 (30%), KCl, KH_2PO_4, NaCl, Na_2CO_3, $CaCl_2$, and Na_2HPO_4 were purchased from Hangzhou High Crystal Special Chemicals Co., Ltd. Standard soil samples were obtained from Beijing Century Aoke Biotechnology Co., Ltd.

A carbonate bicarbonate buffer solution (pH 9.6) was used as a diluent for ELISA antigens. Phosphate-buffered saline (PBS, pH 7.4) was used for wash steps. 1% BSA in PBS (pH 7.4) was used to block the unbound sites of antigens. All other reagents were of analytical grade and used as received. The water used in all experiments was purified by a Milli-Q water system.

2.3. Extraction and Characterization of Antigens. Keratin was extracted from natural wool by an alkali hydrolysis method. Briefly, an alkaline solution was prepared by mixing 4% (w/w) NaOH with an equal volume of 0.6% (w/w) H_2O_2 in advance. Then, natural wool was shredded and immersed in the mixed solution at a bath ratio of 1 : 50 at 50°C. After stirring at 200 rpm for 4 h, the wool was completely dissolved, and a clear liquid was obtained. Next, the resultant keratin solution was dialyzed for 72 h with a molecular weight cutoff of 3500 Da and freeze-dried. Then, the obtained keratin was analyzed by SDS-PAGE and FTIR.

All possible interfering antigens, that is, silk fibroin, sericin, cotton fibre, and hemp fibre, were extracted using the previously reported procedures [28].

FIGURE 1: Digital images of ancient samples unearthed in Kazakhstan: (a) camel hair, unearthed in the western Kazakhstan in 1970s; (b) white wool and (c) black wool, unearthed from the Shymkent region in the South Kazakhstan; (d) wool fabrics, unearthed from the Almaty region in Kazakhstan; (e) wool soil sample and (f) wool fibre, unearthed from the Small River Cemetery, Xinjiang, China. These images were taken by a Canon EOS700D digital camera in micromode.

2.4. Preparation of Anti-Keratin Primary Antibody.

An anti-keratin antibody was produced by animal immunization as follows. The immunogen (keratin) was first mixed with an equal volume of complete Freund's adjuvant to form an emulsion. For the initial immunization, New Zealand white rabbits (14–16 weeks old) were subcutaneously injected multiple times into their thighs with $100\,\mu L$ of the above mixture. Then, incomplete Freund's adjuvant was substituted for the complete Freund's adjuvant as an emulsifier for subsequent boosters every 2 weeks. The titre of the antiserum was measured by indirect ELISA ten days after each immunization. Antiserum was collected 6 weeks after the initial immunization and purified by affinity chromatography. The resultant anti-keratin antibody was stored (3.15 mg/mL) at $-20°C$ before use.

All animal experiments were carried out in accordance with the national standard "Laboratory Animal-Requirements of Environment and Housing Facilities" (GB 14925–2001) and the guidelines issued by the Ethical Committee of Zhejiang Sci-Tec University.

2.5. Indirect ELISA Test.

ELISA tests are performed in an indirect format as shown in Figure 2. Samples were first dissolved in a carbonate bicarbonate buffer solution (pH 9.6) and diluted to $10\,\mu g/mL$. Then, $100\,\mu L$ of the sample solution was added to each well of the 96-well microplate and incubated at $4°C$ overnight. After the solution was removed, the wells were washed 3 times with PBS (7.4). Next, $100\,\mu L$ of the blocking solution was added to each well and incubated at $37°C$ for 1 h to occupy the unbound sites, followed by 3 washings with PBS. Then, $100\,\mu L$ of the anti-keratin antibody was added to the wells and incubated at $37°C$ for 1 h, followed by washing. Next, $100\,\mu L$ of secondary antibody (goat anti-rabbit IgG $(H + L)$ HRP-conjugated) was added, followed by incubation at $37°C$ for 1 h. After washing with PBS 3 times, $100\,\mu L$ of 3, 3', 5, 5'-tetramethylbenzidine (TMB) was added and incubated in darkness for 10 min. Finally, $100\,\mu L$ of 1 mol/L H_2SO_4 (stopping solution) was added to terminate the colour reaction, and the optical density (OD) of each sample at $\lambda = 450\,nm$ was measured using a microplate reader (Model 550, Bio Rad).

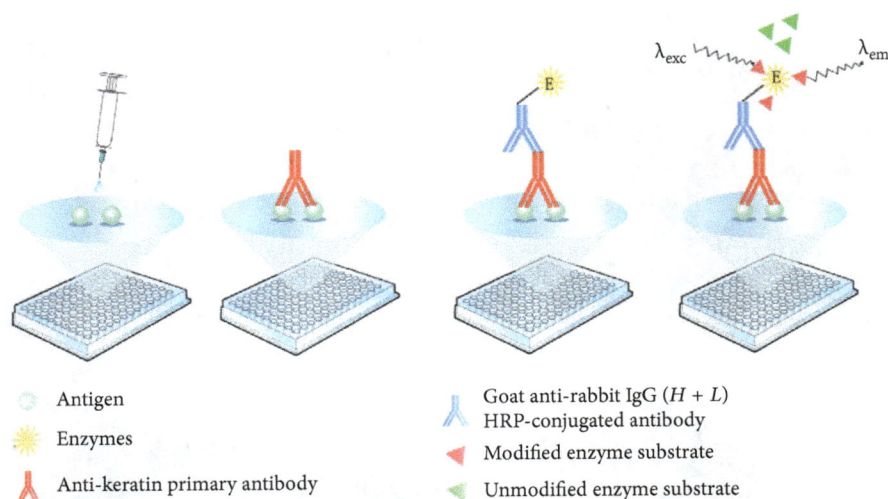

- ○ Antigen
- ✳ Enzymes
- ⅄ Anti-keratin primary antibody
- ⅄ Goat anti-rabbit IgG (*H* + *L*) HRP-conjugated antibody
- ◀ Modified enzyme substrate
- ◀ Unmodified enzyme substrate

FIGURE 2: Schematic diagram of the indirect ELISA used for the microtrace detection of ancient wool.

Optimal antibody and antigen dilutions corresponding to the best specificity and sensitivity of the ELISA method were obtained by panel titrations. To get the best dilution ratios of antibody and antigen, detailed operation steps were set as follows. The keratin powders were dissolved in a carbonate bicarbonate buffer (pH 9.6) solution that was formulated at concentrations of 1,000 μg/mL, 100 μg/mL, 10 μg/mL, and 1 μg/mL. Then, the anti-keratin primary antibody was diluted by 1% BSA solution at dilution ratios of 1 : 200, 1 : 500, 1 : 1000, 1 : 3,000, 1 : 5,000, and 1 : 10,000, respectively. Finally, the mean OD values of different keratin concentrations with different dilutions of primary antibody were tested by a microplate reader (Model 550, Bio Rad). The optical density of the samples at 450 nm was abbreviated as $OD_{450\,nm}$.

A series of controls were set up to ensure the specificity of the indirect ELISA test. PBS replaced sample solutions as a coating antigen for the negative control, and only experiments that presented negative results were considered to be reliable. Keratin was employed as a positive control. For the blank control, the anti-keratin primary antibody was replaced with PBS to ensure that the HRP-conjugated secondary antibody did not react with the coating antigens.

2.6. Preparation of the Immunochromatographic Strip. The detecting principle of the immunochromatographic strip is based on competitive immunoreactions. There are four parts of the immunochromatographic strip: a sample pad, a colloidal gold pad, a nitrocellulose membrane, and an absorbent pad. The colloidal gold-labelled antibody was sprayed onto a polyester fibre pad by a metal spraying machine. Next, wool keratin (1.2 mg/mL) was sprayed onto a nitrocellulose membrane, as the test line, at 1 μg/cm, while the goat anti-rabbit IgG (*H* + *L*) HRP-conjugated secondary antibody (1.5 mg/mL) was sprayed as the control line. Both the polyester fibre pad and the nitrocellulose membrane were dried at 37°C for 2 h. Finally, all four parts were fitted together, cut into 3.5 mm wide strips, and put into a PVC case. When sample solutions were added dropwise onto the sample pad, they flowed chromatographically along the nitrocellulose strip and gave positive or negative reactions.

2.7. Pretreatment of Archaeological Samples for ICS Detection. The extraction of the ancient samples can be summarized as follows. Two milligrams of each sample was added into 100 mL of extracted solutions (including 2% (w/w) sodium hydroxide and 0.3% (w/w) hydrogen peroxide) and incubated for 1 h. Then, the supernatant of each extracted solution was collected and transferred individually into centrifuge tubes for ICS detection.

2.8. Statistical Analysis. Five parallel samples were used for each sample and control and were run simultaneously. All values are expressed as the mean ± standard deviation.

3. Results and Discussion

3.1. Characterization of Keratin. The molecular weight distribution of the resultant keratin was determined by SDS-PAGE. In terms of the molar mass and sulphur content, four fractions, that is, the low sulphur fraction (LSF, Mw: 45–60 kDa), the high sulphur fraction (HSF, Mw: 14–28 kDa), ultrahigh sulphur fraction (USF, Mw: 37 kDa), and the high Gly/Tyr fraction (HGT, Mw: 9–13 kDa) can be extracted from wool keratin. As shown in Figure 3(a), two bands that match the distribution characteristics of the typical keratin bands, including the HSF chain at approximately 28 kDa and the HGT at approximately 12 kDa, were observed, indicating that the resultant keratins were mainly the HSF and HGT fractions. The chemical structure of the resultant keratin was also characterized via FTIR. As shown in Figure 3(b), the sample spectra show the characteristic peaks of amide bonds, specifically a sharp peak at 1655 cm^{-1} (amide I band), a peak at 1536 cm^{-1} (amide II band), and a peak at 1237 cm^{-1} (amide III band). In particular, a peak at 1044 cm^{-1} appears in the spectrum, which is assigned to the

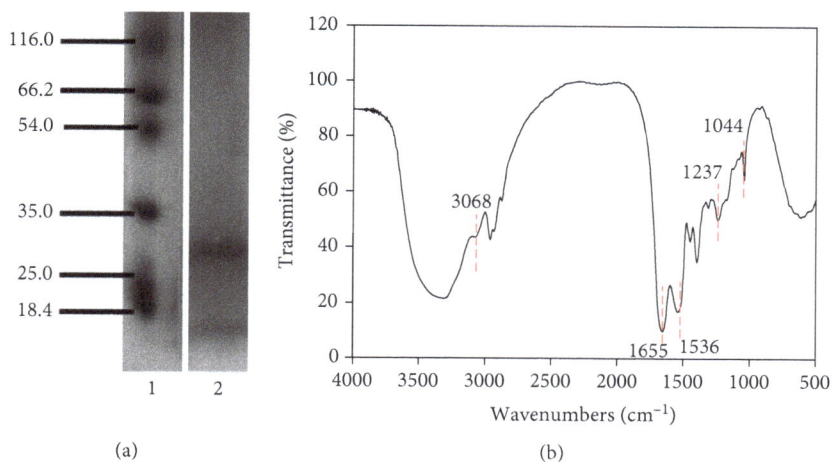

(a)

(b)

FIGURE 3: (a) SDS-PAGE analysis of keratin extracted from wool: lane 1: molecular weight marker; lane 2: keratin. (b) FTIR spectrum of keratin.

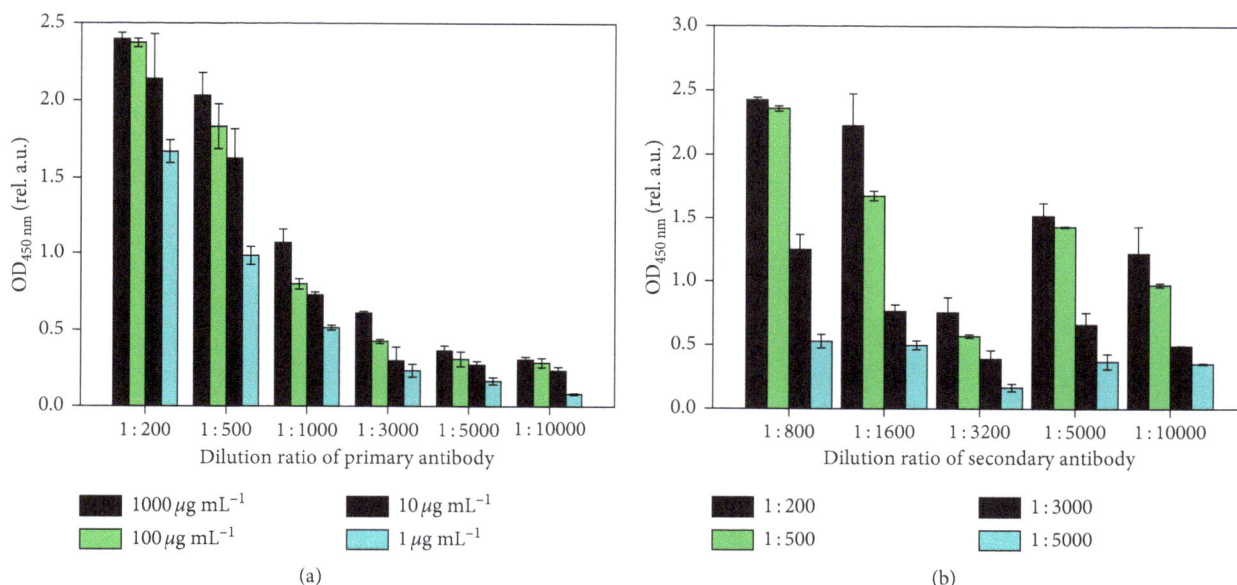

(a)

(b)

FIGURE 4: ELISA titration results in different conditions. (a) The dilution ratio of anti-keratin primary antibody is 1 : 200, 1 : 500, 1 : 1,000, 1 : 3,000, 1 : 5,000, and 1 : 10,000, while the concentration of keratin is 1,000 μg/mL, 100 μg/mL, 10 μg/mL, and 1 μg/mL, respectively. (b) The secondary antibody dilution ratio is 1 : 800, 1 : 1,600, 1 : 3,200, 1 : 5,000, and 1 : 10,000, while the anti-keratin primary antibody dilution ratio is 1 : 200, 1 : 500, 1 : 3,000, and 1 : 5,000, respectively. Each column of the figure represents the mean ± SD (standard deviation) of $n = 5$ assays.

characteristic peak of cysteine sulfenate produced by oxidation of disulfide bond.

3.2. The Optimal Dilution Multiple of Antibody and Antigen.
Investigation of the ELISA procedure for recognition of ancient wool began with an antibody panel titration of standard solutions of keratin. As shown in Figure 4(a), with increasing dilution of the primary antibody, the $OD_{450 nm}$ values decreased. Obviously, when the dilution ratio was 1 : 200 or 1 : 500, the $OD_{450 nm}$ values were larger at various concentrations of keratin. In addition, if the $OD_{450 nm}$ values of keratin ranged from 1.5 relative

arbitrary units to 2.0 relative arbitrary units, subsequent ELISA tests would show much higher accuracy. Considering the cost of the primary antibody, the dilution ratio of the anti-keratin primary antibody was set at 1 : 500, and the chosen concentration of keratin was 100 μg/mL for the following experiments. To achieve the optimized dilution of the secondary antibody, the concentration of keratin was set at 100 μg/mL, and the anti-keratin primary antibody and the HRP-conjugated goat anti-rabbit IgG $(H + L)$ secondary antibody were diluted to various dilution ratios with 1% BSA solution. As shown in Figure 4(b), the $OD_{450 nm}$ values decreased with the increase of the dilution ratios of primary antibody and secondary antibody.

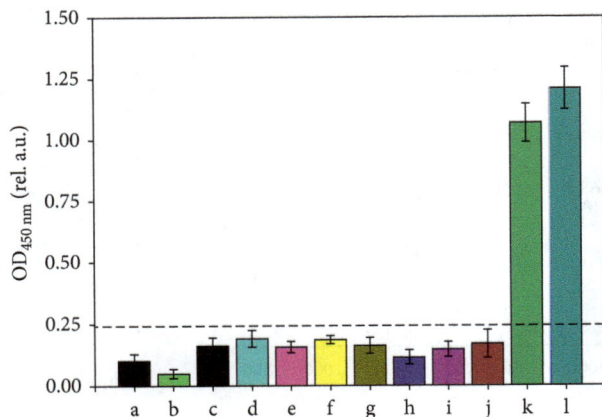

FIGURE 5: ELISA results for interfering antigens to test the specificity of anti-keratin primary antibody: (a) PBS; (b) negative control; (c) BSA; (d) HSA; (e) ovalbumin; (f) collagen; (g) sericin; (h) fibroin; (i) cotton extraction; (j) hemp extraction; (k) wool extraction; (l) keratin. Each column of the figure represents the mean $\pm$ SD (standard deviation) of $n = 5$ assays. The dashed line shows the test criterion (the mean OD_{PBS} plus three times the corresponding standard deviation). If $OD_{450\,nm}$ is above the dashed line, the result is positive, and vice versa.

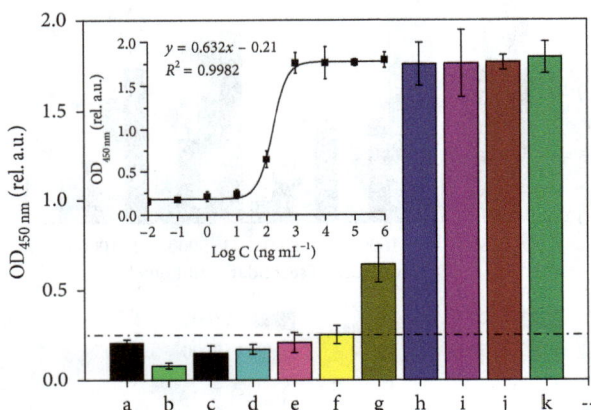

FIGURE 6: The limit of detection (LOD) of the ELISA test for keratin: (a) PBS; (b) negative control; (c–k) keratin with different concentrations of 10^{-2}, 10^{-1}, 1, 10, 10^2, 10^3, 10^4, 10^5, and 10^6 ng/mL.

Paradoxically, under the dilution ratio of 1 : 3,200 for the secondary antibody, the $OD_{450\,nm}$ values were much smaller than the others. After the exclusion of experimental error, it is speculated that it may result from decreasing immune response of secondary antibody, though further research is needed. When the dilution ratio of secondary antibody was 1 : 1600 or 1 : 5,000, the $OD_{450\,nm}$ values ranged from 1.5 relative arbitrary units to 2.0 relative arbitrary units. Considering the antibody titre and production costs, the dilution ratio of the secondary antibody was set as 1 : 5,000. To summarize, the optimal dilution factors of the anti-keratin primary antibody and the secondary antibody were 1 : 500 and 1 : 5,000, respectively.

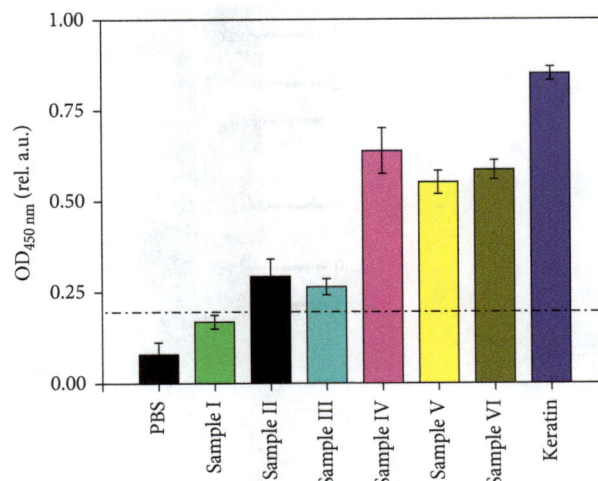

FIGURE 7: ELISA results for ancient samples.

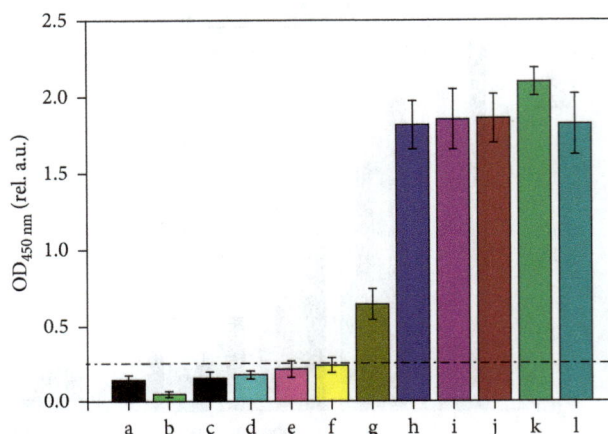

FIGURE 8: The limit of detection (LOD) of the ELISA test for ancient wool samples: (a) PBS control; (b) negative control; (c–k) ancient wool samples with different concentrations of 10^{-2}, 10^{-1}, 1, 10, 10^2, 10^3, 10^4, 10^5, and 10^6 ng/mL; (l) positive control.

3.3. The Specificity of Primary Antibody.

Next, the specificity of the anti-keratin primary antibody was further investigated by checking cross-reactivity with other interfering antigens. The limit of detection (LOD) was set as the mean $OD_{450\,nm}$ value of negative control plus three times the corresponding standard deviation. As shown in Figure 5, a series of interfering antigens, including BSA, HSA, ovalbumin, collagen, sericin, fibroin, cotton fibre extraction, and hemp fibre extraction (100 μg/mL), were tested. All of the above samples showed clearly negative results, while modern wool extraction and keratin showed positive results. These results indicated that the anti-keratin primary antibody is an appropriate choice for detecting ancient wool due to its high specificity.

3.4. The Effectiveness and Sensitivity of Primary Antibody.

An indirect ELISA was then employed to determine the LOD of the primary antibody for keratin. The results are presented

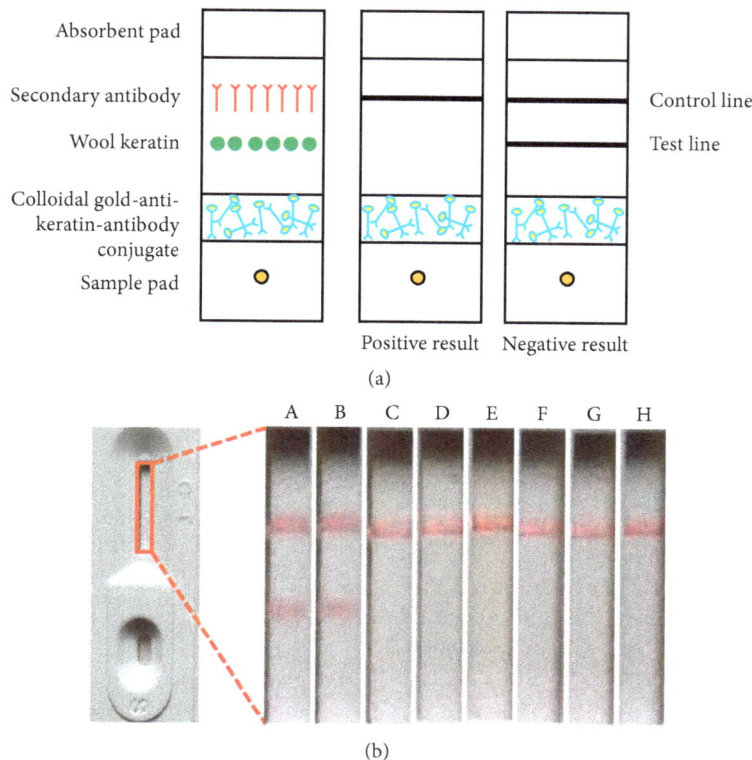

FIGURE 9: (a) Schematic diagram of the colloidal gold-based immunochromatographic assay; (b) ICS detection results for negative control, positive control, and ancient samples: (A) PBS, (B) sample I, (C) sample II, (D) sample III, (E) sample IV, (F) sample (V) (g) sample VI, and (H) keratin.

in Figure 6. Keratin was diluted to 10^{-2}, 10^{-1}, 10^0, 10^1, 10^2, 10^3, 10^4, 10^5, and 10^6 ng/mL with a carbonate bicarbonate buffer (pH 9.6). The results clearly demonstrate that keratin can be detected using the ELISA method when its concentration is above 10 ng/mL and cannot be detected below 10 ng/mL. This finding suggests that the LOD of the primary antibody is 10 ng/mL. Meanwhile, the standard curve, which sets the logarithm of keratin concentration as the abscissa and the $OD_{450\,nm}$ value as the ordinate, is also shown in Figure 6. The relationship between the logarithm of keratin concentration and the $OD_{450\,nm}$ value matches the S-type curve model. In particular, the logarithm of keratin concentration and the OD value showed a linear relationship when concentrations of keratin ranged from 10^1 to 10^3 ng/mL (OD values ranged from 0.435 to 1.758). Additionally, the linear regression equation was $y = 0.632x - 0.21$, $R^2 = 0.973$. Within this range, the samples could be tested quantitatively by indirect ELISA. These findings proved the effectiveness and high sensitivity of the anti-keratin primary antibody.

3.5. Immunodetection of Archaeological Samples with Indirect ELISA Method.
As the ELISA protocol had been optimized for the identification of keratin, it was employed to detect keratin in archaeological samples. To ensure the reliability of the ELISA tests, PBS was used as a negative control, and keratin served as a positive control. Archaeological samples

were extracted using a similar procedure to keratin and were then coated using a carbonate bicarbonate buffer (pH 9.6) solution. As shown in Figure 7, the negative control and sample I showed negative results, while all other ancient samples and positive control (keratin, 100 μg/mL) showed positive results. Because sample I is from camel hair, it could not be detected with this tailored primary antibody. Overall, the ELISA method correctly identified the existence of keratin in ancient wool and wool fabrics, even though they were derived from different species of sheep. The results also proved the effectiveness and specificity of the anti-keratin primary antibody in the ELISA test.

3.6. The Limit of Detection (LOD) of Ancient Wool Samples by Indirect ELISA.
To confirm the ELISA results and decrease the risk of false positives, microtrace detection of ancient wool was carried out. An ancient wool sample was chosen and diluted to 10^{-2}, 10^{-1}, 10^0, 10^1, 10^2, 10^3, 10^4, 10^5, and 10^6 ng/mL with carbonate bicarbonate buffer (pH 9.6). The indirect ELISA protocol was then employed to test the LOD for the ancient wool samples. The results are presented in Figure 8. It demonstrates that the LOD of the ancient wool sample was approximately 10 ng/mL. Therefore, if the concentrations of the ancient wool samples were higher than 10 ng/mL, they could be detected directly with the ELISA method. The LOD was the same as that of the modern samples. Although ancient wool samples unearthed from

Kazakhstan have been buried for a long time and not only the composition but also the colour has undergone great changes and deterioration, the ELISA method still showed high sensitivity. In fact, as long as the targeted site (epitope) of the antigen remained intact, the anti-keratin primary antibody was able to identify the corresponding protein in ancient wool. Therefore, the deterioration of samples did not have a noticeable impact on the ELISA test.

3.7. Immunodetection of Archaeological Samples with ICS Method. The principle of ICS is based on competitive immunoreactions. Wool keratin and HRP-conjugated goat anti-rabbit IgG $(H + L)$ (secondary antibody) were immobilized on the test line and control line, respectively (Figure 9(a)). As the sample solutions flow chromatographically along the nitrocellulose strip, they will pass the conjugate pad, test line, and control line successively. If both the control line and test line become red, it means that the wool keratin is below the LOD (negative result); if only the control line becomes red, this indicates that the wool keratin is above the LOD (positive result) [32, 33]. As shown in Figure 9(b), six archaeological samples were analyzed using the ICS method, while PBS and keratin were used as negative and positive controls, respectively. For PBS and sample I (camel hair), both the control line and the test line exhibited red colour. In contrast, no red band appeared in the test line for other ancient samples and keratin. The results clearly indicated that the ICS strip could identify wool from other types of archaeological samples. These results were consistent with the ELISA results, confirming the high reliability and practicability of immunoassay.

4. Conclusions

This study focused on the microtrace detection of ancient wool with immunological techniques. A specific antibody was developed directly through immunizing rabbits with complete antigen (keratin). Then, antibody-based immunoassays, ELISA, and ICS, were established and conducted to qualitatively identify the corresponding protein in ancient wool. Optimal antibody dilutions corresponding to the best specificity and sensitivity of the ELISA test were obtained by panel titrations. These immunological methods correctly identified the existence of keratin in ancient wool and wool fabrics, even though the samples were from different species of sheep and had been buried underground for a long period. The LOD of both keratin and ancient wool was 10 ng/mL, and no cross-reactions with other possible interference antigens have been noted. Considering the practicality and flexibility of ELISA and ICS, immunological techniques have the potential to become a particularly useful analytical tool for archaeological proteinaceous materials.

Conflicts of Interest

The authors declare that there are no conflicts of interest regarding the publication of this article.

Acknowledgments

This work was supported by the National Natural Science Foundation of China (51603188), Public Technology Research Plan of Zhejiang Province (2016C33175), Outstanding Young Research Program of Science and Technology for the Protection of Cultural Relics (no. 2015-294), and Special Funds from the Administration of Cultural Heritage of Zhejiang Province (2014013 and 2017014).

References

[1] M. Gleba, "From textiles to sheep: investigating wool fibre development in pre-Roman Italy using scanning electron microscopy (SEM)," *Journal of Archaeological Science*, vol. 39, no. 12, pp. 3643–3661, 2012.

[2] E. Karpova, V. Vasiliev, V. Mamatyuk, N. Polosmak, and L Kundo, "Xiongnu burial complex: a study of ancient textiles from the 22nd Noin-Ula barrow (Mongolia, first century AD)," *Journal of Archaeological Science*, vol. 70, pp. 15–22, 2016.

[3] J. Liu, D. Guo, Y. Zhou et al., "Identification of ancient textiles from Yingpan, Xinjiang, by multiple analytical techniques," *Journal of Archaeological Science*, vol. 38, no. 7, pp. 1763–1770, 2011.

[4] I. C. C. von Holstein, P. W. Rogers, O. E. Craig, K. E. H. Penkman, J. Newton, and M. J. Collins, "Provenancing archaeological wool textiles from Medieval Northern Europe by light stable isotope analysis (delta C-13, delta N-15, delta H-2)," *Plos One*, vol. 11, no. 10, article e0162330, 2016.

[5] C. Solazzo, J. Wilson, J. M. Dyer et al., "Modeling deamidation in sheep alpha-keratin peptides and application to archeological wool textiles," *Analytical Chemistry*, vol. 86, no. 1, pp. 567–575, 2014.

[6] S. Clerens, C. D. Cornellison, S. Deb-Choudhury, A. Thomas, J. E. Plowman, and J. M. Dyer, "Developing the wool proteome," *Journal of Proteomics*, vol. 73, no. 9, pp. 1722–1731, 2010.

[7] J. E. Plowman, "The proteomics of keratin proteins," *Journal of Chromatography B*, vol. 849, no. 1-2, pp. 181–189, 2007.

[8] E. Wojciechowska, A. Włochowicz, and A. Wesełucha-Birczyńska, "Application of Fourier-transform infrared and Raman spectroscopy to study degradation of the wool fiber keratin," *Journal of Molecular Structure*, vol. 511–512, no. 99, pp. 307–318, 1999.

[9] K. M. Frei, I. V. Berghe, R. Frei, U. Mannering, and H. Lyngstrøm, "Removal of natural organic dyes from wool–implications for ancient textile provenance studies," *Journal of Archaeological Science*, vol. 37, no. 9, pp. 2136–2145, 2010.

[10] J. M. Cardamone, "Investigating the microstructure of keratin extracted from wool: Peptide sequence (MALDI-TOF/TOF) and protein conformation (FTIR)," *Journal of Molecular Structure*, vol. 969, no. 1-3, pp. 97–105, 2010.

[11] K. M. Frei, R. Frei, U. Mannering, M. Gleba, M. L. Nosch, and H. LyngstrØM, "Provenance of ancient textiles-a pilot study evaluating the strontium isotope system in wool," *Archaeometry*, vol. 51, no. 2, pp. 252–276, 2009.

[12] C. Solazzo, S. Clerens, J. E. Plowman, J. Wilson, E. E. Peacock, and J. M. Dyer, "Application of redox proteomics to the study

of oxidative degradation products in archaeological wool," *Journal of Cultural Heritage*, vol. 16, no. 6, pp. 896–903, 2015.

[13] M. Pechníková, D. Porta, D. Mazzarelli et al., "Detection of metal residues on bone using SEM-EDS. Part I: blunt force injury," *Forensic Science International*, vol. 223, no. 1-3, pp. 87–90, 2012.

[14] S. Bruni, L. E. De, V. Guglielmi, and F. Pozzi, "Identification of natural dyes on laboratory-dyed wool and ancient wool, silk, and cotton fibers using attenuated total reflection (ATR) Fourier transform infrared (FT-IR) spectroscopy and Fourier transform Raman spectroscopy," *Applied Spectroscopy*, vol. 65, no. 9, pp. 1017–1023, 2011.

[15] E. Wojciechowska, M. Rom, A. Włochowicz, M. Wysocki, and A. Wesełucha-Birczyńska, "The use of Fourier transform-infrared (FTIR) and Raman spectroscopy (FTR) for the investigation of structural changes in wool fibre keratin after enzymatic treatment," *Journal of Molecular Structure*, vol. 704, no. 1-3, pp. 315–321, 2004.

[16] E. Wojciechowska, A. Włochowicz, M. Wysocki, A. Pielesz, and A. Wesełucha-Birczyńska, "The application of Fourier-transform infrared (FTIR) and Raman spectroscopy (FTR) to the evaluation of structural changes in wool fibre keratin after deuterium exchange and modification by the orthosilicic acid," *Journal of Molecular Structure*, vol. 614, no. 1-3, pp. 355–363, 2002.

[17] M. S. Hoemberger, C. G. Wilson, and D. Kern, "Probing an ancient protein's dynamics with NMR," *Biophysical Journal*, vol. 106, no. 2, p. 657a, 2014.

[18] R. J. Ward, H. A. Willis, G. A. George et al., "Surface analysis of wool by X-ray photoelectron spectroscopy and static secondary ion mass spectrometry," *Textile Research Journal*, vol. 63, no. 6, pp. 362–368, 1993.

[19] A. Haji and S. S. Qavamnia, "Response surface methodology optimized dyeing of wool with cumin seeds extract improved with plasma treatment," *Fibers and Polymers*, vol. 16, no. 1, pp. 46–53, 2015.

[20] M. Palmieri, M. Vagnini, L. Pitzurra, B. G. Brunetti, and L. Cartechini, "Identification of animal glue and hen-egg yolk in paintings by use of enzyme-linked immunosorbent assay (ELISA)," *Analytical and Bioanalytical Chemistry*, vol. 405, no. 19, pp. 6365–6371, 2013.

[21] M. Gambino, F. Cappitelli, C. Cattò et al., "A simple and reliable methodology to detect egg white in art samples," *Journal of Biosciences*, vol. 38, no. 2, pp. 397–408, 2013.

[22] M. Palmieri, M. Vagnini, L. Pitzurra et al., "Development of an analytical protocol for a fast, sensitive and specific protein recognition in paintings by enzyme-linked immunosorbent assay (ELISA)," *Analytical and Bioanalytical Chemistry*, vol. 399, no. 9, pp. 3011–3023, 2011.

[23] C. Laura, V. Manuela, P. Melissa et al., "Immunodetection of proteins in ancient paint media," *Accounts of Chemical Research*, vol. 43, no. 6, pp. 867–876, 2010.

[24] W. Hu, H. Zhang, and B. Zhang, "Identification of organic binders in ancient Chinese paintings by immunological techniques," *Microscopy and Microanalysis*, vol. 21, no. 5, pp. 1278–1287, 2015.

[25] A. Heginbotham, V. Millay, and M. Quick, "The use of immunofluorescence microscopy and enzyme-linked immunosorbent assay as complementary techniques for protein identification in artists' materials," *Journal of the American Institute for Conservation*, vol. 45, no. 2, pp. 89–105, 2006.

[26] L. Kockaert, "Detection of ovalbumin in paint media by immunofluorescence," *Studies in Conservation*, vol. 34, no. 4, p. 183, 1989.

[27] J. Pavelka, L. Šmejda, and L. Kovačiková, "The determination of domesticated animal species from a Neolithic sample using the ELISA test," *Comptes Rendus Palevol*, vol. 10, no. 1, pp. 61–70, 2011.

[28] Q. Zheng, X. Wu, H. Zheng, and Y. Zhou, "Development of an enzyme-linked-immunosorbent-assay technique for accurate identification of poorly preserved silks unearthed in ancient tombs," *Analytical and Bioanalytical Chemistry*, vol. 407, no. 13, pp. 3861–3867, 2015.

[29] Q. You, Q. Li, H. Zheng, Z. Hu, Y. Zhou, and B. Wang, "Discerning silk produced by Bombyx mori from those produced by wild species using an enzyme-linked immunosorbent assay combined with conventional methods," *Journal of Agricultural and Food Chemistry*, vol. 65, no. 35, pp. 7805–7812, 2017.

[30] Y. Liu, Y. Li, R. Chang et al., "Species identification of ancient leather objects by the use of the enzyme-linked immunosorbent assay," *Anal Methods*, vol. 8, no. 42, pp. 7689–7695, 2016.

[31] Y. Zhou, B. Wang, M. Sui, F. Zhao, and Z. Hu, "Detection of proteinaceous binders in ancient Chinese textiles by enzyme-linked immunosorbent assay," *Studies in Conservation*, vol. 60, no. 6, pp. 368–374, 2015.

[32] M. M. Liu, Y. Li, H. L. Zheng, Y. Zhou, B. Wang, and Z. W. Hu, "Development of a gold-based immunochromatographic strip assay for the detection of ancient silk," *Analytical Methods*, vol. 7, no. 18, pp. 7824–7830, 2015.

[33] Q. You, M. Liu, Y. Liu et al., "Lanthanide-labeled immunochromatographic strip assay for the on-site identification of ancient silk," *ACS Sensors*, vol. 2, no. 4, pp. 569–575, 2017.

11

Development of Global Chemical Profiling for Quality Assessment of *Ganoderma* Species by ChemPattern Software

Hui Zhang, Huijie Jiang, Xiaojing Zhang, Shengqiang Tong, and Jizhong Yan ⓘ

College of Pharmaceutical Science, Zhejiang University of Technology, No. 18 Chaowang Road, Hangzhou 310014, China

Correspondence should be addressed to Jizhong Yan; science5555@163.com

Academic Editor: Luca Campone

Triterpenoids are the major secondary metabolites and active substances in *Ganoderma*, considered as the "marker compounds" for the chemical evaluation or standardization of *Ganoderma*. A response surface methodology was used to optimize the ultrasonic-assisted extraction of triterpenoids. The extraction rate was 7.338 ± 0.150 mg/g under the optimum conditions: 87% ethanol, ratio of solid to liquid (w : v) 1 : 28, and ultrasound extraction time 36 min. Based on the high sensitivity and selectivity of HPLC-LTQ-Orbitrap-MS^n, 24 components of triterpenoids were tentatively identified in the negative mode. Then, the global chemical profiling consisting of HPLC and TLC fingerprints generated by ChemPattern™ software was developed for evaluation of *Ganoderma* species. For fingerprint analysis, 11 peaks of triterpenoids were selected as the characteristic peaks to evaluate the similarities of different samples. The correlation coefficients of similarity were greater than 0.830. The cluster analysis showed a clear separation of three groups, and 11 peaks played key roles in differentiating these samples. The developed global chemical profiling method could be applied for rapid evaluation, quality control, and authenticity identification of *Ganoderma* and other herbal medicines.

1. Introduction

Ganoderma, a popular edible and medicinal mushroom, is commonly used as dietary supplements and touted as a remedy to promote health and longevity [1]. So far, more than 200 species of *Ganoderma* have been found in the world, and *Ganoderma lucidum* (Leyss. ex Fr.) Karst. and *Ganoderma sinense* Zhao, Xu et Zhang are officially recorded in Chinese pharmacopoeia [2]. Previous studies have demonstrated that *Ganoderma* possesses various biological properties, such as antitumor, antiaging, antioxidant, hyperglycemic, and regulating immunity [3–5]. Owing to its satisfactory clinical effects, more and more *Ganoderma* products as health foods or medicines have appeared at the market. However, it is hard to say whether its quality is good or bad, and it is also difficult to identify whether the raw materials are authentic or adulterant.

For *Ganoderma*, triterpenoids are the major secondary metabolites and active substances [6]. The pioneering has isolated and identified more than 300 triterpenoids from the spores, fruiting bodies, and cultured mycelia of *Ganoderma* [7–9]. Triterpenoids could be considered as the "marker compounds" for the chemical evaluation or standardization of *Ganoderma*. Professor Guo and his team's researches concluded that the content and composition of triterpenoids vary significantly due to difference in the strain, geographic origin, cultivation method, extraction process, and other factors [10, 11]. Besides, scholars have made a lot of exploratory work on the chromatographic fingerprint of *Ganoderma* products [12, 13]. However, the traditional fingerprints were always processed by the fingerprint similarity software (2004 or 2012 version), only limited to HPLC profiles without TLC. After that, researchers always apply other statistical software (such as SPSS and SAS) to process the data for cluster analysis or principal component analysis, which is relatively cumbersome and time-consuming. Therefore, it is essential to develop a global chemical profiling method for rapidly evaluating the quality of *Ganoderma* to ensure the efficacy.

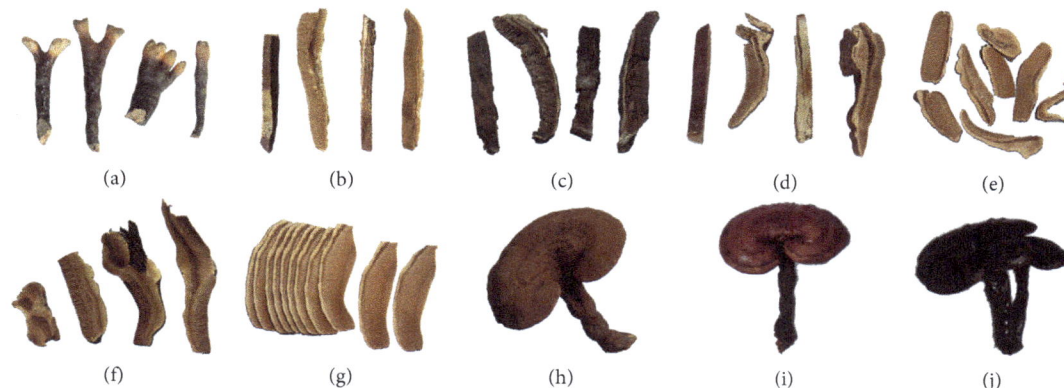

FIGURE 1: Different origins of *Ganoderma*: (a) Anhui: (b) Anhui: (c) America: (d) Anhui: (e) Jilin: (f) Anhui: (g) Anhui: (h) Henan: (i) Anhui: (j) Guangxi.

Chemical profile reflects the totality of intrinsic chemical compounds of herbal medicines and emphasizes the integral characterization of a complex system [14, 15]. Here, the global chemical profiling method contained the establishment of HPLC and TLC fingerprints, the characterization of common peaks and statistical analysis. For processing the large scale of physical properties characteristic in the complex herbal extract, an advanced chemometric and chemical fingerprinting software was developed by Chemmind Technology Co., Ltd. ChemPattern is an advanced chemometric and chemical fingerprinting software, which endeavors to provide solutions for qualitative and quantitative quality evaluation and characteristic analysis. The software was employed to calculate the correlation coefficients between different chromatographic profiles, as well as to generate the representative standard fingerprint by mean simulation.

In the present work, triterpenoids were extracted from different *Ganoderma* samples under the ultrasonic-assisted condition optimized by response surface methodology (RSM) with the Box-Behnken design (BBD) [16]. The extract was analyzed by high-performance liquid chromatography coupled with linear ion trap-Orbitrap mass spectrometry (HPLC-LTQ-Orbitrap-MS^n) for a comprehensive study of the multiple chemical constituents. In addition, the chromatographic data of HPLC and TLC were submitted into the professional Chem-Pattern software for establishing global chemical profiling and evaluating similarity. Furthermore, *Ganoderma* samples from different regions could be distinguished by clustering analysis and principal component analysis of fingerprint data.

2. Experimental

2.1. Materials and Reagents.
Ganoderma collected from different regions is shown in Figure 1. Samples were pulverized into powder and kept in a vacuum dryer. The standard sample of *Ganoderma* was identified by National Institutes for Food and Drug Control. Cellulase and oleanolic acid were obtained from Aladdin. Ganoderic acid A was purchased from Chengdu Must Bio-technology Co., Ltd. Acetonitrile and methanol were of HPLC grade and obtained from Anhui Tedia High Purity Solvents Co., Ltd.

Ethanol, vanillin, sulfuric acid, acetic acid, perchloric acid, phosphoric acid, formic acid, petroleum ether, ethyl acetate, and other chemicals were of analytical grade.

2.2. Heating Reflux Extraction.
Sample powder (2.0 g) and 95% ethanol (30 mL) were extracted under reflux at 75°C for 30 min. Then, the extraction solution was filtered through a filter paper and evaporated to dryness at 60°C.

2.3. Ultrasonic Extraction of Total Triterpenoids.
Sample powder (2.0 g) and 95% ethanol (30 mL) were placed in an ultrasonic bath (300 W) at 75°C for 30 min. The suspension was cooled to room temperature and filtered. The filtrate was vacuum-dried at 60°C.

2.4. Determination of the Total Content of Triterpenoids.
The total content of triterpenoids was determined according to the method of Hou with some modifications [17]. The oleanolic acid (2.0 mg) was dissolved in methanol (10 mL) to produce a standard solution. $0 \mu L$, $100 \mu L$, $200 \mu L$, $300 \mu L$, $400 \mu L$, $500 \mu L$, and $600 \mu L$ standard solutions were added into a test tube, respectively, and evaporated in a water bath. 5% vanillin-acetic acid reagent ($400 \mu L$) and perchloric acid ($1000 \mu L$) were added, and the tube was placed in a water bath for 30 min at 65°C. When the reaction solution was cooled, acetic acid (5 mL) was added. The absorbance was determined at 546 nm using a microplate reader (Infinite®200 Pro NanoQuant, Tecan, Switzerland). The sample ($100 \mu L$) was determined following the abovementioned method. Then, the weight of triterpenoids was calculated according to the standard curve. The extraction yield of triterpenoids was calculated as follows: yield (mg/g, w/w) = weight of triterpenoids/weight of raw materials. All determinations were performed in triplicates. The standard curve was $y = 223.2x + 4.473$, $R^2 = 0.9950$. The oleanolic acid weight in 20–120 μg range showed a good linear relationship.

2.5. Experimental Design of RSM for Ultrasonic Extraction.
Three independent factors of ultrasonic extraction were

investigated using the RSM of BBD, including liquid-solid ratio (A: 15, 20, and 35 mL/g), extraction time (B: 35, 45, and 60 min), and ethanol concentration (C: 75, 85, and 95%), as shown in Table S1. The three levels were designated as −1, 0, and +1 for low, intermediate, and high values.

In order to predict the conditions of ultrasound extraction, experimental data were analyzed using the software Design-Expert version 8.06 and explained using the following nonlinear computer-generated quadratic model [18]:

$$R = \beta_0 + \sum_{i=1}^{k} \beta_i x_i + \sum_{i=1}^{k} \beta_{ii} x_i^2 + \sum_{i=1} \sum_{j=i+1} \beta_{ij} x_i x_j + \varepsilon, \quad (1)$$

where β_0 is the constant coefficient, β_i, β_{ii}, and β_{ij} are the coefficients for the linear, quadratic, and interaction effect, x_i and x_j are the independent variables, and ε is the error.

The adequacy of the model was tested through analysis of variance (ANOVA). The coefficients of determination R^2 and adj R^2 expressed the quality of fit of the resultant polynomial model, and the statistical significance was checked by F-value and lack of fit [19].

2.6. LC-LTQ-Orbitrap-MS^n Conditions. For accurate mass measurements, an Agilent 1290 HPLC instrument was coupled with a LTQ Orbitrap Velos mass spectrometer (Thermo Scientific, Hemel Hempstead, UK) equipped with an ESI source. An Eclipse plus C_{18} (50 mm × 4.6 mm, 1.8 μm) column was used for chromatographic separation. The column temperature was kept at 35°C. The mobile phase consisted of acetonitrile (A) and 0.03% phosphoric acid solution (B). The gradient elution was as follows: 75–68% B at 0–15 min, 68–60% B at 15–20 min, 60–40% B at 20–25 min, 40–0% B at 25–40 min, and 0% B at 40–125 min. The DAD was set at 254 nm. The injection volume was 10 μL, and the flow rate was 0.6 mL/min.

The operation parameters of mass spectrometry were as follows: source voltage, 4.0 kV; sheath gas, 20 (arbitrary units); auxiliary gas, 12 (arbitrary units); sweep gas, 2 (arbitrary units); and capillary temperature, 350°C. Default values were used for most other acquisition parameters: Fourier transformation (FT) automatic gain control (AGC) target 5×10^5 for the MS mode and 5×10^4 for the MS^n mode. Perfusion samples were analyzed in the data-dependent scan mode at a resolving power of 60,000 at m/z 400. The most intense ions were selected, and parent ions were fragmented by high-energy C-trap dissociation (HCD) with a normalized collision energy of 45% and an activation time of 100 ms. The maximum injection time was set to 100 ms with two microscans for the MS mode and to 1000 ms with one microscan for the MS^n mode. The mass range was from m/z 100 to 1500. Each sample was analyzed both in negative and positive modes. Data were analyzed using Xcalibur software version 2.2 (Thermo Fisher Scientific).

2.7. HPLC Chromatographic Fingerprint Analysis Conditions. HPLC chromatographic fingerprint analysis was conducted on a liquid chromatography system (1260, Agilent, America) equipped with a quaternary solvent deliver system,

an autosampler, and a DAD (Agilent Technologies). The mobile phase consisted of acetonitrile (A) and 0.03% phosphoric acid solution (B) using a gradient elution of 75–68% B at 0–40 min, 68–60% B at 40–60 min, 60–40% B at 80–120 min, 40–0% B at 80–120 min, and 0% B at 120–125 min. Chromatographic separation was carried out at an Agilent Zorbax Extend-C_{18} column (4.6 mm × 250 mm, 5 μm) with a solvent flow rate of 1.0 mL/min at a temperature of 35°C. The wavelength was set at 254 nm. The injection volume was 10 μL.

2.8. TLC Chromatographic Conditions. Ganoderic acid A solution and samples a to k (5 μL) were spotted using a microinjector on a 20 × 20 cm silica gel plate (GF254, Qingdao, China). The silica gel was activated in an oven at 80°C for 30 min before use. The mobile phase consisting of petroleum ether : ethyl acetate : formic acid (1 : 1 : 0.02, v/v/v) was added into a twin-trough chamber and saturated for 10 min. The plate in the chamber was developed upward over a path of 15 cm and sprayed with 1% vanillin-sulfuric acid solution. The plate was placed in an oven at 80°C for 10 min until the color of the triterpenoid spots was distinct. The image of TLC was reverse-phase processed, and the information of TLC spots was turned into the gray curve by software.

2.9. Statistical Analysis. All experiments were performed at least in triplicate. The values were expressed as means ± standard deviation (SD).

3. Results and Discussion

3.1. Comparison of Ultrasonic-Assisted Extraction and Reflux Extraction. Comparing heating reflux extraction (6.404 mg/g) with ultrasonic-assisted extraction (6.869 mg/g), the extraction yield of triterpenoids was increased significantly by ultrasonic-assisted extraction in the same extraction time with simple operation. The ultrasonic wave produced a strong cavitation, mechanical crushing, and thermal effect, which dissolved the active ingredients into the solvent more adequately and saved more energy. So, the ultrasonic-assisted extraction was selected for further optimization.

3.2. RSM for Optimization of Ultrasonic-Assisted Extraction. The main factors which affected the ultrasonic extraction yield of triterpenoids were investigated by single-factor experiments, including the temperature, liquid-solid ratio, extraction time, and ethanol concentration. When the procedures were conducted at 30 min with a concentration of 95% ethanol, the maximum yield of triterpenoids was 0.6906% at the liquid-solid ratio 25 mL/g (Figure 2(a)). The liquid-solid ratio and concentration of ethanol were fixed at 25 mL/g and 95%. The result showed that the extraction efficiency increased to the maximum amount of 0.7097% at 45 min (Figure 2(b)). The yield of triterpenoids was significantly increased with the ethanol concentration varying from 15% to 95%, and the optimal ethanol concentration was from 75% to 95%

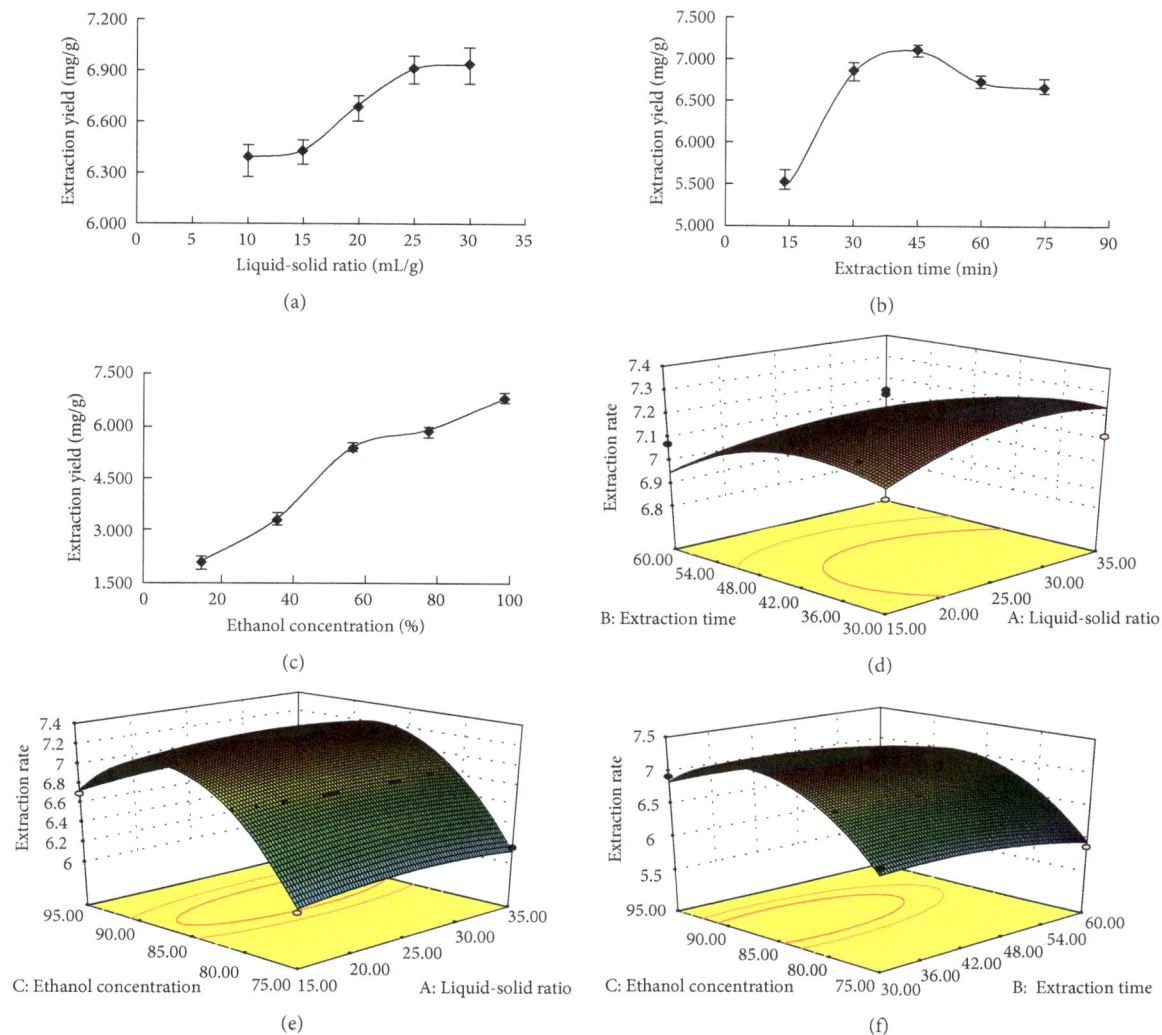

FIGURE 2: Effects of independent parameters: liquid-solid ratio (a), extraction time (b), and ethanol concentration (c) on the extraction yield of triterpenoids. Interaction effects between the liquid-solid ratio and extraction time (d), liquid-solid ratio and ethanol concentration (e), and extraction time and ethanol concentration (f) on the yield of triterpenoids.

(Figure 2(c)). Therefore, the liquid-solid ratio 25 mL/g, extraction time 45 min, and 85% ethanol were selected as the center points for each factor in the RSM experiments. The single-factor experiment of temperature showed that the extraction rate of triterpenoids was almost unchanged with increasing temperature after 50°C. So, the extraction temperature was not considered as the experimental factor of BBD and set at 50°C. An experimental program for optimizing the extraction of triterpenoids using RSM with BBD is shown in Table S2. The predicted values were obtained from the model fitting technique using the software Design-Expert version 8.06.

ANOVA was applied to optimize the extraction conditions of ultrasonic-assisted extraction for the triterpenoid yield and evaluate the relationship between response and variables. ANOVA for the response surface quadratic regression model showed that the F-value of model was 30.56 and the P value of model was smaller than 0.0001 (Table S3), suggesting the model was significant [16]. The coefficient of determination (R^2) of the model was 0.9752, and the adjusted determination coefficient (adj R^2) was 0.9433, which indicated good agreement between the experimental parameters and the predicted values of triterpenoids. The sequence of three factors influencing the triterpenoid yield was the ethanol concentration (C), extraction time (B), and liquid-solid ratio (A).

By statistically processing, the multiple second-order equation for the extraction yield of triterpenoids was obtained as follows:

$$Y = + 7.23 + 0.015A - 0.13B + 0.32C - 0.051AB$$
$$+ 0.012AC + 4.250 \times 10^{-3}BC - 0.075A^2 \qquad (2)$$
$$- 0.11B^2 - 0.72C^2,$$

where Y is the extraction yield of triterpenoids, A, B, and C are the coded values of the liquid-solid ratio, extraction time, and ethanol concentration, respectively.

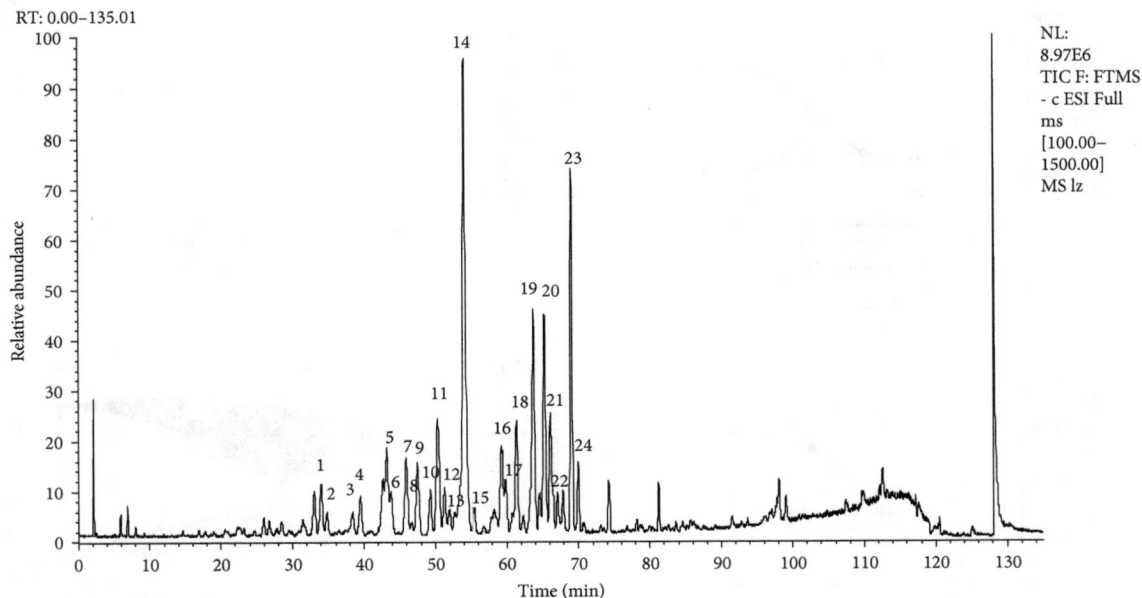

FIGURE 3: The total ion chromatograms (TICs) of the extract from *Ganoderma* by LC-LTQ-Orbitrap-MS^n in negative ion mode.

The extraction yield and the interaction of different variables could be predicted from the three-dimensional (3D) response surface (Figures 2(d)–2(f)). The steeper slope represented that the factor had a more significant effect on the extraction yield. The steeper slope of ethanol concentration implied that it had greater effect on the extraction yield of triterpenoids than on the extraction time and liquid-solid ratio.

According to the regression equation, the optimal parameters were the liquid-solid ratio 28.32 mL/g, extraction time 35.64 min, and ethanol concentration 87.25%. The theoretical highest yield of triterpenoids was 7.3118 mg/g predicted by the model. In order to validate the adequacy of the model, verification experiments were carried out by slightly modified conditions: liquid-solid ratio 28 mL/g, extraction time 36 min, and ethanol concentration 87%. The yield of triterpenoids of 7.338 ± 0.150 mg/g could be attained, which was 6.83% higher than the previous ultrasonic extraction method.

3.3. Characterization and Identification of Triterpenoids by LC-LTQ-Orbitrap-MS^n.
Qualitative analysis of triterpenoids was performed on the HPLC-LTQ-Orbitrap-MS^n system. ESI-MS spectra in both negative and positive modes were examined in this study. Negative-mode ESI was found to be sensitive for triterpenoids. All triterpenoids gave [M – H]$^-$ ions in their negative ion mass spectra. The total ion chromatograms (TICs) of triterpenoids in the negative ion mode by LC-LTQ-Orbitrap-MS^n are shown in Figure 3. The fragmentation pathway of triterpenoids is summarized by using ganoderic acid A as the standard compound. The mass spectrum of ganoderic acid A and its major fragmentation pathways are given in Figure 4. In the negative ion mode, the prominent fragmentation pathways begin with the prominent losses of H_2O or CO_2; then, the cleavages took place on the A, C, and D rings. The [M – H]$^-$ ion at m/z 515.30 ($C_{30}H_{43}O_7{}^-$) of ganoderic acid A produced

a prominent ion at m/z 497.34 by eliminating a molecule of H_2O (18 Da). The m/z 497.34 ion was further subjected to produce signals at m/z 479.32 or 453.36 by the sequential losses of H_2O or CO_2 (44 Da). The [M – H]$^-$ ion at m/z 515.30 produced an ion at m/z 417.32 ($C_{24}H_{33}O_6{}^-$) by direct cleaving on ring A. The ion at m/z 355.29 could be obtained by the process of losing H_2O and CO_2, followed by cleavage of ring A. Ganoderic acid A was also cleaved on ring C to give the product ion at m/z 249.12 ($C_{15}H_{21}O_3{}^-$). The cleavage of ring D could be observed in ganoderic acid A, besides the cleavage of rings A and C. The [M – H]$^-$ ion at m/z 515.30 produced an ion at m/z 301.25 ($C_{19}H_{23}O_3{}^-$). The m/z ion 301.25 then underwent losses of CH_3 (15 Da) to generate an ion at m/z 285.31.

According to the nontarget compound identification strategy based on the accurate mass measurement (<5 ppm), MS/MS fragmentation patterns, diagnostic product ions, and different chromatographic behaviors, 24 compounds were unambiguously identified from triterpenoids [9, 10]. Table 1 summarizes the retention times (t_R), molecular formula, [M – H]$^-$ and CAS number of each compound, and MS/MS ions. The structures of 24 compounds are shown in Figure S1. These results provided the critical information for constructing chemical fingerprints of triterpenoids.

3.4. Validation of HPLC and TLC Methods.
Prior to the establishment of the HPLC fingerprint, the precision, repeatability, and stability were chosen to validate the reliability of HPLC, which were expressed by the relative standard deviations (RSDs) of the retention time (t_R) and peak area (Pa). For the precision test, the working solutions were analyzed in triplicate, and RSD values of t_R and Pa were lower than 0.2% and 4% (Table 2). To confirm the repeatability, five different working solutions prepared from

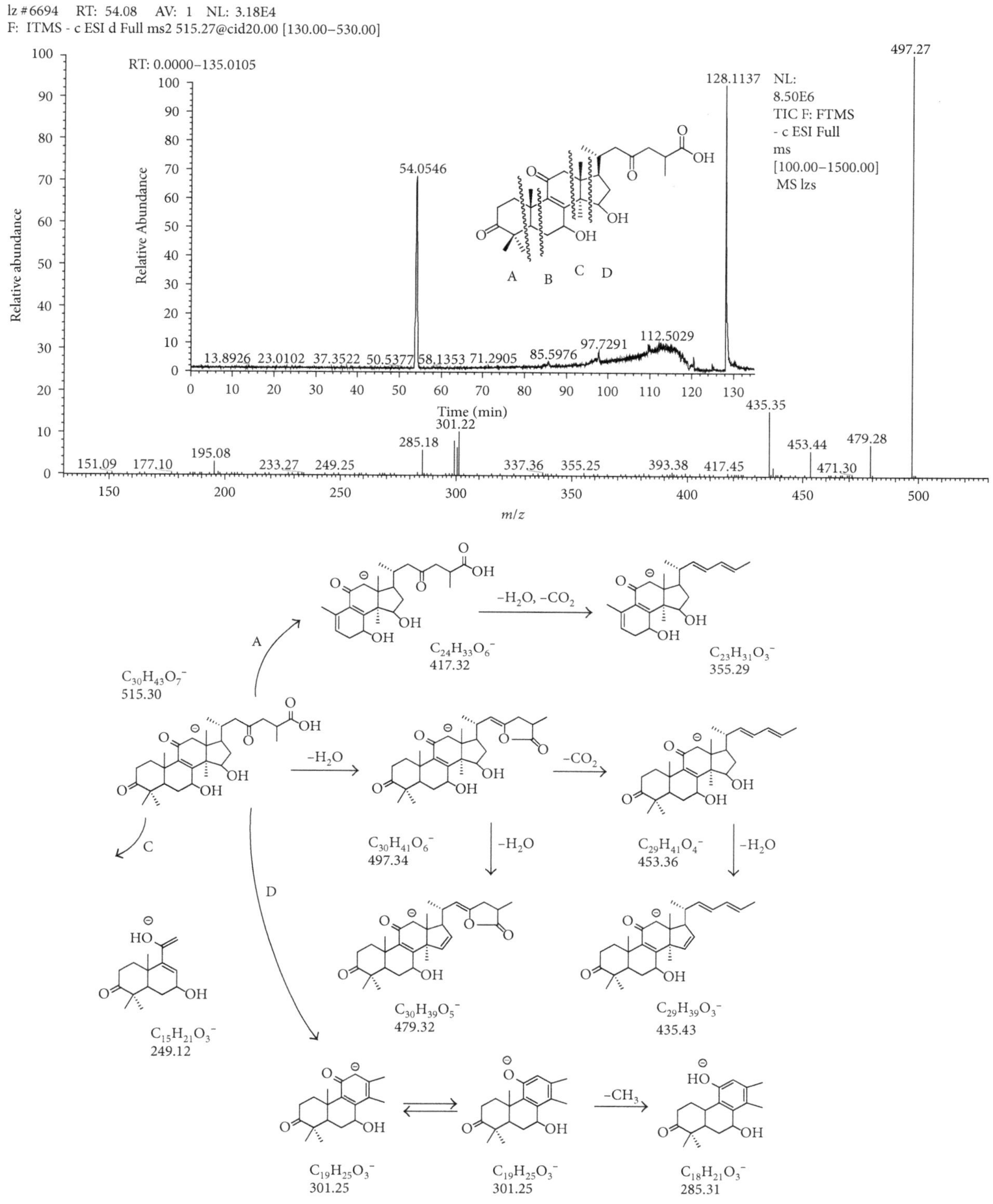

FIGURE 4: The mass spectrum and cleavage pathway of ganoderic acid A.

the same batch of the sample were analyzed. The repeatability test (Table 2) demonstrated that the developed assay was reproducible (RSD < 5%). Stability of 11 analytes was detected at 2, 4, 8, 12, and 24 h, respectively. RSD values of Pa for stability tests were 1.46–4.81% (Table 2), which indicated that the sample was stable in 24 h.

In contrast with HPLC, TLC fingerprints were more cost-effective and provided a vivid colorful image for parallel comparison [20]. At present, there are few literatures about TLC fingerprints applied to *Ganoderma*. In this study, 10 batches of *Ganoderma* and the standard herb were further investigated by TLC fingerprints. In the preliminary

TABLE 1: Compounds identified in triterpenoids from *Ganoderma*.

Number	t_R^a (min)	$[M - H]^-$ (m/z)	MS/MS main fragment ions	Compound	Formula	Weight[b]	CAS number
1	33.88	517.3179	499.3361, 481.3423, 455.5135, 437.4728, 422.2257	Ganoderic acid C_2	$C_{30}H_{46}O_7$	518.68	103773-62-2
2	34.72	475.2600	457.33624, 439.5103	Lucidenic acid C	$C_{27}H_{40}O_7$	476.60	95311-96-9
3	38.40	459.2756	441.3874, 423.3755	Lucidenic acid LM_1	$C_{27}H_{40}O_6$	460.60	364622-33-3
4	39.49	529.2814	481.2692, 467.3972, 437.4240, 407.4240	Ganoderic acid C_6	$C_{30}H_{42}O_8$	530.65	105742-76-5
5	43.11	531.2983	495.4055, 469.3644, 454.3719, 436.3896, 407.4450, 379.1423	Ganoderic acid G	$C_{30}H_{44}O_8$	532.67	98665-22-6
6	43.69	513.2871	495.4676, 477.3961, 469.3892, 451.4328, 437.3875	Ganoderenic acid B	$C_{30}H_{42}O_7$	514.65	100665-41-6
7	45.78	515.3024	479.3840, 453.3503, 438.3453, 409.3114, 391.4153	Ganoderic acid B	$C_{30}H_{44}O_7$	516.67	81907-61-1
8	47.53	515.2659	497.3039, 473.3125, 455.1528, 410.9863	Lucidenic acid E	$C_{29}H_{40}O_8$	516.62	98665-17-9
9	47.73	513.2869	495.4192, 469.4687, 451.3153, 436.2914	Ganoderic acid AM_1	$C_{30}H_{42}O_7$	514.65	149507-55-1
10	49.28	571.2925	553.4242, 538.6849, 511.4037	Ganoderic acid H	$C_{32}H_{44}O_9$	572.69	98665-19-1
11	50.28	513.2868	498.3824, 495.3860, 469.3463, 439.3085, 424.2103, 406.3151	Ganoderenic acid A	$C_{30}H_{42}O_7$	514.65	100665-40-5
12	51.28	527.2643	509.4469, 481.3122, 452.3277, 390.8892	Elfvingic acid A	$C_{30}H_{40}O_8$	528.63	433284-49-2
13	53.29	473.2554	437.3959, 425.3581, 411.2774, 393.3146, 375.2134	Lucidenic acid B	$C_{27}H_{38}O_7$	474.59	95311-95-8
14	54.09	515.4518	497.3640, 479.3914, 435.3200, 417.2299	Ganoderic acid A	$C_{30}H_{44}O_7$	516.30	81907-62-2
15	55.28	499.3073	481.4785, 479.3914, 437.4701, 419.3591	Ganolucidic acid A	$C_{30}H_{44}O_6$	500.67	98665-21-5
16	59.28	457.2603	442.3622, 439.2677, 421.2575, 413.4738, 395.4018	Lucidenic acid A	$C_{27}H_{38}O_6$	458.59	95311-94-7
17	60.57	455.2449	437.3401, 425.2315, 411.5940, 393.3439, 383.2032, 365.0397	Lucidenic acid F	$C_{27}H_{36}O_6$	456.57	98665-18-0
18	61.32	513.2871	478.2901, 463.2901, 449.2825, 434.4029	Ganoderenic acid D	$C_{30}H_{40}O_7$	512.63	100665-43-8
19	63.77	495.2764	477.3036, 451.3039, 436.3244, 407.4024, 365.2903	Ganoderic acid D	$C_{30}H_{42}O_7$	514.65	108340-60-9
20	65.45	513.2506	495.3201, 471.2714, 453.1822, 425.5560, 396.4717	Lucidenic acid D	$C_{29}H_{38}O_8$	514.61	98665-16-8
21	66.29	511.2711	493.4020, 467.4334, 449.3544, 434.2898	Ganoderic acid E	$C_{30}H_{40}O_7$	512.27	98665-14-6
22	67.26	499.3074	481.4776, 437.3365, 419.4566	Ganolucidic acid D	$C_{30}H_{44}O_6$	500.67	102607-22-7
23	69.07	569.2768	521.3403, 509.2697, 465.3616, 447.4046	Ganoderic acid F	$C_{32}H_{42}O9$	570.67	98665-15-7
24	70.18	513.2867	495.3300, 471.3816, 451.3528, 436.4189, 421.2986	Ganoderic acid J	$C_{30}H_{42}O_7$	514.65	100440-26-4

[a]Retention time; [b]relative molecular weight.

experiment, developing solvent, sample concentration, chromogenic reagent, chromogenic temperature, and chromogenic time were optimized to achieve the optimum effect of separation and coloration. Owing to the large polarity of triterpenoids, a small amount of formic acid was added in the developing agent. Then, 1% vanillin-sulfuric acid solution was chosen as a chromogenic agent to color the compounds of triterpenoids. In order to validate the reliability of the TLC method, precision, repeatability, and stability were determined. The operation methods of working solutions were the same as HPLC. As shown in Table 3, RSDs of the flow rate value (R_f) ranged from 0.33% to 2.48%, which proved that the TLC experiments were reliable. However, RSDs of the peak area value (Pa) of the main characteristic peaks were from 2.32% to 7.79% (Table 3), inferring that a certain amount of error was produced by manual spotting.

3.5. Establishment of Global Chemical Profile by ChemPattern Software.
Both common patterns of HPLC and TLC fingerprints for triterpenoids were generated to represent the characteristic peaks of this authenticated herbal medicine by ChemPattern software (Chemmind Technologies, Beijing, China). The similarity and clustering analyses on different kinds of chromatographic fingerprint data comprehensively evaluated the types and quantities of triterpenoids by ChemPattern software.

The HPLC common pattern of 10 batches of samples is shown in Figures 5(a) and 5(b). Sample c from America was not suitable to establish the common model of the HPLC fingerprint because of its significant differences in chemical composition. 11 peaks existing in 10 batches of samples were found as common peaks. According to the compound database of triterpenoids obtained from HPLC-LTQ-Orbitrap-MS^n, peaks 1 to 11 were identified as lucidenic acid LM_1, ganoderic

TABLE 2: Precision, repeatability, and stability of eleven analytes in HPLC.

Analyte	Mean of t_R[a]	Precision ($n = 3$)		Repeatability ($n = 5$)		Stability ($n = 5$)						
		RSD (%) of t_R	RSD (%) of Pa[b]	RSD (%) of t_R	RSD (%) of Pa	RSD (%) of t_R	Pa					RSD (%) of Pa
							2 h	4 h	8 h	12 h	24 h	
1	22.13	0.15	2.22	0.05	3.02	0.14	992	1004	1018	1054	1074	3.34
2	25.66	0.19	3.40	0.05	4.90	0.21	2022	2193	2118	2092	2277	4.57
3	27.74	0.13	3.33	0.05	3.98	0.14	2191	2124	2147	2132	2268	2.73
4	29.02	0.08	1.14	0.06	3.89	0.13	2209	2059	2192	2297	2333	4.81
5	32.15	0.09	3.37	0.05	2.85	0.06	4570	4687	4851	4783	4817	2.40
6	35.78	0.11	0.44	0.05	3.93	0.07	9187	9299	9480	9510	9681	1.88
7	41.51	0.07	2.96	0.06	4.63	0.16	2040	2128	2090	2283	2193	4.40
8	44.57	0.04	2.44	0.03	2.98	0.11	3941	4003	3984	4056	3905	1.46
9	47.32	0.02	0.77	0.02	2.68	0.07	4703	4923	4957	4924	4993	2.32
10	49.67	0.12	3.02	0.05	2.89	0.19	3919	3822	3913	3723	3827	2.09
11	55.67	0.10	3.87	0.03	2.68	0.13	3656	3803	3748	3756	3861	1.92

[a]Retention time of the analytes; [b]peak area of the analytes.

TABLE 3: Precision, repeatability, and stability of nine analytes in TLC.

Analyte	Mean of R_f[a]	Precision ($n = 3$)		Repeatability ($n = 5$)		Stability ($n = 5$)						
		RSD (%) of R_f	RSD (%) of Pa[b]	RSD (%) of R_f	RSD (%) of Pa	RSD (%) of R_f	Pa					RSD (%) of Pa
							2 h	4 h	8 h	12 h	24 h	
1	0.03	0.75	6.81	0.54	4.88	0.86	72.88	75.74	73.45	79.74	72.44	4.03
2	0.19	0.40	3.69	1.61	3.93	2.09	68.49	59.77	66.73	70.52	69.87	6.46
3	0.23	0.87	3.74	1.05	5.44	0.56	88.74	84.91	90.11	85.77	90.45	2.87
4	0.27	1.52	6.29	1.34	3.74	2.48	121.01	125.74	128.44	131.34	130.22	3.24
5	0.33	1.01	6.60	1.01	3.81	1.11	64.41	64.88	66.86	70.11	70.29	4.15
6	0.54	0.37	4.76	0.84	7.64	0.52	31.22	28.87	30.21	33.47	34.49	7.30
7	0.72	1.11	6.10	1.03	7.79	0.39	155.70	149.38	145.06	156.63	149.74	3.18
8	0.80	0.68	7.33	1.43	3.14	1.07	135.75	133.96	141.01	142.13	139.54	2.52
9	0.90	0.94	4.18	0.33	3.78	0.97	117.25	131.49	134.84	129.71	128.56	2.32

[a]R_f value of the analytes; [b]peak area of the analytes.

acid G, ganoderic acid B, lucidenic acid E, ganoderenic acid A, ganoderic acid A, lucidenic acid A, ganoderenic acid D, ganoderic acid D, lucidenic acid D, and ganoderic acid F, respectively. The relative retention time (t_R' = retention time of the characteristic peak/retention time of the marker peak) and relative peak area (RPA = peak area of the characteristic peak/peak area of the marker peak) of the common peaks are shown in Tables S4 and S5. Peak 6, identified as ganoderic acid A, was selected as the reference peak. The results indicated that t_R' of 11 common peaks (between 0.13% and 0.57%) was invariable between samples, and t_R' was a valid parameter for constituent identification. However, the RPA (from 29.19% to 135.20%) showed significant differences between 10 batches of samples, indicating that the content of triterpenoids from various sources was different.

The similarity and clustering analyses of HPLC fingerprints were also analyzed by ChemPattern software. The advanced chemical pattern recognition module of ChemPattern could forecast the classification of complex samples. The similarities of samples a, b, d, e, f, g, h, and i were greater than 0.830 (Figure 5(c)). However, the profiles of samples j and k were different from the common pattern, suggesting that these two samples belonged to Ganoderma sinense. This method could distinguish Ganoderma sinense and Ganoderma lucidum quickly. When the clustering distance was extended to 1.0, the samples from different regions could be divided into three groups. As shown in Figure 5(d), samples f, d, h, b, and a belonged to class I. These five samples were collected from Anhui Province. Samples k, j, g, and e were classified into class II, for the reason that the triterpenoid content and species of these samples were similar. Sample i was classified into class III individually, due to the different categories of triterpenoids. These results illustrated that the origins of Ganoderma could be distinguished via clustering analysis. The HPLC fingerprints could not only reflect the chemistry information of Ganoderma but also distinguish the Ganoderma species from different geographical origins.

The image of TLC taken by a digital camera was reverse-phase processed before imported into ChemPattern software (Figure S2). Twelve tracks were set manually, and the information of TLC spots was turned into the gray curve. TLC fingerprints of different triterpenoid extracts and the common model were obtained as shown in Figures 6(a) and 6(b). The

(a)

(b)

(c)

(d)

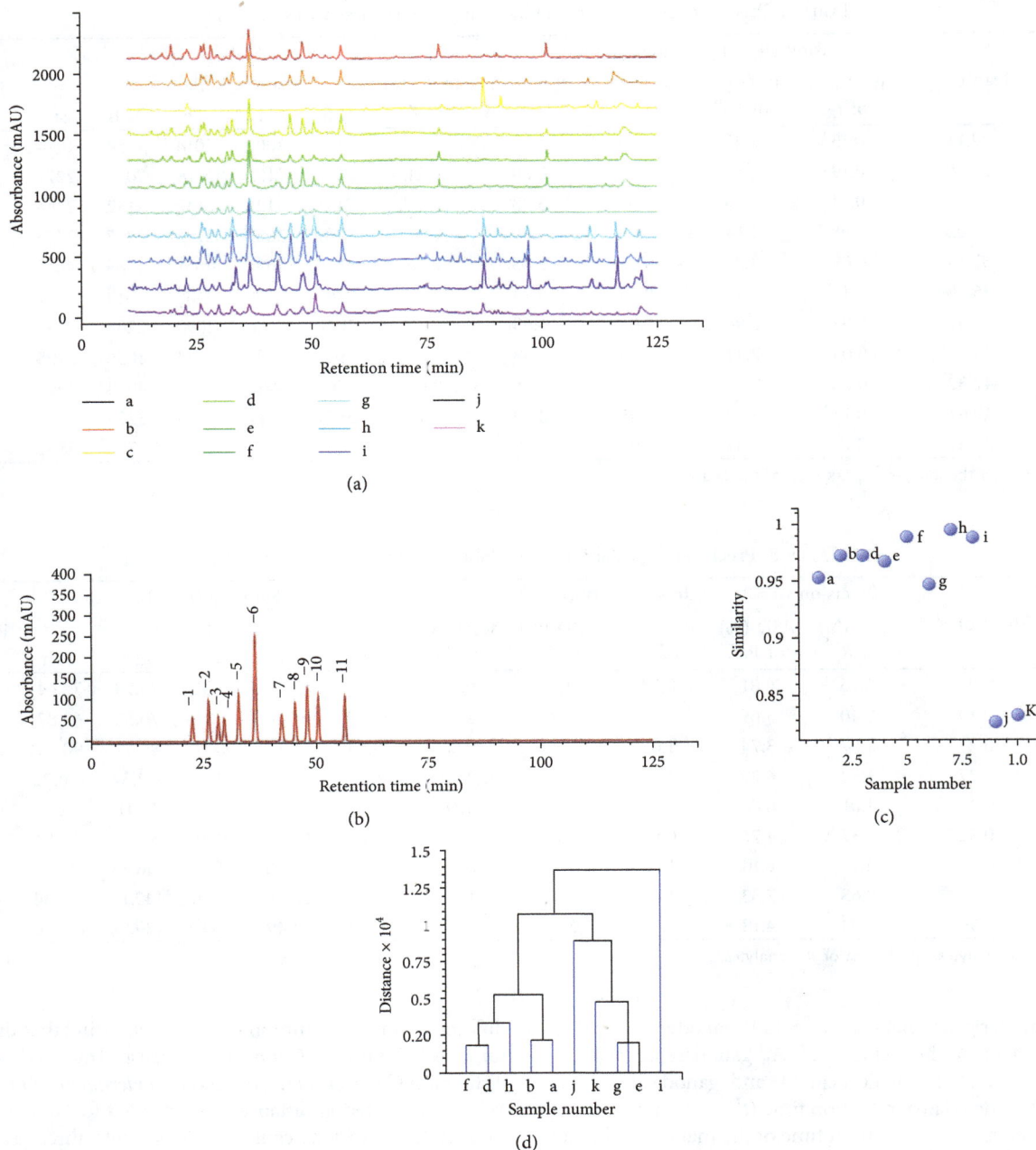

FIGURE 5: HPLC fingerprints of triterpenoids obtained from eleven batches of samples (a) and the common pattern generated by ChemPattern software (b). The similarity analysis (c) and clustering analysis (d) of triterpenoids.

fingerprints exhibited that nine common peaks were likely to represent the major constituents of triterpenoids. The similarity and clustering analyses of TLC by ChemPattern software are shown in Figures 6(c) and 6(d). The similarities of the samples a, b, d, e, f, g, h, i, j, and k were all greater than 0.900, while samples j and k were relatively low, which was consistent with the result of the HPLC fingerprint (Figure 6(c)). The clustering analysis showed that the samples could be divided into three clusters if the Euclidean distance was equal to 25 (Figure 6(d)). The samples e, d, b, f, i, h, and a belonged to class I. Among these samples, only samples e and h were not collected from Anhui. Sample k was classified into class II individually, while samples j and g were classified into class III. The cluster analysis of 10 batches of *Ganoderma* showed a clear separation of the three groups, and common peaks played key roles in differentiating these samples. Both HPLC and TLC fingerprint classifications could provide a simple reference standard for quality identification of *Ganoderma*.

3.6. Discussion. In this article, RSM with the BBD method was successfully applied to optimize the factors for the ultrasonic

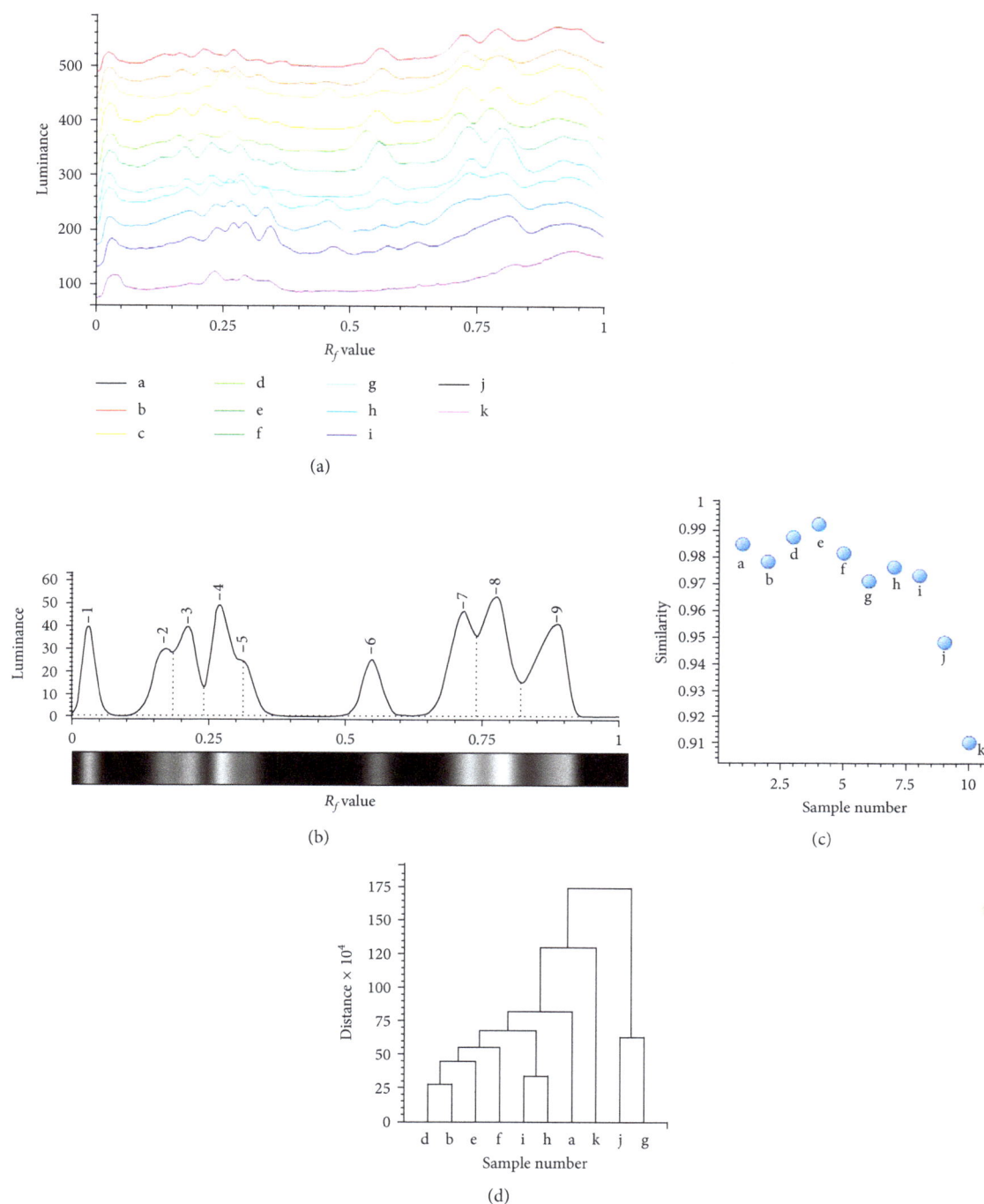

FIGURE 6: TLC fingerprints of triterpenoids obtained from different *Ganoderma* species (a) and their common pattern (b). The similarity analysis (c) and clustering analysis (d) of triterpenoids.

extraction of triterpenoids. The suitable conditions were as follows: liquid-solid ratio 28 mL/g, ethanol concentration 87%, and extraction time 36 min at 50°C. The ultrasonic extraction technology has the advantage of accelerating the extraction time, lowering the temperature, causing less damage to the structure of plant materials, and increasing the extraction yield. It is more suitable for the extraction of triterpenoids. RSM is an effective statistical method useful for optimizing

a complex process and evaluating the interaction between multiple parameters [21]. The yield of triterpenoids of 7.338 ± 0.150 mg/g could be attained, which was consistent with the theoretical predicted value (7.3118 mg/g). Therefore, the model was considered to be reliable, and RSM could be used for predicting the ultrasonic extraction yield of triterpenoids.

HPLC-LTQ-Orbitrap-MS^n technique was applied to identify the chemical structure of triterpenoids in *Ganoderma* with higher sensitivity. It provided high resolution and abundant structural information for not only the pseudo-molecular ions but also the fragment species. The application of HPLC-LTQ-Orbitrap-MS^n could provide a large amount of information related to the chemical structure and make up the deficiency of UV and DAD detectors [8]. Based on the accurate mass measurement, MS/MS fragmentation patterns, and diagnostic product ions provided by HPLC-LTQ-Orbitrap-MS^n and literatures, 24 compounds were identified, which was one of the major tools for the study of the chemical substance of *Ganoderma*.

To investigate triterpenoids further, the chromatographic fingerprint method was applied to analyze the complex composition of *Ganoderma*. The fingerprint analysis technology is different from traditional analysis methods because it analyzes objects from the perspective of whole component information. The chromatographic fingerprint method was established through importing HPLC and TLC profiles into ChemPattern software for the comprehensive quality control of *Ganoderma*. The common pattern of fingerprints provided the critical information for constructing chemical fingerprints of triterpenoids. In addition, the fingerprint method combined with the stoichiometry could be applied to identify the species and origins of *Ganoderma*. The results obtained by clustering analysis of HPLC and TLC fingerprints were slightly different due to the fact that HPLC possesses high resolution and reproducibility. The correct classification percentages of the pattern recognition method of HPLC fingerprint identification were higher than those of TLC. However, in contrast with the HPLC fingerprint, the TLC fingerprint was more cost-effective and provided a vivid colorful image for parallel comparison. In summary, the chemical fingerprint method could be adopted as a reliable tool for the authentication and quality control of *Ganoderma*.

4. Conclusion

In the present study, the chemical composition of triterpenoids was clarified by HPLC-LTQ-Orbitrap-MS^n, and the global chemical profile consisting of HPLC and TLC fingerprints was established. Eleven triterpenoid peaks which differed significantly in all the analyzed samples were used as markers for origin identification and authenticity establishment of *Ganoderma*. This work suggested that the developed global chemical profiling method could provide a convenient approach, which might be applied for rapid evaluation, quality control, and authenticity establishment of *Ganoderma* products. In the future, the chemical constituents and pharmacological activities of *Ganoderma* would be explored in depth through the multidimensional fingerprints combined with chemometric methods, molecular biology, and pattern recognition techniques.

Conflicts of Interest

The authors declare that they have no conflicts of interest.

Acknowledgments

This research was supported by the National Natural Science Foundation of China (Grant no. 81603249) and Scientific Research Project of Science Technology Department of Zhejiang Province (no. 2016C33076).

References

[1] B. S. Sanodiya, G. S. Thakur, R. K. Baghel, G. B. Prasad, and P. S. Bisen, "*Ganoderma lucidum*: a potent pharmacological macrofungus," *Current Pharmaceutical Biotechnology*, vol. 10, no. 8, pp. 717–742, 2009.

[2] National Commission of Chinese Pharmacopoeia, *Pharmacopoeia of Peoples Republic of China*, Part 1, China Medical Science and Technology Press, Beijing, China, 2015.

[3] Z. B. Lin and H. N. Zhang, "Antitumor and immunoregulatory activities of *Ganoderma lucidum* and its possible mechanisms," *Acta Pharmacologica Sinica*, vol. 25, no. 11, pp. 1387–1395, 2004.

[4] L. Y. Liu, H. Chen, C. Liu et al., "Triterpenoids of *Ganoderma theaecolum* and their hepatoprotective activities," *Fitoterapia*, vol. 98, pp. 254–259, 2014.

[5] J. Ćilerdžić, J. Vukojević, M. Stajić, T. Stanojković, and J. Glamočlija, "Biological activity of *Ganoderma lucidum* basidiocarps cultivated on alternative and commercial substrate," *Journal of Ethnopharmacology*, vol. 155, no. 1, pp. 312–319, 2014.

[6] J. Q. Liu, C. F. Wang, Y. Li, H. R. Luo, and M. H. Qiu, "Isolation and bioactivity evaluation of terpenoids from the medicinal fungus *Ganoderma sinense*," *Planta Medica*, vol. 78, no. 4, pp. 368–376, 2012.

[7] Q. Xia, H. Z. Zhang, X. F. Sun et al., "A comprehensive review of the structure elucidation and biological activity of triterpenoids from *Ganoderma* spp.," *Molecules*, vol. 19, no. 11, pp. 17478–17535, 2014.

[8] M. Yang, X. Wang, S. Guan et al., "Analysis of triterpenoids in *Ganoderma lucidum* using liquid chromatography coupled with electrospray ionization mass spectrometry," *Journal of the American Society for Mass Spectrometry*, vol. 18, no. 5, pp. 927–939, 2007.

[9] C. R. Cheng, M. Yang, Z. Y. Wu et al., "Fragmentation pathways of oxygenated tetracyclic triterpenoids and their application in the qualitative analysis of *Ganoderma lucidum* by multistage tandem mass spectrometry," *Rapid Communications in Mass Spectrometry*, vol. 25, no. 9, pp. 1323–1335, 2011.

[10] X. M. Wang, M. Yang, S. H. Guan et al., "Quantitative determination of six major triterpenoids in *Ganoderma lucidum* and related species by high performance liquid chromatography," *Journal of Pharmaceutical and Biomedical Analysis*, vol. 41, no. 3, pp. 838–844, 2006.

[11] J. Da, W. Y. Wu, J. J. Hou et al., "Comparison of two officinal Chinese pharmacopoeia species of *Ganoderma* based on chemical research with multiple technologies and chemometrics analysis," *Journal of Chromatography A*, vol. 1222, no. 2, pp. 59–70, 2012.

[12] Y. Chen, Y. Yan, M. Y. Xie et al., "Development of a chromatographic fingerprint for the chloroform extracts of *Ganoderma lucidum* by HPLC and LC-MS," *Journal of Pharmaceutical and Biomedical Analysis*, vol. 47, no. 3, pp. 469–477, 2008.

[13] Y. Chen, S. B. Zhu, M. Y. Xie et al., "Quality control and

original discrimination of *Ganoderma lucidum* based on high-performance liquid chromatographic fingerprints and combined chemometrics methods," *Analytica Chimica Acta*, vol. 623, no. 2, pp. 146–156, 2008.

[14] D. Custers, P. N. Van, P. Courselle, S. Apers, and E. Deconinck, "Chromatographic fingerprinting as a strategy to identify regulated plants in illegal herbal supplements," *Talanta*, vol. 164, pp. 490–502, 2017.

[15] X. H. Fan, Y. Y. Cheng, Z. L. Ye, R. C. Lin, and Z. Z. Qian, "Multiple chromatographic fingerprinting and its application to the quality control of herbal medicines," *Analytica Chimica Acta*, vol. 555, no. 2, pp. 217–224, 2006.

[16] K. Pan, Q. G. Jiang, G. Q. Liu, X. Y. Miao, and D. W. Zhong, "Optimization extraction of *Ganoderma lucidum* polysaccharides and its immunity and antioxidant activities," *International Journal of Biological Macromolecules*, vol. 55, no. 2, pp. 301–306, 2013.

[17] M. N. Hou and J. Liu, "*Ganoderma* triterpenoids extraction and determination of the total triterpenoid," *Research & Practice on Chinese Medicines*, vol. 24, pp. 70-71, 2010.

[18] Z. Liu, X. Ma, B. Deng, Y. Huang, R. Bo, and Z. Gao, "Development of liposomal *Ganoderma lucidum* polysaccharide: formulation optimization and evaluation of its immunological activity," *Carbohydrate Polymers*, vol. 117, pp. 510–517, 2015.

[19] T. Hu, Y. Y. Guo, Q. F. Zhou et al., "Optimization of ultrasonic-assisted extraction of total saponins from *Eclipta prostrasta* L. using response surface methodology," *Journal of Food Science*, vol. 77, no. 9, pp. 975–982, 2012.

[20] S. Agatonovic-Kustrin and C. M. Loescher, "Qualitative and quantitative high performance thin layer chromatography analysis of *Calendula officinalis*, using high resolution plate imaging and artificial neural network data modeling," *Analytica Chimica Acta*, vol. 798, no. 18, pp. 103–108, 2013.

[21] K. N. Prasad, F. A. Hassan, B. Yang et al., "Response surface optimisation for the extraction of phenolic compounds and antioxidant capacities of underutilised *Mangifera pajang* Kosterm. peels," *Food Chemistry*, vol. 128, no. 4, pp. 1121–1127, 2011.

12

Optimization of Vacuum Microwave-Mediated Extraction of Syringoside and Oleuropein from Twigs of *Syringa oblata*

Xiangping Liu[ID]**, Xuemin Jing**[ID]**, and Guoliang Li**[ID]

College of Animal Science and Veterinary Medicine, Heilongjiang Bayi Agricultural University, Daqing 163319, China

Correspondence should be addressed to Guoliang Li; liguoliang3885@163.com

Academic Editor: Giuseppe Ruberto

A vacuum microwave-mediated method was used to extract syringoside and oleuropein from *Syringa oblata* twigs. The optimal extraction conditions were an ethanol volume fraction of 40%, a liquid-solid ratio of 17 mL/g, 1 h of soaking time, −0.08 MPa of vacuum, a microwave irradiation power of 524 W, and a microwave irradiation time of 8 min. Under optimal parameters, the maximum yields of syringoside (5.92 ± 0.24 mg/g) and oleuropein (4.02 ± 0.18 mg/g) were obtained. The proposed method is more efficient than conventional methods for extracting syringoside and oleuropein from *Syringa oblata*. Moreover, less energy and time were required. The results implied that vacuum microwave-mediated extraction is a suitable method for the extraction of thermosensitive glycosides such as syringoside and oleuropein.

1. Introduction

Recently, many studies have been done on the utilization of microwave-assisted extraction (MAE) to obtain biologically active analytes such as glycosides [1–4], phenolcarboxylic acids [5, 6], procyanidins [7], stilbenes [8], coumarins [9], alkaloid [10], pectin [11], and essential oil [12]. MAE is a process by which microwave energy is used to heat the sample and the solvent, leading to fast movement of the target component molecules from the sample into the extracting solvent. Extraction using microwaves has advantages over traditional extraction techniques, including greater efficiency, less solvent consumption, and reduced extraction times. However, the reaction temperature of MAE is typically near the boiling point of the extraction solvent at atmospheric pressure, which causes the degradation or oxidation of some thermolabile and oxygen-sensitive secondary metabolites in traditional MAE systems [13, 14]. Vacuum microwave-mediated extraction (VMME) is a good alternative technique to prevent degradation of thermal-sensitive and oxygen-sensitive compounds at high extraction temperatures in open MAE [15–19]. Additionally, the exclusion of air from the extraction system can increase the extraction yield as well [17, 20, 21].

Syringa is a genus of the Oleaceae family and is widely cultivated in many warm-temperate countries [22]. China is a rich natural resource center for *Syringa* [23]. Of the *Syringa* species, *Syringa oblata* is widely grown in regions of northern China. In addition to being widely grown for landscape greenery, *S. oblata* has been used as folk medicine for treating various diseases for many centuries [24].

Syringoside and oleuropein are the important active ingredients of *S. oblata* [25, 26]. Syringoside (Figure 1) is commonly regarded as a significant phenylpropanoid glycoside [27]. Pharmacology studies have demonstrated that syringoside has anti-inflammatory and pain-suppressing activities [28], immune regulation activity [29], neuroprotective effects [30], and activity in reducing injury from ultraviolet radiation [31], and it promotes insulin secretion [32, 33]. Oleuropein is a secoiridoid glucoside [27] (Figure 1). Oleuropein possesses significant antioxidant activity and has various pharmacological effects, including anti-inflammatory, antitumor, antibacterial, antiviral, antiarteriosclerosis, hypolipidemic, hypoglycemic, and hepatoprotective activities [34, 35].

Heat reflux extraction (HRE) has been traditionally used for the extraction of glycosides from plants. However, degradation by hydrolysis isomerization, condensation, or

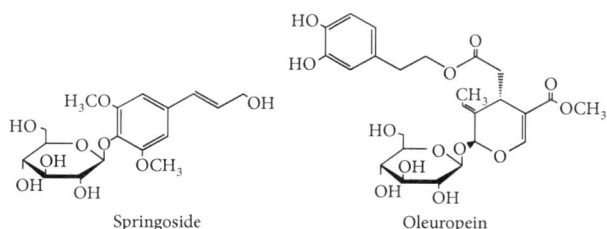

FIGURE 1: The chemical structures of syringoside and oleuropein.

oxidation during HRE contributes to the loss of glycosides. These problems make this technique inefficient and lead to low recovery of these compounds.

In this study, an effective and environmentally friendly vacuum microwave-mediated extraction (VMME) method was used to extract syringoside and oleuropein from twigs of *S. oblata*. VMME parameters were optimized. Moreover, the extraction efficiency of the VMME technique was compared with the traditional HRE technique for the extraction of syringoside and oleuropein from *S. oblata*.

2. Experimental

2.1. Materials and Reagents.
Twigs were manually picked from similar nearly 2.5 m, 10-year-old cultivated *S. oblata* trees in September 2017 in the Daqing suburbs, Heilongjiang Province, China. The twigs were dried at room temperature to a constant weight, and the cut pieces were ground using a laboratory mill and sieved (50-mesh screen). Powdered materials were stored in large brown bottles until use. Reference syringoside and oleuropein were obtained from Sigma-Aldrich (Shanghai, China).

2.2. VMME Apparatus.
A modified microwave-assisted extraction unit (WP700, Galanz Enterprise Co, Ltd., Guangdong, China) was utilized for the experiment. A water condenser coated with PTFE was connected to the top of microwave [36]. Between the condenser and the flask, the extraction system was connected to a vacuum pump (SHB-III, Shanghai Yuezhong Equipment Co., Ltd, Shanghai, China) to generate a vacuum. The extraction system provided an inner microwave volume of $21.5 \times 35 \times 33$ cm³. It operates at a frequency of 2540 Hz and can continuously transmit microwave energy to the reactor.

2.3. VMME Procedure.
The powdered samples (1.0 g) were placed in 100 mL flasks. Then, the extraction agent was added depending on the experimental design. The flask was linked to the condenser through a hole on the microwave oven. The air in the flask was evacuated by a vacuum pump until the required vacuum was acquired. Before extraction, the material was soaked in the solvent for specific times to improve microwave absorption. VMME was conducted under various experimental conditions. The influences of ethanol volume fraction, soaking time, liquid-solid ratio, degree of vacuum, microwave irradiation power, and time on the extraction yields of syringoside and oleuropein were

systematically investigated. After each extraction, the supernatant was filtered through a 0.45 μm membrane and then applied to RP-HPLC.

2.4. Optimization of VMME by RSM.
To investigate the optimal reaction parameters of the VMME, Design Expert (Version 8.0.6) was used to optimize the operating conditions by RSM. An experiment model based on a response pattern was established through the Box–Behnken design (BBD) in Design Expert. Considering the results of preliminary individual factor experiments, microwave irradiation time, liquid-solid ratio, and microwave irradiation power were chosen as key variables and assigned as X_1, X_2, and X_3, respectively. The range of the factors were 4–8 min for the microwave irradiation time, 10–20 mL/g for the liquid-solid ratio, and 230–540 W for the microwave irradiation power, as shown in Table 1. The experiment model designed through BBD included 17 experiments. The extractions were performed in a random order. The three independent variables were related to two dependent variables, the yield of syringoside (Y_1) and oleuropein (Y_2). The complete quadratic equation is as follows:

$$Y = \beta_0 + \sum_{i=1}^{3} \beta_i X_i + \sum_{i=1}^{3} \beta_{ii} X_i^2 + \sum_{i=1}^{2} \sum_{j=i+1}^{3} \beta_{ij} X_i Y_j, \quad (1)$$

where Y represents the estimated response; β_0 is an intercept; β_i, β_{ii}, and β_{ij} represent the regression coefficients for linear, quadratic, and interactive terms, respectively; and X_i and X_j represent the independent variables.

2.5. HPLC Analysis.
HPLC analysis of syringoside and oleuropein was conducted through a Waters Millennium32 system. The Waters HPLC system consisted of a 717 automatic sampler, a 717 automatic column temperature control box, a 1525 pump, and a 2487 UV detector. Chromatographic separation was performed on a Zorbax XDB-C18 column (4.6 mm × 250 mm, 5 μm, Agilent). Methanol with 0.2% phosphoric acid (33 : 67, v/v) was employed as the mobile phase. The wavelength applied to syringoside and oleuropein for the diode detector was 232 nm. Flow rate was 1.0 mL/min, the injection volume was 10 μL, and the column temperature was maintained at 25°C. Syringoside and oleuropein were quantitatively analyzed by a standard analytical method. Calibration curves for syringoside and oleuropein were established in a concentration range of 0.1–0.5 mg/mL. The linear regression equations for syringoside and oleuropein were $Y_{syringoside} = 18914X + 17067$ ($r^2 = 0.9998$, $n = 8$) and $Y_{oleuropein} = 15677X - 72912$ ($r^2 = 0.9993$, $n = 8$), respectively.

3. Results and Discussion

3.1. Individual Factor Tests of VMME Parameters.
The individual factor experiments were conducted to optimize the following parameters: ethanol volume fraction, soaking time, vacuum, liquid-solid ratio, microwave irradiation power, and microwave irradiation time.

TABLE 1: Experimental data and the observed response value with different combinations of microwave irradiation time (X_1), liquid-solid ratio (X_2), and microwave irradiation power (X_3) used in the Box–Behnken design.

Run number	Experimental design			Dependent variables			
	X_1: microwave irradiation time (min)	X_2: liquid-solid ratio (mL/g)	X_3: microwave irradiation power (W)	Yield of syringoside (mg/g)		Yield of oleuropein (mg/g)	
				Predicted yield	Actual yield	Predicted yield	Actual yield
1	4	10	385	3.82	3.82	3.39	3.44
2	8	10	385	4.86	4.94	3.43	3.40
3	4	20	385	3.81	3.73	3.35	3.37
4	8	20	385	5.27	5.27	3.72	3.67
5	4	15	230	3.65	3.63	3.02	2.92
6	8	15	230	4.02	3.92	2.22	2.20
7	4	15	540	3.73	3.83	2.89	2.91
8	8	15	540	5.86	5.88	4.10	4.20
9	6	10	230	4.13	4.14	2.77	2.82
10	6	20	230	4.40	4.51	3.04	3.11
11	6	10	540	5.16	5.05	3.78	3.70
12	6	20	540	5.29	5.28	3.77	3.73
13	6	15	385	5.57	5.71	3.78	3.89
14	6	15	385	5.57	5.57	3.78	3.78
15	6	15	385	5.57	5.52	3.78	3.76
16	6	15	385	5.57	5.39	3.78	3.65
17	6	15	385	5.57	5.68	3.78	3.81

3.1.1. Effect of Ethanol Volume Fraction. Ethanol has low toxicity and is widely used in commercial processes [37]. The extractions were conducted in various ethanol volume fractions from 0 to 80%. The remaining variables were vacuum −0.08 MPa, soaking time 1 h, liquid-solid ratio 20 mL/g, microwave irradiation time 6 min, and microwave irradiation power 385 W. Figure 2(a) shows the significant effect of ethanol volume on extraction yields of syringoside and oleuropein. The extraction yields of syringoside and oleuropein increased with an increasing ethanol volume fraction from 0 to 40%. With further increase in the volume of ethanol, the yield of syringoside and oleuropein decreased gradually. This indicates that using sufficient ethanol enhanced the extraction efficiency for the two compounds, but too much ethanol was disadvantageous for the extraction of syringoside and oleuropein. An appropriate volume of ethanol also contributed to decreasing the viscosity of the extraction solvent and enhancing the permeability of the substrate, which made the target components more easily extracted. However, because syringoside and oleuropein are glycosides, they have similar properties as sugars, so excessive ethanol in the extraction solvent decreased the solubility of syringoside and oleuropein. Based on these data, a 40% ethanol volume fraction was selected for subsequent RSM experiments.

3.1.2. Effect of Soaking Time. To select a proper soaking time, the dry herb powder was soaked in the ethanol solution for 1, 2, 4, 8, 12, and 24 h before VMME. These experiments were performed in a 40% (volume concentration) ethanol solution, and the results are shown in Figure 2(b). The extraction yield of syringoside and oleuropein was highly influenced by

soaking time. To extract target analytes from plant cells, cell contents must be accessible to the solvent. Syringoside and oleuropein yields were increased by soaking because soaking improved penetration of the solvent into the cells, allowing increased solubilization of the target analytes. A 1 h soaking time greatly enhanced the yields of syringoside and oleuropein; however, this improvement was limited with further extension of the soaking time. Therefore, 1 h was chosen as the optimal soaking time.

3.1.3. Effect of the Degree of Vacuum. To study the effect of the degree of vacuum on extraction yields of syringoside and oleuropein, several experiments were conducted at three different levels of vacuum with other extraction parameters set to 40% ethanol as the extraction solvent, 1 h as the soaking time, 20 mL/g as the liquid-solid ratio, 6 min as the microwave irradiation time, and 385 W as the microwave irradiation power. The results are shown in Figure 2(c). A higher degree of vacuum, −0.08 MPa, was appropriate for extraction of syringoside and oleuropein in MAE, and a weaker vacuum, 0.04 MPa, was not helpful for the extraction of these compounds. The reason for the increased yields might be due to the lower solvent boiling point caused by the higher degree of vacuum, which reduced the degradation of syringoside and oleuropein.

3.1.4. Effect of the Liquid-Solid Ratio. The liquid-solid ratio is a key variable that was also investigated to optimize extraction efficiency. A low solvent amount may result in insufficient extraction, whereas large volumes can lead to unnecessary cost and make the procedure difficult. Five different liquid-solid ratios (10, 15, 20, 25, and 30 mL/g) were

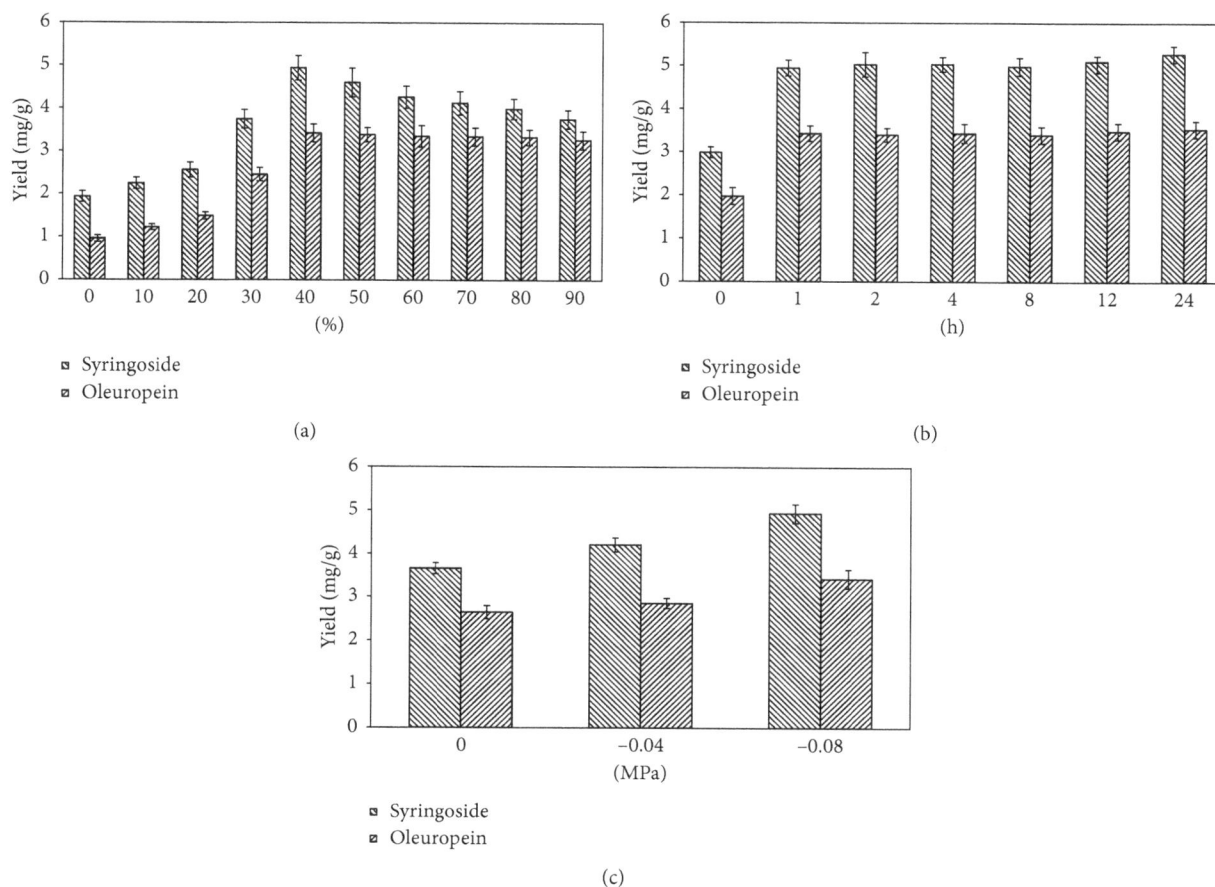

FIGURE 2: Effects of the ethanol volume fraction (a), soaking time (b), and level of vacuum (c) on the extraction yields of the target analytes. Error bars indicate standard deviation ($n = 3$).

tested. The remaining variables used for these experiments were degree of vacuum −0.08 MPa, ethanol volume fraction 40%, microwave power 230 W, microwave time 6 min, and soaking time 1 h. Figure 3(a) shows that the extraction efficiencies of both compounds were enhanced with a liquid-solid ratio of up to 20 mL/g, but increasing the liquid-solid ratio further resulted in no significant additional enhancement of extraction yields. Therefore, 10–20 mL/g was selected as the proper liquid-solid ratio to save solvent and costs.

3.1.5. Effect of Microwave Irradiation Power.
To understand the influence of microwave power on VMME, different power levels were tested. The other variables used in these experiments were degree of vacuum −0.08 MPa, ethanol volume fraction 40%, liquid-solid ratio 20 mL/g, soaking time 1 h, and microwave irradiation time 6 min. As shown in Figure 3(b), the yields of both syringoside and oleuropein reached a peak at 385 W. When the microwave power was 540 W or higher, the yields of both compounds did not increase further. Thus, 230–540 W was selected for the following experiments to achieve optimal reaction parameters of the VMME.

3.1.6. Effect of Microwave Irradiation Time.
Extraction time is a crucial factor that must be studied to guarantee complete extraction of active compounds, regardless of the technique. The impact of different extraction times on the yields of syringoside and oleuropein was evaluated (Figure 3(c)) under the following conditions: degree of vacuum −0.08 MPa, ethanol volume fraction 40%, liquid-solid ratio 20 mL/g, microwave irradiation power 385 W, and soaking time 1 h. With an increase of irradiation time from 2 to 6 min, the extraction efficiency of syringoside improved. However, further increases in microwave irradiation times beyond 6 min did not improve the syringoside extraction efficiency. Additionally, the yield of oleuropein increased from 2 min to 4 min, and the extraction efficiency correlated with the microwave irradiation time. Based on these results, VMME was conducted with an extraction time of 4–8 min in subsequent experiments.

3.2. Optimization of the VMME Parameters Using BBD.
RSM design was used to determine the optimal VMME parameters for extracting syringoside and oleuropein from *S. oblata*. According to the results of the individual factor experiments, three different factors, liquid-solid ratio,

(a)   (b)

(c)

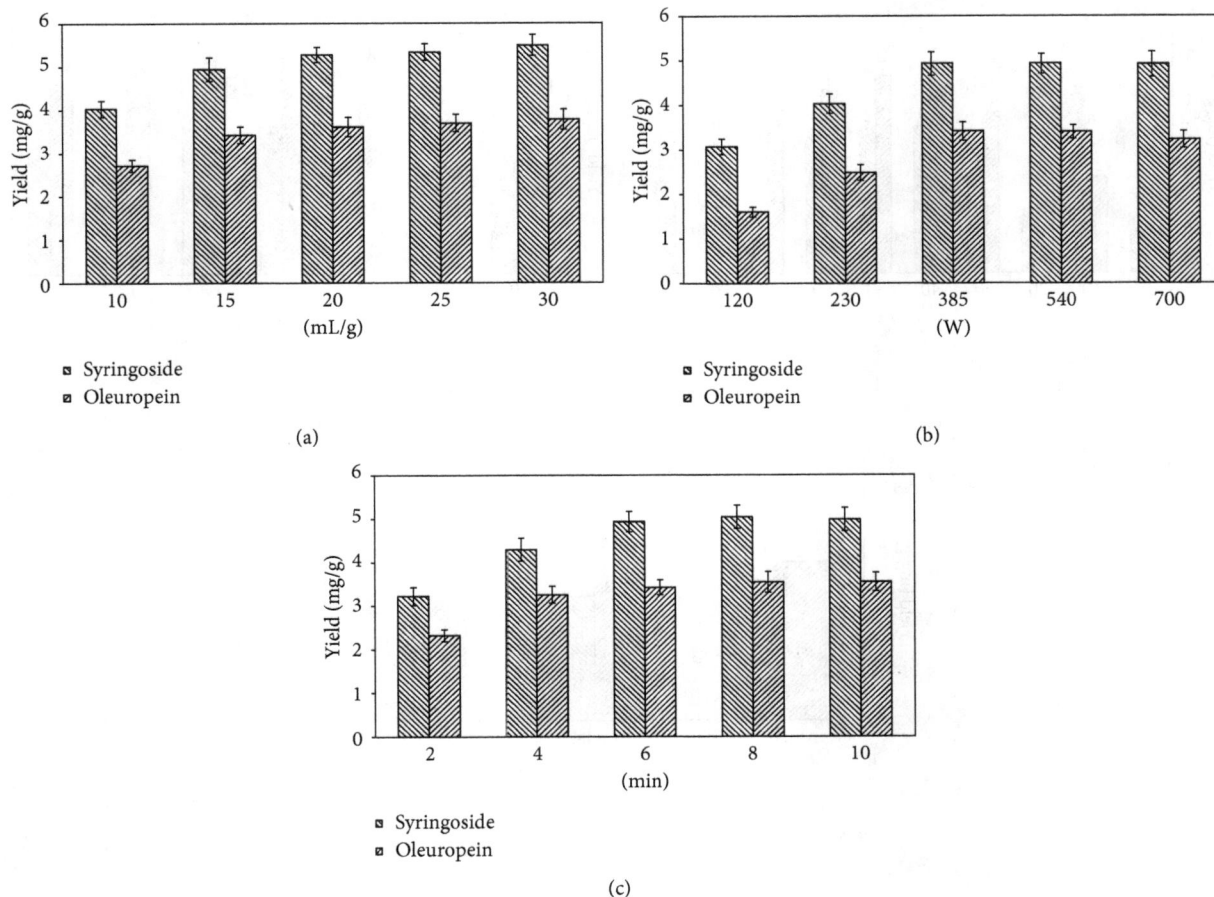

FIGURE 3: Effects of the liquid-solid ratio (a), microwave irradiation power (b), and microwave irradiation time (c) on the extraction yields of the target analytes. Error bars indicate standard deviation ($n = 3$).

microwave irradiation power, and time, were chosen for following optimizing experiments using BBD. The optimal VMME conditions for vacuum, microwave irradiation power, and soaking time in the individual factor experiments were found to be −0.08 MPa, 120 W, and 1 h, respectively, and these values were used in the BBD experiments, which comprised 17 experiments (Table 1). Predicted data were obtained by RSM from mathematical models.

Variance analysis of the RSM was conducted to analyze the significance and suitability of the mathematical models (Table 2). Regression models with P values less than 0.001 indicated that syringoside and oleuropein yields predicted by the two models were adequate. Moreover, P values of "lack of fit" were 0.4263 and 0.2684 for syringoside and oleuropein, respectively, which showed that the "lack of fit" was not significantly correlated with pure error due to statistical noise. There was a significant correlation between the predicted and actual yield for syringoside and oleuropein. The two regression models for syringoside and oleuropein yields, with R^2 values of 0.9885 and 0.9811, respectively, indicated that these models can explain 98.85% and 98.11% of the variation in the response. The following mathematical models were obtained:

$$Y_{\text{syringoside}} = -6.97 + 1.95X_1 + 1.05 \times 10^{-2}X_3$$
$$+ 1.42 \times 10^{-3}X_1X_3 - 1.95 \times 10^{-1}X_1^2 \qquad (2)$$
$$- 1.98 \times 10^{-5}X_3^2,$$

$$Y_{\text{oleuropein}} = 0.85 + 1.77 \times 10^{-1}X_1 + 7.99 \times 10^{-3}X_3$$
$$+ 1.62 \times 10^{-3}X_1X_3 - 7.30 \times 10^{-2}X_1^2 \qquad (3)$$
$$- 4.38 \times 10^{-4}X_2^2 - 1.77 \times 10^{-5}X_3.$$

To research the effects of the three parameters and their interactions on the yields of syringoside and oleuropein, 3D response surface graphs were drawn according to mathematical models (Equation (2) and Equation (3)). The effects of X_1 and X_2 on the extraction yields of two target compounds are shown in Figures 4(a) and 4(d), respectively, at a fixed X_3 of 385 W. At the early stage of extraction, both compounds increased in the extraction yield with extraction time; however, this was followed by slightly decreased extraction yields with further increases in extraction time. These results demonstrated that appropriate microwave irradiation time is necessary for completely extracting the target compounds. Increases in the liquid-solid ratio have

TABLE 2: Estimated regression coefficients for the quadratic polynomial model and ANOVA for the experimental results in the optimization of syringoside and oleuropein extractions[a].

Regression coefficients	Sum of squares		Degree of freedom	Mean square		F value		Probability >F	
	$Y_{oleuropein}$[b]	$Y_{syringoside}$		$Y_{oleuropein}$	$Y_{syringoside}$	$Y_{oleuropein}$	$Y_{syringoside}$	$Y_{oleuropein}$	$Y_{syringoside}$
Model	10.331	3.873	9	1.148	0.430	66.57	40.48	<0.0001***	<0.0001***
X_1[b]	3.125	0.083	1	3.125	0.083	181.24	7.79	<0.0001***	0.0269
X_2	0.085	0.034	1	0.085	0.034	4.95	3.18	0.0614	0.1178
X_3	1.835	1.514	1	1.835	1.514	106.43	142.38	<0.0001***	<0.0001***
X_1X_2	0.045	0.028	1	0.045	0.028	2.62	2.67	0.1494	0.1465
X_1X_3	0.777	1.009	1	0.777	1.009	45.05	94.89	0.0003***	<0.0001***
X_2X_3	0.005	0.018	1	0.005	0.018	0.30	1.66	0.6036	0.2383
X_1^2	2.569	0.359	1	2.569	0.359	148.96	33.75	<0.0001***	0.0007***
X_2^2	0.525	0.001	1	0.525	0.001	30.47	0.05	0.0009	0.8336
X_3^2	0.956	0.760	1	0.956	0.760	55.45	71.48	0.0001***	<0.0001***
Lack of fit	0.056	0.044	3	0.019	0.015	1.17	1.92	0.4263	0.2684

Credibility analysis of the regression equations	Index mark	Standard deviation	Mean	CV %	Press	R^2	Adjust R^2	Predicted R^2	Adequacy precision
	$Y_{oleuropein}$	0.13	4.82	2.73	1.00	0.9885	0.9736	0.9042	21.92
	$Y_{syringoside}$	0.10	3.43	3.00	0.75	0.9811	0.9569	0.8100	23.70

[a]The results were obtained with Design Expert 8.0.6 software; [b]X_1 is microwave irradiation time (min); X_2 is liquid-solid ratio (mL/g); X_3 is microwave irradiation power (W); $Y_{oleuropein}$ and $Y_{syringoside}$ are the yield of oleuropein and syringoside (mg/g); *$p < 0.05$, significant; **$p < 0.01$, highly significant; ***$p < 0.001$, extremely significant.

little effect on the extraction yield of oleuropein. However, for syringoside, the liquid-solid ratio had a positive linear correlation with the yield. A higher liquid to solid ratio is favorable for permeation of the solvent and discharge of target components due to the difference in the target compounds' concentration on either side of the cell membrane. Figures 4(b) and 4(e) illustrate the effect of X_1 and X_3 on the yields of syringoside and oleuropein with a fixed X_2 (15 mL/g). Here, a similar trend for syringoside and oleuropein yields was observed. The maximal syringoside and oleuropein yields were obtained at higher microwave irradiation power and longer microwave irradiation time. Figure 4(c) and 4(f) shows the interaction of X_2 and X_3 at a fixed X_1 (6 min). Microwave irradiation power had a greater effect on the extraction yields of syringoside and oleuropein than the liquid-solid ratio.

The optimal parameters in the extraction of syringoside and oleuropein by VMME predicted by BBD analysis were as follows: microwave irradiation time, 8 min; liquid-solid ratio, 17 mL/g; and microwave irradiation power, 524 W. Using these optimal conditions, the predicted yields of syringoside and oleuropein were 6.0 mg/g and 4.1 mg/g, respectively.

3.3. Verification.
Five extractions were conducted with the optimal conditions (microwave time, 8 min; liquid-solid ratio, 17 mL/g; and microwave power, 524 W) to evaluate the reliability of VMME for extracting syringoside and oleuropein. Using the optimal parameters described above, the actual experimental values of syringoside and oleuropein were 5.92 ± 0.24 mg/g and 4.02 ± 0.18 mg/g, respectively, which were consistent with the predicted values. Therefore, the optimal conditions obtained by the RSM method were reliable and reasonable.

3.4. Comparison of Different Extraction Procedures.
A comparison among VMME, MAE, and HRE is shown in Table 3 for the yields of syringoside and oleuropein from S. oblata twig samples. The extraction yields of analytes in VMME were higher and required a shorter extraction time compared to MAE and HRE using optimal conditions. The superiority of VMME is mainly due to the low pressure and temperature accomplished by introducing vacuum pressure in MAE, which decreases the degradation of thermosensitive and oxidizable compounds. Compared with conventional HRE, VMME and MAE are faster, while obtaining higher extraction yields for the target analytes. The effect of the microwave is to diffuse the inner components, change the internal microscopic structure, and cause the swelling of cells or the breakdown of cell walls when the temperature rises, which enhances the mass transfer of the cell contents [38, 39]. The comparison of VMME to MAE showed that the vacuum was important for increasing the extraction yields. These results are consistent with studies on VMME of vitamin C, vitamin E, and polyphenolic compounds [15, 16]. Therefore, we conclude that the VMME procedure provides a rapid and effective approach for the extraction of thermosensitive glycosides from the plant or other materials.

4. Conclusions

An effective VMME method was presented for extracting syringoside and oleuropein from S. oblata twigs. The optimal conditions for VMME were explored using an RSM method. With the optimal conditions, maximum extraction yields of the two glycosides were obtained. Relative to MAE and HRE, the VMME approach had significantly higher extraction yield. VMME exhibited good extraction efficiencies for extracting syringoside and oleuropein. This method avoids the high temperatures of conventional extraction methods,

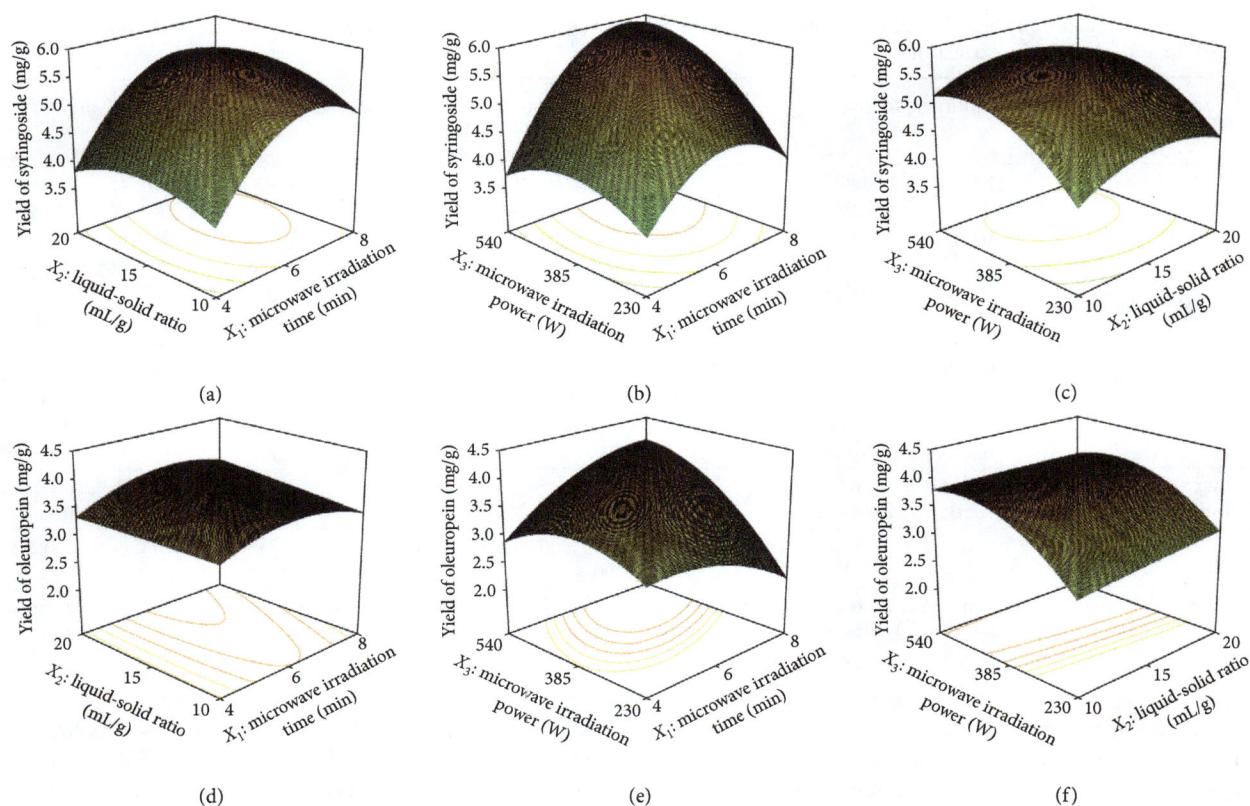

(a) (b) (c)

(d) (e) (f)

FIGURE 4: Response surface plots showing the effects of variables on yields of syringoside and oleuropein: (a) interaction between microwave irradiation time and liquid-solid ratio on the extraction yield of syringoside; (b) interaction between microwave irradiation time and microwave irradiation power on the extraction yield of syringoside; (c) interaction between liquid-solid ratio and microwave irradiation power on the extraction yield of syringoside; (d) interaction between microwave irradiation time and liquid-solid ratio on the extraction yield of oleuropein; (e) interaction between microwave irradiation time and microwave irradiation power on the extraction yield of oleuropein; (f) interaction between liquid-solid ratio and microwave irradiation power on the extraction yield of oleuropein.

TABLE 3: Comparison of extraction by different methods.

Method	Solvent	Soaking time (h)	Heating time consumption (min)	Yield (mean ± SD) (mg/g)	
				Syringoside	Oleuropein
VMME	40% volume fraction ethanol	1	6	5.92 ± 0.24	4.02 ± 0.18
MAE	40% volume fraction ethanol	1	6	3.95 ± 0.14	3.02 ± 0.09
HRE	40% volume fraction ethanol	1	60	2.52 ± 0.08	2.31 ± 0.08

VMME, vacuum microwave-mediated extraction; MAE, microwave-assisted extraction; HRE, hot reflux extraction.

which may also prove useful as a promising extraction method for other glycosides.

Conflicts of Interest

The authors declare that they have no conflicts of interest regarding the publication of this article.

Acknowledgments

This work was supported by the Program for Young Scholars with Creative Talents in Heilongjiang Bayi Agricultural University (no. CXRC2017006) and the National Natural Science Foundation of China (Grant no. NSFC31702169).

References

[1] F. Chen, K. Mo, Q. Zhang, S. Fei, Y. Zu, and L. Yang, "A novel approach for distillation of paeonol and simultaneous extraction of paeoniflorin by microwave irradiation using an

ionic liquid solution as the reaction medium," *Separation and Purification Technology*, vol. 183, pp. 73–82, 2017.

[2] Z. Liu, L. Qiao, H. Gu, F. Yang, and L. Yang, "Development of Brönsted acidic ionic liquid based microwave assisted method for simultaneous extraction of pectin and naringin from pomelo peels," *Separation and Purification Technology*, vol. 172, pp. 326–337, 2017.

[3] X. Li, F. Chen, S. Li, J. Jia, H. Gu, and L. Yang, "An efficient homogenate-microwave-assisted extraction of flavonols and anthocyanins from blackcurrant marc: optimization using combination of Plackett-Burman design and box-behnken design," *Industrial Crops and Products*, vol. 94, pp. 834–847, 2016.

[4] L. Zhao, H. Wang, H. Gu, and L. Yang, "Ionic liquid–lithium salt based microwave pretreatment followed by ultrasonic-assisted extraction of syringin and oleuropein from *Syringa reticulata* var. *mandshurica* branch bark by a dual response surface methodology," *Analytical Methods*, vol. 8, no. 7, pp. 1532–1542, 2016.

[5] Z. Liu, Z. Chen, F. Han, X. Kang, H. Gu, and L. Yang, "Microwave-assisted method for simultaneous hydrolysis and extraction in obtaining ellagic acid, gallic acid and essential oil from *Eucalyptus globulus* leaves using Brönsted acidic ionic liquid [HO$_3$S(CH$_2$)$_4$mim]HSO$_4$," *Industrial Crops and Products*, vol. 81, pp. 152–161, 2016.

[6] S. Li, F. Chen, J. Jia et al., "Ionic liquid-mediated microwave-assisted simultaneous extraction and distillation of gallic acid, ellagic acid and essential oil from the leaves of *Eucalyptus camaldulensis*," *Separation and Purification Technology*, vol. 168, pp. 8–18, 2016.

[7] F. Chen, X. Du, Y. Zu, L. Yang, and F. Wang, "Microwave-assisted method for distillation and dual extraction in obtaining essential oil, proanthocyanidins and poly-saccharides by one-pot process from cinnamomi cortex," *Separation and Purification Technology*, vol. 164, pp. 1–11, 2016.

[8] F. Chen, X. Zhang, Q. Zhang et al., "Simultaneous synergistic microwave–ultrasonic extraction and hydrolysis for preparation of *trans*-resveratrol in tree peony seed oil-extracted residues using imidazolium-based ionic liquid," *Industrial Crops and Products*, vol. 94, pp. 266–280, 2016.

[9] Z. Liu, H. Gu, and L. Yang, "An approach of ionic liquids/lithium salts based microwave irradiation pretreatment followed by ultrasound-microwave synergistic extraction for two coumarins preparation from cortex fraxini," *Journal of Chromatography A*, vol. 1417, pp. 8–20, 2015.

[10] H. Wang, X. Ma, Q. Cheng, X. Xi, and L. Zhang, "Deep eutectic solvent-based microwave-assisted extraction of baicalin from *Scutellaria baicalensis* Georgi," *Journal of Chemistry*, vol. 2018, Article ID 9579872, 10 pages, 2018.

[11] G. Huang, J. Shi, K. Zhang, and X. Huang, "Application of ionic liquids in the microwave-assisted extraction of pectin from lemon peels," *Journal of Analytical Methods in Chemistry*, vol. 2012, Article ID 302059, p. 8, 2012.

[12] A. Filly, X. Fernandez, M. Minuti, F. Visinoni, G. Cravotto, and F. Chemat, "Solvent-free microwave extraction of essential oil from aromatic herbs: from laboratory to pilot and industrial scale," *Food Chemistry*, vol. 150, pp. 193–198, 2014.

[13] A. Liazid, M. Palma, J. Brigui, and C. G. Barroso, "Investigation on ochratoxin a stability using different extraction techniques," *Talanta*, vol. 71, no. 2, pp. 976–980, 2007.

[14] A. Liazid, M. Palma, J. Brigui, and C. G. Barroso, "Investigation on phenolic compounds stability during microwave-assisted extraction," *Journal of Chromatography A*, vol. 1140, no. 1–2, pp. 29–34, 2007.

[15] J. X. Wang, X. H. Xiao, and G. K. Li, "Study of vacuum microwave-assisted extraction of polyphenolic compounds and pigment from Chinese herbs," *Journal of Chromatography A*, vol. 1198-1199, pp. 45–53, 2008.

[16] X. H. Xiao, J. X. Wang, G. Wang, J. Y. Wang, and G. K. Li, "Evaluation of vacuum microwave-assisted extraction technique for the extraction of antioxidants from plant samples," *Journal of Chromatography A*, vol. 1216, no. 51, pp. 8867–8873, 2009.

[17] Z. Huma, M. Abert-Vian, M. Elmaataoui, and F. Chemat, "A novel idea in food extraction field: study of vacuum microwave hydrodiffusion technique for by-products extraction," *Journal of Food Engineering*, vol. 105, no. 2, pp. 351–360, 2011.

[18] H. Gu, F. Chen, Q. Zhang, and J. Zang, "Application of ionic liquids in vacuum microwave-assisted extraction followed by macroporous resin isolation of three flavonoids rutin, hyperoside and hesperidin from *Sorbus tianschanica* leaves," *Journal of Chromatography B*, vol. 1014, pp. 45–55, 2016.

[19] B. Hiranvarachat, S. Devahastin, and S. Soponronnarit, "Comparative evaluation of atmospheric and vacuum microwave-assisted extraction of bioactive compounds from fresh and dried *Centella asiatica* L. leaves," *International Journal of Food Science and Technology*, vol. 50, no. 3, pp. 750–757, 2015.

[20] Y. Hu, Y. Li, Y. Zhang, G. Li, and Y. Chen, "Development of sample preparation method for auxin analysis in plants by vacuum microwave-assisted extraction combined with molecularly imprinted clean-up procedure," *Analytical and Bioanalytical Chemistry*, vol. 399, no. 10, pp. 3367–3374, 2011.

[21] X. Xiao, W. Song, J. Wang, and G. Li, "Microwave-assisted extraction performed in low temperature and *in vacuo* for the extraction of labile compounds in food samples," *Analytica Chimica Acta*, vol. 71, pp. 85–93, 2012.

[22] E. Wallander and V. A. Albert, "Phylogeny and classification of Oleaceae based on *rps*16 and *trnL-F* sequence data," *American Journal of Botany*, vol. 87, no. 12, pp. 1827–1841, 2000.

[23] H. X. Cui, G. M. Jiang, and S. Y. Zang, "The distribution, origin and evolution of *Syringa*," *Bulletin of Botanical Research*, vol. 24, pp. 141–145, 2004.

[24] N. Nenadis, J. Vervoort, S. Boeren, and M. Z. Tsimidou, "*Syringa oblata* Lindl var. *alba* as a source of oleuropein and related compounds," *Journal of the Science of Food and Agriculture*, vol. 87, no. 1, pp. 160–166, 2007.

[25] D. D. Wang, S. Q. Liu, Y. J. Chen, L. J. Wu, J. Y. Sun, and T. R. Zhu, "Studies on the active constituents of *Syringa oblata* Lindl," *Acta Pharmaceutica Sinica*, vol. 17, no. 12, pp. 951–954, 1982.

[26] X. Wei, S. Zhang, W. Zhai, and Z. Wang, "Determination of oleuropein in stem of *Syringa oblata* from different districts by HPLC," *China Journal of Chinese Materia Medica*, vol. 34, no. 3, pp. 304–306, 2009.

[27] A. Lim, N. Subhan, J. A. Jazayeri, G. John, T. Vanniasinkam, and H. K. Obied, "Plant phenols as antibiotic boosters: *in vitro* interaction of olive leaf phenols with ampicillin," *Phytotherapy Research*, vol. 30, no. 3, pp. 503–509, 2016.

[28] J. W. Choi, K. M. Shin, H. J. Park et al., "Anti-inflammatory and antinociceptive effects of sinapyl alcohol and its glucoside syringin," *Planta Medica*, vol. 70, no. 11, pp. 1027–1032, 2004.

[29] J. Y. Cho, K. H. Nam, A. R. Kim et al., "*In-vitro* and *in-vivo* immunomodulatory effects of syringin," *Journal of Pharmacy and Pharmacology*, vol. 53, no. 9, pp. 1287–1294, 2001.

[30] E. J. Yang, S. I. Kim, H. Y. Ku et al., "Syringin from stem bark of *Fraxinus rhynchophylla* protects Aβ$_{(25-35)}$-induced toxicity in neuronal cells," *Archives of Pharmacal Research*, vol. 33, no. 4, pp. 531–538, 2010.

[31] R. J. Zhang, J. K. Qian, G. H. Yang, B. Z. Wang, and X. L. Wen, "Medicinal protection with Chinese herb-compound against radiation damage," *Aviation, Space and Environmental Medicine*, vol. 61, no. 8, pp. 729–731, 1990.

[32] K. Y. Liu, Y. C. Wu, I. M. Liu, W. C. Yu, and J. T. Cheng, "Release of acetylcholine by syringin, an active principle of *Eleutherococcus senticosus*, to raise insulin secretion in Wistar rats," *Neuroscience Letters*, vol. 434, no. 2, pp. 195–199, 2008.

[33] H. S. Niu, F. L. Hsu, and I. M. Liu, "Role of sympathetic tone in the loss of syringin-induced plasma glucose lowering action in conscious Wistar rats," *Neuroscience Letters*, vol. 445, no. 1, pp. 113–116, 2008.

[34] S. H. Omar, "Oleuropein in olive and its pharmacological effects," *Scientia Pharmaceutica*, vol. 78, no. 2, pp. 133–154, 2010.

[35] S. Park, Y. Choi, S. J. Um, S. K. Yoon, and T. Park, "Oleuropein attenuates hepatic steatosis induced by high-fat diet in mice," *Journal of Hepatology*, vol. 54, no. 5, pp. 984–993, 2011.

[36] S. Wang, L. Yang, Y. Zu et al., "Design and performance evaluation of ionic liquids-microwave based environmental-friendly extraction technique for camptothecin and 10-hydroxycamptothecin from samara of *Camptotheca acuminate*," *Industrial & Engineering Chemistry Research*, vol. 50, no. 24, pp. 13620–13627, 2011.

[37] S. Hemwimon, P. Pavasant, and A. Shotipruk, "Microwave-assisted extraction of antioxidative anthraquinones from roots of *Morinda citrifolia*," *Separation and Purification Technology*, vol. 54, no. 1, pp. 44–50, 2007.

[38] L. Yang, X. Sun, F. Yang, C. Zhao, L. Zhang, and Y. Zu, "Application of ionic liquids in the microwave-assisted extraction of proanthocyanidins from *Larix gmelini* bark," *International Journal of Molecular Sciences*, vol. 13, no. 4, pp. 5163–5178, 2012.

[39] C. Ma, L. Yang, Y. Zu, and T. Liu, "Optimization of conditions of solvent-free microwave extraction and study on antioxidant capacity of essential oil from *Schisandra chinensis* (Turcz.) Baill," *Food Chemistry*, vol. 134, no. 4, pp. 2532–2539, 2012.

13

Headspace Solid-Phase Microextraction Coupled to Comprehensive Two-Dimensional Gas Chromatography Time-of-Flight Mass Spectrometry for the Determination of Short-Chain Chlorinated Paraffins in Water Samples

Nan Zhan, Feng Guo, Shuai Zhu, and Zhu Rao (iD)

The Key Laboratory of Eco-Geochemistry, Ministry of Nature Resources and National Research Center for Geoanalysis, Beijing 100037, China

Correspondence should be addressed to Zhu Rao; raozhu@126.com

Academic Editor: Verónica Pino

Short-chain chlorinated paraffins (SCCPs) are a new type of persistent organic pollutants. In this work, a simple and effective method involving headspace solid-phase microextraction (HS-SPME) and comprehensive two-dimensional gas chromatography time-of-flight mass spectrometry (GC × GC-TOF-MS) was developed and optimized for the determination of trace SCCPs in water samples. The key parameters related to extraction and separation efficiency were systematically optimized. The SCCP congener groups were best resolved using an Rxi-5Sil MS (30 m × 0.25 mm × 0.25 μm) column followed by an Rxi-17Sil MS (1.0 m × 0.15 mm × 0.15 μm) column; the optimum extraction conditions were achieved with a 100 μm polydimethylsiloxane SPME fiber, when a 10 mL water sample added with 3.6 g sodium chloride was incubated for 15 min at 90°C and then extracted during 60 min at 90°C and desorption at 260°C for 2 min. The proposed method showed good linearity in the concentration range of 0.2–20.0 μg/L with the determination coefficient greater than 0.995. The detection and quantification limits ranged from 0.06 to 0.13 μg/L and 0.18 to 0.40 μg/L, respectively, which are sufficient to meet the regulatory detection limits as set by most environmental regulations. The accuracy and precision of the method was also good, where the recoveries ranged from 82.5 to 95.4%, and intra- and interday precision was within 7.2% and 14.5%, respectively. The optimized method has been applied to the determination of SCCPs in ten freshwater samples of three different types.

1. Introduction

Short-chain chlorinated paraffins (SCCPs) are complex mixtures of polychlorinated *n*-alkanes with carbon chain lengths from 10 to 13 carbon atoms and with a chlorination degree in the range 30–70%. Because of their excellent physicochemical properties, SCCPs have been widely used in metalworking applications and polyvinyl chloride processing and used as plasticizers or flame retardants in paints, adhesives, sealants, leather fat liquor, lubricants, plastics, rubber textiles, and polymeric materials since the 1930s [1–3]. However, in the last two decades, SCCPs have been found to be toxic towards aquatic organisms and persistent in the environment and have high potential for bioaccumulation and long-distance

transport at relatively low concentrations. Since then, their production and use have been gradually restricted and included in some environmental laws and regulations. In Japan, SCCPs are listed in Pollutant Release and Transfer Register Law and are no longer manufactured [1, 2]; in Canada, SCCPs have been included in the "first Priority Substances List" under the Canadian Environmental Protection Act [1, 2, 4]; and in Europe, SCCPs are no longer permitted for sale or manufacture and have been listed as hazardous priority substances in the European Water Framework Directive [1, 2, 5]. Not long ago, SCCPs were officially included in the Annex A of Stockholm Convention [6] and Annex III of Rotterdam Convention [7], indicating that their production, use, and emission will soon be limited worldwide.

Among various environmental matrices, water is the most important environmental sink for SCCPs by either direct emission or from sewage treatment systems [1, 8]. In most cases, the presence of SCCPs in freshwater or seawater is reported at low concentrations (μg/L level or below) [1, 2]. Hence, suitable sample preparation methods and reliable instrumental analysis methods are much required for the determination of SCCPs in water environment.

Over recent years, liquid-liquid extraction (LLE) [8, 9] and solid-phase extraction (SPE) [10–12] are widely used sample preparation techniques for the preconcentration of SCCPs from water samples. However, LLE [8, 9] is time-consuming and requires large quantities of organic solvent; SPE [10–12] is susceptible to high baseline blanks, sorbent bed blocking, and poor reproducibility problems; and both methods cannot avoid the loss of the sample. The development of solid-phase microextraction (SPME) technique overcame the traditional sample preparation procedure, integrating extraction, concentration, and sample introduction in a single step, thus significantly saving time and labor [13]. Moreover, SPME only requires a few milliliters of water samples and does not require organic solvent, which is an added advantage. With the automation technology, the automated online SPME has realized in situ sample preparation with high efficiency and has been widely used for the analysis of volatile and semivolatile organic compounds in water samples [14–17]. However, the application of the SPME technique for extracting SCCPs from water samples is still limited and scarcely reported [10, 14, 17]. Gandolfi et al. [17] compared SPME in the direct immersion mode (DI-SPME) and headspace mode (HS-SPME) for the study of SCCPs in water samples and found that HS-SPME is preferable than DI-SPME, as it could reduce the extraction time, extend the service life of the extraction fiber, and prevent the matrix effect as well as keep the instrument system (e.g., GC column, MS) cleaner. Therefore, the HS-SPME method was selected in this study based on its numerous advantages.

The instrumental methods for the determination of SCCPs, just like many halogenated compounds, are mainly based on GC with different detectors, especially with low-resolution MS. Since SCCPs have thousands of isomers, enantiomers, and diastereomers in the mixtures, a single capillary GC column is insufficient to resolve all the congeners as individual peaks, always resulting a characterized "hump" peak in the chromatogram as unresolved complex mixture [18–23]. The state of art for SCCP analysis is the use of comprehensive two-dimensional gas chromatography (GC × GC) that significantly improves the peak capacity by using two GC columns with different retention mechanisms, allowing the SCCP congeners to be separated into ordered structures according to the carbon chain length and number of chlorine atoms on the chromatogram [24–26]. Moreover, with improved chromatographic separation, SCCPs can also be better separated from the interference compounds such as polychlorinated biphenyls and from complex matrices at the same time [25, 26]. Xia et al. demonstrated the feasibility of GC × GC for quickly screening and separating SCCPs in a fish sample, not only improving the detection and

separation of SCCPs but also enabling separation of the SCCP congeners from other coeluting halogenated organic compounds [25].

Currently, time-of-flight mass spectrometry (TOF-MS) is the most commonly used mass spectroscopy technique connected to GC × GC, whose fast acquisition rates allow proper reconstruction of 2D chromatogram by producing hundreds of spectra per second. The electron capture negative ionization mode (ECNI) has been extensively used for the quantification of SCCPs [18–23] because of its capability to differentiate the structural isomers of SCCPs by less fragmentation at low ionization energies, thus enhancing the selectivity of the method. However, in the ECNI mode, the response factors depend heavily on the number of chlorine atoms and their position in the carbon chain [1–3, 27], making the detection of the lower chlorinated SCCPs congeners (Cl < 5) difficult, resulting in quantitative errors [1, 3]. Contrary to ECNI, electron impact (EI) ionization is independent of the chlorine content and carbon chain length and thus can overcome the chlorination discrimination flaw; therefore, EI-MS can also detect lower chlorinated SCCPs (Cl < 5). This is important because lower chlorine components accounted for 10–62% mass of the total content in the environmental samples and are not negligible [28, 29]. The EI mode is known to lead to extensive fragmentation, making the mass spectra difficult to interpret. In other words, EI-MS cannot identify congener and homologue patterns; therefore, it is usually used for the detection of total concentration of SCCPs in the sample [1–3, 20].

At present, to the best of our knowledge, online HS-SPME coupled to GC × GC-TOF-MS for the determination of SCCPs in water samples has not been reported. Therefore, the aim of this study was to develop a HS-SPME GC × GC-TOF-MS analytical method for the determination of SCCPs in water samples. The key parameters related to extraction and separation efficiency were first carefully optimized. Then, the performance of the developed method was validated with respect to linearity, limit of detection (LOD), limit of quantification (LOQ), accuracy, and precision. Finally, the proposed method was applied to ten freshwater samples of three different types.

2. Materials and Methods

2.1. Chemicals and Materials. Commercial standards of SCCPs (C_{10-13}, 55.5% chlorination; $100 \, \mu$g/mL in cyclohexane), medium-chain chlorinated paraffins (MCCPs, C_{14-17}, 52% chlorination; $100 \, \mu$g/mL in cyclohexane), and long-chain chlorinated paraffins (LCCPs, C_{18-20}, 49% chlorination; $100 \, \mu$g/mL in cyclohexane) were purchased from Dr. Ehrenstorfer GmbH (Augsburg, Germany). $^{13}C_6$-hexachlorobenzene ($100 \, \mu$g/mL in nonane) used as the internal standard was purchased from Cambridge Isotope Laboratories (Andover, USA). Acetone and cyclohexane of pesticide residue analysis grade and sodium chloride (NaCl) of analytical grade were purchased from J&K Scientific Limit (Beijing, China). Before use, NaCl was purified in a furnace oven at 450°C for 5 h. Ultrapure water was obtained from a Milli-Q purification system (Millipore, Bedford MA, USA).

The stock solutions of SCCPs (C_{10-13}, 55.5% chlorination) were prepared in two different solvents, acetone and cyclohexane, at 10 μg/mL and stored at 4°C until use.

Three commercially available SPME fibers, 65 μm polydimethylsiloxane/divinylbenzene (PDMS/DVB), 85 μm polyacrylate (PA), and a 100 μm polydimethylsiloxane (PDMS), were purchased from Sepelco (Sigma Aldrich, USA). All these fibers were preconditioned in a GC injection port according to the manufacturer's recommendations prior to use. 20 mL headspace SPME vials, polytetrafluoroethylene septum, and crimp caps were obtained from GERSTEL (Mülheim, GER).

2.2. Water Samples and Sample Preparation. Ten water samples including three types of freshwater were collected at different sites in Beijing (China). Two groundwater samples were collected from a tap of a resident and a tap of our laboratory. Four lake water samples and four river water samples were collected from Bayi lake and Kunyu river at different sampling points on October 2017. All the samples were collected in 500 mL amber glass bottles and preserved at 4°C. Before analysis, all water samples were filtered through a 0.45 μm membrane filter (J&K, Beijing, China) to remove particulate matter.

2.3. HS-SPME Procedure. The HS-SPME procedure was performed in a 20 mL SPME vial containing a 10 mL water sample with 20 ng internal standard. The initial experimental conditions were based on the report by Gandolfi et al. [17] and are as follows: water sample was first stabilized in a thermal-static incubator for 10 min at 90°C; then, the SPME fiber was exposed in the headspace above the sample for 80 min at 90°C, followed by desorption on the GC injection port at 260°C for 2 min. After optimizing several key factors of the SPME step, the optimum extraction conditions were obtained: 10 mL water sample with the addition of 3.6 g NaCl was incubated at 90°C for 15 min and then extracted for 60 min at the same temperature, and finally, the fiber was desorbed in the GC injector at 260°C for 2 min. After desorption, fibers were reconditioned for 30 min at 260°C for removing the memory effect.

2.4. GC × GC-TOF-MS Conditions. The GC × GC-TOF-MS system was built from an Agilent 7890B GC (Agilent Technologies, Santa Clara, CA, USA) coupled to a Pegasus 4D TOF-MS (LECO, St. Joseph, MI, USA). The system was equipped with a multipurpose autosampler (GERSTEL, Mülheim, GER), including an SPME conditioning incubator and a temperature-controllable sample tray for 20 mL SPME vials. Liquid nitrogen, automatically filled from a Dewar using a liquid meter, was used to cool the nitrogen gas for cold pulses. Instrument control and data processing were performed by ChromaTOF software, version V4.51 (Leco, St. Joseph, MI, USA).

Based on the previous reports [24–26] and our preliminary experiment, a combination of nonpolar and midpolar column was found to be suitable for the analysis of

SCCPs in the GC × GC system. Therefore, in this study, the first-dimension and second-dimension columns were an Rxi-5Sil MS column (5% phenyl + 95% methylpolysiloxane, Restek) and an Rxi-17Sil MS column ((50%-diphenyl)-dimethyl polysiloxane, Restek, 1.0 m × 0.15 mm × 0.15 μm), respectively. The injection temperature was 260°C in the splitless mode, and helium (purity 99.999%) was used as the carrier gas. The optimum temperature program of the main oven was obtained from the experiments, comprising an initial temperature 100°C for 1 min, a ramp of 20°C/min to 150°C followed by a ramp of 3°C/min to 290°C and held for 1 min. The temperature of the secondary oven was programmed 5°C above the primary oven gradient. The modulator had a 15°C offset above the second oven. The TOF-MS was operated in the electron impact (EI) mode with ionization energy of 70 eV. Mass spectra were collected in the full-scan mode over the *m/z* range 50–500. Ion source and transfer line temperatures were set as 240 and 280°C, respectively. Other instrument parameters such as modulation period, gas flow rate, and MS acquisition rate were obtained from the results of optimized experiments, as described in Section 3.1.

2.5. Identification and Quantification. The identification of SCCPs in real samples was performed by comparing their elution area and retention times with that of standard solutions and by comparing their mass spectra with the reference mass spectra from the NIST (National Institute of Standards and Technology) library. A minimum similarity value of 750 was applied.

For quantification of SCCPs, the seven-point standard calibration curves were constructed at the 0.2, 0.5, 1, 2, 5, 10, and 20 ng/mL concentration levels by spiking the blank water samples with the appropriate amount of the stock solutions and 20 ng internal standard. Since EI-MS uses high energetic electrons to produce fragment ions and thus leads to extensive fragmentation, the molecular characteristic peaks are usually weak. Therefore, the quantification is based on the characteristic fragment ions of low mass-to-charge ratio values (*m/z*). Here, the fragment ion at *m/z* 91 and *m/z* 187 was selected as the quantitative ion for SCCPs and internal standard, respectively. The LODs and LOQs of the method were determined as three and ten times the SD of the average values of seven blank water samples, respectively. The accuracy of this method was tested by the recovery studies of blank water samples spiked with 1.0, 5.0, and 10.0 μg/L SCCP standards in five replicates. The recoveries are calculated by comparing the peak area ratio obtained with a 100 μm PDMS fiber in DI-SPME, which were the optimal conditions established by Gandolfi et al. [17]. The precision of the method was evaluated by intraday and interday reproducibility experiments by analyzing the spiked water sample (10.0 μg/L) for five times on the same day and daily for five times over three different days.

3. Results and Discussion

3.1. Optimization of GC × GC-EI-TOF-MS. The GC × GC-TOF-MS conditions were optimized to achieve good separation of SCCP congeners, by analyzing a SCCP standard solution (C_{10-13}, 55.5% chlorination in cyclohexane) at 10 μg/mL in the splitless mode. Each experiment was performed in triplicate.

Since the column dimension could essentially affect the retention behavior of target compounds (i.e., SCCPs) on GC, the column length and film thickness of the primary column were first optimized. Three Rxi-5Sil MS columns were studied with a certain second column Rxi-17Sil MS (1.0 m × 0.15 mm × 0.15 μm). Accordingly, the modulation period was adjusted to ensure all the SCCP congeners eluting in one modulation period. The results revealed that SCCPs can be resolved into an ordered structure on the 2D chromatogram by all three columns, where the SCCPs are arranged by boiling points on the x-axis and by polarity on the y-axis, far better than that obtained by 1D GC. And this typical "tile-structure" of SCCPs helps them to be identified quickly in the sample. As can be seen from Figure 1, a narrow-bore column (30 m long and 0.15 μm film thickness) and a short narrow-bore column (15 m long and 0.15 μm film thickness) decreased the retention of target compounds on GC and thus accelerated their elution; therefore, a relatively long modulation period (e.g., 6 s or 7 s) was required to avoid the wraparound effect. But using such a long modulation period, the separation efficiency achieved on the primary column decreased, and the peak shape and peak intensity deteriorated too (Figures 1(b) and 1(c)). However, a regular size column (30 m long and 0.25 μm film thickness) could separate SCCPs well in a quite short modulation period such as 3 s (Figure 1(a)). In addition, regarding the anti-interference ability, a thin and/or short GC column is more susceptible to matrix interference than one of the regular dimension columns, especially for matrix-rich environmental samples such as wastewater sample, while a regular dimension column is more adaptable and robust for analyzing various water samples. Therefore, a 30 m long, 0.25 mm internal diameter, and 0.25 μm film thickness Rxi-5Sil MS column was selected in this study.

The GC oven temperature program was then optimized for the optimum extent of the SCCP congeners. In this experiment, heating rates from 2 to 4°C/min were evaluated in the main oven temperature programs. The initial temperature program used is as follows: held for 1 min at 100°C, increased to 150°C at 20°C/min, and finally ramped to 290°C at X°C/min (X = 2, 3, and 4) and held for 1 min. A heating rate of 2°C/min (Figure 2(a)) provided the best separation of SCCP congeners but caused a "wraparound" for some high-boiling-point SCCPs. When the heating rate was increased to 3°C/min (Figure 2(b)), all SCCP congeners eluted in one modulation period with good separation, together with providing a more efficient laboratory output in a relatively short run time (51.5 min) and reducing the liquid nitrogen consumption. At a higher heating rate such as 4°C/min, some neighboring groups of SCCPs overlapped and coeluted, leading to worse separation (Figure 2(c)). Therefore, 3°C/min was selected as the optimum heating rate. For this, the initial temperature of the first oven was programmed at 100°C and held for 1 min, then ramped to 150°C at 20°C/min, and further increased to 290°C at 3°C/min and held for 1 min.

The modulation period, an another key factor in the GC × GC system, needs to preserve the 1D separation and elute all the compounds from two columns with the good peak shape; therefore, a relatively short modulation period is usually preferable. In this experiment, three different modulation periods (2, 3, and 4 s) with 20% hot pulse duration were investigated. At a modulation period of 2 s (Figure 3(a)), most peaks had good shapes, but wraparound was observed for a few less-volatile SCCPs, suggesting that the current modulation period was too short to ensure the injection of all the congeners into the second column. When the modulation period was extended to 3 or 4 s, all the SCCP congeners flew out in one modulation cycle; a modulation period of 3 s (Figure 3(b)) provided better separation and peak shapes than that obtained at 4 s (Figure 3(c)), indicating most peaks with appropriate modulations at a 3 s modulation. Thus, a modulation period of 3 s and hot and cold pulses of 0.6 s and 0.9 s, respectively, were applied in this study.

Finally, the carrier gas flow rate and MS acquisition rate were investigated. Since the two GC columns of different dimensions are linked in series, the same flow rate will affect the separation in both columns differently. To obtain a suitable gas flow rate for both the dimensions, the flow rate ranging from 1.0 to 1.6 mL/min was tested. The results showed that a lower flow rate could slow down the elution of the sample components and improve the separation of neighboring SCCP congener groups. Moreover, a lower flow rate required a relatively low head pressure, even at high temperature, which was a benefit to the GC × GC system. Therefore, a flow rate of 1.0 mL/min was chosen in this study. The MS acquisition rate was based on the peak width of the target compounds. Since most SCCP congeners were extremely narrow of only approximately 0.1-0.2 s wide at the base, corresponding to the acquisition rate of 50–100 Hz, 100 Hz was therefore selected as the MS scan rate in this study.

Under the abovementioned optimized conditions, the separation of SCCP congener groups was achieved. According to the distribution rules of SCCPs [24–26], the SCCPs elution patterns are identified into 9 parallel peak groups (polygons) on the 2D contour plot (Figure 4(a)), where the congeners in each polygon have similar physicochemical properties, retention times, and the same total number of carbon and chlorine atoms. This classification not only makes the identification of SCCPs simple and intuitive but also facilitates the quantification, as the polygon areas also represent the quantitative area for SCCPs. Figure 4(b) shows a mass spectrum of a SCCP congener $C_{10}H_{17}Cl_5$, where the major fragment ions at m/z 67, 75, 91, and 103 have good relative abundance while the abundance of the characteristic ion $[M-Cl]^+$ (m/z 278) is very low. Although the fragment ion at m/z 91 was not the most abundant ion, this pentacyclic positive ion $[C_4H_8Cl]^+$ was considered specific as it was only generated by the rearrangement of polychlorinated n-alkanes during the mass fracture process, and thus, it was selected as the quantitative ion. The other fragment ions at m/z 67, 75, and 103, although they have higher abundance than $[C_4H_8Cl]^+$, can also be produced by other chlorinated compounds, so they were not selected as quantitative ions.

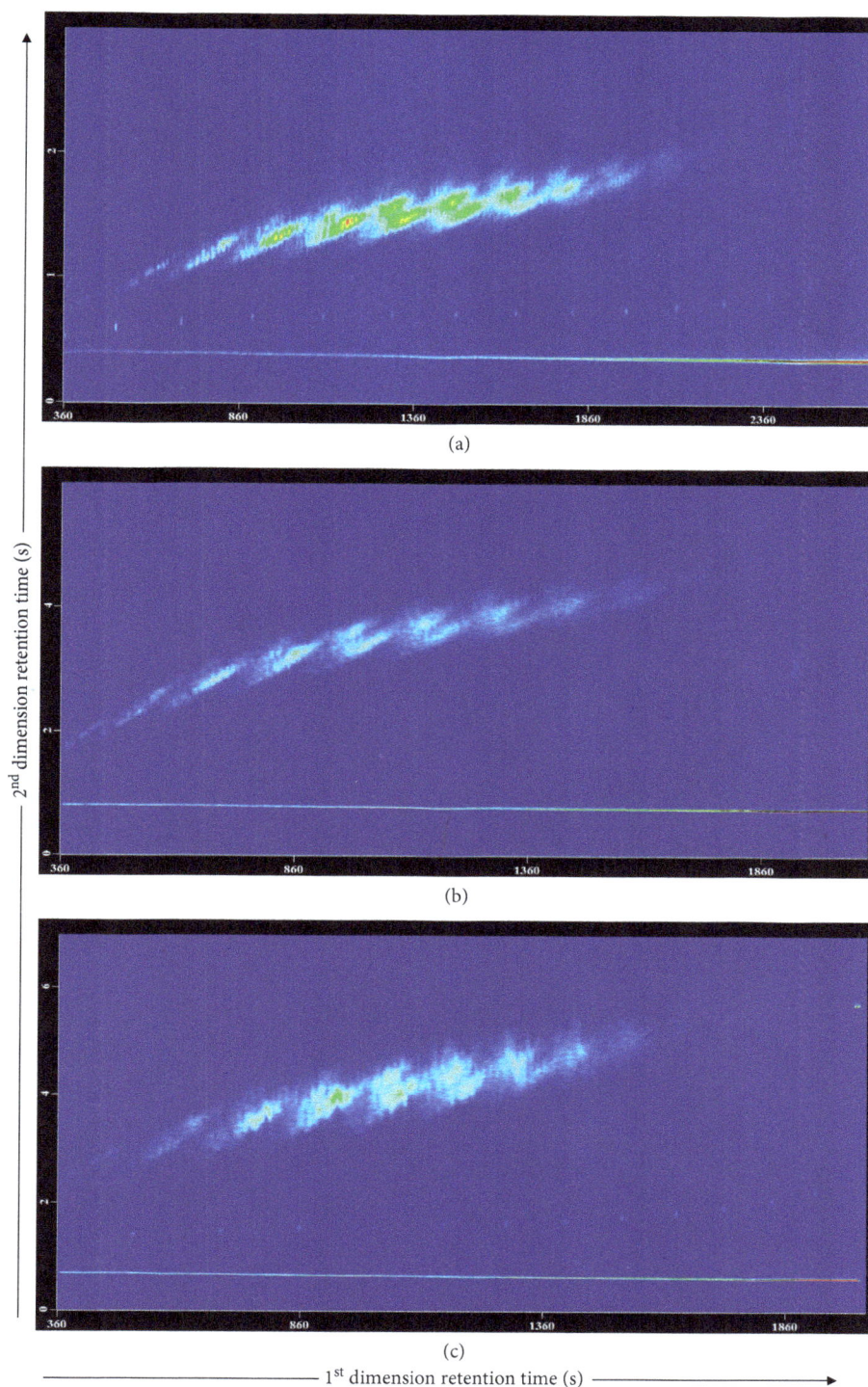

FIGURE 1: The effect of column length and film thickness of the primary column Rxi-5Sil MS on the separation performance of SCCP congeners. GC × GC-TOF-MS chromatograms of a mixed SCCP standard solution (C_{10-13}, containing 55.5% chlorine), acquired using (a) 30 m × 0.25 mm × 0.25 μm, (b) 15 m × 0.25 mm × 0.15 μm, and (c) 30 m × 0.25 mm × 0.15 μm Rxi-5Sil MS columns.

3.2. Fiber Selection and Optimization of HS-SPME Procedure. Initially, three different SPME fibers, including one nonpolar (100 μm PDMS), one semipolar (65 μm PDMS/DBV), and one highly polar (85 μm PA) fibers, were investigated in terms of extraction capacity and reproducibility. Here, the

extraction capacity was assessed by the peak area ratio, which was calculated by summing all the peak areas of SCCPs and dividing by the internal standard peak area; the reproducibility of the fibers was evaluated by the RSD values of peak area ratios. Each experiment was performed in triplicate

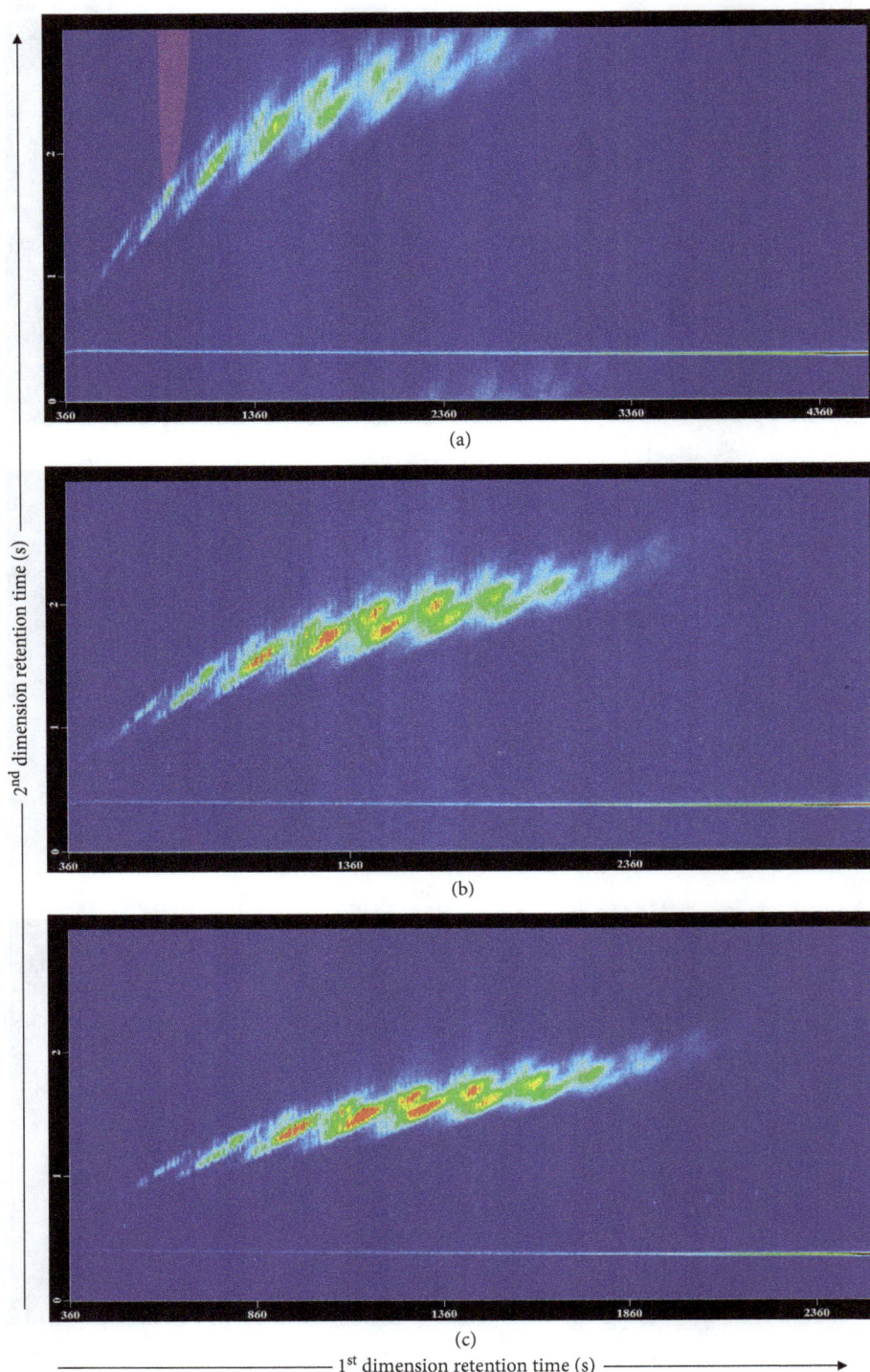

FIGURE 2: The effect of the heating rate on the separation performance of SCCP congeners. Temperature program of the primary oven: 100°C (1 min), at 20°C/min to 150°C, and then at X°C/min to 290°C (1 min). X = (a) 2°C/min, (b) 3°C/min, and (c) 4°C/min.

with 10 mL blank water spiked at 10.0 μg/L with standard mixture of SCCPs (C_{10-13}, 55.5% chlorination in acetone).

As can be seen from Table 1, 100 μm PDMS fiber not only provided the highest sensitivity to SCCPs but also offered the best reproducibility, whereas PA and PDMS/DBV fibers both showed lower sensitivity and poorer

repeatability. These observations were basically consistent with the previous study [10, 13, 17]; therefore, 100 μm PDMS fiber was chosen as the most appropriate fiber for SCCPs and used in the subsequent experiments.

After the selection of the SPME fiber, four key parameters of the extraction step were then optimized [13, 14, 17].

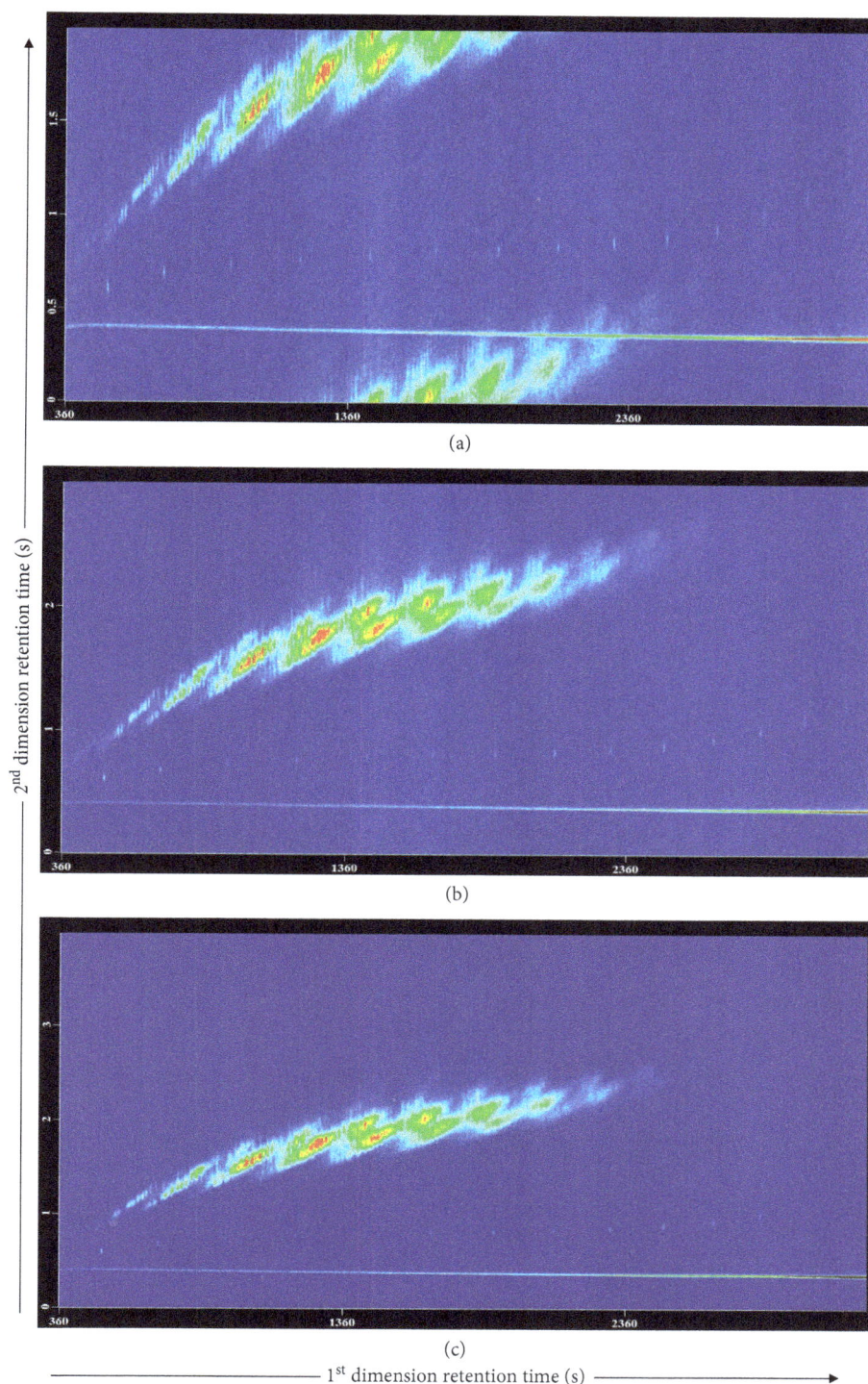

FIGURE 3: The effect of the modulation period on the separation performance of SCCP congeners: modulation period = (a) 2 s, (b) 3 s, and (c) 4 s with 20% hot pulse duration.

First, the incubation times were evaluated between 2 and 25 min. Figure 5(a) shows that that increasing incubation time leads to an increase in SCCPs extraction from 2 to 15 min, but when the incubation time exceeded 15 min, the extraction efficiency was slightly affected. Besides, considering a shorter extraction time would enhance the experimental throughput, and 15 min was chosen as the ideal extraction time.

Extraction temperature plays an important role in the HS-SPME step but in two opposite ways: increasing temperature not only increases the analytes content in the gas phase but also accelerates the desorption of adsorbates on

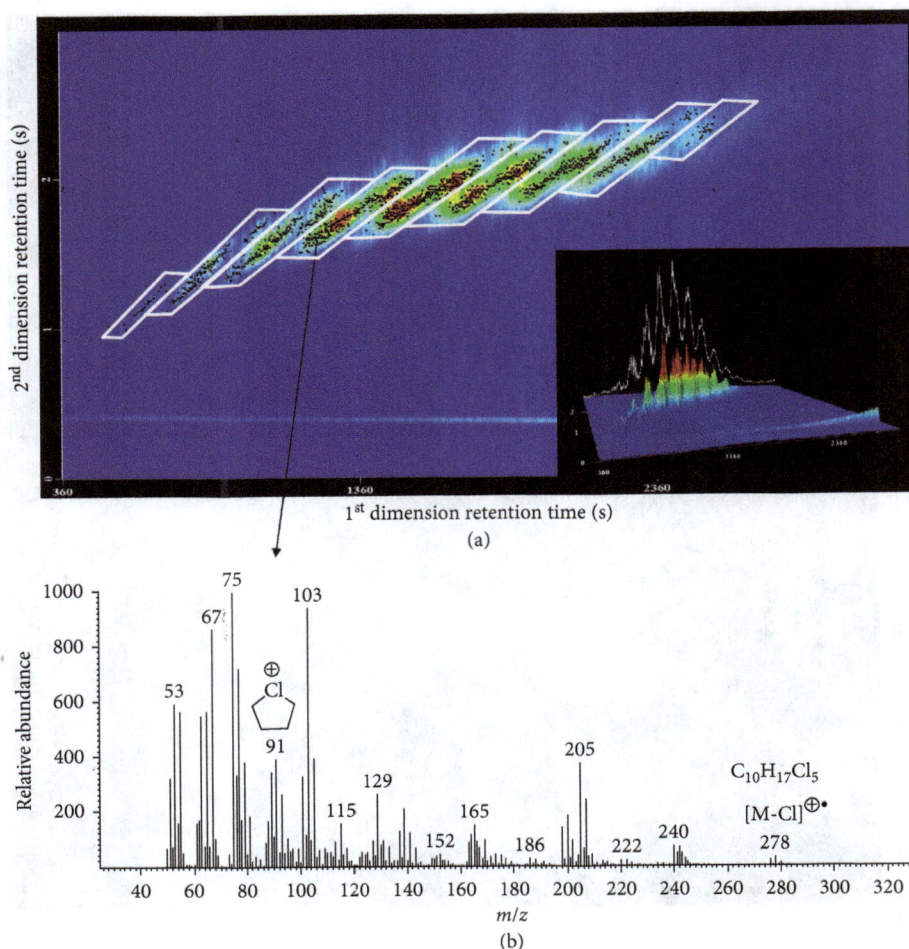

FIGURE 4: (a) GC × GC-TOF-MS chromatogram and elution pattern of a mixed SCCPs standard solution (C_{10-13}, 55.5% chlorine) obtained on Rxi-5Sil MS × Rxi-17Sil MS column combination. Lower right: 3D contour plot. (b) Mass spectrum of a selected SCCP congener $C_{10}H_{17}Cl_5$, where the fragment ion m/z 91 is the quantitative ion.

TABLE 1: Extraction efficiency and reproducibility of three studied SPME fibers for extracting SCCPs.

Fiber type	Peak area ratio				RSD
	Test 1	Test 2	Test 3	Average value	(%, $n = 3$)
100 μm PDMS	22.39	24.16	21.45	22.66	6.06
65 μm PDMS/DBV	17.43	15.63	17.62	16.89	6.51
85 μm PA	17.46	16.45	13.94	15.29	8.30

the fiber at the same time. In this experiment, extraction temperature was studied at 50, 60, 70, 80, 90, and 95°C with a fixed extraction time of 80 min. As can be seen from Figure 5(b), the extraction efficiency first increased with increasing temperature and reached maximum at 90°C and then dropped at 95°C. This is probably because volatilization was the main controlling factor from 50 to 90°C, while desorption became dominant when temperature reached above 90°C. Given the above, 90°C was chosen as the optimum extraction temperature.

Extraction time was further evaluated from 40 to 100 min with an interval of 10 min. In Figure 5(c), the

chromatographic response showed a fast increase from 40 to 60 min and then remained almost unchanged afterwards, which means the SCCPs reached an extraction equilibrium from 60 min. Continuous increasing extraction time did not improve the extraction efficiency but reduces the experimental throughput and increases matrix interference [13], and thus, 60 min was chosen as the optimum extraction time.

Ionic strength, i.e., salt concentration, is also important for the analyte extraction in HS-SPME. Increasing ionic strength usually enhances the extraction efficiency of hydrophobic compounds through salting out phenomenon [13]. In this experiment, quantities of NaCl were added to reach 0, 0.1, 0.2, 0.3, and 0.36 (saturation concentration at room temperature) g/mL NaCl solutions. As shown in Figure 5(d), increasing salt concentration increases extraction and the best values are those obtained at the saturated concentration (i.e., 0.36 g/mL). Therefore, 0.36 g/mL of NaCl was chosen. It is worth noting that this result is a supplement to the result of Gandolfi et al. [17], who did not observe any obvious salting-out effect in the salt concentration range between 0 and 0.035 g/mL. However, in a wider salinity

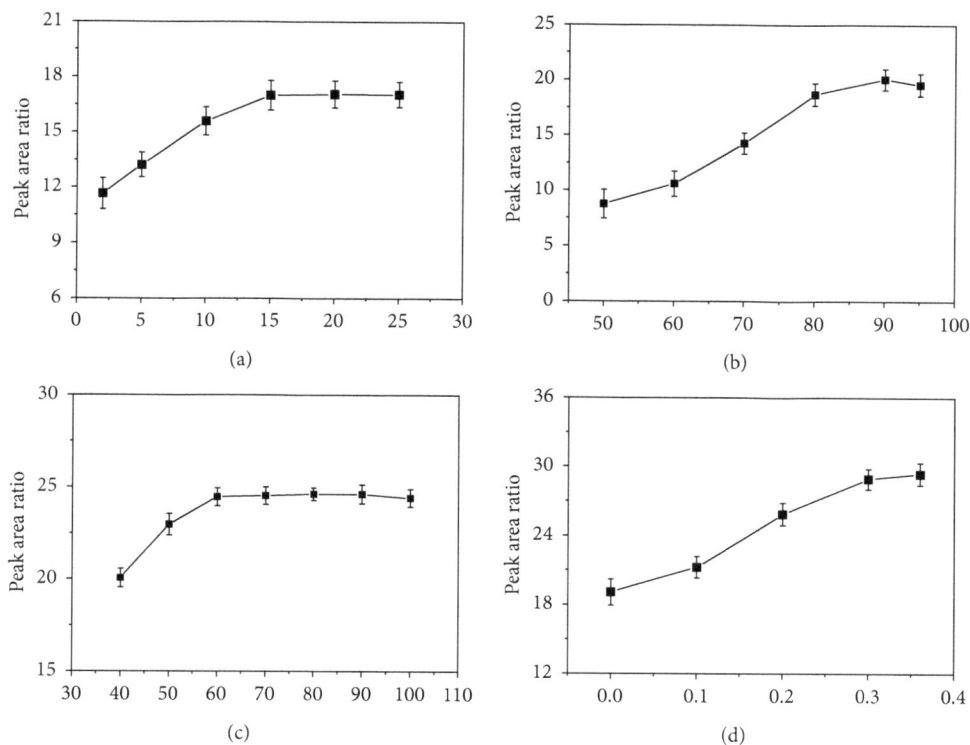

FIGURE 5: Influence of (a) incubation time, (b) extraction temperature, (c) extraction time, and (d) salt concentration on the HS-SPME extraction efficiency of SCCPs by a $100\,\mu$m PDMS fiber.

ranges, the salting-out phenomenon was observed and had a positive effect on the extraction efficiency.

In summary, the best HS-SPME extraction conditions are as follows: 10 mL water sample with the addition of 3.6 g NaCl was stabilized at 90°C for 15 min, then extracted for 60 min at the same temperature, and finally desorbed at 260°C for 2 min. The overall extraction time for each sample was approximately 80 min, which is less time-consuming compared to the traditional methods using LLE (120 min) [9] or SPE (130 min) [10].

3.3. *Influence of MCCPs and LCCPs.* Since MCCPs and LCCPs may also be present in the sample of SCCPs and their presence may interfere with SCCPs analysis, it is necessary to study their influence on SCCPs. To study their effect, 10 mL Milli-Q water spiked with 10 μg/L mixed CPs standard solutions (C_{10-13}, 55.5% chlorination; C_{14-17}, 52% chlorination; and C_{18-20}, 49% chlorination in acetone) were submitted to the analysis under the optimized conditions.

As can be seen from Figure 6, in the sample containing three kinds of CPs, only SCCPs show obvious signals, while MCCPs merely have weak signals and LCCPs have no signals. Although the three types of CPs had the same concentration, MCCPs and LCCPs were hardly detected under the experimental condition. This probably because SPME only extracts volatile or semivolatile components but cannot extract the less volatile components—that is, MCCPs and LCCPs cannot be enriched on the SPME fiber, so they would not interfere with the extraction of SCCPs in the initial extraction step.

Moreover, the samples were compared with those containing only SCCPs (10 μg/L, C_{10-13}, 55.5% chlorination in acetone). Results showed that the total peak areas of SCCPs in two kinds of samples were basically same, and the RSD values of SCCPs peak areas were less than 0.9%, indicating that the presence of MCCPs and LCCPs would not interfere with the subsequent quantification of SCCPs.

3.4. *Method Validation.* The quality parameters of the proposed method were then validated with respect to the linearity, sensitivity, accuracy, and precision to Milli-Q water, lake water, and river water samples. Calibration curves were elaborated using matrix blanks over the range of 0.2–20 μg/L. The water samples used as matrix blanks were also collected from the same lake and river, merely from which SCCPs were not detected by the proposed method. Table 2 summarizes the results.

Matrix-matched calibration curves exhibited good linearity within the calibration range, and all the correlation coefficients R^2 were greater than 0.995. It is worth noting that the linearity range of this method is relatively narrow, mainly because the enrichment ability of the SPME fiber is limited. The strong enrichment ability of the SPME fiber not only pulls down the lowest point of the linearity range but also takes down the highest point of the linearity range.

The LODs and LOQs ranged from 0.06 to 0.13 μg/L and 0.18–0.40 μg/L, respectively. The obtained LODs were better than the one reported when using DI-SPME with GC-NCI-MS (LOD = 0.1 μg/L for Milli-Q water) [10, 14], but worse

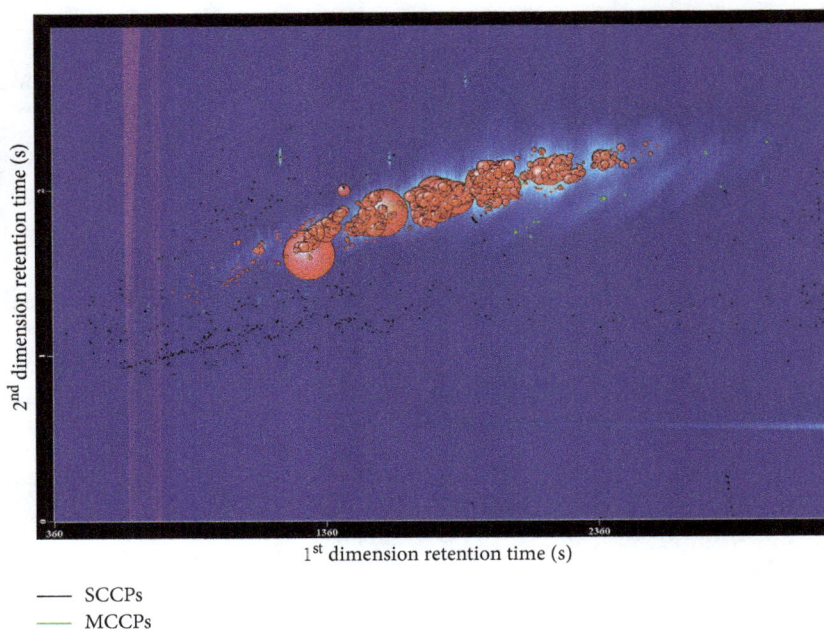

—— SCCPs
—— MCCPs

FIGURE 6: Influence of MCCP and LCCPs on the analysis of SCCPs. GC×GC-TOF-MS bubble chart obtained from a 10 μg/L mixed CPs standard solution (C$_{10-13}$, 55.5% chlorine; C$_{14-17}$, 52% chlorine; C$_{18-20}$, 49% chlorine) under the optimized conditions.

TABLE 2: Analytical features of the proposed HS-SPME GC×GC-TOF-MS method for determination of SCCPs in water samples.

Water type	Linear range (μg/L)	R^2	LOD (μg/L)	LOQ (μg/L)	Recovery (%, $n = 5$)			Precision (RSD%, $n = 5$)	
					1 μg/L	5 μg/L	10 μg/L	Intraday	Interday
Milli-Q water	0.2–20.0	0.999	0.06	0.18	90.1	97.3	104	3.6	6.7
Lake water	0.2–20.0	0.996	0.10	0.31	85.4	95.8	92.6	5.8	12
River water	0.2–20.0	0.995	0.13	0.40	82.5	84.9	88.5	7.2	15

than the one obtained from HS-SPME with GC-ECNI-MS (LOD = 0.004 for Milli-Q water) [17], mainly because of less sensitivity of EI-MS than ENCI-MS. Fortunately, the quality standards of SCCPs in most environmental regulations are at the level of μg/L, such as 2.4 μg/L for water in the Canadian Environmental Protection Act [4] and 0.4 μg/L (annual average quality standard) for surface water in European Water Framework Directive [5], so the sensitivity of our method is sufficient to fulfil these criteria for the detection of SCCPs at low levels in water samples.

The accuracy of the method was evaluated by recovery experiments using the spiked sample at three levels (1, 5, and 10 μg/L) in five replicates. The recoveries ranged from 82.5% to 104.2% in three types of water samples, demonstrating the good accuracy of the method.

The method precision, expressed as the RSDs of the peak areas of SCCPs at 10 μg/L, was also satisfactory, with the RSD values less than 7.2% and 15% for the intra- and interday experiments, respectively.

3.5. Analysis of Real Samples. To confirm the applicability of this method, the proposed method was applied to ten real water samples, including two underground water samples,

four lake water samples, and four river water samples. All samples were measured in duplicate. Table 3 summarizes the average concentration of total SCCPs in the investigated water samples, and the concentration are shown as the mean ± SD. SCCPs were not detected in two groundwater water samples but detected in the lake samples (0.91–1.52 μg/L) and river samples (2.98–5.07 μg/L). The concentration of SCCPs in the detected river and lake samples (#3–#10) was generally greater than the values reported in the freshwater of Britain [30], Spain [14], Japan [31], and Canada [32]. Two lake samples (#3–#4) and four river samples (#7–#10) exceeded the maximum acceptable concentration quality standards of European Water Framework Directive (1.4 μg/L) [5], probably because the sampling site was in the urban areas and the stock of SCCPs in China water environment was heavier than in these countries [1, 2, 33]. Considering that the mobility of SCCPs may threaten groundwater, even the trace amount of SCCPs in the water environment should be given more attention.

4. Conclusions

The present study is the first report which propose a green, sensitive, and reliable method for the determination of

TABLE 3: Occurrence and concentration levels of SCCPs in different water samples.

Sample code	Water type	Concentration (μg/L) ± SD
#1	Groundwater	n.d.
#2	Groundwater	n.d.
#3	Lake water	1.52 ± 0.08
#4	Lake water	1.40 ± 0.05
#5	Lake water	0.91 ± 0.03
#6	Lake water	1.05 ± 0.03
#7	River water	4.1 ± 0.2
#8	River water	2.98 ± 0.08
#9	River water	5.1 ± 0.1
#10	River water	3.36 ± 0.04

n.d., not detected (below the method detection limit).

SCCPs in water samples by online HS-SPME and GC × GC-TOF-MS. Compared to the existing analytical methods, this method not only simplifies the operation process saving both time and labor but also yields good accuracy, precision, and sensitivity. The LODs (0.06–0.13 μg/L) and LOQs (0.18–0.40 μg/L) were relatively low and could meet the requirement of quality detection on SCCPs in water samples as mentioned in most international conventions and regulations. Besides, the interference from MCCPs and LCCPs can be excluded in the initial extraction step because these compounds cannot be enriched on the SPME fiber. Based on these observations, we believe that this method can be used for the routine analysis of SCCPs in water samples.

Conflicts of Interest

The authors declare that there are no conflicts of interest regarding the publication of this paper.

Acknowledgments

The authors would like to acknowledge the engineers from Leco Corporation for their kind support in this study. Comments from anonymous reviewers are deeply appreciated. This work was funded by the China National Research Center for Geoanalysis (no. 20020123500170042512), China Geological Survey (no. 12120110500017250102 and no. 12120110500017270106), and China National Science Foundation (no. 21507017).

References

[1] S. Bayen, O. J. P. Obbard, and G. O. Thomas, "Chlorinated paraffins: a review of analysis and environmental occurrence," *Environment International*, vol. 32, no. 7, pp. 915–929, 2006.

[2] L. M. van Mourik, P. E. G. Leonards, C. Gaus, and J. de Boer, "Recent developments in capabilities for analysing chlorinated paraffins in environmental matrices: a review," *Chemosphere*, vol. 136, pp. 259–272, 2015.

[3] L. M. van Mourik, C. Gaus, P. E. G. Leonards, and J. de Boer, "Chlorinated paraffins in the environment: a review on their production, fate, levels and trends between 2010 and 2015," *Chemosphere*, vol. 155, pp. 415–428, 2016.

[4] Canadian Environmental Protection Act, 1999 Federal Environmental Quality Guidelines Chlorinated Alkanes, May 2018, http://www.ec.gc.ca/ese-ees/C4148C43-C35E-44EA-87A7-866E5907C42C/FEQG_Chlorinated%20Alkanes_EN.pdf.

[5] Directive 2000/60/EC of the European Parliament and of the Council Establishing a Framework for the Community Action in the Field of Water Policy.

[6] UNEP/POPS/POPRC.12/11/Add.3, *United Nations Environmental Programme*, Stockholm Convention on Persistent Organic Pollutants, Rome, Italy, 2016.

[7] UNEP/FAO/RC/COP.8/12/Add.1, *Rotterdam Convention on the Prior Informed Consent Procedure for Certain Hazardous Chemicals and Pesticides*, International Trade, Geneva, Switzerland, 2017.

[8] L. X. Zeng, H. J. Li, T. Wang et al., "Behavior, fate, and mass loading of short chain chlorinated paraffins in an advanced municipal sewage treatment plant," *Environmental Science and Technology*, vol. 47, no. 2, pp. 732–740, 2013.

[9] S. Geiß, J. W. Einax, and S. P. Scott, "Determination of the sum of short chain polychlorinated n-alkanes with a chlorine content of between 49 and 67% in water by GC-ECNI-MS and quantification by multiple linear regression," *CLEAN—Soil, Air, Water*, vol. 38, no. 1, pp. 57–76, 2010.

[10] P. Castells, F. J. Santos, and M. T. Galceran, "Solid phase extraction versus solid phase microextraction for the determination of chlorinated paraffins in water using gas chromatography–negative chemical ionisation mass spectrometry," *Journal of Chromatography of A*, vol. 1025, no. 2, pp. 157–162, 2004.

[11] M. Coelhan, "Levels of chlorinated paraffins in water," *CLEAN—Soil, Air, Water*, vol. 38, no. 5-6, pp. 452–456, 2010.

[12] X. D. Ma, C. Chen, H. J. Zhang et al., "Congener-specific distribution and bioaccumulation of short-chain chlorinated paraffins in sediments and bivalves of the Bohai Sea, China," *Marine Pollution Bulletin*, vol. 79, no. 1-2, pp. 299–304, 2014.

[13] J. Płotka-Wasylka, N. Szczepańska, M. de la Guardia, and J. Namieśnik, "Miniaturized solid phase extraction techniques," *TrAC-Trends in Analytical Chemistry*, vol. 73, pp. 19–38, 2015.

[14] P. Castells, F. J. Santos, and M. T. Galceran, "Solid phase microextraction for the analysis of short-chain chlorinated paraffins in water samples," *Journal of Chromatography A*, vol. 984, no. 1, pp. 1–8, 2003.

[15] E. Passeport, A. Guenne, T. Culhaoglu et al., "Design of experiments and detailed uncertainty analysis to develop and validate a solid phase microextraction/gas chromatography-mass spectrometry method for the simultaneous analysis of 16 pesticides in water," *Journal of Chromatography A*, vol. 1217, no. 33, pp. 5317–5327, 2010.

[16] S. F. Yuan, Z. H. Liu, H. X. Lian et al., "Simultaneous determination of eleven estrogenic and odorous chloro- and bromo-phenolic compounds in surface water through an automated online HS SPME followed by on-fiber derivatization coupled with GC-MS," *Analytical Methods*, vol. 9, no. 33, pp. 4819–4827, 2017.

[17] F. Gandolfi, L. Malleret, M. Sergent, and P. Doumenq, "Parameters optimization using experimental design for HS solid phase micro-extraction analysis of short-chain chlorinated paraffins in waters under the European water framework

directive," *Journal of Chromatography of A*, vol. 1406, pp. 59–67, 2015.

[18] E. Eljarrat and D. Barcelo, "Quantitative analysis of polychlorinated *n*-alkanes in environmental samples," *TrAC-Trends in Analytical Chemistry*, vol. 25, no. 4, pp. 421–434, 2006.

[19] G. T. Tomy, G. A. Stern, D. C. G. Muir et al., "Quantifying C10-C13 polychloroalkanes in environmental samples by high-resolution gas chromatography electron capture negative ion high-resolution mass spectrometry," *Analytical Chemistry*, vol. 69, no. 14, pp. 2762–2771, 1997.

[20] Z. Zencak, A. Borgen, M. Reth, and M. Oehme, "Evaluation of four mass spectrometric methods for the gas chromatographic analysis of polychlorinated *n*-alkanes," *Journal of Chromatography A*, vol. 1067, no. 1-2, pp. 295–301, 2005.

[21] M. Reth, Z. Zencak, and M. Oehme, "New quantification procedure for the analysis of chlorinated paraffins using electron capture negative ionization mass spectrometry," *Journal of Chromatography A*, vol. 1081, no. 2, pp. 225–231, 2005.

[22] B. Yuan, Y. W. Wang, J. J. Fu, Q. H. Zhang, and G. B. Jiang, "An analytical method for chlorinated paraffins and their determination in soil samples," *Chinese Science Bulletin*, vol. 55, no. 22, pp. 2396–2402, 2010.

[23] W. Gao, J. Wu, Y. W. Wang, and G. B. Jiang, "Quantification of short- and medium-chain chlorinated paraffins in environmental samples by gas chromatography quadrupole time-of-flight mass spectrometry," *Journal of Chromatography A*, vol. 1452, pp. 98–106, 2016.

[24] P. Korytár, J. Parera, P. E. G. Leonards et al., "Characterization of polychlorinated *n*-alkanes using comprehensive two-dimensional gas chromatography-electron-capture negative ionisation time-of-flight mass spectrometry," *Journal of Chromatography A*, vol. 1086, no. 1-2, pp. 71–82, 2005.

[25] D. Xia, L. Gao, S. Zhu, and M. H. Zheng, "Separation and screening of short-chain chlorinated paraffins in environmental samples using comprehensive two-dimensional gas chromatography with micro electron capture detection," *Analytical and Bioanalytical Chemistry*, vol. 406, no. 29, pp. 7561–7570, 2014.

[26] D. Xia, L. R. Gao, M. H. Zheng et al., "A novel method for profiling and quantifying short- and medium-chain chlorinated paraffins in environmental samples using comprehensive two-dimensional gas chromatography–electron capture negative ionization high-resolution time-of-flight mass spectrometry," *Environmental Science and Technology*, vol. 50, no. 14, pp. 7601–7609, 2016.

[27] M. Reth and M. Oehme, "Limitations of low resolution mass spectrometry in the electron capture negative ionization mode for the analysis of short- and medium-chain chlorinated paraffins," *Analytical and Bioanalytical Chemistry*, vol. 378, no. 7, pp. 1741–1747, 2004.

[28] S. Moore, L. Vromet, and B. Rondeau, "Comparison of metastable atom bombardment and electron capture negative ionization for the analysis of polychloroalkanes," *Chemosphere*, vol. 54, no. 4, pp. 453–459, 2004.

[29] Y. Gao, H. J. Zhang, L. L. Zou et al., "Quantification of short-chain chlorinated paraffins by deuterodechlorination combined with gas chromatography-mass spectrometry," *Environmental Science and Technology*, vol. 50, no. 7, pp. 3746–3753, 2016.

[30] C. Nicholls, C. Allchin, and R. Law, "Levels of short and medium chain length polychlorinated *n*-alkanes in environmental samples from selected industrial areas in England and Wales," *Environmental Pollution*, vol. 114, no. 3, pp. 415–430, 2001.

[31] F. Iino, T. Takasuga, K. Senthikumar, N. Nakamura, and J. Nakanishi, "Risk assessment of short-chain chlorinated paraffins in Japan based on the first market basket study and species sensitivity distributions," *Environmental Science and Technology*, vol. 39, no. 3, pp. 859–866, 2005.

[32] M. Houde, D. C. G. Muir, G. T. Tomy et al., "Bioaccumulation and trophic magnification of short- and medium-chain chlorinated paraffins in food webs from Lake Ontario and Lake Michigan," *Environmental Science and Technology*, vol. 42, no. 10, pp. 3893–3899, 2008.

[33] W. Y. H. Jiang, T. Huang, X. X. Mao et al., "Gridded emission inventory of short-chain chlorinated paraffins and its validation in China," *Environmental Pollution*, vol. 220, pp. 132–141, 2017.

14

A Turn-On Fluorescent Probe for Sensitive Detection of Cysteine in a Fully Aqueous Environment and in Living Cells

Xiaohua Ma,[1,2] **Guoguang Wu** (ID),[1] **Yuehua Zhao,**[3] **Zibo Yuan,**[3] **Yu Zhang,**[3] **Ning Xia** (ID),[3] **Mengnan Yang,**[3] **and Lin Liu** (ID)[3]

[1]*School of Chemical Engineering and Technology, China University of Mining and Technology, Xuzhou, Jiangsu 221116, China*
[2]*Henan Key Laboratory of Biomolecular Recognition and Sensing, College of Chemistry and Chemical Engineering, Shangqiu Normal University, Shangqiu, Henan 476000, China*
[3]*Key Laboratory of New Optoelectronic Functional Materials (Henan Province), College of Chemistry and Chemical Engineering, Anyang Normal University, Anyang, Henan 455000, China*

Correspondence should be addressed to Guoguang Wu; tb12040004@cumt.edu.cn, Ning Xia; xianing82414@163.com, and Lin Liu; liulin@aynu.edu.cn

Academic Editor: Chih-Ching Huang

We reported here a turn-on fluorescent probe (1) for the detection of cysteine (Cys) by incorporating the recognition unit of 2,4-dinitrobenzenesulfonyl ester (DNBS) to a coumarin derivative. The structure of the obtained probe was confirmed by NMR and HRMS techniques. The probe shows a remarkable fluorescence off-on response (~52-fold) by the reaction with Cys in 100% aqueous buffer. The sensing mechanism was verified by the HPLC test. Probe 1 also displays high selectivity towards Cys. The detection limit was calculated to be 23 nM. Moreover, cellular experiments demonstrated that the probe is highly biocompatible and can be used for monitoring intracellular Cys.

1. Introduction

Cysteine (Cys), a kind of critical biothiols, plays many crucial physiological roles, such as maintaining biological redox homeostasis, participation in enzymatic reactions, and sequestering inimical metal ions [1–4]. The abnormal levels of Cys are associated with many syndromes and diseases, including growth retarding, muscle loss, skin lesions, liver damage, severe neurotoxicity, and cardiovascular diseases [5–7]. Therefore, it is highly desired to develop effective Cys assays for application in biological systems, which would be very helpful to further elucidate its biological functions and reveal its relevance to certain diseases.

Analytical methods for the detection of Cys include capillary electrophoresis (CE) [8–10], highperformance liquid chromatography (HPLC) [11–13], electrochemical methods [14], and colorimetric and fluorescent assays [3, 15–18]. Among them, the fluorescence assay based on optical probes has gained tremendous attentions due to its inherent advantages of high sensitivity and selectivity, simplicity of implementation, high spatiotemporal resolution, and good compatibility for biosamples [19–26]. Up to now, some fluorescent probes have been synthesized for the detection of Cys by exploiting mechanisms of Michael addition, cleavage of the selenium-nitrogen bond and of disulfides, cyclization with aldehydes, cleavage of sulfonamide and sulfonate esters, and metal complex replacement of ligands [27–40]. However, many of these developed probes have drawbacks of low sensitivity, complicated synthetic process, and/or the use of high-content organic solvent. Thus, developing facile and reliable fluorescent Cys probes is still highly desired. Herein, we report a highly sensitive fluorogenic Cys probe (1) by installing the recognition moiety of 2,4-dinitrobenzenesulfonyl ester (DNBS) onto a coumarin fluorophore. Coumarin and its derivatives are popular fluorescent reporters due to their high photostability, excellent biocompatibility, and high quantum yield [41–43]. Upon the target-mediated cleavage of 2,4-dinitrobenzenesulfonyl ester and release of the coumarin fluorophore, probe 1 exhibits efficient turn-on fluorescent response towards Cys. Moreover, the proposed probe 1 displays good water solubility, high sensitivity and selectivity, and low cytotoxicity and can be used for imaging intracellular Cys.

2. Experimental Section

2.1. General Procedure for Analysis. All spectral measurements were performed in the aqueous phosphate buffer (pH 7.4, 10 mM). Stock solution of probe 1 (0.1 mM) was prepared in the same phosphate buffer solution. The following solutions (10.0 mM) were prepared in deionized water: amino acids (Cys, Hcy, GSH, Gly, Ser, Val, Leu, Tyr, His, Trp, Arg, Glu, Pro, Asp, Thr, Asn, and Phe), ascorbic acid (AA), and glutathione (GSH). Test solutions were prepared by placing 300.0 μL of stock solution 1 (0.1 mM), an appropriate aliquot of each analyte stock solution into a 5.0 mL centrifugal tube, and diluting the solution to 3.0 mL with the phosphate buffer (pH 7.4, 10 mM). The solution was mixed for a given time at the room temperature. Then, the fluorescence and UV absorption spectra were recorded. For fluorescence assays, the excitation and emission slit width are both 5 nm.

2.2. Synthesis of Probe 1. Synthesis procedures for probe 1 were displayed in Scheme 1. Compound 3 was obtained according to literature methods [44, 45].

Compound 2, compound 3 (12.8 g, 50 mmol), 2,3,6,7-tetrahydro-8-hydroxy-1H, and 5H-benz[i, j]quinolizine-9-carboxaldehyde (8.26 g, 50 mmol) were added to toluene (0.1 L), and the mixture was refluxed for 10 h. Then, the formed solid product was filtered and washed with hexanes. The obtained precipitation was further dried under vacuum giving a white solid (8.7 g, 76%). ^{1}H NMR (400 MHz, DMSO) δ: 11.78 (s, 1H), 7.19 (d, J = 31.3 Hz, 1H), 5.22 (s, 1H), 3.23 (s, 4H), 2.70 (s, 4H), and 1.87 (s, 4H) (Figure S1). ^{13}C NMR (100 MHz, DMSO): δ 167.06 (s), 163.29 (s), 151.46 (s), 146.45 (s), 120.35 (s), 117.78 (s), 105.80 (s), 103.53 (s), 86.25 (s), 49.67 (s), 49.14 (s), 27.42 (s), 21.46 (s), and 20.58 (d, J = 2.3 Hz) (Figure S2). HRMS: m/z, calcd. $[M + H]^+$ 258.1130; found 258.1126 (Figure S3).

Probe 1 was prepared by reacting compound 2 with 2,4-dinitrobenzenesulfonyl chloride. In brief, compound 2 (2.57 g, 10 mmol), 2,4-dinitrobenzenesulfonyl chloride (2.67 g, 10 mmol), and triethylamine (1.21 g, 12 mmol) were added in anhydrous CH_2Cl_2 (0.1 L) at 0°C. After stirring for 1 h, the mixture was gradually warmed to the room temperature and reacted for another 2 h. Then, the reaction mixture was evaporated to dryness and purified by column chromatography (silica, DCM-EtOAc as eluent, 2 : 1, v/v) yielded 1 as a yellow solid (12.82 g, 58%). ^{1}H NMR (400 Hz, CDCl$_3$): δ 8.71 (s, 1H), 8.60 (d, J = 8.1 Hz, 1H), 8.41 (d, J = 8.3 Hz, 1H), 7.16 (s, 1H), 5.87 (s, 1H), 3.30 (s, 4H), 2.79 (d, J = 35.0 Hz,

4H), and 1.96 (s, 4H) (Figure S4). ^{13}C NMR (101 MHz, CDCl$_3$) δ 161.98 (s), 158.41 (s), 151.30 (s), 151.19 (s), 148.87 (s), 134.00 (s), 133.36 (s), 126.96 (s), 120.79 (s), 120.06 (s), 119.25 (s), 96.54 (s), 77.34 (s), 77.23 (s), 77.03 (s), 76.71 (s), 50.11 (s), 49.67 (s), 27.54 (s), 21.06 (s), 20.31 (s), and 20.13 (s) (Figure S5). HRMS: m/z, calcd. $[M + H]^+$ 488.0764; found 488.0759 (Figure S6).

3. Results and Discussion

3.1. Design and Synthesis. The probe 1 was devised by exploiting 2 as the fluorophore and DNBS as the reaction moiety. The coumarin derivative (compound 2) was selected here because of its high emission efficiency, facile preparation procedure, excellent water solubility, and biocompatibility. DNBS group has been exploited as a good reaction moiety for fluorescent biothiols probes. Scheme 1 illustrates the synthesis procedures for probe 1. Compound 2 was prepared via refluxing malonate ester with 2,3,6,7-tetrahydro-8-hydroxy-1H and 5H-benz[i, j]quinolizine in toluene. Furthermore, coupling 2 with 2,4-dinitrobenzenesulfonyl chloride in CH_2Cl_2 afforded 1. The structures of compound 2 and probe 1 were confirmed by NMR and HRMS (Supporting Information).

3.2. Spectral Characteristics of Probe 1 and Its Optical Responses towards Cys. The spectroscopic characteristics of probe 1 were inspected with or without Cys (10.0 equiv) (Figure 1). 1 alone displayed an absorption band at about 415 nm (ε = 1.57 × 10^4 M^{-1}·cm^{-1}) and nonemissivity (curve a). With the addition of Cys (10.0 equiv), the absorbance at 415 nm decreased significantly, and a new absorption band centered at 347 nm (ε = 2.93 × 10^4 M^{-1}·cm^{-1}) appeared (curve b). Meanwhile, the emission of the probe solution increased remarkably (λ_{em} = 413 nm). These obvious spectral responses imply that probe 1 is capable of monitoring Cys.

To study the response time of probe 1 for Cys, time-dependent fluorescence response of probe 1 towards Cys with different concentrations was investigated (Figure 2(a)). The peak emission intensity of probe 1 did not obviously change in the absence of Cys during the time course of testing, indicating the high stability of the probe in the aqueous buffer solution under the neutral condition. And the emission intensity was observed to increase in the presence of Cys in a concentration-dependent fashion. Higher concentration of Cys (ca. 10.0 equiv) afforded a quicker and more dramatic fluorescent response. The pseudofirst-order rate of the reaction is found to be 1.4 × 10^{-2}·s^{-1} (Figure S7). And 1 h was set as the optimized reaction time as the fluorescence intensity reached a plateau within 1 h at each inspected concentration of Cys.

The effect of pH on the response of 1 toward Cys was studied. Without Cys, the fluorescence intensity of the probe remained unchanged with pH ≤8 and increased significantly with the pH value over 8, indicating that probe 1 is stable under the neutral condition and prone to hydrolysis under

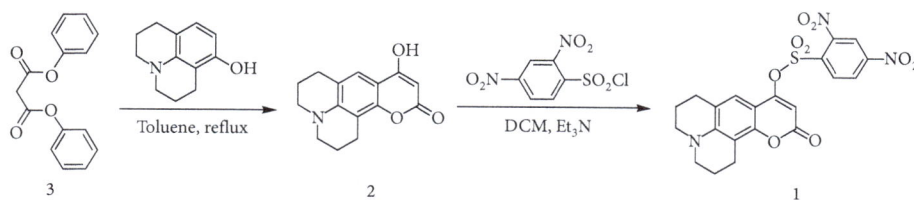

SCHEME 1: Synthetic route for probe 1.

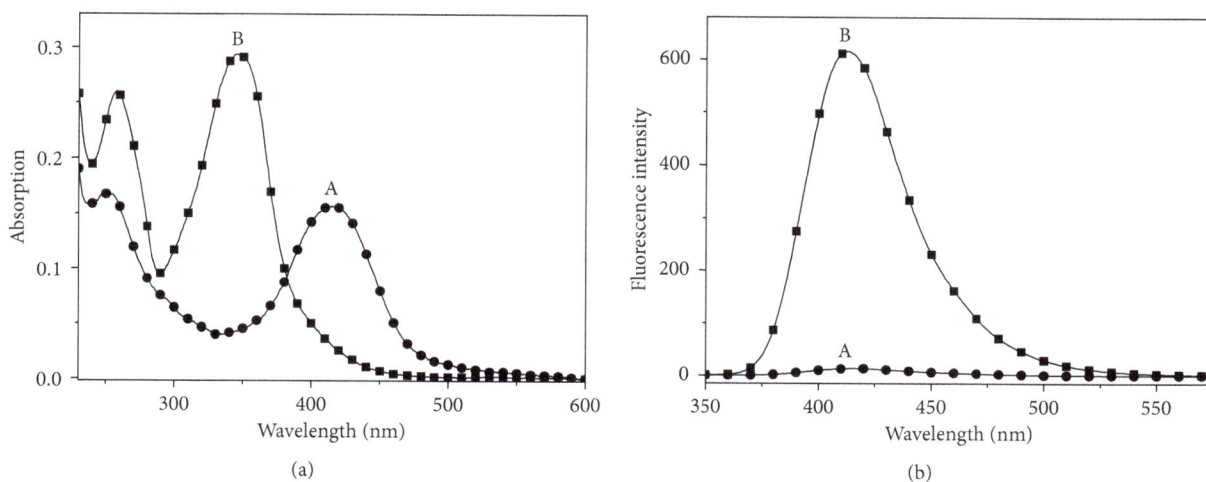

(a)

(b)

FIGURE 1: (a) Absorption and (b) emission spectra of probe 1 (10 μM) in the absence (A) and presence (B) of Cys (100 μM) in a solution of the phosphate buffer (pH 7.4, 10 mM). λ_{ex} = 347 nm.

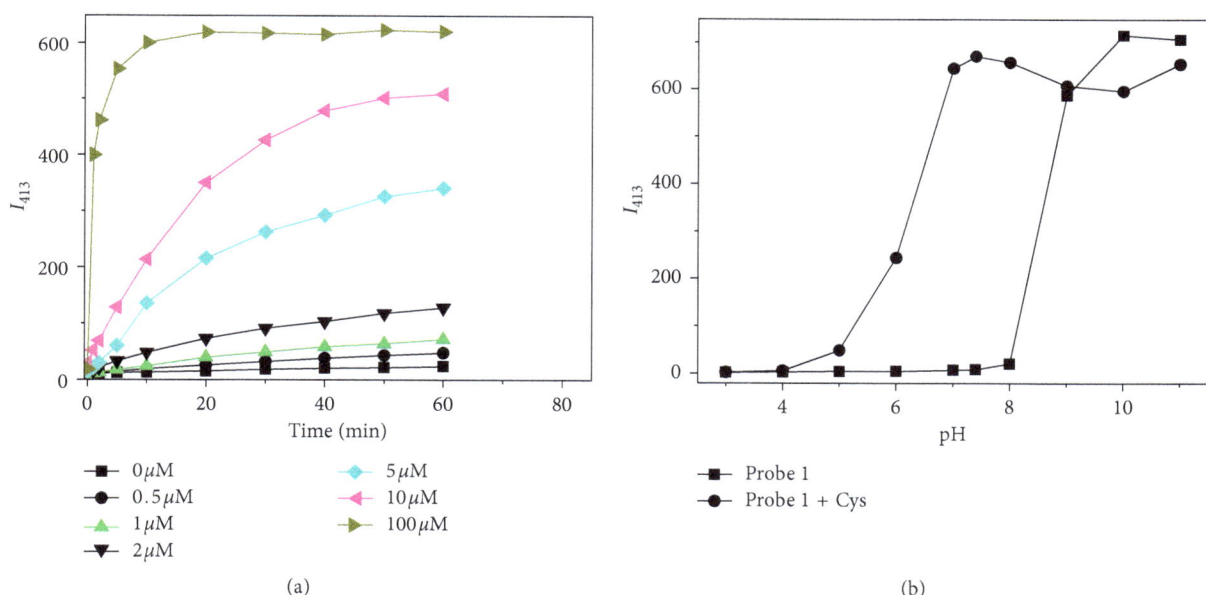

(a)

(b)

FIGURE 2: (a) Time-dependent fluorescence intensity changes of probe 1 (10 μM) at 413 nm in the presence of various concentrations of Cys (0, 0.5, 1, 2, 5, 10, and 100 μM); (b) effect of pH on the fluorescence response of probe 1 (10 μM) towards Cys (100 μM). λ_{ex} = 347 nm.

the alkaline condition (Figure 2(b)). With addition of Cys, the fluorescence was gradually increased in the region of pH 4.0–7.0 and reached the maximum at pH 7.4. These results demonstrated that 1 responds well to Cys at round physiological pH.

3.3. *Sensitivity and Selectivity.* The quantitative response ability of probe 1 towards Cys was inspected via fluorescence titration. The fluorescence intensity gradually increases with increment of Cys contents and reaches a plateau with the Cys concentration up to 30 μM (Figure 3). And there is a good

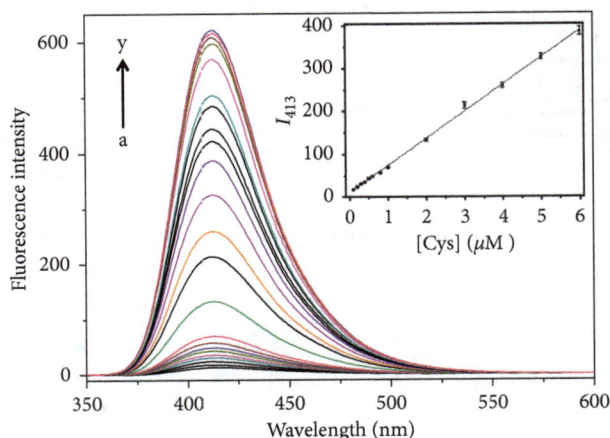

FIGURE 3: Fluorescence spectra of probe 1 (10 μM) in the presence of Cys with various concentrations: 0, 0.1, 0.2, 0.3, 0.4, 0.5, 0.6, 0.8, 1.0, 2.0, 3.0, 4.0, 5.0, 6.0, 7.0, 8.0, 9.0, 10.0, 20.0, 30.0, 40.0, 50.0, and 100.0 μM (from a to y). Inset shows the standard curve. λ_{ex} = 347 nm.

linear correlation between emission intensity at 413 nm and the Cys concentration in the range of 0.1–6 μM. Linear equation can be expressed as I = 10.81 + 63.16 × [Cys]/μM (R^2 = 0.998). The detection limit was estimated to be 23 nM (3σ). The analytical performances of probe 1 were also compared with other reported fluorescent Cys probes using 2,4-dinitrobenzenesulfonyl ester (DNBS) as the recognition moiety [46–55] (Table S1).

To evaluate the specificity of the assay, the spectral response of probe 1 towards various biologically relevant analytes including twenty natural protein amino acids, AA, Hcy, and GSH was recorded. Cys generated a significant enhancement of fluorescence intensity. Another biothiol, Hcy, also created an obvious fluorescent emission for the probe solution, indicating that probe 1 can response to both Cys and Hcy but showing a higher reactivity for Cys. Other analytes, including GSH, did not show any significant changes (Figure 4(a)). The higher reactivity of the probe toward Cys over GSH may be ascribed to the bulkiness of GSH and the significant steric hindrance around its thiol group. The spectral response of probe 1 for various metal ions (Al^{3+}, Ca^{2+}, Cd^{2+}, CO^{2+}, Cu^{2+}, Fe^{2+}, Fe^{3+}, K^+, Mg^{2+}, Mn^{2+}, Na^+, Ni^{2+}, Pb^{2+}, and Zn^{2+}) was also inspected (Figure S8), indicating that probe 1 is nonresponsive for these metal ions. Furthermore, competition experiments also indicated that the coexistence of other interfering species did not influence the reactivity of probe 1 for Cys (Figure 4(b)).

3.4. Sensing Mechanism. The presented fluorescent Cys probe (1) was obtained by incorporating the DNBS functional group (a well-known recognition moiety for the biothiols) onto the coumarin-based fluorophore. Probe 1 is nonfluorescent due to the quenching effect of DNBS unit via the electron-transfer process. The introduced Cys can first react with DNBS of the probe via the nucleophilic aromatic substitution and form a unstable negative-charged intermediate, which further involved the intramolecular rearrangement to yield the sulfur dioxide, 2,4-dinitrophenyl cysteine, and compound 2 (a highly-emissive

fluorophore) (Scheme 2). HPLC analysis was performed to verify this proposed sensing mechanism. 1 alone exhibited a single chromatographic peak at 2.00 min (curve a in Figure 5). After incubation probe 1 (10 μM) with Cys (5 μM), a new peak at 0.56 min appeared, which can be ascribed to compound 2 (curves b and d in Figure 5). Incubating probe 1 (10 μM) with high concentration of Cys (100 μM) resulted in the disappearance of the peak at 2.00 min and leaded to a chromatographic profile identical to that of compound 2, which indicated that probe 1 can be completely converted to compound 2 upon the Cys-induced thiolysis process.

3.5. Cellular Imaging. The good water solubility and high selectivity inspired us to use 1 for the bioimaging application. Firstly, cellular cytotoxicity of probe 1 was inspected (Figure S9). The high survival rates of all these three kinds of cells with different concentrations of probe 1 indicated that the probe was highly biocompatible. Then, cellular imaging experiments were conducted. HeLa cells incubated with probe 1 alone displayed no intracellular fluorescence (Figure 6(b)). However, cells incubated with 1 and consequently with Cys (100 μM) exhibited strong blue fluorescent emission (Figure 6(e)). These imaging results indicated that probe 1 is living cell membrane permeable and can be used to monitor intracellular Cys.

4. Conclusions

In conclusion, we developed a turn-on fluorescent probe for Cys based on a coumarin-derived fluorophore. The sensing mechanism involved the Cys-induced cleavage of the DNBS group and the follow-up release of the coumarin fluorophore, which was confirmed by HPLC and spectral results. Probe 1 displayed high selectivity for Cys and a low detection limit of 23 nM. The proposed probe also features excellent water solubility and biocompatibility and has been successfully utilized for imaging Cys in living cells.

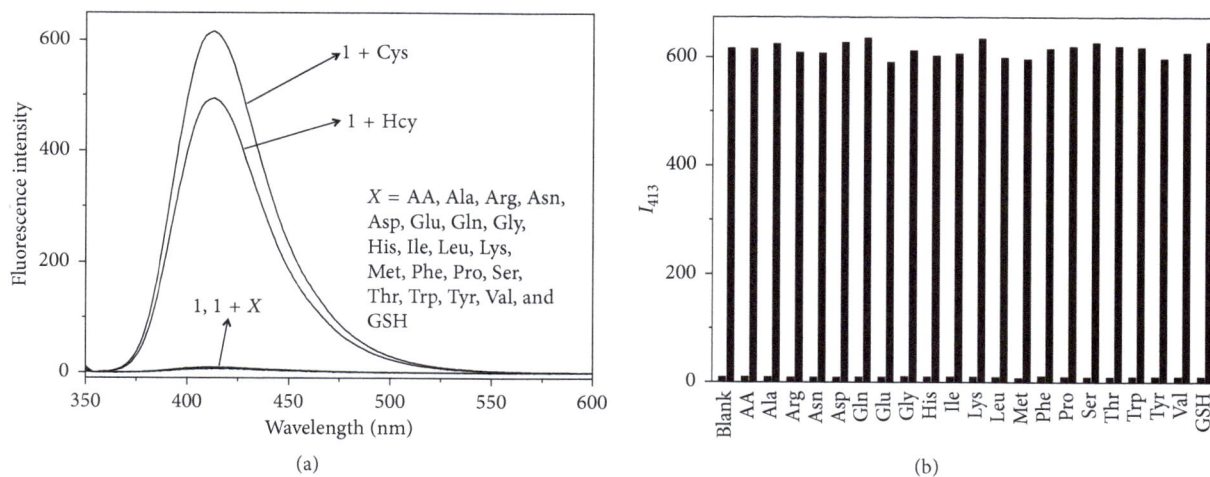

(a)

(b)

FIGURE 4: (a) Fluorescence spectra of probe **1** (10 μM) towards different amino acids, AA and GSH (100 μM). (b) Fluorescence intensities of probe **1** (10 μM) at 413 nm upon the addition of different interfering species (100 μM) (low bars), followed by addition of Cys (100 μM) (high bars). λ_{ex} = 347 nm.

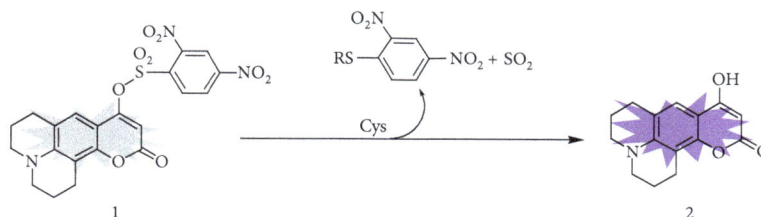

SCHEME 2: Sensing mechanism of probe **1** for Cys.

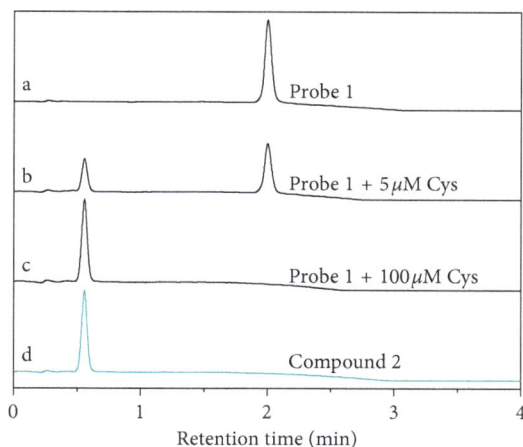

FIGURE 5: Reversed-phase HPLC chromatograms of (a) probe **1** (10 μM), (b) probe **1** (10 μM) and Cys (5 μM), (c) probe **1** (10 μM) and Cys (100 μM), and (d) compound **2**.

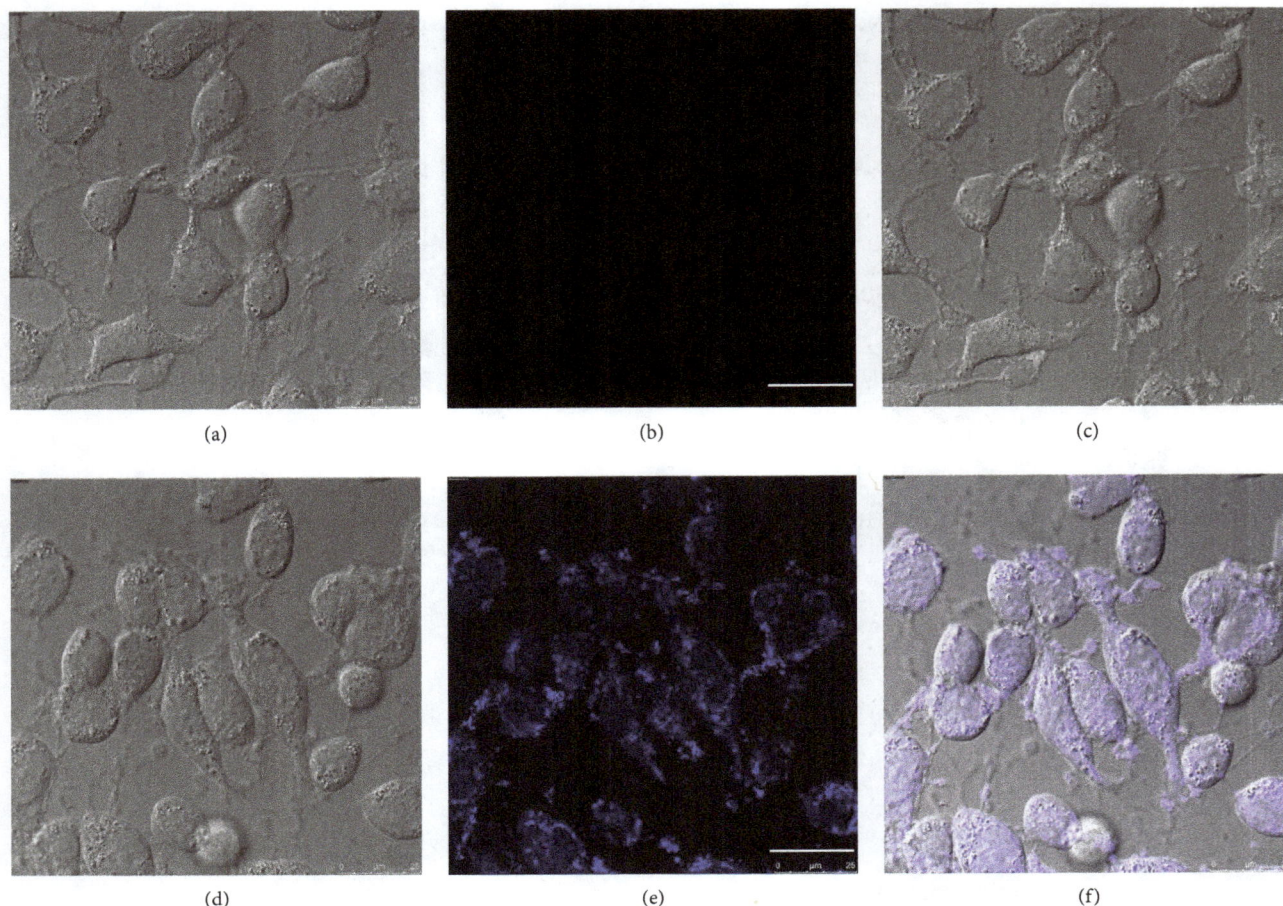

FIGURE 6: Confocal fluorescence images of living HeLa cells. Bright-field image (a) and fluorescence image (b) of cells incubated with probe 1 (10 μM) for 1 h; (c) overlay of the images of (a) and (b); bright-field image (d) and fluorescence image (e) of cells incubated with probe 1 (10 μM) for 1 h and subsequent treatment with Cys (100 μM) for another 1 h; (f) overlay of the images of (d) and (e). λ_{ex} = 405 nm; scare bar = 25 μm.

Conflicts of Interest

The authors declare that there are no conflicts of interest regarding the publication of this article.

Acknowledgments

We are grateful to the Program for Science and Technology Innovation Talents at the University of Henan Province (18HASTIT005), the Fund Project for Young Scholar sponsored by Henan Province (2016GGJS-122), and the Program for Innovative Research Team of Science and Technology in the University of Henan Province (18IRT-STHN004) for support.

Supplementary Materials

Supplementary Description Part 1: experimental materials, instrumentations, and experimental procedures for the HPLC test and cell viability assay/imaging. Part 2: characterization of compound 2 and probe 1. Figure S1: 1H NMR chemical shifts of compound 2. Figure S2: 13C NMR chemical shifts of compound 2. Figure S3: high-resolution mass spectrum (HRMS) of compound 2. Figure S4: 1H NMR chemical shifts of probe 1. Figure S5: 13C NMR chemical shifts of probe 1. Figure S6: high-resolution mass spectrum (HRMS) of probe 1. Part 3: kinetic study of 1 to Cys. Figure S7: the kinetic study of the response of probe 1 to Cys (10 equiv) under pseudofirst-order conditions based on the time course of the emission intensity at 413 nm. Part 4: comparison of DNBS-based fluorescent probes for Cys. Table S1: comparison of DNBS-based fluorescent probes for Cys. Part 5: spectral responses of probe 1 for various metal ions. Figure S8: fluorescence intensities of probe 1 (10 μM) at 413 nm upon the addition of Cys (100 μM) and different metal ions (100 μM); λ_{ex} = 347 nm. Part

6: cell cytotoxicity of probe **1**. Figure S9: cell cytotoxicity of probe **1** against HeLa, A549, and MDA-MB-231 cells upon 24 h of incubation. (*Supplementary Materials*)

References

[1] K. G. Reddie and K. S. Carroll, "Expanding the functional diversity of proteins through cysteine oxidation," *Current Opinion in Chemical Biology*, vol. 12, no. 6, pp. 746–754, 2008.

[2] T. Dudev and C. Lim, "Metal binding affinity and selectivity in metalloproteins: insights from computational studies," *Annual Review of Biophysics*, vol. 37, no. 1, pp. 97–116, 2008.

[3] X. Chen, Y. Zhou, X. Peng, and J. Yoon, "Fluorescent and colorimetric probes for detection of thiols," *Chemical Society Reviews*, vol. 39, no. 6, pp. 2120–2135, 2010.

[4] C. E. Paulsen and K. S. Carroll, "Cysteine-mediated redox signaling: chemistry, biology, and tools for discovery," *Chemical Reviews*, vol. 113, no. 7, pp. 4633–4679, 2013.

[5] W. Dröge, H. P. Eck, and S. Mihm, "HIV-induced cysteine deficiency and T-cell dysfunction—a rationale for treatment with N-acetylcysteine," *Immunology Today*, vol. 13, no. 6, pp. 211–214, 1992.

[6] M. W. Lieberman, A. L. Wiseman, Z. Z. Shi et al., "Growth retardation and cysteine deficiency in gamma-glutamyl transpeptidase-deficient mice," *Proceedings of the National Academy of Sciences*, vol. 93, no. 15, pp. 7923–7926, 1996.

[7] J. A. McMahon, T. J. Green, C. M. Skeaff, R. G. Knight, J. I. Mann, and S. M. Williams, "A controlled trial of homocysteine lowering and cognitive performance," *New England Journal of Medicine*, vol. 354, no. 26, pp. 2764–2772, 2006.

[8] A. V. Ivanov, E. D. Virus, B. P. Luzyanin, and A. A. Kubatiev, "Capillary electrophoresis coupled with 1,1′-thiocarbonyldiimidazole derivatization for the rapid detection of total homocysteine and cysteine in human plasma," *Journal of Chromatography B*, vol. 1004, pp. 30–36, 2015.

[9] J. Lacna, F. Foret, and P. Kuban, "Capillary electrophoresis in the analysis of biologically important thiols," *Electrophoresis*, vol. 38, no. 1, pp. 203–222, 2017.

[10] K. Y. Liu, H. Wang, J. L. Bai, and L. Wang, "Home-made capillary array electrophoresis for high-throughput amino acid analysis," *Analytica Chimica Acta*, vol. 622, no. 1-2, pp. 169–174, 2008.

[11] L. Y. Zhang, F. Q. Tu, X. F. Guo, H. Wang, P. Wang, and H. S. Zhang, "A new BODIPY-based long-wavelength fluorescent probe for chromatographic analysis of low-molecular-weight thiols," *Analytical and Bioanalytical Chemistry*, vol. 406, no. 26, pp. 6723–6733, 2014.

[12] L. J. Zhang, B. Q. Lu, C. Lu, and J. M. Lin, "Determination of cysteine, homocysteine, cystine, and homocystine in biological fluids by HPLC using fluorosurfactant-capped gold nanoparticles as postcolumn colorimetric reagents," *Journal of Separation Science*, vol. 37, no. 1-2, pp. 30–36, 2014.

[13] D. Tsikas, J. Sandmann, M. Ikic, J. Fauler, D. O. Stichtenoth, and J. C. Frolich, "Analysis of cysteine and N-acetylcysteine in human plasma by high-performance liquid chromatography at the basal state and after oral administration of N-acetylcysteine," *Journal of Chromatography B*, vol. 708, no. 1-2, pp. 55–60, 1998.

[14] P. T. Lee, D. Lowinsohn, and R. G. Compton, "Simultaneous detection of homocysteine and cysteine in the presence of ascorbic acid and glutathione using a nanocarbon modified electrode," *Electroanalysis*, vol. 26, no. 7, pp. 1488–1496, 2014.

[15] Y. Q. Hao, D. D. Xiong, L. Q. Wang, W. S. Chen, B. B. Zhou, and Y. N. Liu, "A reversible competition colorimetric assay for the detection of biothiols based on ruthenium-containing complex," *Talanta*, vol. 115, pp. 253–257, 2013.

[16] D. Y. Lee, G. M. Kim, J. Yin, and J. Yoon, "An aryl-thioether substituted nitrobenzothiadiazole probe for the selective detection of cysteine and homocysteine," *Chemical Communications*, vol. 51, no. 30, pp. 6518–6520, 2015.

[17] L. Cui, Y. Baek, S. Lee, N. Kwon, and J. Yoon, "An AIE and ESIPT based kinetically resolved fluorescent probe for biothiols," *Journal of Materials Chemistry C*, vol. 4, no. 14, pp. 2909–2914, 2016.

[18] B. Babur, N. Seferoglu, M. Ocal, G. Sonugur, H. Akbulut, and Z. Seferoglu, "A novel fluorescence turn-on coumarin-pyrazolone based monomethine probe for biothiol detection," *Tetrahedron*, vol. 72, no. 30, pp. 4498–4502, 2016.

[19] N. Xia, B. Zhou, N. Huang, M. Jiang, J. Zhang, and L. Liu, "Visual and fluorescent assays for selective detection of beta-amyloid oligomers based on the inner filter effect of gold nanoparticles on the fluorescence of CdTe quantum dots," *Biosensors and Bioelectronics*, vol. 85, pp. 625–632, 2016.

[20] Y. Q. Hao, Y. T. Zhang, K. H. Ruan et al., "A naphthalimide-based chemodosimetric probe for ratiometric detection of hydrazine," *Sensors and Actuators B-Chemical*, vol. 244, pp. 417–424, 2017.

[21] W. Q. Chen, X. X. Yue, H. Zhang et al., "Simultaneous detection of glutathione and hydrogen polysulfides from different emission channels," *Analytical Chemistry*, vol. 89, no. 23, pp. 12984–12991, 2017.

[22] X. H. Li, X. H. Gao, W. Shi, and H. M. Ma, "Design strategies for water-soluble small molecular chromogenic and fluorogenic probes," *Chemical Reviews*, vol. 114, no. 1, pp. 590–659, 2014.

[23] L. Yuan, W. Y. Lin, K. B. Zheng, L. W. He, and W. M. Huang, "Far-red to near infrared analyte-responsive fluorescent probes based on organic fluorophore platforms for fluorescence imaging," *Chemical Society Reviews*, vol. 42, no. 2, pp. 622–661, 2013.

[24] M. La, Y. Q. Hao, Z. Y. Wang, G. C. Han, and L. B. Qu, "Selective and sensitive detection of cyanide based on the displacement strategy using a water-soluble fluorescent probe," *Journal of Analytical Methods in Chemistry*, vol. 2016, Article ID 1462013, 6 pages, 2016.

[25] Y. Jung, J. Jung, Y. Huh, and D. Kim, "Benzo-g-coumarin-based fluorescent probes for bioimaging applications," *Journal of Analytical Methods in Chemistry*, vol. 2018, Article ID 5249765, 11 pages, 2018.

[26] L. Liu, Y. Chang, J. Yu, M. S. Jiang, and N. Xia, "Two-in-one polydopamine nanospheres for fluorescent determination of beta-amyloid oligomers and inhibition of beta-amyloid aggregation," *Sensors and Actuators B-Chemical*, vol. 251, pp. 359–365, 2017.

[27] H. S. Jung, X. Q. Chen, J. S. Kim, and J. Yoon, "Recent progress in luminescent and colorimetric chemosensors for detection of thiols," *Chemical Society Reviews*, vol. 42, no. 14, pp. 6019–6031, 2013.

[28] L. Y. Niu, Y. Z. Chen, H. R. Zheng, L. Z. Wu, C. H. Tung, and Q. Z. Yang, "Design strategies of fluorescent probes for selective detection among biothiols," *Chemical Society Reviews*, vol. 44, no. 17, pp. 6143–6160, 2015.

[29] H. Chen, Y. H. Tang, and W. Y. Lin, "Recent progress in the fluorescent probes for the specific imaging of small molecular weight thiols in living cells," *Trac-Trends in Analytical Chemistry*, vol. 76, pp. 166–181, 2016.

[30] C. Yin, F. Huo, J. Zhang et al., "Thiol-addition reactions and their applications in thiol recognition," *Chemical Society Reviews*, vol. 42, no. 14, pp. 6032–6059, 2013.

[31] Y. Geng, H. Tian, L. Yang, X. Liu, and X. Song, "An aqueous methylated chromenoquinoline-based fluorescent probe for instantaneous sensing of thiophenol with a red emission and a large Stokes shift," *Sensors and Actuators B: Chemical*, vol. 273, pp. 1670–1675, 2018.

[32] X. Ren, H. Tian, L. Yang et al., "Fluorescent probe for simultaneous discrimination of Cys/Hcy and GSH in pure aqueous media with a fast response under a single-wavelength excitation," *Sensors and Actuators B: Chemical*, vol. 273, pp. 1170–1178, 2018.

[33] S. Zhou, Y. Rong, H. Wang, X. Liu, L. Wei, and X. Song, "A naphthalimide-indole fused chromophore-based fluorescent probe for instantaneous detection of thiophenol with a red emission and a large Stokes shift," *Sensors and Actuators B: Chemical*, vol. 276, pp. 136–141, 2018.

[34] Y. Wang, L. Wang, E. Jiang et al., "A colorimetric and ratiometric dual-site fluorescent probe with 2,4-dinitrobenzenesulfonyl and aldehyde groups for imaging of aminothiols in living cells and zebrafish," *Dyes and Pigments*, vol. 156, pp. 338–347, 2018.

[35] Y. Wang, M. Zhu, E. Jiang, R. Hua, R. Na, and Q. X. Li, "A simple and rapid turn on ESIPT fluorescent probe for colorimetric and ratiometric detection of biothiols in living cells," *Scientific Reports*, vol. 7, no. 1, p. 4377, 2017.

[36] R. Na, M. Zhu, S. Fan et al., "A simple and effective ratiometric fluorescent probe for the selective detection of cysteine and homocysteine in aqueous media," *Molecules*, vol. 21, no. 8, p. 1023, 2016.

[37] L. Xia, Y. Zhao, J. Huang, Y. Gu, and P. Wang, "A fluorescent turn-on probe for highly selective detection of cysteine and its bioimaging applications in living cells and tissues," *Sensors and Actuators B: Chemical*, vol. 270, pp. 312–317, 2018.

[38] P. Wang, Q. Wang, J. Huang, N. Li, and Y. Gu, "A dual-site fluorescent probe for direct and highly selective detection of cysteine and its application in living cells," *Biosensors and Bioelectronics*, vol. 92, pp. 583–588, 2017.

[39] Q. Wang, H. Wang, J. Huang, N. Li, Y. Gu, and P. Wang, "Novel NIR fluorescent probe with dual models for sensitively and selectively monitoring and imaging Cys in living cells and mice," *Sensors and Actuators B: Chemical*, vol. 253, pp. 400–406, 2017.

[40] P. Wang, Y. Wang, N. Li, J. Huang, Q. Wang, and Y. Gu, "A novel DCM-NBD conjugate fluorescent probe for discrimination of Cys/Hcy from GSH and its bioimaging applications in living cells and animals," *Sensors and Actuators B: Chemical*, vol. 245, pp. 297–304, 2017.

[41] Y. Hao, W. Chen, L. Wang et al., "A retrievable, water-soluble and biocompatible fluorescent probe for recognition of Cu(II) and sulfide based on a peptide receptor," *Talanta*, vol. 143, pp. 307–314, 2015.

[42] D. T. Gryko, J. Piechowska, and M. Gałęzowski, "Strongly emitting fluorophores based on 1-azaperylene scaffold," *Journal of Organic Chemistry*, vol. 75, no. 4, pp. 1297–1300, 2010.

[43] R. Nazir, A. J. Stasyuk, and D. T. Gryko, "Vertically π-expanded coumarins: the synthesis and optical properties," *Journal of Organic Chemistry*, vol. 81, no. 22, pp. 11104–11114, 2016.

[44] X. Shi, F. Huo, J. Chao, and C. Yin, "A ratiometric fluorescent probe for hydrazine based on novel cyclization mechanism and its application in living cells," *Sensors and Actuators B: Chemical*, vol. 260, pp. 609–616, 2018.

[45] J. Liu, Y.-Q. Sun, Y. Huo et al., "Simultaneous fluorescence sensing of Cys and GSH from different emission channels," *Journal of the American Chemical Society*, vol. 136, no. 2, pp. 574–577, 2014.

[46] H. Maeda, H. Matsuno, M. Ushida, K. Katayama, K. Saeki, and N. Itoh, "2,4-Dinitrobenzenesulfonyl fluoresceins as fluorescent alternatives to ellman's reagent in thiol-quantification enzyme assays," *Angewandte Chemie International Edition*, vol. 44, no. 19, pp. 2922–2925, 2005.

[47] J. Bouffard, Y. Kim, T. M. Swager, R. Weissleder, and S. A. Hilderbrand, "A highly selective fluorescent probe for thiol bioimaging," *Organic Letters*, vol. 10, no. 1, pp. 37–40, 2008.

[48] M. Wei, P. Yin, Y. Shen et al., "A new turn-on fluorescent probe for selective detection of glutathione and cysteine in living cells," *Chemical Communications*, vol. 49, no. 41, pp. 4640–4642, 2013.

[49] Y. Liu, K. Xiang, B. Tian, and J. Zhang, "A fluorescein-based fluorescence probe for the fast detection of thiol," *Tetrahedron Letters*, vol. 57, no. 23, pp. 2478–2483, 2016.

[50] X.-D. Jiang, J. Zhang, X. Shao, and W. Zhao, "A selective fluorescent turn-on NIR probe for cysteine," *Organic & Biomolecular Chemistry*, vol. 10, no. 10, pp. 1966–1968, 2012.

[51] J. Shao, H. Sun, H. Guo et al., "A highly selective red-emitting FRET fluorescent molecular probe derived from BODIPY for the detection of cysteine and homocysteine: an experimental and theoretical study," *Chemical Science*, vol. 3, no. 4, pp. 1049–1061, 2012.

[52] J. Shao, H. Guo, S. Ji, and J. Zhao, "Styryl-BODIPY based red-emitting fluorescent OFF–ON molecular probe for specific detection of cysteine," *Biosensors and Bioelectronics*, vol. 26, no. 6, pp. 3012–3017, 2011.

[53] W. Qu, L. Yang, Y. Hang, X. Zhang, Y. Qu, and J. Hua, "Photostable red turn-on fluorescent diketopyrrolopyrrole chemodosimeters for the detection of cysteine in living cells," *Sensors and Actuators B: Chemical*, vol. 211, pp. 275–282, 2015.

[54] C. Yin, W. Zhang, T. Liu, J. Chao, and F. Huo, "A near-infrared turn on fluorescent probe for biothiols detection and its application in living cells," *Sensors and Actuators B: Chemical*, vol. 246, pp. 988–993, 2017.

[55] S. Chen, H. Li, and P. Hou, "Imidazo[1,5-α]pyridine-derived fluorescent turn-on probe for cellular thiols imaging with a large Stokes shift," *Tetrahedron Letters*, vol. 58, no. 27, pp. 2654–2657, 2017.

15

Investigation of Interactions between Thrombin and Ten Phenolic Compounds by Affinity Capillary Electrophoresis and Molecular Docking

Qiao-Qiao Li,[1] Yu-Xiu Yang,[1] Jing-Wen Qv,[2] Guang Hu ⓘ,[3] Yuan-Jia Hu ⓘ,[2] Zhi-Ning Xia,[1] and Feng-Qing Yang ⓘ[1]

[1]School of Chemistry and Chemical Engineering, Chongqing University, Chongqing 401331, China
[2]State Key Laboratory of Quality Research in Chinese Medicine, Institute of Chinese Medical Sciences, University of Macau, Macau, China
[3]School of Pharmacy and Bioengineering, Chongqing University of Technology, Chongqing 400054, China

Correspondence should be addressed to Yuan-Jia Hu; yuanjiahu@umac.mo and Feng-Qing Yang; fengqingyang@cqu.edu.cn

Academic Editor: Chih-Ching Huang

Thrombin plays a vital role in blood coagulation, which is a key process involved in thrombosis by promoting platelet aggregation and converting fibrinogen to form the fibrin clot. In the receptor concept, drugs produce their therapeutic effects via interactions with the targets. Therefore, investigation of interaction between thrombin and small molecules is important to find out the potential thrombin inhibitor. In this study, affinity capillary electrophoresis (ACE) and in silico molecular docking methods were developed to study the interaction between thrombin and ten phenolic compounds (p-hydroxybenzoic acid, protocatechuic acid, vanillic acid, gallic acid, catechin, epicatechin, dihydroquercetin, naringenin, apigenin, and baicalein). The ACE results showed that gallic acids and six flavonoid compounds had relative strong interactions with thrombin. In addition, the docking results indicated that all of optimal conformations of the six flavonoid compounds were positioned into the thrombin activity centre and had interaction with the HIS57 or SER195 which was the key residue to bind thrombin inhibitors such as argatroban. Herein, these six flavonoid compounds might have the potential of thrombin inhibition activity. In addition, the developed method in this study can be further applied to study the interactions of other molecules with thrombin.

1. Introduction

Thrombosis persists as a leading cause of death and incapacity worldwide [1]. The formation of thrombosis is a very complex pathological process involving the platelets and blood coagulation components. Thrombin plays a vital role in blood coagulation by promoting platelet aggregation and by converting fibrinogen to form the fibrin clot in the final step of the coagulation cascade [2]. Thrombin is composed of two polypeptide chains of 36 (A chain) and 259 (B chain) residues that are covalently linked through a disulfide bond, and the B chain carries the functional epitopes of the enzyme [3]. As same as all chymotrypsin-like serine proteases, thrombin has a conserved active centre located inside the molecule and contains amino acid residues

of HIS57, ASP102, and SER195, which are called the catalytic triad [4]. Except for its active centre, thrombin possesses two exosites (1 and 2), positively charged domains located at opposite poles of the enzyme, among them, exosite 1 is utilized to dock on the substrates such as fibrinogen, and exosite 2 serves as the heparin-binding domain [5].

The current antithrombotic therapies include heparin (unfractionated heparin and low-molecular-weight heparins), fondaparinux, vitamin K antagonists, factor Xa inhibitors, and direct thrombin inhibitors [6]. Direct thrombin inhibitors (DTIs), such as argatroban, dabigatran, lepirudin, desirudin, and bivalirudin, which bind to thrombin and block its enzymatic activity, are widely and effectively used in the treatment of thromboembolic diseases; however, dabigatran is not orally available due to its high polarity [7].

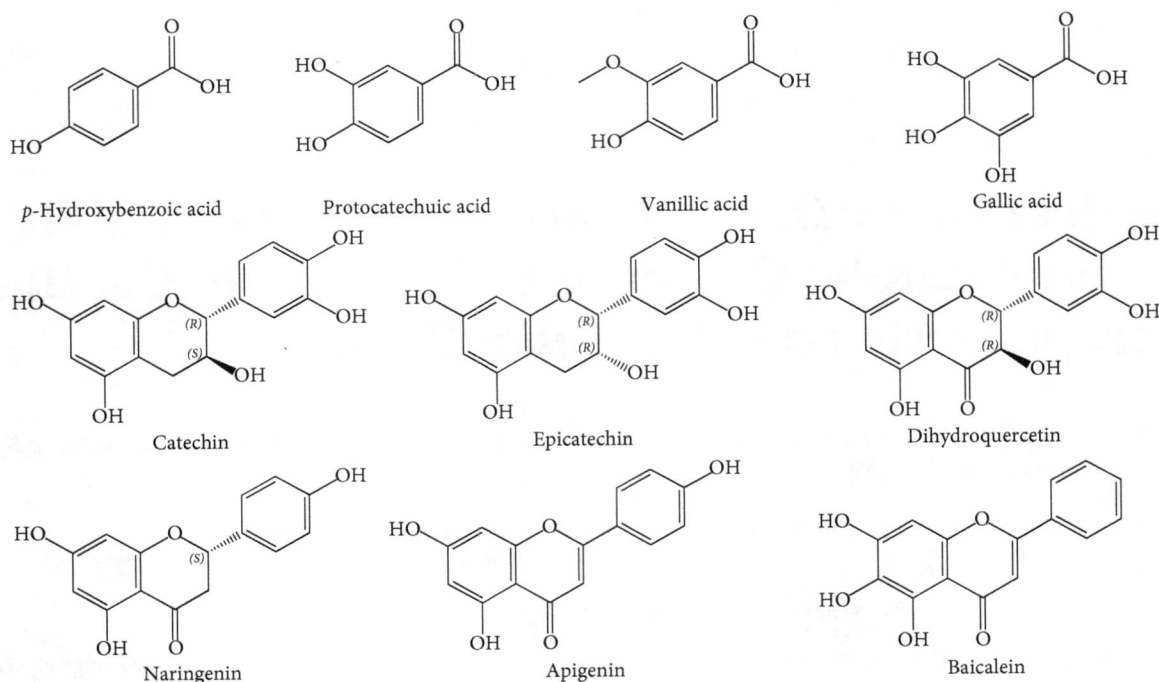

FIGURE 1: The chemical structures of ten investigated phenolic compounds.

Therefore, dabigatran etexilate as prodrug was developed to facilitate gastrointestinal absorption by adding an ethyl group at the carboxylic acid group and a hexyloxycarbonyl side chain at the amidine group [8]. Compared with heparins, DTIs do not require antithrombin as a cofactor and do not bind to plasma proteins; therefore, they produce a more predictable anticoagulant effect, and variability of patient response is low relative to other drug classes [9]. In reality, they still present limitations such as a narrow therapeutic window, and bleeding and anaphylaxis as side effects [10]. Therefore, alternative antithrombotic therapies are under extensive investigation, and many entities from natural products are being isolated and studied to counteract these side effects [11]. Liu et al. [12] evaluated a series of natural flavonoids as potential thrombin inhibitors by optimized method of thrombin time and found that myricetin and quercetin were the best thrombin inhibitors. Bijak et al. [4] showed that cyanidin, quercetin, and silybin changed thrombin proteolytic activity, while cyanidin and quercetin caused a strong response in the interaction with immobilized thrombin by BIAcore analyses. Thus, their results suggested that polyphenolic compounds might be potential structural bases and source to find and project nature-based, safe, orally bioavailable direct thrombin inhibitors.

In the receptor concept, drugs produce their therapeutic effects via interactions with the targets [13]. Therefore, evaluation of the interaction between thrombin and phenolic compounds is important for studying of drug action mechanism and drug discovery. Affinity capillary electrophoresis (ACE) is one of the predominant methods for interaction studies. Comparing with traditional methods

such as equilibrium dialysis, ultrafiltration, ultracentrifugation, liquid chromatography, spectroscopic methods (UV-visible, fluorescence, infrared, nuclear magnetic resonance, optical rotatory dispersion, and circular dichroism), and isothermal titration calorimetry [14], ACE has lots of advantages including high separation efficiency, short analysis duration, low sample and reagent volumes, low purity requirement, ease of automation, and the ability to work under near-physiological conditions [14, 15]. At present, ACE had usually been applied in evaluating the noncovalent interaction between protein and ligands [15–17] or cell and ligands [13, 18]. In addition, in silico molecular docking had recently been presented as an assistant technology to obtain information about the interactions between thrombin and small molecules [19–21].

The aim of this study was to investigate the interaction between thrombin and phenolic compounds including p-hydroxybenzoic acid, protocatechuic acid, vanillic acid, gallic acid, catechin, epicatechin, dihydroquercetin, naringenin, apigenin, and baicalein (the chemical structures of the investigated compounds are presented in Figure 1) by ACE and in silico molecular docking. Firstly, ACE was used to calculate the binding constants (K_b) of the interactions between thrombin and phenolic compounds in experimental environment. Then, in silico molecular docking as assistant technology was also utilized to further explore the interactions between phenolic compounds and thrombin, including the binding sites and positioning of the ligand into the binding site. Finally, the interaction of compounds with enzyme activity centre, which might be potential thrombin inhibitors, could be obtained by the results of ACE and molecular docking.

2. Materials and Methods

2.1. Reagents. Bovine thrombin was obtained from Sigma-Aldrich (Shanghai) Trading Co., Ltd. (Shanghai, China). *p*-Hydroxybenzoic acid, vanillic acid, naringenin, apigenin, baicalein, catechin, and epicatechin were purchased from Chengdu Biopurify Phytochemicals Ltd. (Chengdu, China). Protocatechuic acid and dihydroquercetin were purchased from Push Bio-Technology Co., Ltd. (Chengdu, China). Gallic acid was the product of Aladdin Reagent Co., Ltd. (Shanghai, China). Tris(hydroxymethyl)aminomethane (Tris) was obtained from Sangon Biotech (Shanghai) Co., Ltd. (Shanghai, China). HCl and acetone were analytical grade reagents and purchased from Chengdu Kelong Chemical Reagent Factory (Chengdu, China).

2.2. Apparatus. The capillary electrophoresis experiments were performed on an Agilent 7100 3D CE system (Agilent Technologies, Palo Alto, CA, USA) equipped with a diode array detector and Agilent ChemStation software. The bare fused-silica capillary (Yongnian Ruifeng Chromatographic Device Co., Ltd., Hebei, China) was 75 μm id and had a total length of 50 cm and an effective length of 41.8 cm. The water used for all the experiments was purified by water purification system (ATS-H20, Antesheng Environmental Protection Equipment Co., Ltd., Chongqing, China). Background electrolytes buffer and phenolic compound solutions were ultrasonicated in a KQ-100B ultrasonic cleaner (Kunshan Ultrasonic Instruments Co., Ltd., Kunshan, China). The pH of running buffer was measured by a FE28 pH meter (Mettler-Toledo Instruments, Shanghai, China).

2.3. Sample and Buffer Preparation. The running buffer containing 50 mol/L Tris was adjusted to pH 9.0 with 1 mol/L HCl. 10,000 U of thrombin was solved in 20 mL water, and the enzyme activity was 500 U/mL. The running buffers containing different concentrations of thrombin (0.4 U/mL, 0.8 U/mL, 1.2 U/mL, 1.6 U/mL, and 2.0 U/mL) were prepared by diluting the 500 U/mL thrombin stock solutions with running buffer. The samples of phenolic compounds (about 0.2 g/L) were prepared by dissolving each compound in buffer solution containing 5% (*v/v*) acetone as neutral EOF marker.

2.4. Electrophoresis Conditions. The voltage applied across the capillary was 20 kV. The sample was injected into the capillary under the pressure of 50 mbar for 5 s. The temperature of the capillary cartridge was set at 25°C. The conditioning between two successive runs was done according to the following protocol: 1 mol/L NaOH solution for 2 min, water for 2 min, and buffer solution for 2 min under 930 mbar.

2.5. Calculation of K_b by ACE. A schematic diagram of calculating K_b is shown in Figure 2. In ACE, K_b can be determined from the variation of mobility shifts of the sample at running buffer containing different concentrations of thrombin. The binding constant K_b could be calculated by the Scatchard equation [22] as follows:

$$\frac{\mu_i - \mu_f}{[L]} = -K_b(\mu_i - \mu_f) + K_b(\mu_c - \mu_f), \qquad (1)$$

where K_b is the binding constant, $[L]$ is the equilibrium concentration of uncomplexed ligand, μ_f and μ_c are the electrophoretic mobility of free and complexed solute, and μ_i is the solute mobility measured at ligand concentration $[L]$.

In this work, the mobility ratio M was inducted to eliminate the effects of the variation of running buffer, as defined in the following equation:

$$M = \frac{\mu_{net}}{\mu_{eo}} = \frac{\mu}{\mu_{eo}} + 1 = \frac{l_c l_d / (Vt)}{l_c l_d / (Vt_{eo})} + 1 = \frac{t_{eo}}{t} + 1, \qquad (2)$$

where l_c is the total length of the capillary, l_d is the effective length of the capillary, μ_{net} is the net mobility measured for the solute, μ_{eo} is the mobility due to EOF, μ is the inherent mobility of the solute, t is the measured analyte migration time, t_{eo} is the migration time of the neutral marker, and V is the operating voltage.

Analysis that can be performed using M is thus independent of the capillary length and voltage; the expression depends solely on the migration time of the phenolic compound relative to acetone. Equations (1) and (2) were integrated into (3). K_b could be obtained as follows:

$$\frac{M_i - M_f}{[L]} = -K_b(M_i - M_f) + K_b(M_c - M_f), \qquad (3)$$

where M_i is the solute mobility ratio measured at the ligand concentration $[L]$ and M_f and M_c are the electrophoretic mobility ratio of free and complexed solute.

2.6. In Silico Molecular Docking. An in silico molecular docking study was performed to validate the binding potency of the phenolic compounds to thrombin by using AutoDock 4.2 program [23]. The molecular dockings were conducted by using the crystal structure of the thrombin-argatroban complex (PDB ID = 1DWC) at 1.53 Å resolution [24], where the ligand argatroban was deleted using UCSF Chimera. Besides, polar hydrogen atoms were added, and the crystal water was remained. The three-dimensional chemical structures of compounds were drawn by ChemOffice and minimized energy, with outputting in PDB format.

The cubic grid box was set to $60 \times 60 \times 60$ points with a spacing of 0.375 Å. The catalytic site of the grid box was centralized using the following coordinates ($x = 35.887$; $y = 19.178$; $z = 18.856$). To find the best orientations and conformations of the ligands in the protein binding sites, the Lamarckian genetic algorithm was selected, with an initial population size of 150, a maximum number of evaluations of 2.5×10^6 (medium), maximum number of generations of 27,000, gene mutation rate of 0.02, crossover rate of 0.8, and number of GA runs equal to 50 [25]. The interaction figures were generated, and the results of docking were recorded with binding potency and bonded residues. Additionally, the

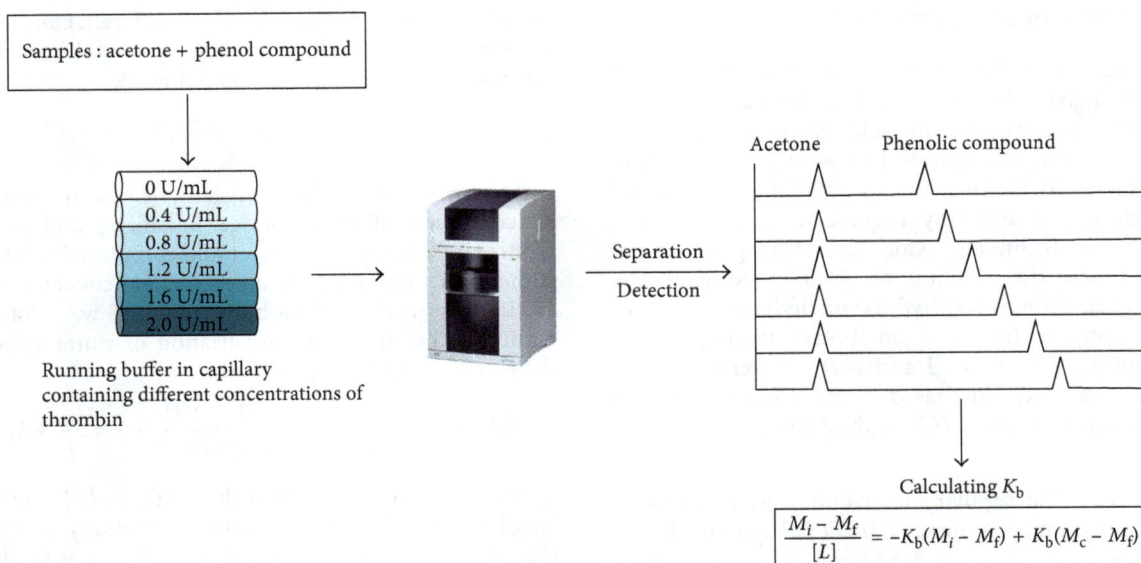

FIGURE 2: Schematic diagram of the steps of this study.

3. Results and Discussion

2D interaction diagrams also were produced by Discovery Studio 4.5 to obtain specific interaction analysis including the functional groups, bonded residues, and interaction force.

3.1. ACE Analysis. Through preliminary investigation of thrombin concentration in running buffer, the six suitable concentrations were arranged as 0 U/mL, 0.4 U/mL, 0.8 U/mL, 1.2 U/mL, 1.6 U/mL, and 2.0 U/mL for ACE analyses. Considering about some flavonoid compounds were sparingly soluble in buffer at pH 7.4, and the activity of thrombin presents a maximum around pH 9.5 [26]. Therefore, the pH 9.0 of running buffer was used.

It was attempted to calculate the K_b value of argatroban and thrombin. However, the migration time of argatroban was same as argatroban-thrombin complex which did not meet the basic requirement of the ACE method [14]. So the K_b value of interaction between argatroban and thrombin could not be obtained in this study. Baicalein as sample containing 5% (v/v) acetone was analyzed in running buffer at six concentrations. As shown in Figure 3, the migration of acetone delayed due to the effect of the variation of running buffer. But the introduced mobility ratio eliminated the error. Then, the binding constant was calculated by the variation of mobility shifts of baicalein at running buffer containing different concentrations of thrombin. The K_b values of other compounds were calculated by the same procedure (electrophoregrams of other compounds were provided in Supplementary Figure S1). The migration time of p-hydroxybenzoic acid, vanillic acid, and protocatechuic acid delayed, and mobility ratio became smaller with increasing the thrombin concentration in running buffer which indicated that thrombin had interaction with them. However, there was no nonlinear correlation between $(M_f - M_i)/[L]$ and $(M_f - M_i)$. According to the previous

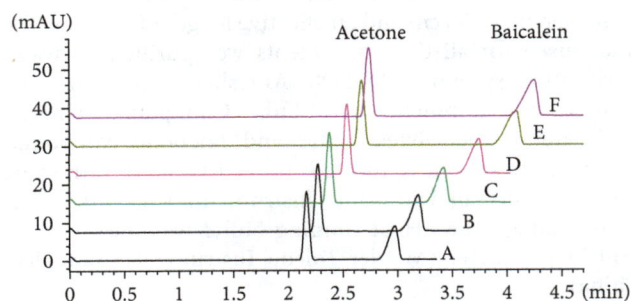

FIGURE 3: Electrophoregrams of baicalein and acetone in running buffers containing different concentrations of thrombin. Thrombin concentration in running buffer: 0 U/mL (a), 0.4 U/mL (b), 0.8 U/mL (c), 1.2 U/mL (d), 1.6 U/mL (e), and 2.0 U/mL (f).

study [26], the binding modes of small molecules and biomacromolecules include site-specific and nonspecific binding. Furthermore, the K_b value could only be calculated by the Scatchard equation when the site-specific binding mode was dominant. Therefore, the reason of K_b value of interaction of thrombin and p-hydroxybenzoic acid, vanillic acid, and protocatechuic acid could not be obtained in this study might be that the site-specific binding force was weak. The other phenolic compounds had relative stronger affinity with thrombin, and the K_b values are shown in Table 1.

In general, the affinities of thrombin binding with flavonoids were stronger than phenolic acids according to the K_b values obtained by ACE. Gallic acid possessed more OH group than other three phenolic acid compounds (p-hydroxybenzoic acid, vanillic acid, and protocatechuic acid), which may relate to its stronger affinity with thrombin. Compared with the K_b values of other flavonoid compounds, baicalein had the strongest affinity with thrombin, which is possibly contributed to the existence of three OH groups (5th, 6th, and 7th positions) at A-ring. Liu et al. [12] demonstrated that more OH groups in the A-ring will

TABLE 1: Interactions of ten investigated compounds with thrombin evaluated by ACE.

Compounds	Molecule weight	pKa	Detection wavelength (nm)	Regression equation of $(M_f - M_i)/[L]$ and $(M_f - M_i)$	Binding constant (mL/U)
p-Hydroxybenzoic acid	138.12	4.57 ± 0.10	250	—	—
Protocatechuic acid	154.12	4.45 ± 0.10	250	—	—
Vanillic acid	168.15	4.45 ± 0.10	250	—	—
Gallic acid	170.12	4.33 ± 0.10	250	$y = -0.184x + 0.105$ $(R^2 = 0.95)$	0.184
Naringenin	272.25	7.52 ± 0.40	280	$y = -0.238x + 0.117$ $(R^2 = 0.99)$	0.238
Apigenin	270.24	6.53 ± 0.40	280	$y = -0.302x + 0.147$ $(R^2 = 0.99)$	0.302
Baicalein	270.24	6.31 ± 0.40	280	$y = -0.508x + 0.093$ $(R^2 = 0.94)$	0.508
Catechin	290.27	9.54 ± 0.10	280	$y = -0.353x + 0.080$ $(R^2 = 0.98)$	0.353
Epicatechin	290.27	9.54 ± 0.10	280	$y = -0.297x + 0.043$ $(R^2 = 0.99)$	0.297
Dihydroquercetin	304.25	7.39 ± 0.60	250	$y = -0.389x + 0.075$ $(R^2 = 0.99)$	0.389

increase thrombin inhibition activity. Previous reported results also indicated that hydroxyl groups at C-3′ and C-4′ positions in the B-ring and hydroxyl group at C-3 position in the C-ring played key roles in the thrombin inhibitory activity [27]. And the present results of ACE showed that naringenin had the lowest affinity with thrombin due to lack of OH groups. According to the calculated K_b values of naringenin and apigenin, the presence of C2=C3 was also important for the affinity with thrombin. Additionally, spatial structure difference could result in different binding strength to thrombin such as catechin and epicatechin.

3.2. Molecular Docking Results. The essence of molecular docking was a recognition process of two or more molecules, involving the space and energy matching between them. The major tasks of the docking procedure were characterization of the binding site, positioning of the ligand into the binding site, and evaluating the strength of interaction for a specific ligand-receptor complex [28]. The semiflexible docking was applied in this study. In the docking, the thrombin conformation kept constant, but the compound conformation could change slightly to get the most stable phenolic compound-thrombin complex also called optimal conformation.

Clustering the docking conformations of thrombin-phenolic acid complexes, the result showed that their optimal conformation was not present in the activity centre, meaning that the four phenolic acid compounds could bind with other place of the enzyme. Therefore, this study only discussed the docking results of thrombin and flavonoid compounds. The 2D interaction diagram of baicalein with residues of thrombin can be observed in Figure 4. Three OH groups (5th, 6th, and 7th positions) at A-ring of baicalein have interaction with ASP189 and SER214 of thrombin by hydrogen bonds which are marked by deep green dotted lines on the 2D interaction diagram. The 2D interaction diagrams of other compounds with residues of thrombin are shown in Supplementary Figure S2, and the docking results and residue interactions are demonstrated in Table 2.

It was recognized that the region with binding energy below −5.0 kcal/mol could be regarded as the "Potential Targets" [23]. The docking results showed that the binding energies of six flavonoid compounds were similar and below −7.0 kcal/mol. A possible explanation is that their highly

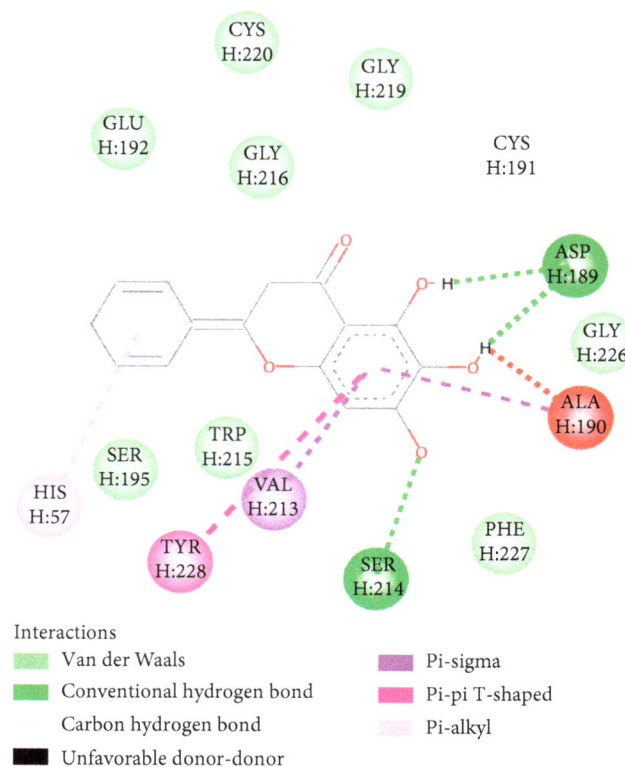

FIGURE 4: The 2D interaction diagram of baicalein with residues of thrombin.

similar structures lead to the approximate docking results. Therefore, all of six compounds had relative strong affinity with thrombin. In addition, the binding positions of six flavonoid compounds were alike with argatroban as shown in Figures 5(a) and 5(b). It is worth mentioning that all of them located in the activity centre and had interaction with the residue HIS57 or SER195, which indicated that these compounds likely had thrombin inhibitory activity.

4. Conclusion

In this work, quick and simple methods (ACE and in silico molecular docking) were developed to investigate the interactions between thrombin and ten phenolic compounds.

TABLE 2: The docking results and residue interactions of the complexes of argatroban and six flavonoids with thrombin.

Compounds	Molecule weight	NC	NP	BE (kcal/mol)	H-bond	EI	VDW
Argatroban	508.63	26	7	−8.93	**HIS57**, ASP189, ALA190, **SER195**, GLY219	TYR60A, TRP60D, LEU99, TRP215	CYS60F, CYS191, GLU192, GLY193, VAL213, SER214, GLY216, GLU217, CYS220, GLY226, PHE227
Baicalein	270.24	1	50	−7.31	ASP189, CYS191, SER214	**HIS57**, ALA190, VAL213, TYR228	GLU192, **SER195**, TRP215, GLY216, GLY219, CYS220, GLY226, PHE227
Apigenin	270.24	1	50	−7.39	ASP189, GLU192, GLY226, PHE227	ALA190, VAL213, TYR228	CYS191, **SER195**, SER214, TRP215, GLY216, GLY219, CYS220
Naringenin	272.25	1	50	−7.80	ASP189, SER214, GLY216, GLY219	ALA190, CYS191, CYS220	**HIS57**, LEU99, GLU192, **SER195**, VAL213, TRP215, GLU217, GLY226, PHE227, TRY228
Dihydroquercetin	304.25	4	42	−7.67	ASP189, SER214, GLY216, GLY219	ALA190, CYS220	**HIS57**, LEU99, CYS191, GLU192, **SER195**, VAL213, TRP215, GLU217, ASP221, GLY226, PHE227, TYR228
Catechin	290.27	4	27	−7.59	ASP189, **SER195**, SER214, GLY216, CYS220, TYR228	ALA190, VAL213, TYR228	**HIS57**, CYS191, GLU192, TRP215, GLY219, GLY226, PHE227
Epicatechin	290.27	3	45	−8.09	ASP189, GLU192, **SER195**, SER214, GLY216, GLY219, CYS220, TYR228	ALA190, VAL213, TRP215	**HIS57**, CYS191, ASP194, GLY226, PHE227

NC: number of clusters; NP: number of poses on the lowest energy cluster; BE: binding energy; H-bond: hydrogen bond; EI: electrostatic interactions; VDW: van der Waals.

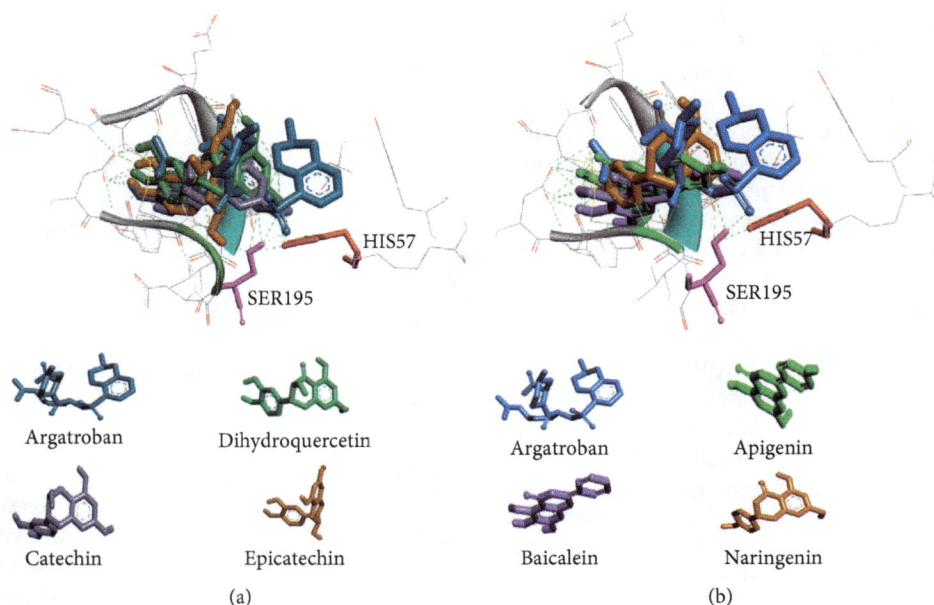

FIGURE 5: Comparison of 3D structures of six flavonoids docked with the thrombin catalytic site. (a) The 3D structures of argatroban, dihydroquercetin, catechin, and epicatechin; (b) the 3D structures of argatroban, apigenin, baicalein, and naringenin.

As compared with conventional methods for interaction study, combination of ACE and molecular docking could greatly decrease the analyses time and sample consumption, and the mechanisms of drug effect could be explained by specific interaction with residues of proteins. Furthermore, the results of interaction study in actual (experimental) and simulation environment are mutually complementary or verified.

The results of present study indicated that the interactions between thrombin and phenolic acids were weak except for gallic acids. On the other hand, in docking results, phenolic acids could only interact with residues other than activity centre due to lack of space matching. So these phenolic acids (except gallic acid) possibly have not thrombin inhibition effects. The calculated K_b values by ACE in actual environment indicated that gallic acid and six flavonoid compounds had relative strong affinity with thrombin, and the affinity was positively correlated with the number of OH groups in A-ring specially. Furthermore, OH groups at C-3' and C-4' position in the B-ring and OH group at C-3 position in the C-ring, and the presence of C2=C3 also played key roles for the interaction with thrombin. The docking results showed that all of optimal conformations of the investigated flavonoid compounds were positioned into the thrombin activity centre and had interactions with HIS57 or SER195, which indicated that these flavonoids possibly had the thrombin inhibition activity. In a word, combination of ACE and molecular docking is an effective approach for studying the interaction between thrombin and small molecules, and characterizing of the binding site, which is an important process for drug discovery.

Conflicts of Interest

The authors declared that they have no conflicts of interest.

Acknowledgments

This work was supported by the National Natural Science Foundation of China (21275169 and 81703687) and the Natural Science Foundation Project of CQ CSTC (cstc2015jcyjA10044). The authors sincerely acknowledge the Science and Technology Development Fund of Macao SAR and the University of Macau for financial support by the projects FDCT013-2015-A1 and MYRG2016-00144-ICMS-QRCM for this research.

References

[1] J. W. Yau, K. K. Singh, Y. Hou et al., "Endothelial-specific deletion of autophagy-related 7 (ATG7) attenuates arterial thrombosis in mice," *Journal of Thoracic and Cardiovascular Surgery*, vol. 154, no. 3, pp. 978–988, 2017.

[2] A. M. Tanaka-Azevedo, K. Morais-Zani, R. J. Torquato, and A. S. Tanaka, "Thrombin inhibitors from different animals," *Journal of Biomedicine & Biotechnology*, vol. 2010, Article ID 641025, 9 pages, 2010.

[3] E. DiCera, Q. D. Dang, and Y. M. Ayala, "Molecular mechanisms of thrombin function," *Cellular and Molecular Life Sciences*, vol. 53, no. 9, pp. 701–730, 1997.

[4] M. Bijak, R. Ziewiecki, J. Saluk et al., "Thrombin inhibitory activity of some polyphenolic compounds," *Medicinal Chemistry Research*, vol. 23, no. 5, pp. 2324–2337, 2013.

[5] J. I. Weitz and M. Crowther, "Direct thrombin inhibitors," *Thrombosis Research*, vol. 106, no. 3, pp. V275–V284, 2002.

[6] M. K. Dabbous, F. R. Sakr, and D. N. Malaeb, "Anticoagulant therapy in pediatrics," *Journal of Basic and Clinical Pharmacy*, vol. 5, no. 2, pp. 27–33, 2014.

[7] J. Stangier and A. Clemens, "Pharmacology, pharmacokinetics, and pharmacodynamics of dabigatran etexilate, an oral direct thrombin inhibitor," *Clinical and Applied Thrombosis-Hemostasis*, vol. 15, no. 1, pp. 9S–16S, 2009.

[8] J. Stangier, K. Rathgen, H. Staehle, D. Gansser, and W. Roth, "The pharmacokinetics, pharmacodynamics and tolerability of dabigatran etexilate, a new oral direct thrombin inhibitor, in healthy male subjects," *British Journal of Clinical Pharmacology*, vol. 64, no. 3, pp. 292–303, 2007.

[9] E. A. Nutescu, N. L. Shapiro, A. Chevalier, and A. N. Amin, "A pharmacologic overview of current and emerging anticoagulants," *Cleveland Clinic Journal of Medicine*, vol. 72, no. 1, p. S2, 2005.

[10] Y. Kong, H. Chen, Y. Q. Wang, L. Meng, and J.-F. Wei, "Direct thrombin inhibitors: patents 2002–2012 (review)," *Molecular Medicine Reports*, vol. 9, no. 5, pp. 1506–1514, 2014.

[11] L. de Andrade Moura, A. C. Marqui de Almeida, T. F. Domingos et al., "Antiplatelet and anticoagulant effects of diterpenes isolated from the marine alga, *Dictyota menstrualis*," *Marine Drugs*, vol. 12, no. 12, pp. 2471–2484, 2014.

[12] L. Liu, H. Ma, N. Yang et al., "A series of natural flavonoids as thrombin inhibitors: structure-activity relationships," *Thrombosis Research*, vol. 126, no. 5, pp. e365–378, 2010.

[13] F. Q. Wang, Q. Zhang, C. H. Li et al., "Evaluation of affinity interaction between small molecules and platelets by open tubular affinity capillary electrochromatography," *Electrophoresis*, vol. 37, no. 5-6, pp. 736–743, 2016.

[14] K. Vuignier, J. Schappler, J. L. Veuthey, P.-A. Carrupt, and S. Martel, "Drug-protein binding: a critical review of analytical tools," *Analytical and Bioanalytical Chemistry*, vol. 398, no. 1, pp. 53–66, 2010.

[15] M. Mozafari, S. Balasupramaniam, L. Preu et al., "Using affinity capillary electrophoresis and computational models for binding studies of heparinoids with p-selectin and other proteins," *Electrophoresis*, vol. 38, no. 12, pp. 1560–1571, 2017.

[16] M. Nachbar, M. Mozafari, F. Krull et al., "Metal ion–dehydrin interactions investigated by affinity capillary electrophoresis and computer models," *Journal of Plant Physiology*, vol. 216, pp. 219–228, 2017.

[17] Y. Xu, T. Hong, X. Chen, and Y. Ji, "Affinity capillary electrophoresis and fluorescence spectroscopy for studying enantioselective interactions between omeprazole enantiomer and human serum albumin," *Electrophoresis*, vol. 38, no. 9-10, pp. 1366–1373, 2017.

[18] P. Yakufu, H. Qi, M. Li, X. Ling, W. Chen, and Y. Wang, "CCR4 expressing cells cultured adherently on a capillary wall and formaldehyde fixed as the stationary phase for ligand screening by ACE," *Electrophoresis*, vol. 34, no. 4, pp. 531–540, 2013.

[19] X. Wang, Y. Zhang, Y. Yang, X. Wu, H. Fan, and Y. Qiao, "Identification of berberine as a direct thrombin inhibitor from traditional Chinese medicine through structural, functional and binding studies," *Scientific Reports*, vol. 7, p. 44040, 2017.

[20] Q. Zhang, Y. X. Yang, S. Y. Li et al., "An ultrafiltration and high performance liquid chromatography coupled with diode array detector and mass spectrometry approach for screening and characterizing thrombin inhibitors from Rhizoma Chuanxiong," *Journal of Chromatography B*, vol. 1061-1062, pp. 421–429, 2017.

[21] Y. X. Yang, S. Y. Li, Q. Zhang, H. Chen, Z.-N. Xia, and F.-Q. Yang, "Characterization of phenolic acids binding to thrombin using frontal affinity chromatography and molecular docking," *Analytical Methods*, vol. 9, no. 35, pp. 5174–5180, 2017.

[22] K. L. Rundlett and D. W. Armstrong, "Examination of the origin, variation, and proper use of expressions for the estimation of association constants by capillary electrophoresis," *Journal of Chromatography A*, vol. 721, no. 1, pp. 173–186, 1996.

[23] J. Lu, H. P. Song, P. Li, P. Zhou, X. Dong, and J. Chen, "Screening of direct thrombin inhibitors from Radix Salviae Miltiorrhizae by a peak fractionation approach," *Journal of Pharmaceutical and Biomedical Analysis*, vol. 109, pp. 85–90, 2015.

[24] D. W. Banner and P. Hadvary, "Crystallographic analysis at 3.0-Å resolution of the binding to human thrombin of 4 active site-directed inhibitors," *Journal of Biological Chemistry*, vol. 266, no. 30, pp. 20085–20093, 1991.

[25] R. Pereira, A. Lourenço, L. Terra et al., "Marine diterpenes: molecular modeling of thrombin inhibitors with potential biotechnological application as an antithrombotic," *Marine Drugs*, vol. 15, no. 3, p. 79, 2017.

[26] S. Leborgne and M. Graber, "Amidase activity and thermal-stability of human thrombin," *Applied Biochemistry and Biotechnology*, vol. 48, no. 2, pp. 125–135, 1994.

[27] Z. H. Shi, N. G. Li, Y. P. Tang et al., "Metabolism-based synthesis, biologic evaluation and SARs analysis of O-methylated analogs of quercetin as thrombin inhibitors," *European Journal of Medicinal Chemistry*, vol. 54, pp. 210–222, 2012.

[28] M. Mozzicafreddo, M. Cuccioloni, A. M. Eleuteri, E. Fioretti, and M. Angeletti, "Flavonoids inhibit the amidolytic activity of human thrombin," *Biochimie*, vol. 88, no. 9, pp. 1297–1306, 2006.

16

Synchronized Survey Scan Approach Allows for Efficient Discrimination of Isomeric and Isobaric Compounds during LC-MS/MS Analyses

Keabetswe Masike and Ntakadzeni Madala ⓘD

Department of Biochemistry, University of Johannesburg, P.O. Box 524, Auckland Park 2006, South Africa

Correspondence should be addressed to Ntakadzeni Madala; emadala@uj.ac.za

Academic Editor: Josep Esteve-Romero

Liquid chromatography-mass spectrometry- (LC-MS-) based multiple reaction monitoring (MRM) methods have been used to detect and quantify metabolites for years. These approaches rely on the monitoring of various fragmentation pathways of multiple precursors and the subsequent corresponding product ions. However, MRM methods are incapable of confidently discriminating between isomeric and isobaric molecules and, as such, the development of methods capable of overcoming this challenge has become imperative. Due to increasing scanning rates of recent MS instruments, it is now possible to operate MS instruments both in the static and dynamic modes. One such method is known as synchronized survey scan (SSS), which is capable of acquiring a product ion scan (PIS) during MRM analysis. The current study shows, for the first time, the use of SSS-based PIS approach as a feasible identification feature of MRM. To achieve the above, five positional isomers of dicaffeoylquinic acids (diCQAs) were studied with the aid of SSS-based PIS method. Here, the MRM transitions were automatically optimized using a 3,5-diCQA isomer by monitoring fragmentation transitions common to all five isomers. Using the mixture of these isomers, fragmentation spectra of the five isomers achieved with SSS-based PIS were used to identify each isomer based on previously published hierarchical fragmentation keys. The optimized method was also used to detect and distinguish between diCQA components found in *Bidens pilosa* and their isobaric counterparts found in *Moringa oleifera* plants. Thus, the method was shown to distinguish (by differences in fragmentation patterns) between diCQA and their isobars, caffeoylquinic acid (CQA) glycosides. In conclusion, SSS allowed the detection and discrimination of isomeric and isobaric compounds in a single chromatographic run by producing a PIS spectrum, triggered in the automatic MS/MS synchronized survey scan mode.

1. Introduction

Plants produce a myriad of organic compounds referred to as secondary metabolites (natural products) which differ in their structure and biosynthetic origins. These metabolites undergo chemical modifications, such as conjugation [1, 2], and isomerization (positional and geometrical) [3–6], which further contribute to the high complexity of the plant metabolome. For instance, secondary metabolites such as hydroxycinnamic acid (HCA) derivatives have the potential to form conjugates with organic acids such as isocitric acid [1], tartaric acid [2, 7–9], and quinic acid [4, 10, 11], thus forming hydroxycinnamoyl-isocitric acid [1], hydroxycinnamoyl-tartaric acid [2, 7–9], and hydroxycinnamoyl-quinic acid [4, 10, 11], respectively. The most common HCA derivatives include caffeic acid, ferulic acid, and *p*-coumaric acid, to name a few.

In addition, these HCA derivatives undergo isomerization to produce positional isomers such as di-acylated hydroxycinnamoyl-quinic acid derivatives like 1,3-dicaffeoylquinic acid (1,3-diCQA) or 1,5-diCQA [3–6]. As such, some HCA conjugates have been found to result in isobaric compounds which produce similar mass spectrometry (MS) fragmentation patterns [1], and this renders identification challenging from an analytical perspective. For instance, diCQA positional isomers, apart from being isobaric constituents of each other, also produce similar fragmentation patterns to structurally related molecules such as caffeoylquinic acid (CQA) glycosides [12–14], making identification in different plant species very challenging.

From the above, it can be surmised that secondary metabolites are diverse and, in some cases, are unique to specific plant species and, as such, can be used as chemotaxonomic markers [15–18]. Therefore, the unambiguous detection and identification of these metabolites using analytical techniques such as liquid chromatography-mass spectrometry (LC-MS) is of paramount importance. For targeted analysis, multiple reaction monitoring (MRM) using a triple quadrupole LC-MS/MS system can be employed to selectively screen for and detect and quantify metabolites of interest [19–22]. However, studies dedicated to providing specific fragmentation patterns that discriminate between isomeric and isobaric secondary metabolites during MRM analyses are limited [23, 24]. As such, a novel method/approach referred to as synchronized survey scan (SSS) is proposed as a possible way of discriminating structurally similar metabolites during MRM analyses, especially when they produce similar fragmentation transitions. Using a SSS function, we have demonstrated that positional isomers of diCQAs and their isobaric compounds, CQA glycosides, can be distinguished in a single chromatographic run. Here, authentic standards and plant extracts of *Bidens pilosa* [25] and *Moringa oleifera* were employed, since these plant species are reported to, respectively, accumulate/produce these compounds. Thus, the overall aim of the current study was to use the SSS approach as an orthogonal identification component of the MRM method for efficient discrimination of structurally related plant metabolites.

2. Materials and Methods

2.1. Materials. Authentic standards (with the purity of above 99.6%) of dicaffeoylquinic acids (1,3-diCQA, 1,5-diCQA, 3,4-diCQA, 3,5-diCQA, and 4,5-diCQA) were purchased from Phytolab (Vestenbergsgreuth, Germany). Mass spectrometry grade (99.9%) methanol was purchased from Romil Pure Chemistry (Cambridge, UK). Mass spectrometry grade formic acid (with the purity of above 96%) was obtained from Sigma-Aldrich (St. Louis, MO, USA). The analytical column used was a reverse-phase Raptor biphenyl (2.1 × 100 mm, 3 μm) column purchased from Restek (Bellefonte, PA, USA).

2.2. Methods

2.2.1. Sample Preparation. A 1 mg/mL solution for each diCQA positional isomer was prepared with 100% methanol. The solution (for each positional isomer) was diluted 10× with 100% methanol. Furthermore, equal amounts (e.g., 40 μL) were taken from each positional isomer sample to prepare a mixed sample (e.g., of a final volume of 200 μL). The samples (individuals and the mixture) were placed in amber vials and subjected to HPLC-PDA analyses.

2.2.2. Metabolite Extraction. The dried leaves of *B. pilosa* and *M. oleifera* were pulverized, respectively, using a clean and dry quartz mortar and pestle. Extraction was conducted using an organic solvent-based extraction. The respective amounts of samples (0.2 g) were mixed with 2 mL of 50% aqueous methanol, and these extracts were placed (with the lids of the tubes closed to avoid evaporation) in a heating block at 60°C for 2 h. The samples were sonicated for 30 min using an ultrasonic bath and then centrifuged at 9740 ×*g* for 10 min at 4°C. The resulting supernatants were subjected to UHPLC-MS analyses.

2.2.3. Ultra-High Performance Liquid Chromatography Tandem Mass Spectrometry (UHPLC-MS/MS) Analysis. Once prepared, samples were analyzed on a Shimadzu Nexera 8050 UHPLC (Kyoto, Japan) fitted with a Raptor biphenyl analytical column, with the column temperature set at 40°C. A binary solvent mixture consisting of MilliQ water made up of 0.1% formic acid (eluent A) and methanol made up of 0.1% formic acid (eluent B) at a constant flow rate of 0.4 mL/min was used to analyze 2 μL of the injected samples. For the gradient elution, the following conditions were used: isocratic 5% eluent B from 0 min to 1 min, linear 5%–20% eluent B from 1 min to 5 min, linear 20%–90% eluent B from 5 min to 40 min, and isocratic 90% eluent B from 40 min to 45 min. At the end of analysis, conditions were changed to the initial conditions (5% eluent B) from 45 min to 48 min and finally, the column was reequilibrated with isocratic 5% eluent B from 48 min to 52 min. The data were acquired using a UV detector set at 325 nm and 330 nm.

For the MS analysis, the chromatographic effluent was introduced to a MS source and ionized by electrospray (ESI). ESI conditions were as follows: the interface voltage was set at 3.0 kV (in the negative ESI mode), the source temperature was 300°C, nitrogen was used as the drying gas at the flow rate of 10.00 L/min, and as a nebulizing gas at a flow rate of 3.00 L/min. Argon was used as a collision gas with a pressure of ±230 kPa in the collision cell. Sensitive and qualitative analysis of isomeric and isobaric plant metabolites was achieved by developing a MRM and MRM-dependent product ion scan (PIS) method. The MRM transitions were developed or optimized using 3,5-diCQA as the sample of choice based on the work done by Clifford et al. [3]. According to Clifford and colleagues, 3,5-diCQA contains product ions (e.g., *m/z* 353 representing a caffeoylquinic acid moiety, *m/z* 179 representing a caffeic acid moiety, and *m/z* 191 representing a quinic acid moiety) characteristic of all diCQA isomers, as well as CQA glycosides [13, 14]. The MRM transition parameters were automatically optimized to produce the transitions shown in Table 1. The dwell time for all the MRM transitions was 30 ms.

A synchronized survey scan (SSS) function was selected which automatically performed MS/MS analysis when the precursor threshold peak intensity exceeded 2,000,000. This resulted in a combined MRM and a MRM-dependent PIS, both of which were produced in a single analysis. For the PIS mode, ions were collected at a mass range 100–1000 Da with a continuous scan time of 1 sec, at a collision energy of 25 eV.

TABLE 1: MRM transitions automatically optimized using 3,5-diCQA.

Precursor (m/z)	Transitions (m/z)	Collision energy (eV)
515	353	18
515	191	40
515	179	28

3. Results and Discussion

3.1. LC-MS/MS Method Optimization. In this study, LC-MS analysis was used to sensitively and qualitatively analyze isomeric and isobaric plant metabolites by developing a MRM-dependent product ion scan (PIS) method. Samples (authentic standards and plant samples) containing positional isomers of dicaffeoylquinic acids (diCQAs) and caffeoylquinic acid (CQA) glycosides were analyzed under reverse-phase chromatographic conditions using a Raptor biphenyl column with methanol as part of the binary solvent mixture. To identify the respective diCQA positional isomers, a sample made up of a mixture of the positional isomers (authentic standards) was analyzed, and it produced a chromatogram showing well-resolved peaks representing the five isomers (Figure 1; Table 2). The retention times (Rts) of the respective peaks were compared with the Rts of the individual (nonmixed) authentic standards, and the elution order under the abovementioned conditions was noted as 1,3-diCQA (Figure 1, **A**), 3,4-diCQA (Figure 1, **B**), 3,5-diCQA (Figure 1, **C**), 1,5-diCQA (Figure 1, **D**), and 4,5-diCQA (Figure 1, **E**). Furthermore, the respective product ion scan (PIS) spectra, triggered in the automatic MS/MS synchronized survey scan mode, were also referred to for the analysis of the fragmentation patterns of the diCQA positional isomers for further identification (Figure 2). Although the MRM transition parameters were automatically optimized using 3,5-diCQA, all five diCQA isomers were detected, since these compounds have been shown to share similar product ions [3, 11].

3.2. MRM and MRM-Dependent Product Ion Scan Analysis of Plant Extracts. After optimization, *Bidens pilosa* and *Moringa oleifera* plant extracts were analyzed under the abovementioned conditions (Section 3.1). When the MRM transition parameters, which were automatically optimized using 3,5-diCQA, were used to analyze extracts of *B. pilosa* and *M. oleifera* plant samples, several peaks were detected. Briefly, three peaks were detected in *B. pilosa*, and only two peaks were detected in *M. oleifera*. When compared, the two peaks detected from *M. oleifera* samples showed an earlier elution profile (Rt = 4.89 and 7.32) and were thus observed to be more hydrophilic than the three peaks detected in *B. pilosa* samples, which showed a later elution profile (Rt = 21.24, 22.05, and 24.46). These observations are summarized in Table 2. Such chromatographic observations are important in scenarios whereby the analyzed plant sample contains both diCQAs and CQA glycosides, thus circumstances whereby the plant extract contains both the isomeric and isobaric compounds [11, 13, 14, 26].

The Rts of the peaks (Rt = 21.29, 22.05, and 24.61) found in *B. pilosa* samples were compared with those observed with the diCQA authentic standard samples, and the respective peaks were identified as 3,4-diCQA (Figure 3(a)), 3,5-diCQA (Figure 3(b)), and 4,5-diCQA (Figure 3(c)) (Figure 3; Table 2), since the fragmentation pattern of these peaks was identical to the fragmentation pattern observed in Figures 2(b), 2(c), and 2(e), respectively. For the peak representing 3,4-diCQA, the observed fragment ions were (Figure 3(a)) at m/z 353 ([caffeoylquinic acid-H]⁻) due to the loss of a caffeic acid moiety [3], at m/z 335 ([caffeoylquinic acid-H_2O-H]⁻) due to the dehydration of the caffeoylquinic acid moiety which subsequently results in a lactone group on the quinic acid [27], at m/z 191 ([quinic acid-H]⁻) as a result of a loss of a caffeic acid moiety from a caffeoylquinic acid group, at m/z 179 ([caffeic acid-H]⁻) ascribed to the loss of the caffeoylquinic acid moiety (or the loss of a quinic acid moiety from a caffeoylquinic acid group), at m/z 173 ([quinic acid-H_2O-H]⁻) (base peak, bp) due to the dehydration of a quinic acid moiety, at m/z 161 ([caffeoylquinic acid-quinic acid moiety-H_2O-2H]⁻) as a result of a loss of a dehydrated quinic acid moiety from a caffeoylquinic acid lactone, and at m/z 135 ([caffeic acid-CO_2-H]⁻) due to the decarboxylation of a caffeic acid moiety (Figures 2(b) and 3(a); Table 2). The peak representing 3,5-diCQA showed fragment ions (Figure 3(b)) at m/z 353 ([caffeoylquinic acid-H]⁻), 191 ([quinic acid-H]⁻) (bp), 179 ([caffeic acid-H]⁻), and 135 ([caffeic acid-CO_2-H]⁻). Lastly, the peak identified as 4,5-diCQA showed product ions (Figure 3(c)) at m/z 353 ([caffeoylquinic acid-H]⁻) (bp), 191 ([quinic acid-H]⁻), 179 ([caffeic acid-H]⁻), and 173 ([quinic acid-H_2O-H]⁻). It is worth noting that in the absence of authentic standards, the ion at m/z 173 is diagnostic of HCA derivatives acylated at position 4 on the quinic acid (e.g., 3,4-diCQA and 4,5-diCQA), as it has been noted in the work done by Clifford et al. [3, 10]. However, to distinguish between 3,4-diCQA and 4,5-diCQA an ion at m/z 335 is noteworthy, as its presence signifies the formation of a lactone group between a carboxylic group at position 1 and a hydroxyl group at position 5 on the quinic acid [27]. Thus, position 5 on the quinic acid need not be acylated to allow the formation of the lactone group as shown elsewhere [27]. Therefore, 3,4-diCQA and 4,5-diCQA can be distinguished based on the presence of the ion at m/z 335 on the MS² spectra representing 3,4-diCQA [10].

The two peaks (Rt = 4.87 and 7.30) detected in the *M. oleifera* plant extracts were tentatively characterized as 3-caffeoylquinic acid (CQA) glycoside and 4-CQA glycoside (Table 2) [26]. The peak (Rt = 4.87) representing 3-CQA glycoside showed fragment ions (Figure 3(d)) at m/z 353 ([caffeoylquinic acid-H]⁻) due to the neutral loss of a glycoside residue (162 Da), 341 ([caffeoyl glycoside-H]⁻) as a result of a loss of a quinic acid moiety, 191 ([quinic acid-H]⁻) ascribed to the loss of a caffeoyl glycoside moiety, and 179 ([caffeic acid-H]⁻) (bp) due to the loss of the quinic acid and glycosyl moieties from the caffeoylquinic acid and caffeoyl glycoside groups, respectively. The peak (Rt = 7.30) representing 4-CQA glycoside showed product ions (Figure 3(e)) at m/z 353 ([caffeoylquinic acid-H]⁻), at m/z 341 ([caffeoyl glycoside-H]⁻), at m/z 191 ([quinic acid-H]⁻), at m/z 179 ([caffeic acid-H]⁻), and ion at m/z 173 ([quinic acid-H_2O-H]⁻) (bp) due to the dehydration of

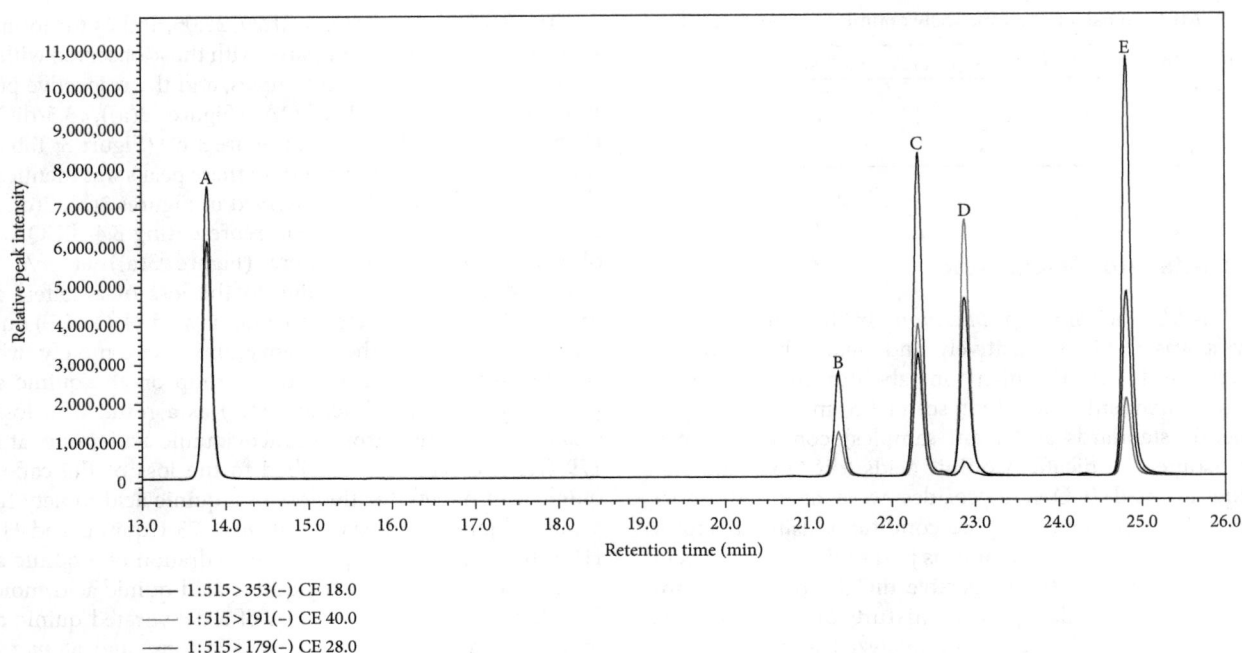

FIGURE 1: Chromatogram showing differences in the relative abundance of the MRM transitions of a sample containing a mixture of dicaffoylquinic acids (diCQAs) authentic standards: **A** = 1,3-diCQA, **B** = 3,4-diCQA, **C** = 3,5-diCQA, **D** = 1,5-diCQA, and **E** = 4,5-diCQA.

TABLE 2: MRM transitions and MRM-dependent product ion scan of isomeric and isobaric compounds from different sample types.

Sample type	Compound	Retention time (Rt) (min)	Precursor (m/z)	Optimal transitions	Collision energy (eV)	MRM-dependent product ion
	1,3-O-dicaffeoylquinic acid	13.71	515	515 > 353 515 > 179/191	18 28/40	515→353, 335, 191 (bp), 179, 161, 135
	3,4-O-dicaffeoylquinic acid	21.34	515	515 > 179/353 515 > 191	28/18 40	515→353, 335, 191, 179, 173 (bp), 161, 135
Standard	3,5-O-dicaffeoylquinic acid	22.25	515	515 > 353 515 > 191/179	18 40/28	515→353, 191 (bp), 179, 135
	1,5-O-dicaffeoylquinic acid	22.82	515	515 > 191 515 > 353	40 18	515→353, 191 (bp)
	4,5-O-dicaffeoylquinic acid	24.75	515	515 > 353 515 > 179	18 28	515→353 (bp), 191, 179, 173, 135
Bidens pilosa	3,4-O-dicaffeoylquinic acid	21.24	515	515 > 179/353 515 > 191	28/18 40	515→353, 335, 191, 179, 173 (bp), 161, 135
	3,5-O-dicaffeoylquinic acid	22.05	515	515 > 353 515 > 191/179	18 40/28	515→353, 191 (bp), 179, 135
	4,5-O-dicaffeoylquinic acid	24.61	515	515 > 353 515 > 179	18 28	515→353 (bp), 191, 179, 173, 135
Moringa oleifera	3-O-(4′-O-caffeoyl glucosyl) quinic acid	4.89	515	515 > 179 —	28 —	515→353, 341, 191, 179 (bp)
	4-O-(4′-O-caffeoyl glucosyl) quinic acid	7.32	515	515 > 179 515 > 353	28 18	515→353, 341, 191, 179, 173 (bp)

a quinic acid moiety. The presence and abundance of the peak at m/z 173 allowed characterization of the peak with the Rt of 7.30 as 4-CQA glycoside [10]. For both the peaks, the presence of the ion at m/z 353 and 341 suggests that the caffeoyl moiety forms an ester bond with the quinic acid (to produce an ion at m/z 353) and an ether bond with the glucose group (to produce an ion at m/z 341). However, where the glycosyl group is connected on the caffeic acid catechol group (C-3′ or C-4′) is not clear. According to Jaiswal et al., an intense ion (bp) at

m/z 323 ([caffeoyl glycoside- H_2O-H]$^-$), due to the dehydration of a caffeoyl glycoside moiety, on the MS^2 spectra, suggests connectivity at C-3′ and an intense ion (bp) at m/z 353 on the MS^2 spectra is characteristic of connectivity at C-4′ [13, 14]. In this study, both ions were secondary ions and not base peaks, and the intensity of both peaks could have been influenced by the differences in the MS conditions/parameters [28]. Thus, due to the low abundance/absence of the ion at m/z 323 on the MS^2 spectra (Figures 3(d) and 3(e)), the CQA glycosides were

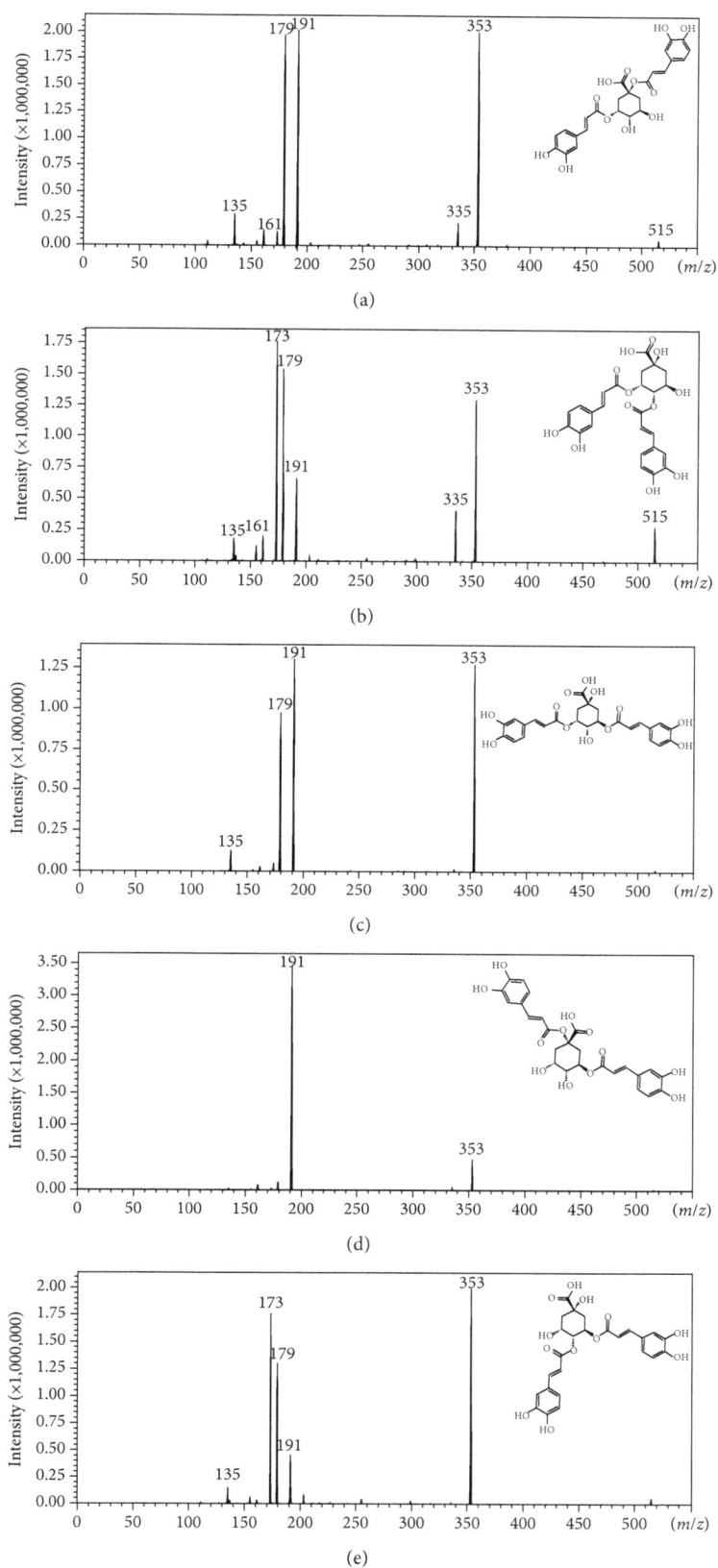

Figure 2: Product ion scan (PIS) spectra and fragmentation pathways of (a) 1,3-diCQA, (b) 3,4-diCQA, (c) 3,5-diCQA, (d) 1,5-diCQA, and (e) 4,5-diCQA authentic standards.

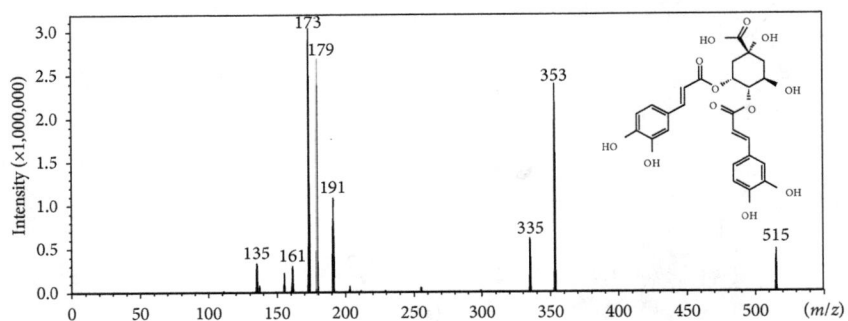

(a)

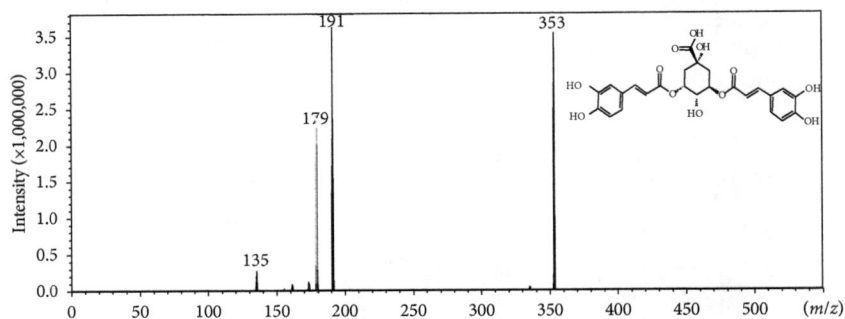

(b)

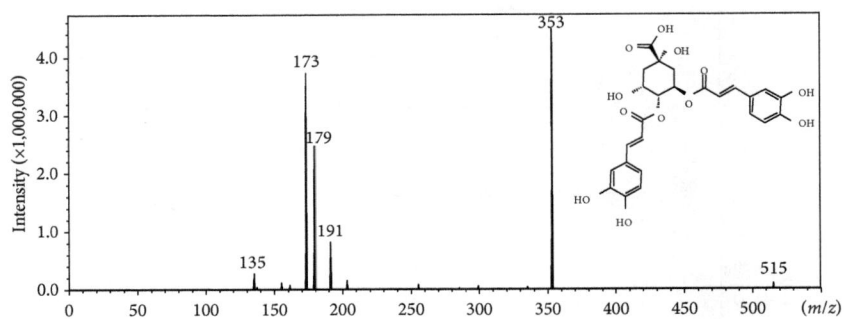

(c)

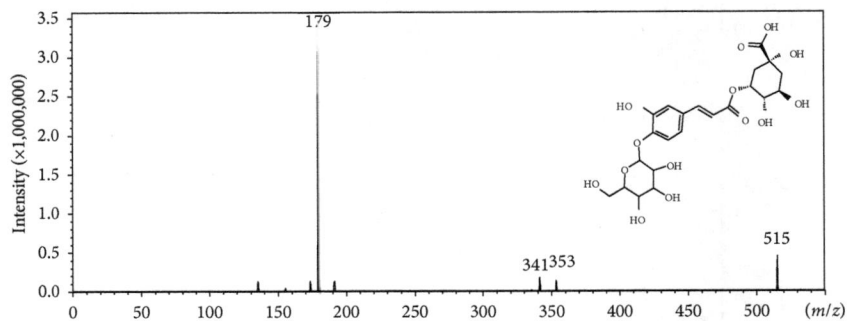

(d)

FIGURE 3: Continued.

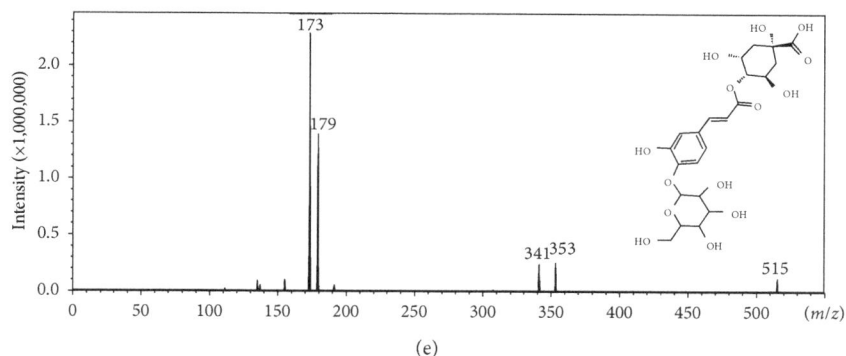

FIGURE 3: Product ion spectra and fragmentation pathway of diCQA positional isomers: (a) 3,4-diCQA, (b) 3,5-diCQA, and (c) 4,5-diCQA detected in *Bidens pilosa* plant extracts. Product ion spectra and fragmentation pathway of CQA glycoside positional isomers: (d) 3-*O*-(4′-*O*-caffeoyl glucosyl) quinic acid and (e) 4-*O*-(4′-*O*-caffeoyl glucosyl) quinic acid detected in *Moringa oleifera* plant extracts.

putatively annotated as 3-*O*-(4′-*O*-caffeoyl glucosyl) quinic acid and 4-*O*-(4′-*O*-caffeoyl glucosyl) quinic acid (Figures 3(d) and 3(e); Table 2).

3.3. Synchronized Survey Scan (SSS). From the above, it is apparent that in the absence of the triggered product ion scan (PIS), the optimized MRM method would have detected the diCQA and the CQA glycosides as these compounds share similar transitions. However, the method would fail to differentiate between the structurally similar compounds and thus lead to misidentification. Thus, the novel approach, SSS, allowed the detection and differentiation of all five diCQA positional isomers (Figure 2) as well as two CQA glycoside positional isomers (Figures 3(d) and 3(e)) in a single chromatographic analysis. Furthermore, the method allowed the discrimination of isobaric compounds such as diCQA and CQA glycosides. Here, an ion at *m/z* 341 was found to be present on the PIS MS/MS spectra of CQA glycosides. This ion represents caffeoyl glycoside [11] and, as seen from our results, it was only observed on the PIS spectra but not the MRM spectra. This is an indication that SSS dependent PIS allows other diagnostic ions to be used for the differentiation of closely related molecules, which otherwise would be impossible if only MRM transitions are relied upon. Thus, SSS produces an automatically performed MS/MS analysis which allows the simultaneous identification of these isomeric and isobaric compounds, without any ambiguity.

4. Conclusion

For the first time, we have successfully demonstrated synchronized survey scan (SSS) to be an efficient approach to detect and discriminate isomeric and isobaric plant metabolites. Briefly, this approach allowed the discrimination and identification of (1) diCQA positional isomers, (2) CQA glycosides positional isomers, and (3) isobaric compounds, diCQAs and CQA glycosides, in a single chromatographic run, albeit the developed MRM transition parameters were automatically optimized using the positional isomer 3,5-diCQA. Our results show that this method can be further applied

in any method where isomers are expected. Furthermore, in phytochemistry, where identification is the key, SSS is expected to add a novel orthogonal feature during identification.

Conflicts of Interest

The authors declare that they have no conflicts of interest.

Acknowledgments

The authors would like to thank the University of Johannesburg and the NRF for the financial support. The authors would also like to thank RESTEK for the RASP grant used to purchase the chromatographic columns. Dr. Riaan Meyer, Dr. Charles Yates, and Mr. Darryl Harris from Shimadzu, South Africa, are thanked for their technical assistance.

References

[1] K. Masike, M. I. Mhlongo, S. P. Mudau et al., "Highlighting mass spectrometric fragmentation differences and similarities between hydroxycinnamoyl-quinic acids and hydroxycinnamoyl-isocitric acids," *Chemistry Central Journal*, vol. 11, no. 29, pp. 1–7, 2017.

[2] K. Masike, F. Tugizimana, N. Ndlovu et al., "Deciphering the influence of column chemistry and mass spectrometry settings for the analyses of geometrical isomers of L-chicoric acid," *Journal of Chromatography B*, vol. 1052, pp. 73–81, 2017.

[3] M. N. Clifford, S. Knight, and N. Kuhnert, "Discriminating between the six isomers of dicaffeoylquinic acid by LC-MS^n," *Journal of Agricultural and Food Chemistry*, vol. 53, no. 10, pp. 3821–3832, 2005.

[4] M. N. Clifford, J. Kirkpatrick, N. Kuhnert, H. Roozendaal, and P. R. Salgado, "LC-MS^n analysis of the *cis* isomers of chlorogenic acids," *Food Chemistry*, vol. 106, no. 1, pp. 379–385, 2008.

[5] M. M. Makola, P. Steenkamp, I. A. Dubery, M. M. Kabanda, and N. E. Madala, "Preferential alkali metal adduct formation

by *cis* geometrical isomers of dicaffeoylquinic acids allows for efficient discrimination from their *trans* isomers during ultra-high-performance liquid chromatography/quadrupole time-of-fight mass spectrometry," *Rapid Communications in Mass Spectrometry*, vol. 30, no. 8, pp. 1011–1018, 2016.

[6] K. Masike, I. Dubery, P. Steenkamp, E. Smit, and E. Madala, "Revising reverse-phase chromatographic behavior for efficient differentiation of both positional and geometrical isomers of dicaffeoylquinic acids," *Journal of Analytical Methods in Chemistry*, vol. 2018, Article ID 8694579, 11 pages, 2018.

[7] J. Lee and C. F. Scagel, "Chicoric acid: chemistry, distribution, and production," *Frontiers in Chemistry*, vol. 1, no. 40, pp. 1–17, 2013.

[8] J. Lee and C. F. Scagel, "Chicoric acid levels in commercial basil (*Ocimum basilicum*) and *Echinacea purpurea* products," *Journal of Functional Foods*, vol. 2, no. 1, pp. 77–84, 2010.

[9] B. Khoza, S. Gbashi, P. Steenkamp, P. Njobeh, and N. Madala, "Identification of hydroxylcinnamoyl tartaric acid esters in *Bidens pilosa* by UPLC-tandem mass spectrometry," *South African Journal of Botany*, vol. 103, pp. 95–100, 2016.

[10] M. N. Clifford, K. L. Johnston, S. Knight, and N. Kuhnert, "Hierarchical scheme for LC-MS^n identification of chlorogenic acids," *Journal of Agricultural and Food Chemistry*, vol. 51, no. 10, pp. 2900–2911, 2003.

[11] E. N. Ncube, M. I. Mhlongo, L. A. Piater, P. A. Steenkamp, I. A. Dubery, and N. E. Madala, "Analyses of chlorogenic acids and related cinnamic acid derivatives from *Nicotiana tabacum* tissues with the aid of UPLC-QTOF-MS/MS based on the in-source collision-induced dissociation method," *Chemistry Central Journal*, vol. 8, no. 66, pp. 1–10, 2014.

[12] M. N. Clifford, W. Wu, J. Kirkpatrick, and N. Kuhnert, "Profiling the chlorogenic acids and other caffeic acid derivatives of herbal Chrysanthemum by LC-MS^n," *Journal of Agricultural and Food Chemistry*, vol. 55, no. 3, pp. 929–936, 2007.

[13] R. Jaiswal, S. Deshpande, and N. Kuhnert, "Profiling the chlorogenic acids of *Rudbeckia hirta, Helianthus tuberosus, Carlina acaulis* and *Symphyotrichum novae-angliae* leaves by LC-MS^n," *Phytochemical Analysis*, vol. 22, no. 5, pp. 432–441, 2011.

[14] R. Jaiswal, E. A. Halabi, M. G. E. Karar, and N. Kuhnert, "Identification and characterisation of the phenolics of *Ilex glabra* L. Gray (Aquifoliaceae) leaves by liquid chromatography tandem mass spectrometry," *Phytochemistry*, vol. 106, pp. 141–155, 2014.

[15] L. M. Calabria, V. P. Emerenciano, M. J. Ferreira, M. T. SCotti, and T. J. Mabry, "A phylogenetic analysis of tribes of the Asteraceae based on phytochemical data," *Natural Product Communications*, vol. 2, pp. 277–285, 2007.

[16] C. Makita, L. Chimuka, P. Steenkamp, E. Cukrowska, and E. Madala, "Comparative analyses of flavonoid content in *Moringa oleifera* and *Moringa ovalifolia* with the aid of UHPLC-qTOF-MS fingerprinting," *South African Journal of Botany*, vol. 105, pp. 116–122, 2016.

[17] M. E. P. Martucci, R. C. de Vos, C. A. Carollo, and L. Gobbo-Neto, "Metabolomics as a potential chemotaxonomical tool: application in the genus *Vernonia* schreb," *PloS One*, vol. 9, no. 4, Article ID e93149, 2014.

[18] A. Vallverdú-Queralt, A. Medina-Remon, M. Martínez-Huélamo, O. Jáuregui, C. Andres-Lacueva, and R. M. Lamuela-Raventos, "Phenolic profile and hydrophilic antioxidant capacity as chemotaxonomic markers of tomato varieties," *Journal of Agricultural and Food Chemistry*, vol. 59, no. 8, pp. 3994–4001, 2011.

[19] O. Aizpurua-Olaizola, J. Omar, P. Navarro, M. Olivares, N. Etxebarria, and A. Usobiaga, "Identification and quantification of cannabinoids in *Cannabis sativa* L. plants by high performance liquid chromatography-mass spectrometry," *Analytical and Bioanalytical Chemistry*, vol. 406, no. 29, pp. 7549–7560, 2014.

[20] N. R. Kitteringham, R. E. Jenkins, C. S. Lane, V. L. Elliott, and B. K. Park, "Multiple reaction monitoring for quantitative biomarker analysis in proteomics and metabolomics," *Journal of Chromatography B*, vol. 877, no. 13, pp. 1229–1239, 2009.

[21] M. Martínez-Huélamo, S. Tulipani, X. Torrado, R. Estruch, and R. M. Lamuela-Raventós, "Validation of a new LC-MS/MS method for the detection and quantification of phenolic metabolites from tomato sauce in biological samples," *Journal of Agricultural and Food Chemistry*, vol. 60, no. 11, pp. 4542–4549, 2012.

[22] X. Su, R. Lin, S. Wong, S. Tsui, and S. Kwan, "Identification and characterisation of the Chinese herb Langdu by LC-MS/MS analysis," *Phytochemical Analysis*, vol. 14, no. 1, pp. 40–47, 2003.

[23] L. Abrankó and B. Szilvássy, "Mass spectrometric profiling of flavonoid glycoconjugates possessing isomeric aglycones," *Journal of Mass Spectrometry*, vol. 50, no. 1, pp. 71–80, 2015.

[24] M. E. Fridén and P. J. Sjöberg, "Strategies for differentiation of isobaric flavonoids using liquid chromatography coupled to electrospray ionization mass spectrometry," *Journal of Mass Spectrometry*, vol. 49, no. 7, pp. 646–663, 2014.

[25] S. Gbashi, P. Njobeh, P. Steenkamp, H. Tutu, and N. Madala, "The effect of temperature and methanol–water mixture on Pressurized Hot Water Extraction (PHWE) of anti-HIV analogous from *Bidens pilosa*," *Chemistry Central Journal*, vol. 10, no. 37, pp. 1–12, 2016.

[26] C. Makita, L. Chimuka, E. Cukrowska et al., "UPLC-qTOF-MS profiling of pharmacologically important chlorogenic acids and associated glycosides in *Moringa ovalifolia* leaf extracts," *South African Journal of Botany*, vol. 108, pp. 193–199, 2017.

[27] L. Dawidowicz and R. Typek, "Transformation of chlorogenic acids during the coffee beans roasting process," *European Food Research and Technology*, vol. 243, no. 3, pp. 379–390, 2016.

[28] N. E. Madala, P. A. Steenkamp, L. A. Piater, and I. A. Dubery, "Collision energy alteration during mass spectrometric acquisition is essential to ensure unbiased metabolomic analysis," *Analytical and Bioanalytical Chemistry*, vol. 404, no. 2, pp. 367–372, 2012.

17

The Simultaneous Voltammetric Determination of Aflatoxins B_1 and M_1 on a Glassy-Carbon Electrode

G. B. Slepchenko ⓘ, T. M. Gindullina, M. A. Gavrilova, and A. Zh. Auelbekova

National Research Tomsk Polytechnic University, Tomsk 634050, Russia

Correspondence should be addressed to G. B. Slepchenko; slepchenkogb@mail.ru

Academic Editor: Luca Campone

For the first time, the possibility of using stripping voltammetry for the simultaneous determination of aflatoxins B_1 and M_1 on a glassy-carbon electrode has been shown. The influence of various factors ($E_э$, $\tau_э$, w, and the nature of the background electrolyte) on the potential and magnitude of the oxidation current of mycotoxins has been estimated. Working conditions for voltammetric determination and reproducibility of analytical signals for two mycotoxins have been selected. The mutual influence of aflatoxins B_1 and M_1 on the value of analytical signals in their simultaneous presence has been studied. It has been found that, in the range of their detectable contents, the presence of aflatoxin B_1 reduces the analytical signal of aflatoxin M_1 by 45–50%, but the linearity of the calibration dependence is preserved. The content of aflatoxin M_1 in determination of aflatoxin B_1 does not exert a significant effect in the range of 10–15%. Based on the results obtained, a procedure has been proposed for determining the content of aflatoxins B_1 and M_1 in their joint presence in milk by voltammetry in the concentration ranges $2 \times 10^{-3} \div 2 \times 10^{1}$ mg/dm^3 and $2 \times 10^{-4} \div 2 \times 10^{2}$ mg/dm^3, respectively (Sr not more than 18%).

1. Introduction

Aflatoxin M_1 is a metabolite of aflatoxin B_1, a product of life activity of *Aspergillus* microscopic fungi. In natural conditions, aflatoxin B_1 contaminates cereals, legumes, various nuts, oil seeds, cocoa and coffee, animal feed, and other food products. It can be converted into aflatoxin M_1 in the body of animals and is present in meat [1]. In [2], the information on admissible levels of contents of the specified kinds of mycotoxins in flour obtained from various kinds of grains used for cooking human food, as well as in the composition of feed of various animal species, is presented. With contaminated feed, aflatoxins enter the body of animals, and their residual quantities are found in meat, milk, and eggs. Aflatoxins are the only mycotoxins that are strictly regulated in markets such as the EU and US [3]. The information on tolerable levels of mycotoxins taken in Australia, China, Guatemala, India, Ireland, Kenya, and Taiwan is reported [4]. Aflatoxin M_1 is a hydroxylated metabolite present in human milk and animals exposed to aflatoxin B_1. Like its precursor, aflatoxin B_1, aflatoxin M_1 already in low concentrations poses a serious threat to the health of animals and humans. Aflatoxin B_1 is found not only in whole milk (including reconstituted milk) but also in cottage cheese, cheese, and yoghurt. Dairy products contaminated with aflatoxin M_1 are environmentally hazardous to humans. In adult food and baby food, aflatoxins are not allowed. Figure 1 shows the structural formula of aflatoxins B_1 and M_1.

Aflatoxin B_1 is (6aR-*cis*)(2,3,6a,9a)-tetrahydro-4-methoxycyclopenta[c]furo[2,3-h][1]benzopyran-1,11-dione with a molecular weight of 312. Aflatoxin M_1 is (2,3,6a,9a)-tetrahydro-9a-hydroxy-4-methoxycyclopenta[c]furo[2,3-h][1]benzopyran-1,11-dione with a molecular weight of 328.

There are known methods of simultaneous quantitative determination of aflatoxins B_1 and M_1, aflatoxins B_1, B_2, G_1, G_2, and M_1, and ochratoxin A by the method of high-performance liquid chromatography with fluorescent detection in breast milk [5–8] with detection limits between 0.5 and 0.25 μg/L [7], of 5 ng/L [7], and from 0.005 to 0.03 ng/mL [8].

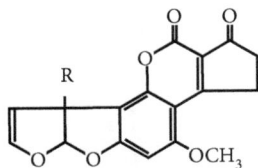

FIGURE 1: Structural formula of aflatoxins B_1 (R=H) and M_1 (R=OH).

The method of high-performance liquid chromatography was used to determine aflatoxin M_1 in eggs [9], and together with tandem mass-spectrometry, it was used to determine aflatoxins B_1 and M_1 in fresh and dry milk after ultrasonic extraction; the detection limit was 0.05 μg/kg, and the limit of quantification was 0.1 μg/kg [10]. The detection limits of aflatoxins B_1, B_2, G_1, G_2, and M_1 and ochratoxin A in food products of animal origin were in the range of 0.07–0.59 μg/kg [11]. In case of simultaneous determination of six aflatoxins (B_1, B_2, G_1, G_2, M_1, and M_2) by this method in peanut, the detection limits are in the range of 0.03 to 0.26 ng/g and 0.1 to 0.88 ng/g[12]. The use of chromatography methods is complicated by the duration and the need to use expensive equipment and highly toxic solvents as a mobile phase.

The possibility of indirect competitive enzyme-linked immunosorbent assay for the determination of aflatoxin M_1 in various objects was demonstrated in [13]. At present, highly sensitive, inexpensive, and easy-to-use electrochemical methods, in particular voltammetry, are becoming increasingly popular for the determination of a number of organic substances including aflatoxins. In literature, there is a rather large number of works on the individual determination of aflatoxins B_1 [14–19] and M_1 [20–23] using amperometric and voltammetric immunosensors. The use of enzymes and nanomaterials to design sensors provides high sensitivity and selectivity for detection. At the same time, the analysis of numerous publications in the databases of Science Direct, Scopus, Web of Science, and so on shows that, at the moment, there is no research work on the simultaneous quantification of aflatoxins B_1 and M_1 by the method of voltammetry.

The purpose of the work consists of studying the possibility of simultaneous voltammetric determination of aflatoxins B_1 and M_1 on a glassy-carbon electrode (GCE), selecting working conditions for measurements and developing a method for their determination in whole milk.

2. Methods and Materials

In this study, the voltammetric analyzer "STA" (Russia) consisting of electronic and measuring units and an IBM-compatible personal computer with the installed software package "STA" was used. As the indicator electrode, a glassy-carbon electrode (GCE) was used, and the conventional silver chloride electrode (CSE) served as an auxiliary and reference electrode.

The measurements were carried out in a constant-current sweep mode with the speed $w = 30$ mV/s in the potential range from 0.0 to +1.1 V. To mix the analyzed solution, vibration of the electrodes without removal of dissolved oxygen was used.

The working solutions of aflatoxins B_1 and M_1 were prepared from standard samples of aflatoxin B_1 (GSO 7936-2001) with a concentration of 10.0 mg/dm^3 and aflatoxin M_1 (GSO 7935-2001) with a concentration of 1.0 mg/dm^3 in the volume of 1.0 dm^3 mixture of benzene and acetonitrile (in the 9 : 1 ratio), followed by diluting 10 times in ethyl alcohol. As background electrolytes, solutions with different pH: 0.1 M Na_2HPO_4, 0.1 M $C_6H_5O_7(NH_4)_3$, 0.1 M $(NH_4)_2SO_4$, 0.1 M Na_3PO_4, 0.1 M K_2HPO_4, 0.1 M Li_2CO_3, and 0.1 M $ZnSO_4$, were used.

3. Preparation of the Sample of Whole Milk

When preparing for analysis, a sample of whole milk 25.00 g is taken in a conical flask with the capacity of 100 cm^3, and 1.0 cm^3 of hydrochloric acid with a concentration of 6–7 mol/dm^3 is added in portions of 0.2 cm^3. The mixture is slightly stirred and left for 15 minutes, poured into centrifuge tubes, and then centrifuged at 15,000 rpm within 15 minutes.

The centrifugate is poured into a conical flask, and 5-6 g of ammonium sulfate ((NH_2)$_2SO_4$) is added in portions of 2 to 3 grams, each time stirring the contents of the flask with a glass rod until the salt dissolves. The flask is left for 20 minutes, after which the contents of the flask are poured into centrifuge tubes and centrifuged within 15 minutes at the speed of 6000 rpm. The centrifugate is filtered into a clean cup with the capacity of 30 cm^3 through the double-layered filter paper (blue tape). The resulting filtrate is a prepared sample. For analysis, an aliquot of the prepared sample of 5.0 cm^3 is taken.

4. Results and Discussion

Studies on the effect of the background electrolyte composition on the analytical signals of aflatoxins B_1 and M_1 on a glassy-carbon electrode under working conditions previously developed for the determination of aflatoxin B_1 were conducted [24]. Experiments on the choice of the background electrolyte showed that the value of the analytical signal aflatoxin M_1 on background electrolytes: 0.1 M Na_3PO_4, 0.1 M Na_2HPO_4, 0.1 M K_2HPO_4, and 0.1 M $ZnSO_4$, was found to be low, and it was high on background electrolytes: 0.1 M $(NH_4)_2SO_4$ and 0.1 M Li_2CO_3; the maximum current of its electric oxidation was obtained against the background of 0.1 M $C_6H_5O_7(NH_4)_3$. Changing the cation-anion composition and pH of the background electrolyte may negatively shift the peak potential of the electric oxidation peak of aflatoxin M_1 in the range (0.6 $\pm$ 0.08) V. The effect of the background electrolyte pH on the analytical signals of these aflatoxins was studied, and it was shown that it was preferable to use neutral or weak acidic solutions as background ones, since mycotoxins decompose into nontoxic or low-toxic compounds in the alkaline medium, and the use of background electrolytes with pH > 6.5 is impractical. In Figure 2, calibration curves of the peak current of electric oxidation of aflatoxins B_1 and M_1 in

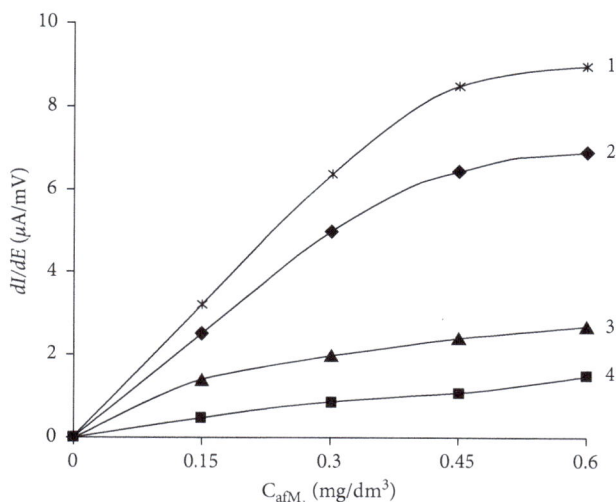

FIGURE 2: Calibration dependences of electric oxidation of aflatoxins M_1 (1 and 3) and B_1 (2 and 4) on various background electrolytes: (1, 2) 0.1 M $C_6H_5O_7(NH_4)_3$; (3, 4) 0.1 M $(NH_4)_2SO_4$.

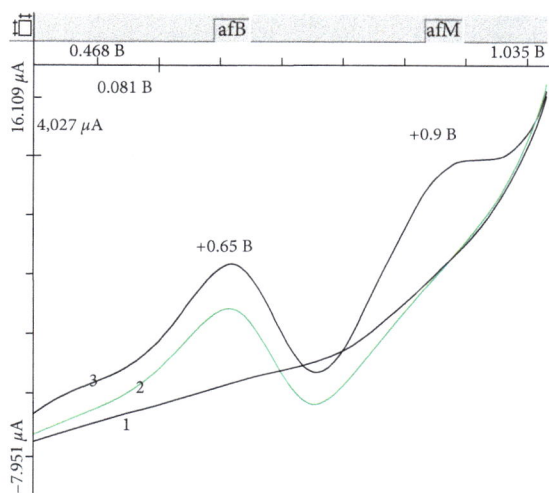

FIGURE 3: Voltammograms of aflatoxins B_1 and M_1 electric oxidation with the joint presence on the GCE: (1) background electrolyte 0.1 M $C_6H_5O_7(NH_4)_3$; (2) $C_{afB_1} = 2 \times 10^{-3}$ mg/dm^3 and $C_{afM_1} = 0$; (3) $C_{afB_1} = 2 \times 10^{-3}$ mg/dm^3 and $C_{afM_1} = 2 \times 10^{-4}$ mg/dm^3.

various background electrolytes are presented. According to the calibration curves, two background electrolytes were selected: 0.1 M 3-substituted ammonium citrate and 0.1 M ammonium sulfate solution, providing a high detection sensitivity coefficient in the range of determined contents $2 \times 10^{-4} \div 0.6$ mg/dm^3.

In Figure 3, voltammograms of the electric oxidation of aflatoxins B_1 and M_1 on the GCE in the selected background electrolyte are shown. Analytical signals are well separated and reproduced.

Both electrolytes can be used for the joint quantification of mycotoxins, but 0.1 M $C_6H_5O_7(NH_4)_3$ was selected as the working background electrolyte providing sufficient resolution and satisfactory reproducibility of the analytical signal.

Figure 4 shows the dependences of the current derivatives of the peak of aflatoxins B_1 (curve 1) and M_1 (curve 2) on the accumulation potential of the GCE in the selected background electrolyte. It can be seen in Figure 4 that the maximum values of the electric oxidation currents of aflatoxins are observed at the potential of 0.0 V that was selected as the accumulation potential for further studies.

The mutual influence of aflatoxins B_1 and M_1 in their simultaneous determination on a glassy-carbon electrode was studied. For this purpose, the current derivative of the peak of aflatoxin B_1 electric oxidation was obtained as a function of the concentration of aflatoxin M_1 in the solution (Figure 5) and the calibration curves of aflatoxin M_1 in the presence of aflatoxin B_1 (Figure 6).

In Figure 5, it can be seen that, in the concentration range studied, the effect of aflatoxin M_1 on the aflatoxin B_1 current is practically negligible at the ratio $C_{a\Phi B_1} : C_{a\Phi M_1} = 1:1$ in the presence of a two- or threefold excess of aflatoxin M_1, the peak current of aflatoxin B_1 decreases by 10–15% (curve 1), and the potential of the peak remains unchanged. It is shown that the systematic error in determination of aflatoxin B_1 in the presence of aflatoxin M_1 at the ratio $C_{B_1}/C_{M_1} \leq 1:40$ does not exceed 20%.

In Figure 6, it is seen that, in the presence of aflatoxin B_1, the derivative of the peak current of aflatoxin M_1 decreases almost 1.5 times and the peak potential shifts to the anode region from +0.85 V to +0.95 V, but the linearity of the calibration dependence remains in a wide range which proves the possibility of their simultaneous determination.

Based on the conducted studies, the working conditions of the simultaneous voltammetric determination of aflatoxins B_1 and M_1 were proposed (Table 1).

On the basis of the obtained data of the electrochemical behavior of aflatoxins, an algorithm for quantifying these toxic substances in order to effectively control the detection of their minimum acceptable amounts in whole milk was developed. The algorithm for quantification of mycotoxins in whole milk includes the following steps:

(1) Taking a sample

(2) Acid hydrolysis with concentrated HCl and centrifugation

(3) Precipitation of proteins with ammonium salts of sulfate $(NH_4)_2SO_4$ and centrifugation

(4) Filtration of the obtained precipitate

(5) Quantitative determination of the aflatoxins content by the method of differential voltammetry

Verification of the correctness of the proposed procedure was carried out by the "introduced-found" method (Table 2).

The data in Table 2 show that the voltammetric joint for determination of quantities of aflatoxins B_1 and M_1 is possible with the measurement error of 15–20% in the concentration ranges $2 \times 10^{-3} \div 2 \times 10^{-1}$ mg/dm^3 and $2 \times 10^{-4} \div 2 \times 10^{-2}$ mg/dm^3, respectively.

The proposed method is simple, and it does not require a lot of reagents and labor. The range of detectable

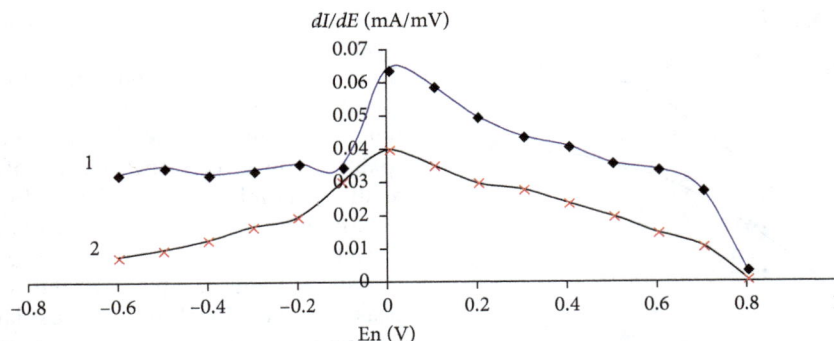

FIGURE 4: Dependences of the current derivatives of the aflatoxins B_1 and M_1 peak on the accumulation potential of the GCE. The background electrolyte is 0.1 M $C_6H_5O_7(NH_4)_3$; $\tau_{\ni} = 30$ s; $w = 30$ mV/s; (1) $C_{afM_1} = 2 \times 10^{-3}$ mg/dm^3; (2) $C_{afM_1} = 2 \times 10^{-4}$ mg/dm^3.

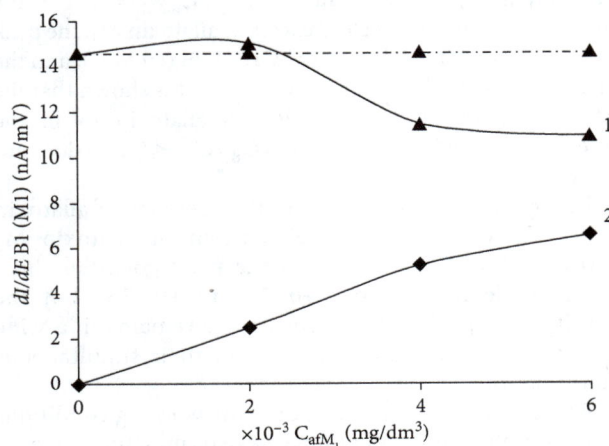

FIGURE 5: Dependences of the current derivatives of the peak of aflatoxins B_1 (1) and M_1 (2) on the aflatoxin M_1 content in the background electrolyte on the GCE: the background electrolyte is 0.1 M $C_6H_5O_7(NH_4)_3$; (1) $C_{afB_1} = 2 \times 10^{-3}$ mg/dm^3 and $C_{afM_1} = $ var (on 2×10^{-3} mg/dm^3); (2) dependence of $I_{p \cdot afM_1}$ on C_{afM_1}.

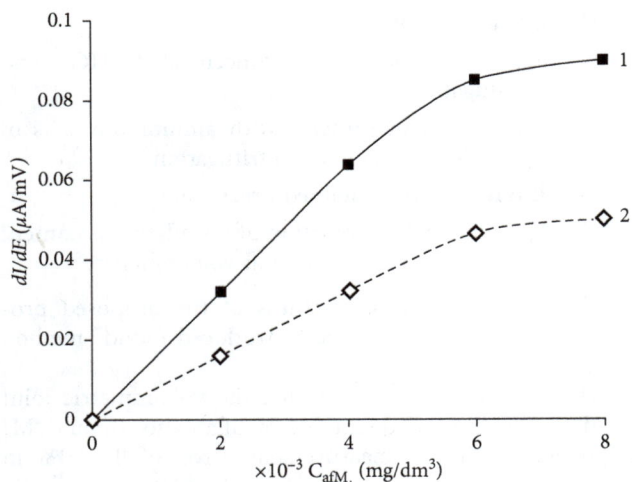

FIGURE 6: Calibration dependences of aflatoxin M_1 on the GCE: the background electrolyte is 0.1 M $C_6H_5O_7(NH_4)_3$; $E_e = 0,0$ B; $\tau_e = 30$ s; (1) $C_{afB_1} = 0$; (2) $C_{afB_1} = 2 \times 10^{-3}$ mg/dm^3.

TABLE 1: Working conditions of the simultaneous voltammetric determination of aflatoxins B_1 and M_1.

Parameters of votammetric determination of aflatoxins	Parameters values	
	B_1	M_1
The system used	3-electrode	
Electrodes		
(i) Indicator	GCE	
(ii) Comparison/auxiliary	CSE/CSE	
Background electrolyte	0.1 M $C_6H_5O_7(NH_4)_3$	
Electrolysis potential $E_{\ni}$ (V)	0,0	
Potential sweeping range (V)	$0,0 \div +1,1$	
Potential changing speed w (mV/s)	30	
Registration mode	Differential	
Peak potential E_{II} (V)	$+0.65 \pm 0.05$	$+0.90 \pm 0.05$

TABLE 2: Verification of correctness of the voltammetric method for determination of the content of aflatoxins B_1 and M_1 in model samples of whole milk by the "introduced-found" method ($P = 0.95$ and $n = 5$).

Object	Component	Content of aflatoxins $B_1(10^3)$ and $M_1(10^4)$ (mg/dm^3)		
		In samples	Introduced	Found
Cow's milk	B_1	2.79 ± 0.41	2.00	4.81 ± 0.72
Fat content: 3.8%	M_1	3.01 ± 0.45	2.00	5.43 ± 0.81
Cow's milk	B_1	1.94 ± 0.35	2.00	3.95 ± 0.65
Fat content: 1.5%	M_1	2.12 ± 0.32	2.00	4.21 ± 0.98
Goat's milk	B_1	3.92 ± 0.58	4.00	7.85 ± 1.12
Fat content: 4.4%	M_1	2.08 ± 0.31	4.00	5.95 ± 0.91
Sour milk	B_1	13.5 ± 1.9	10.0	22.6 ± 3.1
Fat content: 2.5%	M_1	15.3 ± 2.1	10.0	26.1 ± 3.7

concentrations is from 0.001 to 0.12 mg/dm^3. The relative standard deviation (Sr) is not more than 30%.

5. Conclusion

Thus, the possibility of the simultaneous voltammetric determination of aflatoxins B_1 and M_1 on the GCE in the background electrolyte 0.1 M $C_6H_5O_7(NH_4)_3$ has been shown.

When determining aflatoxins, the method of "soft" sample preparation has been used for separating the matrix by hydrolysis and salting out proteins followed by their separation by centrifugation or filtration which reduces the analysis time to less than one hour as compared with thin-layer and high-performance chromatography (GOST 30711-2001). The developed technique has a number of advantages in comparison with the already known methods of analysis. The algorithm of the technique is characterized by the express analysis (the analysis time does not exceed 1 hour), sensitivity (the range of the determined contents is not inferior, and in the case of aflatoxin M_1, it exceeds the capabilities of chromatographic methods), and the equipment cheapness. The technique is characterized by simplicity of execution, minimal consumption of reagents, and improved metrological characteristics.

Conflicts of Interest

The authors declare that they have no conflicts of interest for this research.

Acknowledgments

This work was financed by the State Task (Science) 1.1343.2014.

References

[1] Codex Stan 193-1995, *General Standard for Pollutant Impurities and Toxins in Foodstuffs and Feeds (Codex Stan 193-1995)*, 1995.

[2] F. Wu, "Mycotoxin risk assessment for the purpose of setting international regulatory standards," *Environmental Science and Technology*, vol. 38, no. 15, pp. 4049–4055, 2004.

[3] The Commission of the European, *Commission regulation (EC) No. 1881/2006*, The Commission of the European, Brussels, Belgium, 2006.

[4] J. B. Coulter, S. M. Lamplugh, G. I. Suliman, M. I. Omer, and R. G. Hendrickse, "Aflatoxins in human breast milk," *Annals of Tropical Paediatrics*, vol. 4, no. 2, pp. 61–66, 1984.

[5] P. T Scaglioni, T. Becker-Algeri, D. Drunkler, and E. Badiale-Furlong, "Aflatoxin B1 and M1 in milk," *Analytica Chimica Acta*, vol. 829, pp. 68–74, 2014.

[6] O. Adejumo, O. Atanda, A. Raiola, Y. Somorin, R. Bandyopadhyay, and A. Ritieni, "Correlation between aflatoxin M1 content of breast milk, dietary exposure to aflatoxin B1 and socioeconomic status of lactating mothers in Ogun State, Nigeria," *Food and Chemical Toxicology*, vol. 56, pp. 171–177, 2013.

[7] A. Gürbay, S. A. Sabuncuoğlu, G. Girgin et al., "Exposure of newborns to aflatoxin M1 and B1 from mothers' breast milk in Ankara, Turkey," *Food and Chemical Toxicology*, vol. 48, no. 1, pp. 314–319, 2010.

[8] P. D. Andrade, J. L. Gomes da Silva, and E. D. Caldas, "Simultaneous analysis of aflatoxins B1, B2, G1, G2, M1 and ochratoxin A in breast milk by high-performance liquid chromatography/fluorescence after liquid–liquid extraction with low temperature purification (LLE–LTP)," *Journal of Chromatography A*, vol. 1304, pp. 61–68, 2013.

[9] A. P. Wacoo, D. Wendiro, P. C. Vuzi, and J. F. Hawumba, "Methods for detection of aflatoxins in agricultural food crops," *Journal of Applied Chemistry*, vol. 2014, Article ID 706291, 15 pages, 2014.

[10] S. Fan, Q. Li, L. Sun, Y. Du, J. Xia, and Y. Zhang, "Simultaneous determination of aflatoxin B1 and M1 in milk, fresh milk and milk powder by LC-MS/MS utilising online turbulent flow chromatography," *Food Additives & Contaminants: Part A*, vol. 32, no. 7, pp. 1175–1184, 2015.

[11] C. Dongmei, C. Xiaoqin, T. Yanfei et al., "Development of a sensitive and robust liquid chromatography coupled with tandem mass spectrometry and a pressurized liquid extraction for the determination of aflatoxins and ochratoxin A in animal derived foods," *Journal of Chromatography A*, vol. 1253, pp. 110–119, 2012.

[12] A. V. Sartori, J. S. de Mattos, Y. P. Souza, R. Pereira dos Santos, M. H. P. de Moraes, and A. W. da Nóbrega, "Determination of aflatoxins M1, M2, B1, B2, G1 and G2 in peanut by modified QuEChERS method and ultra-high performance liquid chromatography-tandem mass spectrometry," *Vigilância Sanitária em Debate*, vol. 3, pp. 115–121, 2015.

[13] A. Radoi, M. Targa, B. Prieto-Simon, and J.-L. Marty, "Enzyme-linked immunosorbent assay (ELISA) based on superparamagnetic nanoparticles for aflatoxin M1 detection," *Talanta*, vol. 77, no. 1, pp. 138–143, 2008.

[14] Z. Linting, L. Ruiyi, L. Zaijun, X. Qianfang, F. Yinjun, and L. Junkang, "An immunosensor for ultrasensitive detection of aflatoxin B1 with an enhanced electrochemical performance based on graphene/conducting polymer/gold nanoparticles/ the ionic liquid composite film on modified gold electrode with electrodeposition," *Sensors and Actuators B: Chemical*, vol. 174, pp. 359–365, 2012.

[15] L. Masoomi, O. Sadeghi, M. H. Banitaba, A. Shahrjerdi, and S. S. H. Davarani, "A non-enzymatic nanomagnetic electro-immunosensor for determination of aflatoxin B1 as a model antigen," *Sensors and Actuators B: Chemical*, vol. 177, pp. 1122–1127, 2013.

[16] H. Maa, J. Suna, Y. Zhanga, C. Biana, S. Xiaa, and T. Zhen, "Label-free immunosensor based on one-step electrodeposition of chitosan-gold nanoparticles biocompatible film on Au microelectrode for the determination of aflatoxin B1 in maize," *Biosensors and Bioelectronics*, vol. 80, pp. 222–229, 2016.

[17] A. Sharma, A. Kumar, and R. Khan, "Electrochemical immunosensor based on poly(3,4-ethylenedioxythiophene) modified with gold nanoparticle to detect aflatoxin B1," *Materials Science and Engineering: C*, vol. 76, pp. 802–809, 2017.

[18] K. Abnous, N. M. Danesh, M. Alibolandi et al., "A new amplified π-shape electrochemical aptasensor for ultrasensitive detection of aflatoxin B1," *Biosensors and Bioelectronics*, vol. 94, pp. 374–379, 2017.

[19] S. Mavrikoua, E. Flampouria, D. Iconomoub, and S. Kintziosa, "Development of a cellular biosensor for the detection of aflatoxin B1, based on the interaction of membrane engineered Vero cells with anti-AFB1 antibodies on the surface of gold nanoparticle screen printed electrodes," *Food Control*, vol. 73, pp. 64–70, 2017.

[20] N. Paniel, A. Radoi, and J.-L. Marty, "Development of an electrochemical biosensor for the detection of aflatoxin M1 in milk," *Sensors*, vol. 10, no. 10, pp. 9439–9448, 2010.

[21] C. O. Parker, Y. H. Lanyon, M. Manning, D. W. Arrigan, and I. E. Tothill, "Electrochemical immunochip sensor for aflatoxin M1 detection," *Analytical Chemistry*, vol. 81, no. 13, pp. 5291–5298, 2009.

[22] C. O. Parker and I. E. Tothill, "Development of an electro-chemical immunosensor for aflatoxin M1 in milk with focus on matrix interference," *Biosensors and Bioelectronics*, vol. 24, no. 8, pp. 2452–2457, 2009.

[23] G. Istambouliéa, N. Paniela, L. Zaraa et al., "Development of an impedimetric aptasensor for the determination of aflatoxin M1 in milk," *Talanta*, vol. 146, pp. 464–469, 2016.

[24] M. A. Gavrilova, G. B. Slepchenko, E. V. Mikheeva, and V. I. Deryabina, "Voltammetric determination of aflatoxin B1," *Procedia Chemistry*, vol. 10, pp. 114–119, 2014.

18

A Method to Determination of Lead Ions in Aqueous Samples: Ultrasound-Assisted Dispersive Liquid-Liquid Microextraction Method based on Solidification of Floating Organic Drop and Back-Extraction Followed by FAAS

Çiğdem Arpa ⓘ and Itır Aridaşir

Chemistry Department, Hacettepe University, Beytepe, 06800 Ankara, Turkey

Correspondence should be addressed to Çiğdem Arpa; carpa@hacettepe.edu.tr

Academic Editor: Valdemar Esteves

Ultrasound-assisted dispersive liquid-liquid microextraction method based on solidification of floating organic drop and back-extraction (UA-DLLME-SFO-BE) technique was proposed for preconcentration of lead ions. In this technique, two SFODME steps are applied in sequence. The classical SFODME was applied as the first step and then the second (back-extraction) step was applied. For the classical SFODME, Pb ions were complexed with Congo red at pH 10.0 and then extracted into 1-dodecanol. After this stage, a second extraction step was performed instead of direct determination of the analyte ion in the classical method. For this purpose, the organic phase containing the extracted analyte ions is treated with $1.0 \, \text{mol·L}^{-1}$ HNO_3 solution and then exposed to ultrasonication. So, the analyte ions were back-extracted into the aqueous phase. Finally, the analyte ions in the aqueous phase were determined by FAAS directly. Owing to the second extraction step, a clogging problem caused by 1-dodecanol during FAAS determination was avoided. Some parameters which affect the extraction efficiency such as pH, volume of extraction solvent, concentration of complexing agent, type, volume, and concentration of back-extraction solvent, effect of cationic surfactant addition, effect of temperature, and so on were examined. Performed experiments showed that optimum pH was 10.0, 1-dodecanol extraction solvent volume was $75 \, \mu\text{L}$, back-extraction solvent was $500 \, \mu\text{L}$, $1.0 \, \text{mol·L}^{-1}$ HNO_3, extraction time was 4 min, and extraction temperature was 40°C. Under optimum conditions, the enhancement factor, limit of detection, limit of quantification, and relative standard deviation were calculated as 81, $1.9 \, \mu\text{g·L}^{-1}$, $6.4 \, \mu\text{g·L}^{-1}$, and 3.4% (for $25 \, \mu\text{g·L}^{-1}$ Pb^{2+}), respectively.

1. Introduction

Lead is highly toxic and potentially dangerous to living organisms because of the accumulation ability in the vital organs. Long-term exposure to lead causes cumulative poisoning which leads to central nervous system damages [1, 2]. On account of this, even at low concentration, determination and controlling of lead content are crucial [3]. However, because of their complex matrix and trace concentration, determination of heavy metals in many real samples is a difficult analytical task and requires sensitive detection techniques and probably a preconcentration procedure [4].

Nowadays, for determination of trace heavy metals, atomic absorption spectrometry equipped with flame (FAAS) or graphite furnace (ETAAS) and inductively coupled plasma emission spectrometry (ICP) [5–7] are widely used analytical techniques. Among these techniques, although flame atomic absorption spectrometry is commonly used for heavy metal determination, its sensitivity is not enough to direct determination of trace and ultra-trace amount of heavy metals in real samples; therefore, some pretreatment procedures, such as separation and preconcentration, are needed [8]. Several sample preparation methods for the preconcentration and separation of heavy metals have been developed. Solid-phase extraction (SPE) [9], solid-phase microextraction (SPME) [10], dispersive liquid-liquid microextraction (DLLME) [11], and cloud point extraction (CPE) [12, 13] are some of these methods [14].

Recently, the solidified floating organic drop micro-extraction (SFODME) [15–17] has been attracted more attention and properly utilized as an effective method for heavy metal extraction and preconcentration. In this technique, a free microdrop of the organic solvent which has a low melting point (in the range of 10–30°C) and a density lower than water is gently dropped to the stirred sample solution. Under the appropriate stirring conditions, the organic microdrop floats on the aqueous sample and species, which have a hydrophobic character, are extracted into this phase. After completion of the extraction, the sample vial is placed in an ice bath, and the floated drop solidifies. After solidification, the organic drop is easily transferred to a conical tube and melted there. Finally, the analyte in the extraction phase is determined by using FAAS. The SFODME method needs very small volumes of the harmful organic solvent, and this makes the method simple, fast, cheap, and environmental friendly [18]. Application of ultrasonic energy to a solution causes acoustic cavitation. Cavitation is the formation and then immediate implosion of bubbles in a liquid. When high temperatures and pressures are produced at the interface of the bubble and other phases, an increase in chemical reactivity occurs [19]. Hence, combination of microextraction processes with ultrasound energy accelerates the extraction step and improves the efficiency of preconcentration. This combined technique is called as ultrasound-assisted emulsification-microextraction (USAEME) and bears all the advantages of both methods.

Besides all these advantages mentioned above, SFODME has some important disadvantages: selected extracting solvent should not be volatile and toxic, and its melting point should be close to or below room temperature [20]. Having a melting point near room temperature also means that the extracting solvent freezes easily at room temperature. After solidification, the microdrop is transferred to a conical vial, is diluted with an appropriate solvent to a certain volume, and is then analyzed for determination of heavy metal concentration. But, unfortunately, after detection of the analyte, the solvent residues freeze again in the capillaries, tubes, and pipes of the FAAS instrument, and this causes clogging of the sample introduction path and drainage path of the instrument. (Especially, when the room temperature is slightly below 24°C in winter). Remedy of clogging in the system requires washing the pipes with plenty of hot water and alcohol solution several times, which is a time- and chemical-consuming process. In addition to clogging problems, 1-dodecanol has a background signal, which interferes with the analyte signal. To eliminate this problem, the back-extraction step can be applied [1].

In this ultrasound-assisted dispersive liquid-liquid microextraction method based on solidification of floating organic drop and back-extraction (UA-DLLME-SFO-BE) technique, two solidified floating organic drop microextraction (SFODME) steps are performed in sequence. In the first step, the following conventional SFODME procedure is applied: the extraction solvent is added into the solution including heavy metal ions that form hydrophobic complexes with suitable complexing agent. After applying ultrasound energy, extraction occurs, and hydrophobic

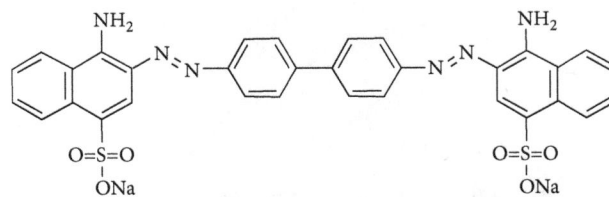

FIGURE 1: Molecular structure of CR.

analyte species are transferred into the organic solvent. At this stage, instead of direct analysis of the extraction phase, a second SFODME round is performed, in which the extraction phase is treated with diluted acid solution, and after applying ultrasound, interested analyte is back-extracted into the aqueous phase. At the end of second SFODME, the obtained aqueous solution, which includes the analyte, is injected to the FAAS. The second extraction step provides the elimination of the clogging problem of the organic extraction solvent.

In this study, we developed a UA-DLLME-SFO-BE method for the determination of Pb^{2+} ions in aqueous samples. For this purpose, Congo red (CR) is used as a complexing agent to obtain hydrophobic Pb^{2+} complexes. CR-Pb^{2+} complexes are extracted into 1-dodecanol and back-extracted into aqueous nitric acid solution, which eliminates the drawbacks of the organic extraction phase as mentioned above. CR (3,3'-[[1,1' biphenyl]-4,4'-diylbis-(azo)] bis[4-amino-1-naphthalene-sulfonic acid] disodium salt) is a sulfonated bis-azo dye [21]. As seen from Figure 1, CR has azo, sulfonate, and amino groups, and these groups provide two different donor atoms: one is N from azo or amino groups and other one is O from sulfonate groups. Complexation between metals and CR may occur through these donor atoms [22, 23].

There are two special features at the heart of the study: (i) implementation of ultrasound energy shortens the extraction time and improves the extraction efficiency; (ii) by performing a second extraction step, drawbacks of 1-dodecanol in conventional SFODME (clogging of pipes and tubings of instrument and background signal) are eliminated. In order to fully characterize the proposed method, several parameters were explored and optimized.

2. Experimental

2.1. Reagents and Standard Solutions. $1000\ mg\cdot L^{-1}$ stock standard solution of lead was prepared by dissolving a necessary mass of $Pb(NO_3)_2$ in a small amount of deionized water and diluting it to the appropriate volume. All solutions were prepared with deionized water ($18.1\ M\Omega m^{-1}$) obtained from a Millipore Simplicity UV purification system. Working standard solutions were prepared daily by serial dilutions of the stock solution with deionized water just before analysis. The chelating agent, 0.1% CR solution, was prepared by dissolving the appropriate amount of CR (purchased from BDH, Poole, England) in acetone. Na_2HPO_4/NaOH (purchased from Merck, Darmstadt, Germany) buffer system was used to adjust the

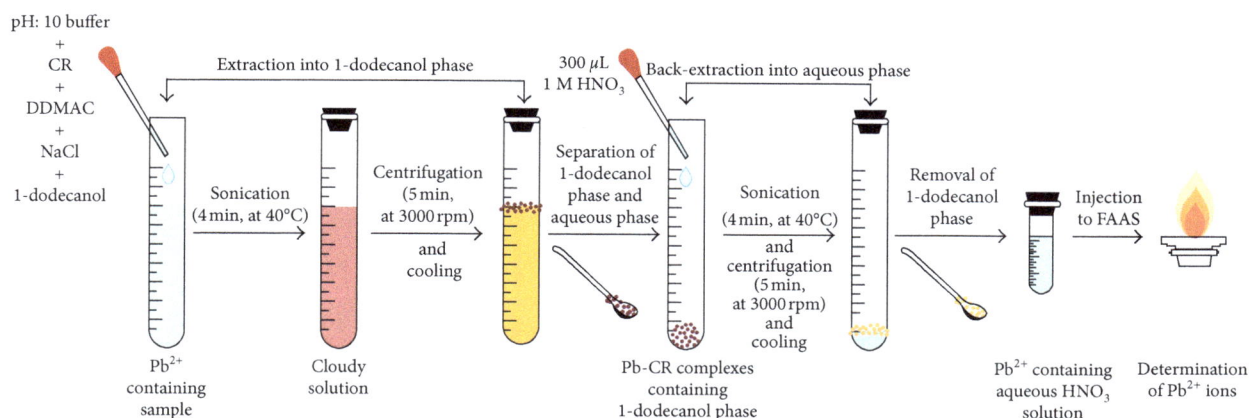

FIGURE 2: Schematic representation of UA-DLLME-SFO-BE procedure.

sample pH to 10.0. 1-dodecanol, used as an extraction solvent, was obtained from Merck, and didecyldimethyl ammoniumchloride (DDMAC) was purchased from Sigma-Aldrich. A solution of 8% (v/v) DDMAC was prepared by dissolving proper amount of DDMAC in water. All reagents used were of analytical reagent grade. All the solutions were prepared by using deionized water. In order to eliminate possible contamination, all glassware was immersed into 10% hydrochloric acid for at least 24 h and then rinsed three times with deionized water.

2.2. Instrument. A Perkin Elmer Analyst atomic absorption spectrometer equipped with a lead hollow cathode lamp (Perkin Elmer) as the radiation source was used for determination of lead. The analytical wavelength (283.3 nm), slit width (0.7 nm), and lamp current (28 mA) were used as recommended by the manufacturer. All pH measurements were carried out with an Isolab Laborgeräte GmbH digital pH meter equipped with a pH electrode. A Hettich Eba 21 model centrifuge was used to accelerate the phase separation. A 53 kHz, 100 W temperature-controlled ultrasonic bath (Kudos SK3310LHC) was used in the ultrasound-assisted emulsification process of the method.

2.3. Procedure for UA-DLLME-SFO-BE. The UA-DLLME-SFO-BE method includes two SFODME steps. In the first step, 25.0 mL Pb^{2+} solution or water samples, 2 mL of Na_2HPO_4/NaOH buffer solution, 1 mL of 0.5% (w/v) CR solution, 100 μL 8% (v/v) DDMAC, and 0.1 g NaCl were mixed. Then, 75 μL of 1-dodecanol was added. The conical tube was sonicated for 4 min at 40°C to ensure complete extraction. At this step, the lead ions reacted with CR and extracted into 1-dodecanol. With the addition of DDMAC, Pb-CR complex gained more hydrophobic character and was more extractable into the 1-dodecanol phase. With the aid of centrifugation (5 min at 3000 rpm), very small droplets of the 1-dodecanol group were brought together and collected at the vicinity of sample solution. After cooling of the test tube in the refrigerator, the 1-dodecanol phase was solidified, and this solidified phase was transferred into

another conical tube. At this stage, instead of adding a diluent and determination of Pb^{2+}, a second SFODME procedure was applied. For this purpose, the 1-dodecanol phase containing Pb-CR complexes was treated with 500 μL of 1 mol·L^{-1} HNO_3 solution. After sonication for 4 min at 40°C, the tube was centrifuged for 5 min at 3000 rpm. At this stage, Pb^{2+} ions were transferred from the 1-dodecanol phase to the aqueous HNO_3 phase. Finally, the supernatant was introduced into FAAS by direct nebulization for Pb^{2+} analysis, and the blank was also treated in the same way. Schematic representation of the proposed procedure is shown in Figure 2.

3. Results and Discussion

In order to obtain the maximum extraction efficiency and the highest enhancement factor, several parameters that influence the metal-ligand formation and the extraction conditions were studied. In the course of optimization experiments, 100 μg·L^{-1} of Pb^{2+} working solutions were utilized. After obtaining optimum conditions, the Pb^{2+} content of certified reference materials and some natural water samples were determined by using these optimum conditions.

3.1. Selection of the Extraction Solvent and Effect of Its Volume on Extraction. In the optimization of the SFODME process, the choice of the proper extraction solvent is one of the important steps. For this method, the extraction solvent must have the following features: (1) being immiscible with water, (2) having low volatility, which provides a stable character during the extraction step, (3) having a considerable extraction efficiency, (4) having a lower density than that of water, (5) having a melting point around room temperature (in the range of 10–30°C), and (6) having a low toxicity level [24]. For this purpose, 1-dodecanol (density: 0.8201–0.8309 g·mL^{-1}; melting point: 24°C) was chosen as the extracting solvent because of its sensitivity, stability, low water solubility, low vapor pressure, and low price.

To investigate the effect of the extraction solvent volume on extraction efficiency, 1-dodecanol volumes were changed in the range of 10–250 μL, while other parameters were kept at optimum values. It was not possible to work with volumes

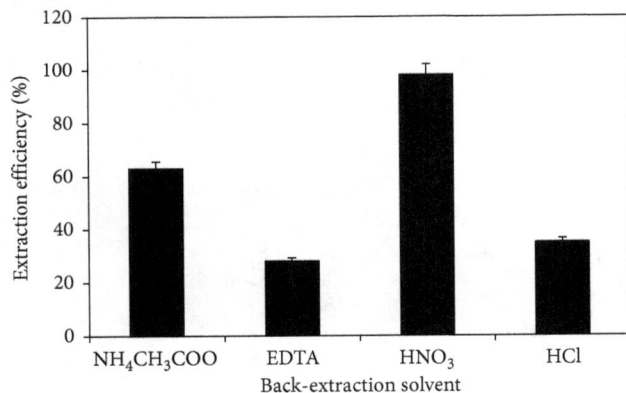

FIGURE 3: Effect of back-extraction solvent type. Conditions: 25 mL of $100\,\mu g \cdot L^{-1}$ Pb^{2+}, 0.5% (w/v) CR, 0.03% (v/v) DDMAC, 0.1 g NaCl, $75\,\mu L$ 1-dodecanol, 4 min extraction time, and 40°C extraction temperature.

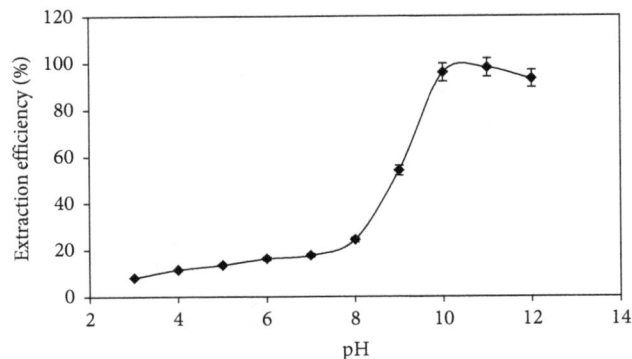

FIGURE 4: Effect of pH on the extraction efficiency. Conditions: 25 mL of $100\,\mu g \cdot L^{-1}$ Pb^{2+}, 0.5% (w/v) CR, 0.03% (v/v) DDMAC, 0.1 g NaCl, $75\,\mu L$ 1-dodecanol, 4 min extraction time, and 40°C extraction temperature.

less than $10\,\mu L$. The extraction efficiency values (calculated from absorbance values obtained various volumes of 1-dodecanol) showed that initially increasing the volume of 1-dodecanol in the range of 10–50 μL causes increase in the extraction efficiency, and then the extraction efficiency remained constant at 98% in the range of 75–100 μL. Finally, extraction efficiencies slightly decreased by the volumes greater than 100 μL. Therefore, in the subsequent studies, 75 μL was chosen as the optimum volume for 1-dodecanol.

3.2. Selection and Effect of Back-Extraction Solvent.

In the second step of the procedure, several back-extraction solvents were tried to extract Pb^{2+} ions from their Pb-CR complexes into the aqueous phase in order to eliminate drawbacks of the 1-dodecanol phase. For this purpose, aqueous solutions of EDTA, NH_4CH_3COO, HCl, and HNO_3 were examined as back-extraction solvents. During this optimization study, other experimental conditions were adjusted to optimum values. Among all these solvents, HNO_3 gave the superior extraction efficiency (>96%) and therefore, was chosen as the back-extraction solvent (Figure 3). After experiments, which were performed to explore the effects of HNO_3 concentration and volume on the extraction efficiency, optimum HNO_3 concentration and volume were found as 1 M and 500 μL, respectively. (During these optimization experiments, concentration and volume of HNO_3 were varied in the range of 0.1 M–2.0 M and $100\,\mu L$–$1000\,\mu L$, respectively, and other parameters were kept at optimum values).

3.3. Effect of pH.

Complex formation interaction between ligand and metal, and hence extraction efficiency, is pH-dependent [25, 26]. The effect of pH on the formation and extraction of Pb-CR complex was investigated at the pH range of 3.0–12.0 by adding appropriate buffer solutions. During these experiments, other parameters were adjusted to their optimum values. Obtained results are shown in Figure 4. As seen in this figure, the extraction efficiency of

lead with CR into 1-dodecanol was maximized and remained nearly constant (>98%) in the pH range of 10.0–11.0. Decreasing extraction efficiencies at low pH values were probably due to the competition of hydrogen ions with lead ions in the complex formation reaction with CR. Therefore, the pH of aqueous solution was adjusted at around pH 10.0.

3.4. Effect of Amount of CR.

The extraction efficiency of the analyte depends on the distribution ratio of the metal chelate between the organic and aqueous phases [27]. At a constant aqueous phase pH, the value of the distribution ratio and therefore the extraction efficiency increases to a certain value with the increasing amount of the chelate. The influence of CR amount on the extraction efficiency of lead was studied using different volumes of 0.1% (w/v) CR solution ranging from 0.1 mL to 2.0 mL, while other parameters were kept at their optimum values. According to the results obtained after optimization experiments, the extraction efficiency was increased with the increase of CR volume up to 0.5 mL and then remained constant at a value greater than 98%. In other words, the extraction efficiency was constant when the CR volume was higher than 0.5 mL, indicating complete complexation. Therefore, 0.5 mL of 0.1% (w/v) CR solution was chosen as the optimum value for subsequent studies.

3.5. Effect of DDMAC Concentration.

In general, as mentioned before, species to be extracted into 1-dodecanol should have a hydrophobic character, and this is the greatest limitation of all the SFODME techniques. Consequently, hydrophilic and polar characters of the extracted compounds lower the extraction efficiency. To cope with this limitation, a cationic surfactant, DDMAC, has been introduced to the sample solution to attain sites with hydrophobic solubilizing properties that would make the extraction of metal chelates with hydrophilic character easier [28]. By adding DDMAC, the Pb-CR complex was possessed of more hydrophobic features, and thus, the complex was more extractable into 1-dodecanol. As a result, a higher extraction efficiency factor was obtained. In order to explore

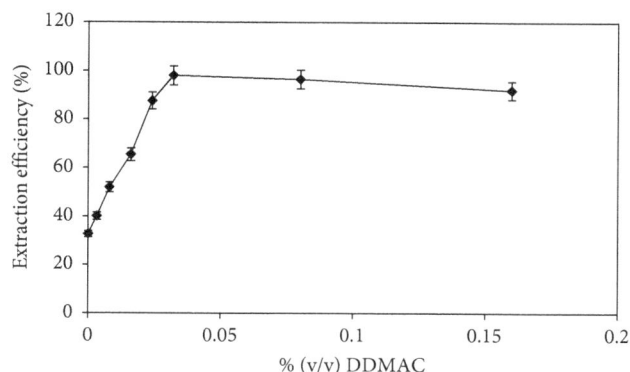

FIGURE 5: Effect of DDMAC concentration on the extraction efficiency. Conditions: 25 mL of $100\,\mu g \cdot L^{-1}\,Pb^{2+}$, pH: 11.0, 0.5% (w/v) CR, 0.1 g NaCl, $75\,\mu L$ 1-dodecanol, 4 min extraction time, and 40°C extraction temperature.

TABLE 1: Tolerance limits of interfering ions in the determination of $50\,\mu g \cdot L^{-1}$ of Pb^{2+}.

M^a	$[Pb^{2+}]/[M]$
Na^+	$>1:5000$
Mg^{2+}	$>1:5000$
Fe^{3+}	$1:500$
Cd^{2+}	$1:1500$
NH_4^+	$>1:5000$
Zn^{2+}	$1:500$
Co^{2+}	$1:100$
Cr^{3+}	$1:1000$
Ni^{2+}	$1:100$
SCN^-	$1:2000$
Mn^{2+}	$1:500$
CO_3^{2-}	$1:1000$
CrO_4^{2-}	$1:2000$
Cu^{2+}	$1:500$
Al^{3+}	$1:100$

^{a}Interfering ion.

the effect of DDMAC concentration on the extraction efficiency, various amounts of DDMAC solution in the range of 0.003% to 0.16% were evaluated. The obtained results given in Figure 5 showed that with the increasing DDMAC concentration, the extraction efficiency increased up to 0.03% (v/v) and then remained constant up to 0.16% (v/v). So, a DDMAC concentration of 0.03% (v/v) was selected in the following experiments. Note that, the extraction efficiencies were about 30% and 98% in the absence and in the presence of DDMAC, respectively. So, it can be concluded that the presence of DDMAC enhances the extraction efficiency by more than 3 folds.

3.6. Effect of Salt Addition.
Increase in ionic strength resulting from salt addition causes two opposite effects on the extraction efficiency: (1) the physical properties of the Nernst diffusion film can change by the presence of dissolved salt in water, and consequently, the diffusion rate of the analyte into the drop can decrease [29]; (2) the salting-out effect can cause an increase in the extraction efficiency [30]. In order to investigate the effect of ionic strength arising from salt addition, on performance of UA-DLLME-SFO-BE, various experiments were performed by adding varying amounts of NaCl from 0% to 4% (w/v). Other experimental parameters were kept constant. The results showed a gradual increase in the extraction efficiency of the Pb ions with increased NaCl amount up to 0.4% (w/v) and then remained constant when salt concentration is between 0.4% and 2%. However, a further increase in the salt concentration caused a decrease in the extraction efficiency. Thus, a concentration of 0.4% (w/v) of NaCl was chosen as the optimum value.

3.7. Effects of Temperature and Duration on Extraction and Back-Extraction.
The temperature affects both the solubility of the organic extraction solvent in the aqueous phase and the emulsification process. Thus, it also affects the mass transfer and the extraction efficiency [28]. If the extraction temperature is lower than 25°C, the selected extraction solvent 1-dodecanol (melting point: 22–24°C) becomes more viscous in the extraction medium and the emulsification does not occur properly. On the contrary, at higher temperatures, the solubility and volatility of 1-dodecanol increase, and this leads to decrease in the extraction efficiency. In addition, higher temperatures cause decomposition of the metal complexes. In order to get a satisfying extraction efficiency, the extraction temperature and incubation time required to complete extraction should be optimized. The effect of the extraction temperature (ultrasonic bath temperature) and duration was explored in the range of 20–60°C and 1–8 min, respectively. It was found that a temperature of 40°C and a time of 4 min for sonication were adequate to achieve an optimum extraction efficiency. Same temperature and time values, 40°C and a 4 min, were used for back-extraction in UA-DLLME-SFO-BE.

3.8. Effect of Interfering Ions.
Other ions except Pb in natural samples can affect the extraction efficiency. In order to explore the effect of coexisting ions, 25 mL solutions containing $50\,\mu g \cdot L^{-1}\,Pb^{2+}$ ions together with foreign ions were processed according to the developed procedure. If an added foreign ion caused ±5% variation in the absorbance value of analyte, then it was considered as interfering species. The obtained results are summarized in Table 1. According to the given results, the method has a good tolerance to coexisting ions.

3.9. Analytical Performance of the Method.
The calibration graphs were linear in the range of $10–500\,\mu g \cdot L^{-1}$ lead under the optimum conditions of the proposed UA-DLLME-SFO-BE procedure. The regression equation for lead determination after microextraction was $A = 4.22 \times 10^{-4}\,C - 2.81 \times 10^{-2}$, where A is the absorbance and C is the metal ion concentration in solution ($\mu g \cdot L^{-1}$). The correlation coefficient of the calibration curve equation was 0.994. The equation obtained by direct aspiration in FAAS without the microextraction procedure was $A = 5.22 \times 10^{-6}\,C + 1.79 \times 10^{-4}$, with linear range between 2,000 and $5,000\,\mu g \cdot L^{-1}$ with the correlation coefficient of 0.999. The enhancement factor calculated as the ratio of the

TABLE 2: Analytical characteristics of the proposed method.

Analytical feature	Value
Enhancement factor (EF)	81
Limit of detection (LOD) (μg·L^{-1}) ($n = 10$)	1.9
Limit of quantitation (LOQ) (μg·L^{-1}) ($n = 10$)	6.4
Linear range (μg·L^{-1})	10–500
Precision (%RSD) ($n = 10$, for 25 μg·L^{-1} Pb^{2+})	3.4

TABLE 3: Determination of lead ions in several water samples ($n = 3$).

Sample	Added (μg·L^{-1})	Found (μg·L^{-1})	Recovery %
Tap water	0	0	—
	25.0	24.1 ± 0.9	96.4
	50.0	48.6 ± 1.3	97.2
	100.0	98.1 ± 1.9	98.1
Lake water	0	7.6 ± 0.2	—
	25.0	31.9 ± 1.1	97.8
	50.0	56.1 ± 1.4	97.4
	100.0	106.2 ± 1.8	98.7

slope of calibration curves of the analyte after microextraction to that of prior microextraction was found as 81.

As an indication of precision, the relative standard deviation (RSD) was calculated for 10 replicate measurements by using 25 μg·L^{-1} Pb^{2+} and was found to be less than 3.4%. The limit of detection (LOD), defined as the concentration ratio of three times the standard deviation of the blank signal and the slope of the calibration graph after preconcentration, was found as 1.9 μg·L^{-1}. The limit of quantification (LOQ) was the lowest level of the analyte that can be accurately and precisely measured. LOQ, defined as the concentration ratio of ten times the standard deviation of the blank signal and the slope of the calibration graph after preconcentration, was found as 6.4 μg·L^{-1}. Table 2 summarizes some analytical figures of the method.

3.10. Accuracy of the Method. In order to prove the performance, the proposed method, a certified reference material (TM-61.2, fortified water), was used and recovery values were calculated. The certified concentration value for Pb(II) ions was 61.4 μg·L^{-1}. Recovery values were calculated as the averages of three parallel experiments and found as 99%. This value proves the satisfying consistency between the obtained results and certified value. These results also show that the developed method was successful for the determination of lead.

3.11. Analysis of Real Samples. In order to evaluate the validation of the method, the recovery experiments were performed by spiking different water samples such as tap (Ankara, Turkey) and lake (Ankara, Turkey) water samples. During these experiments, different amounts of lead were added to these water samples and the optimized method was applied. Obtained results are given in Table 3. As understood from the table, calculated recovery values for spiked water samples were always greater than 96%, and these results verify the validity of the proposed method.

4. Conclusion

An ultrasound-assisted dispersive liquid-liquid microextraction method based on solidification of floating organic drop and back-extraction (UA-DLLME-SFO-BE) technique for lead enrichment in aqueous samples prior to flame atomic absorption determination was reported in this paper. The novelty of the proposed method is the elimination of the clogging problem of the instrument parts caused by frozen 1-dodecanol. In addition, applying ultrasound energy shortens the extraction time and improves the extraction efficiency. Besides known advantages of the conventional SFODME method such as low cost, rapidity, simplicity of operation, high enhancement factor, and extraction efficiency, advantages originated from back-extraction make the proposed method very effective for the determination of trace amounts of lead in natural water samples.

Conflicts of Interest

The authors declare that they have no conflicts of interest.

References

[1] Y. Wang, J. Han, Y. Liu, L. Wang, L. Ni, and X. Tang, "Recyclable non-ligand dual cloud point extraction method for determination of lead in food samples," *Food Chemistry*, vol. 190, pp. 1130–1136, 2016.

[2] M. Falahnejad, H. Zavvar Mousavi, H. Shirkhanloo, and A. M. Rashidi, "Preconcentration and separation of ultra-trace amounts of lead using ultrasound-assisted cloud point-micro solid phase extraction based on amine functionalized silica aerogel nanoadsorbent," *Microchemical Journal*, vol. 125, pp. 236–241, 2016.

[3] S. Nekouei, F. Nekouei, I. Tyagi, S. Agarwal, and V. K. Gupta, "Mixed cloud point/solid phase extraction of lead(II) and cadmium(II) in water samples using modified-ZnO nanopowders," *Process Safety and Environmental Protection*, vol. 99, pp. 175–185, 2016.

[4] S. S. Arain, T. G. Kazi, J. B. Arain, H. I. Afridi, K. D. Brahman, and Naeemullah, "Preconcentration of toxic elements in artificial saliva extract of different smokeless tobacco products by dual-cloud point extraction," *Microchemical Journal*, vol. 112, pp. 42–49, 2014.

[5] M. Tuzen, K. O. Saygi, and M. Soylak, "Solid phase extraction of heavy metal ions in environmental samples on multiwalled carbon nanotubes," *Journal of Hazardous Materials*, vol. 152, no. 2, pp. 632–639, 2008.

[6] J. Cao, P. Liang, and R. Liu, "Determination of trace lead in water samples by continuous flow microextraction combined with graphite furnace atomic absorption spectrometry," *Journal of Hazardous Materials*, vol. 152, no. 3, pp. 910–914, 2008.

[7] A. Szymczycha-Madeja, M. Welna, and P. Pohl, "Comparison and validation of different alternative sample preparation procedures of tea infusions prior to their multi-element analysis by FAAS and ICP-OES," *Food Analytical Methods*, vol. 9, no. 5, pp. 1398–1411, 2016.

[8] Ç. Arpa Şahin and İ. Durukan, "Ligandless-solidified floating organic drop microextraction method for the preconcentration of trace amount of cadmium in water samples," *Talanta*, vol. 85, no. 1, pp. 657–661, 2011.

[9] T. Daşbaşı, Ş. Saçmacı, N. Çankaya, and C. Soykan, "A new synthesis, characterization and application chelating resin for determination of some trace metals in honey samples by FAAS," *Food Chemistry*, vol. 203, pp. 283–291, 2016.

[10] B. Barfi, M. Rajabi, M. M. Zadeh, M. Ghaedi, M. Salavati-Niasari, and R. Sahraei, "Extraction of ultra-traces of lead, chromium and copper using ruthenium nanoparticles loaded on activated carbon and modified with *N,N*-bis-(α-methyl-salicylidene)-2, 2-dimethylpropane-1,3-diamine," *Microchimica Acta*, vol. 182, no. 5-6, pp. 1187–1196, 2015.

[11] M. Ataee, T. Ahmadi-Jouibari, and N. Fattahi, "Application of microwave-assisted dispersive liquid–liquid microextraction and graphite furnace atomic absorption spectrometry for ultra-trace determination of lead and cadmium in cereals and agricultural products," *International Journal of Environmental Analytical Chemistry*, vol. 96, no. 3, pp. 271–283, 2016.

[12] R. Rahnama and M. Najafi, "The use of rapidly synergistic cloud point extraction for the separation and preconcentration of trace amounts of Ni (II) ions from food and water samples coupling with flame atomic absorption spectrometry determination," *Environmental Monitoring and Assessment*, vol. 188, no. 3, pp. 1–9, 2016.

[13] G. Peng, Q. He, G. Zhou et al., "Determination of heavy metals in water samples using dual-cloud point extraction coupled with inductively coupled plasma mass spectrometry," *Analytical Methods*, vol. 7, no. 16, pp. 6732–6739, 2015.

[14] L. Zhang, X. Li, X. Wang, W. Wang, X. Wang, and H. Han, "Preconcentration and determination of chromium species by cloud point extraction-flame atomic absorption spectrometry," *Analytical Methods*, vol. 6, no. 15, pp. 5578–5583, 2014.

[15] M. R. Moghadam, A. M. H. Shabani, and S. Dadfarnia, "Simultaneous spectrophotometric determination of Fe(III) and Al(III) using orthogonal signal correction–partial least squares calibration method after solidified floating organic drop microextraction," *Spectrochimica Acta Part A: Molecular and Biomolecular Spectroscopy*, vol. 135, pp. 929–934, 2015.

[16] C. Arpa Sahin and I. Tokgöz, "A novel solidified floating organic drop microextraction method for preconcentration and determination of copper ions by flow injection flame atomic absorption spectrometry," *Analytica Chimica Acta*, vol. 667, no. 1-2, pp. 83–87, 2010.

[17] H. Fazelirad and M. A. Taher, "Ligandless, ion pair-based and ultrasound assisted emulsification solidified floating organic drop microextraction for simultaneous preconcentration of ultra-trace amounts of gold and thallium and determination by GFAAS," *Talanta*, vol. 103, pp. 375–383, 2013.

[18] B. Barfi, A. Asghari, M. Rajabi, and N. Mirkhani, "Dispersive suspended-solidified floating organic droplet microextraction of nonsteroidal anti-inflammatory drugs: comparison of suspended droplet-based and dispersive-based liquid-phase microextraction methods," *RSC Advances*, vol. 5, no. 129, pp. 106574–106588, 2015.

[19] J. J. Ma, X. Du, J. W. Zhang, J. C. Li, and L. Z. Wang, "Ultrasound-assisted emulsification–microextraction combined with flame atomic absorption spectrometry for determination of trace cadmium in water samples," *Talanta*, vol. 80, no. 2, pp. 980–984, 2009.

[20] J. Xiong, G. Zhou, Z. Guan, X. Tang, Q. He, and L. Wu, "Determination of chlorpyrifos and its main degradation product tcp in water samples by dispersive liquid –liquid microextraction based on solidification of floating organic droplet combined with high-performance liquid chromatography," *Journal of Liquid Chromatography & Related Technologies*, vol. 37, no. 11, pp. 1499–1512, 2014.

[21] T. Sismanoglu, S. Pura, and A. S. Bastug, "Binary and ternary metal complexes of Congo red with amino acids," *Dyes and Pigments*, vol. 70, no. 2, pp. 136–142, 2006.

[22] O. Abollino, M. Aceto, C. Sarzanini, and E. Mentasti, "Behavior of different metal/ligand systems in adsorptive cathodic stripping voltammetry," *Electroanalysis*, vol. 11, no. 12, pp. 870–878, 1999.

[23] G. Alpdogan, N. San, and F. Dinc Zor, "Analysis of some trace metals in fish species after preconcentration with congo red on amberlite XAD-7 resin by flame atomic absorption spectrometry," *Journal of Chemistry*, vol. 2016, 8 pages, 2016.

[24] G. Khayatian and S. Hassanpoor, "Development of ultrasound-assisted emulsification solidified floating organic drop microextraction for determination of trace amounts of iron and copper in water, food and rock samples," *Journal of Iranian Chemical Society*, vol. 10, no. 1, pp. 113–121, 2013.

[25] F. Shemirani, M. Baghdadi, M. Ramezani, and M. R. Jamali, "Determination of ultra trace amounts of bismuth in biological and water samples by electrothermal atomic absorption spectrometry (ET-AAS) after cloud point extraction," *Analytica Chimica Acta*, vol. 534, no. 1, pp. 163–169, 2005.

[26] I. Durukan, M. Soylak, and M. Doğan, "Enrichment and separation of Fe (III), Mn (II), Ni (II), and Zn (II) as their congo red chelates on multiwalled carbon nanotube (MWCNT) disk in food and water samples," *Atomic Spectroscopy*, vol. 34, pp. 20–25, 2013.

[27] S. Dadfarnia, A. M. H. Shabani, and A. Mirshamsi, "Solidified floating organic drop microextraction and spectrophotometric determination of vanadium in water samples," *Turkish Journal of Chemistry*, vol. 35, pp. 625–636, 2011.

[28] İ. Durukan, C. Arpa Sahin, and S. Bektas, "Determination of copper traces in water samples by flow injection-flame atomic absorption spectrometry using a novel solidified floating organic drop microextraction method," *Microchemical Journal*, vol. 98, no. 2, pp. 215–219, 2011.

[29] M. Leonga, C. Changa, M. Fuhb, and S. Huanga, "Low toxic dispersive liquid–liquid microextraction using halosolvents for extraction of polycyclic aromatic hydrocarbons in water samples," *Journal of Chromatography A*, vol. 1217, no. 34, pp. 5455–5461, 2010.

[30] S. Bahar and R. Zakerian, "Determination of copper in human hair and tea samples after dispersive liquid-liquid microextraction based on solidification of floating organic drop (DLLME-SFO)," *Journal of the Brazilian Chemical Society*, vol. 23, no. 6, pp. 1166–1173, 2012.

19

Preparation of Imazethapyr Surface Molecularly Imprinted Polymers for its Selective Recognition of Imazethapyr in Soil Samples

Yanqiang Zhou,[1] Yinhui Yang,[2] Meihua Ma,[1] Zhian Sun,[1] Shanshan Wu,[1] and Bolin Gong ⓘ[1]

[1]College of Chemistry and Chemical Engineering, North Minzu University, Yinchuan 750021, China
[2]College of Chemistry and Chemical Engineering, Beijing Institute of Technology, Beijing 100081, China

Correspondence should be addressed to Bolin Gong; gongbolin@163.com

Academic Editor: Núria Fontanals

A novel strategy based on imazethapyr (IM) molecular-imprinting polymers (MIPs) grafted onto the surface of chloromethylation polystyrene resin via surface-initiated atom transfer radical polymerization (SI-ATRP) for specific recognition and sensitive determination of trace imazethapyr in soil samples was developed. The SI-ATRP was performed by using methanol-water (4 : 1, v/v) as the solvent, acrylamide as the functional monomer, trimethylolpropane trimethacrylate (TRIM) as the cross-linker, imazethapyr as the template, and CuBr/2,2'-bipyridine as the catalyst. The resulting MIPs were characterized by elemental analysis, Fourier-transform infrared spectroscopy (FT-IR), scanning electron microscopy (SEM), and transmission electron microscopy (TEM). Then, the binding selectivity, adsorption capacity, and reusability of the MIPs were evaluated. The results indicated that the prepared MIPs exhibited specific recognition and high selectivity for imazethapyr. The MIPs were further used as solid-phase extraction (SPE) materials coupled with high-performance liquid chromatography (HPLC) for selective extraction and detection of trace imazethapyr from soil samples. The results showed that good linearity was observed in the range of 0.10–5.00 μg/mL, with a correlation coefficient of 0.9995. The limit of detection (LOD) of this method was 15 ng/g, and the extraction recoveries of imazethapyr from real samples were in the range of 91.1–97.5%, which proved applicable for analysis of trace imazethapyr in soils. This work proposed a sensitive, rapid, and convenient approach for determination of trace imazethapyr in soil samples.

1. Introduction

Imazethapyr, a type of imidazolinones herbicide, belongs to the imidazolinones family of herbicides that are being extensively used in a wide range of cropping systems to enhance crop yields and protect crops from damage by weeds and annual grasses in soybean and peanut [1, 2]. The improper or incorrect use of imazethapyr may lead to the residue and pollution of the soil and groundwater near field crops and also may affect nontargeted plants [3, 4]. In recent years, the deeper toxic mechanisms of imazethapyr revealed that the accumulated imazethapyrs were particularly unfriendly toward animals and humans, which may result in acute or chronic toxicity. In addition, it would affect the transcription of photosynthesis-related genes and inhibit the antioxidant system of the plants [5] and affect the chlorophyll synthesis [6].

To avoid its harmful effects on target and nontarget organisms and to better understand the behavior of imazethapyr in the environment, it is essential to propose a sensitive, rapid and, convenient approach for adsorption of the imazethapyr. The development of analytical methods suitable to fulfill the requirement of determining imazethapyr in soil has attracted widespread interest. The methods for determination of herbicide residues in soil include solid-phase extraction [7], dispersive solid-phase extraction [8, 9], magnetic solid-phase extraction (MSPE) [10], multiple monolithic fiber solid-phase microextraction (MMF-SPME) [11], high-performance liquid chromatography (HPLC) [12], high-performance liquid chromatography with mass spectroscopy (HPLC-MS/MS) [13], gas chromatography coupled with an electron capture detector (GC-ECD) [14], and molecular imprinting [15, 16]. Compared with other methods, molecular imprinting has emerged as a powerful technique [17] for the preparation of tailor-made recognition material owing to its high affinity and specificity [18–21]. Molecularly imprinted polymers (MIPs) have been widely applied to sample preparation for the selective separation and enrichment of trace analytes [22–26]. Traditionally, the MIPs were synthesized through precipitation polymerization, bulk polymerization, and emulsion polymerization [27, 28]. In most cases, traditional MIPs were prepared in organic solvents. However, low imprinted efficiency, slow mass transfer rate, and high hydrophilia of some pesticide residues were restricting the application of MIPs. It is clear that the development and application of selective sorbent phases are still in demand, particularly if they are applied in aqueous conditions. Thus, MIP materials with high sensitivity, selectivity, and versatility are still in demand [29]. Furthermore, the recognition sites of MIPs synthesized by traditional methods are located in the interior of the polymers, which have poor dispersion and serious embedding. As a result, the template molecule has a certain residue when eluting. The MIP product can be used after being crushed and sieved, and the shape of the product is not uniform. The subsequent processing work requires a large amount of time [30, 31]. The surface-initiated atom transfers radical polymerization (SI-ATRP) has been proposed as a new class of controlled living radical polymerization, which possessed a wide range of monomer choice, high selectivity, effectively reduced the embedding of recognition sites, improved adsorption capacity and adsorption efficiency, high mass transfer rate, and high imprinting efficiency [32, 33]. The conditions required for the polymerization reaction were simple and easy to operate, and there are many types of functional monomers that can be selected, water can be used as a medium for the reaction, and a block polymer can be synthesized. Meanwhile, it could be well controlled by SI-ATRP reagent to avoid adverse reactions, such as radical coupling or disproportionation action. Particularly, the polymer segment, template molecular density, and film thickness could be effectively controlled, which is beneficial to increase the adsorption capacity and adsorption efficiency of MIPs [34].

In this study, the imazethapyr surface MIPs on the surface of chloromethylation polystyrene resin were successfully prepared in the binary mixture solvents of methanol-water (4 : 1, v/v) via SI-ATRP. The MIPs were characterized by elemental analysis, Fourier-transform infrared spectroscopy (FT-IR), scanning electron microscopy (SEM), and transmission electron microscopy (TEM). Then, MIP polymers were used as SPE adsorbents coupled with high-performance liquid chromatography (HPLC) to detect a low concentration of imazethapyr in soil samples.

2. Materials and Methods

2.1. Reagent and Instrument. Polystyrene resin was purchased from Hebei Cangzhou Baoen Chemical Co., Ltd. (Hebei, China). Copper(I) bromide (CuBr), 2, 2′-bipyridyl (BPy), imazethapyr (IM), imazapyr acid, carbendazim, trimethylolpropane trimethacrylate (TRIM, technical grade), and acrylamide (AM) were purchased from Aladdin Reagent (Shanghai, China). Anhydrous zinc chloride, acetonitrile, sulfuric acid, paraformaldehyde, acetic acid (glacial), acetone, and methanol were purchased from Tianjing Damao Chemical Reagent Factory (Tianjing, China). Soil samples were collected from the field and then mixed, dried, grounded and sifted. All other reagents were of analytical grade, and double distilled water was used throughout the experiment. Before use in HPLC analysis, the solution must be filtered through a 0.45 μm nylon filter.

All chromatographic tests were performed using an LC-20AT chromatographic system (Shimadzu, Japan), including two LC-20AT pumps and an SPD-20A UV–VIS detector. FT-IR was performed on FTIR-8400S (Shimadzu, Japan). SEM was performed on JSM-7500F (JEOL, Japan). TEM was measured on HT7700 (Hitachi Ltd., Japan). An elemental analyzer was purchased from the YiLe Man Element Analysis System Company of German (VarioEL III, German). A constant temperature water bath oscillator was purchased from Shanghai Pudong Physical Optics Instrument Plant (SHZ-C, Shanghai, China). A TG16-WS high-speed centrifuge (Centrifuge Factory, China) was used in this experiment, and a TU-1810-type ultraviolet spectrophotometer was purchased from Beijing general instrument Co., Ltd. (Beijing, China).

2.2. Preparation of Chloromethylation Polystyrene Resin (CMCPS). The chloromethylation polystyrene resin (CMCPS) was prepared by following a previously reported protocol [35]. 15.4 g of polystyrene resin, 9.09 g of paraformaldehyde, 20.4 g of anhydrous zinc chloride, 40 mL of 80% sulfuric acid, and 80 mL of glacial acetic acid were placed in three flasks at 50 to 55°C with magnetic stirring. During the reaction, hydrogen chloride gas was continuously bubbled into the three flasks for 12 h. The obtained resin was first washed with deionized water until the pH of the resin to neutral, and then with acetone for a few times. Lastly, it was dried in the vacuum oven at 60°C.

2.3. Preparation of Imazethapyr-Imprinted Polymer via SI-ATRP. The imazethapyr-imprinted polymer was prepared via SI-ATRP. Briefly, the template, functional monomer, and cross-linker that were used in this study were imazethapyr, acrylamide (AM), and TRIM, respectively. IM (1.10 g), AM (1.30 g), and TRIM (7.50 mL) were dissolved in 20 mL methanol-water (4 : 1, v/v). This mixture solution was stirred for 2 h at room temperature, leading to the formation of a complex of imprint molecules and monomers. Then, it was purged with the gas of N_2 for 30 min and was transferred to a flask containing ATRP initiator CMCPS resin (1.30 g), catalyst-CuBr (0.0258 g), and Bpy

(0.0562 g). This reaction system was incubated at room temperature under the protection of nitrogen for 24 h. The final product were collected and washed with ethylenediamine tetraacetic acid disodium salt (EDTA) solution (0.1 mol/L), with deionized water, several times with acetone, and were dried in a vacuum oven at 50°C temperature overnight. Then, the MIPs were extracted by using methanol-acetic acid (9 : 1, v/v) solution in a Soxhlet extractor for 24 h to remove the template thoroughly. After that, the MIPs were washed with ultrapure water until neutral. The nonimprinted polymers (NIPs) were prepared by using the same procedure mentioned above, except the addition of a template molecule.

2.4. Batch Rebinding Experiment.

The adsorption isotherms were obtained by suspending 0.1 g MIPs in 25 mL methanol solution with different imazethapyr concentrations (1~10 mmol/L). Meanwhile, the adsorption kinetic curves were obtained by detecting the change over time. Adsorption equilibrium of MIPs and NIPs in IM solution was determined by using an initial IM concentration of 8 mmol/L with different adsorption time periods of 1, 2, 3, 4, 5, 6, 8, and 10 h. The binding amount of IM on MIPs was determined as the difference between the total IM amount and the residual amount in the solution by using the UV spectrophotometer at 218 nm. The same experiment steps were carried out for NIPs.

The adsorption capacity Q (mg/g) was calculated according to the following:

$$Q = \frac{(C_0 - C_e)V}{m}, \quad (1)$$

where C_0 (mg/L) and C_e (mg/L) are the initial and final concentrations of IM in the solution, respectively; V (L) is the total volume of the solution; and m (g) is the mass of MIPs.

2.5. Selective Binding Experiment.

To evaluate the selectivity of MIPs, two structural analogues (schemes), imazapyr and carbendazim, were selected to determine their binding capacities on MIPs and NIPs. 0.1 g of MIPs and NIPs were dispersed in 25 mL of methanol solution containing imazethapyr, imazapyr acid, and carbendazim with an initial concentration of 8 mmol/L. After the adsorption is finished, equilibrium concentrations of each analyte were determined by using a UV spectrophotometer (Figure 1).

The selectivity coefficient of MIPs for imazethapyr with respect to the competition species, imazapyr acid and carbendazim, could be obtained from the equilibrium-binding data according to the following equation:

$$k = \frac{K_{d1}}{K_{d2}}, \quad (2)$$

where K_{d1} represents the distribution coefficient (L/g) of imazethapyr, K_{d2} represents the distribution coefficient (L/g) of imazethapyr structural analogues imazapyr and carbendazim, and k is the selectivity coefficient. The value of k allows an estimation of selectivity of MIPs for imazethapyr.

2.6. Desorption Experiment.

Certain amounts of MIP particles adsorbing imazethapyr in a saturated state were packed into a piece of glass pipe with an internal diameter of 0.8 cm. The bed volume (BV) of the packed column was 2.0 mL. The eluent of methanol-acetic acid (9 : 1, v/v) was allowed to flow gradually through the column at a rate of two bed volumes per 2 minutes (2 BV/2 min) in countercurrent manner. The effluent with two-volume (2 BV) interval was collected, and the concentration of imazethapyr was determined by UV spectrophotometry at 218 nm. The dynamic desorption curve was plotted, and elution property of MIPs was examined.

2.7. Preparation of Imazethapyr MIPs Solid-Phase Extraction Column.

An empty polypropylene solid-phase extraction column prepared for the experiment was blocked by a polyethylene sieve plate, and 500 mg of MIP powder was filled into the column, another polyethylene sieve plate was used to compact the MIP powder. The packed columns were washed by methanol-acetic acid (9 : 1, v/v) to remove the template, and the effluent was collected for HPLC detection until IM became undetectable, showing that the template had been cleaned. Then, the residual solvent was removed by methanol and dried for subsequent use. Before each use, 5 mL methanol and deionized water were used to activate the solid phase extraction column. The analytes were eluted by methanol and ammonium hydroxide with a volume ratio of 9 : 1 at each step. The elutions were evaporated until drying and redissolved with 1 mL methanol for further HPLC analysis.

2.8. Analysis of Soil Sample.

Ten grams of soil sample, 25 mL of 0.1 mol/L ammonium chloride, and ammonia buffer (pH 10) were added to a 50 mL centrifuge tube. The mixture was vortex mixed for 5 min, ultrasonicated for 20 min, and then centrifuged at 6000 r/min for 4 min. The supernatant was collected. This procedure was repeated thrice. The pH of the combined supernatant was adjusted to 2.0 with 1.0 mol/L HCl and volatilized to dry with a rotary evaporator at 45°C. Then, the residue was redissolved into 5 mL methanol, diluted with water to 15 mL, and applied to MIP or NIP cartridges. After SPE, the elution was concentrated to 1 mL and analyzed on the HPLC.

3. Results and Discussion

3.1. Preparation of MIPs.

The preparation procedures for MIPs of imazethapyr were described in Figure 2. The functional monomer initially formed complexes with the template molecules. Then, their functional groups were held in position by the highly cross-linked polymeric structure after polymerization. The MIPs were finally grafted onto the surface of the CMCPS resin via SI-ATRP. After template removal, specific binding sites were left in the polymer material.

3.2. Characterization of the MIPs

3.2.1. Elemental Analysis.

Nitrogen, carbon, and hydrogen elemental analysis data of CMCPS resin, MIPs, and NIPs

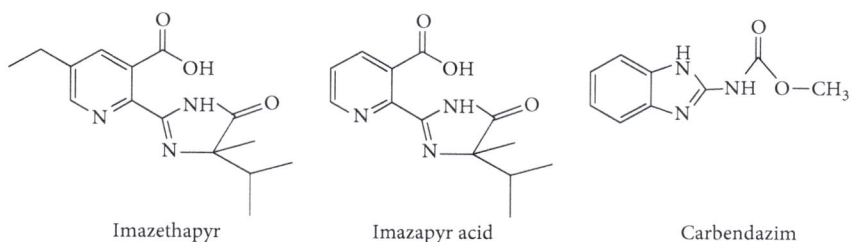

Imazethapyr   Imazapyr acid   Carbendazim

FIGURE 1

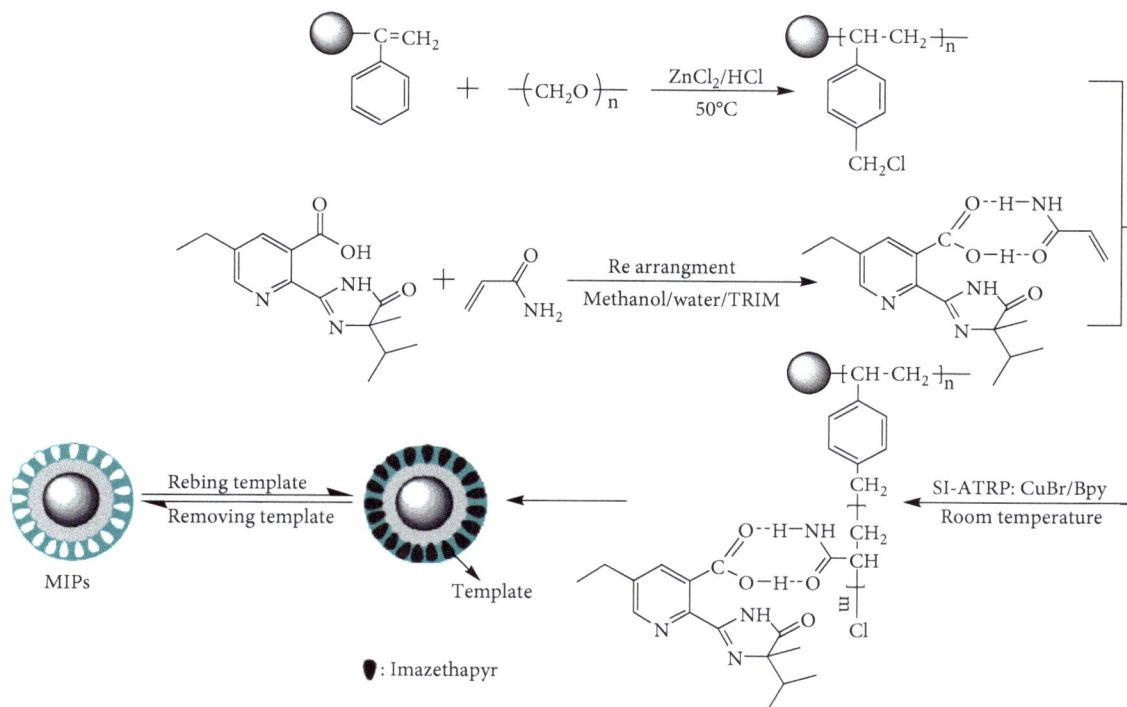

FIGURE 2: Process for preparing MIPs of imazethapyr.

are summarized in Table 1. Obviously, N content had a significant increase after the IM imprinted onto the surface of CMCPS resin. The amounts of the grafted functional monomer and cross-linker on CMCPS resin were calculated based on the elemental analysis data by using the following:

$$GR = \frac{N_i\% \times 10^6}{28 \times (1 - C\% - H\% - N\%) \times S}, \qquad (3)$$

where C%, H%, and N% are the weight fractions of carbon, hydrogen, and nitrogen of MIPs in grafted imazethapyr on CMCPS resin, respectively. $N_i\%$ represents the weight fraction of nitrogen increase over that of CMCPS resin. S is the specific surface area of the chloromethylation polystyrene resin. According to Equation (3), the calculated amount of grafted density was $25.14\,\mu mol/m^2$.

3.2.2. FT-IR Characterization of Polystyrene Resin, CMCPS, and MIPs.
FT-IR spectra of polystyrene resin (a), CMCPS (b), and MIPs (c) are shown in Figure 3. After

chloromethylation, the absorption peak changed significantly, of which $676\,cm^{-1}$ is the characteristic absorption peak of $-CH_2Cl$, $1232\,cm^{-1}$ is the in-plane bending vibration of $-CH$ bond in $-CH_2Cl$, $3016\,cm^{-1}$ is the stretching vibration peak of the $-CH$ bond on the benzene ring, $2958\,cm^{-1}$ and $2927\,cm^{-1}$ are the stretching vibrations of saturated $-CH$ and $-CH_2$, respectively. $1741\,cm^{-1}$ and $1720\,cm^{-1}$ are the vibration absorption peaks of $-C=O$ in the aldehyde group. $1484\,cm^{-1}$ and $1444\,cm^{-1}$ are the in-plane and out-of-plane bending vibration peaks of $-CH$ bond on the benzene ring. $828\,cm^{-1}$ is the out-of-plane bending vibration peak of two adjacent hydrogen atoms on the para-disubstituted benzene ring. FT-IR indicated that the H at the 4th position on the cross-linked polystyrene benzene ring has been replaced by $-CH_2Cl$, which has successfully produced a chloromethylated polystyrene cross-linked resin.

In MIPs, the increase of peak intensity at $1160\,cm^{-1}$ is due to the stretching vibration of $-C-O-C$ in the cross-linker TRIM. The peak intensity of $3527\,cm^{-1}$ is enhanced because

TABLE 1: Element analysis of the MIPs.

Samples	Elemental composition (%)		
	N	C	H
CMCPS resin	0.158	80.5	6.54
MIPs	0.289	80.8	6.65
NIPs	0.173	80.7	6.73

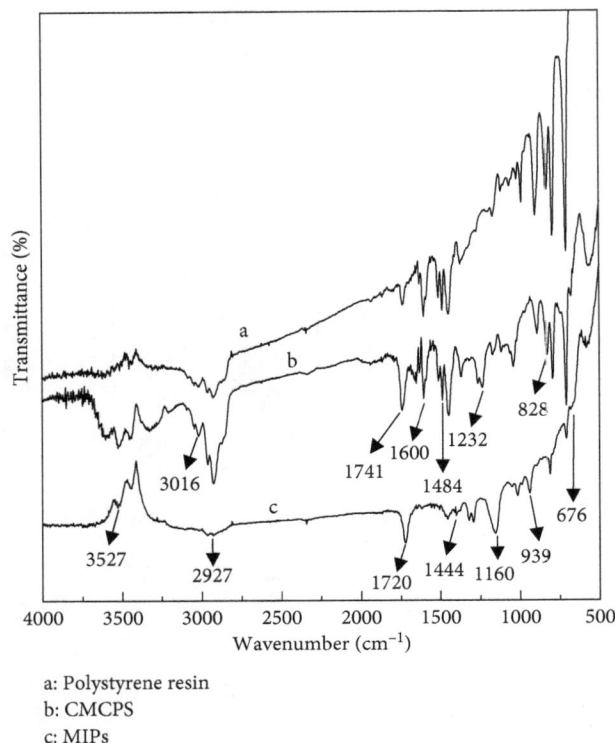

a: Polystyrene resin
b: CMCPS
c: MIPs

FIGURE 3: FT-IR spectra of polystyrene resin (a), CMCPS (b), and MIPs (c).

the end of TRIM can react with the exposed hydroxyl groups on the CMCPS, and the number of free hydroxyl groups increases, indicating that the cross-linker TRIM has been successfully polymerized on the surface of the CMCPS.

3.2.3. SEM and TEM Characterization.

SEM and TEM were used to observe the morphology of CMCPS resin and MIPs. As shown in Figures 4(a) and 4(b), the SEM of CMCPS resin exhibited a very smooth and tight surface before imprinting. After imprinting, the MIPs had a rough surface with morphological features and a large number of mesopores, which was beneficial for molecular adsorption and mass transfer. Then, the TEM of MIPs (Figure 5) shows an obvious nuclear-shell structure of MIPs, and particles adhere to each other. The results showed that the imprinted polymer was grafted onto the surface of the CMCPS resin after polymerization and formed a site of interaction with the template molecule.

3.3. Effect of the Ratio of Template and Monomer on Adsorption.

The ratio of template and monomer is one of the most important factors influencing the adsorption behavior of imazethapyr on the MIPs. From Figure 6, the ratios of the template and monomer at 1 : 2, 1 : 4, 1 : 6, and 1 : 8 were selected to investigate the effects. At first, as the concentration of the monomer increased, the adsorption capacity increased significantly. However, when the ratio of the template and monomer was at 1 : 6, the adsorption capacity gradually decreased. It could be that too much of the amount of monomers will enhance nonspecific adsorption capacity, resulting in the reduction of the specific adsorption capacity of MIPs. In this study, the optimal ratio of the template and monomer was at 1 : 4.

3.4. Scatchard Analysis.

To further estimate the binding property of MIPs, the obtained data were processed with scatchard analysis according to the following equation:

$$\frac{Q}{C_e} = \frac{Q_{max}}{K_d} - \frac{K_d}{Q}, \tag{4}$$

where Q and Q_{max} are the experimental adsorption capacity to the template imazethapyr (mg/g) and the theoretical maximum adsorption capacity of the polymer (mg/g), respectively; C_e is the concentration of imazethapyr in equilibrium solution (mg/L); and K_d is the dissociation constant (mg/L).

The values of K_d and Q_{max} could be calculated from the slope and intercept of the linear line plotted in Q/C_e versus Q. As shown in Figure 7, the two distinct sections of the Scatchard' plot indicate that the binding sites in the MIPs can be classified into two distinct groups with specific binding properties. The respective K_d and Q_{max} values calculated from the slopes and intercepts of the two linear portions are 47.39 mg/L and 78.21 mg/g for higher affinity sites and 833.3 mg/L and 237.32 mg/g for lower affinity sites, respectively.

The linear relevant fitting equations were $y = -0.0211x + 1.6502$ and $R = 0.9982$ for higher affinity sites; $y = -0.0012x + 0.2848$ and $R = 0.9915$ for lower affinity sites. These results indicated that MIPs had high specific recognition to the template imazethapyr.

3.5. Adsorption Kinetics and Selectivity.

Figure 8 shows the rapid rebinding rate of MIPs to the template molecule imazethapyr and a sharp increase in the adsorption capacity on MIPs in the first 3 h due to the presence of a large amount of empty and high-affinity binding sites on the surface of the MIPs. After 3 h, the adsorption rate slowed down, and the equilibrium binding of imazethapyr on the MIPs could be reached within approximately 3 h. Clearly, the surface-imprinted MIPs required a much shorter equilibrium time than the traditionally referenced polymerization method. Without the imprinting process, the functional groups were distributed randomly in NIPs, resulting in the low adsorption ability of imazethapyr. The nonspecific adsorption of imazethapyr was observed in NIPs. The results indicated that the prepared MIPs had high specific recognition to the template imazethapyr.

The adsorption selectivity is an indispensable factor for appreciating capacities of an adsorbent. To evaluate the

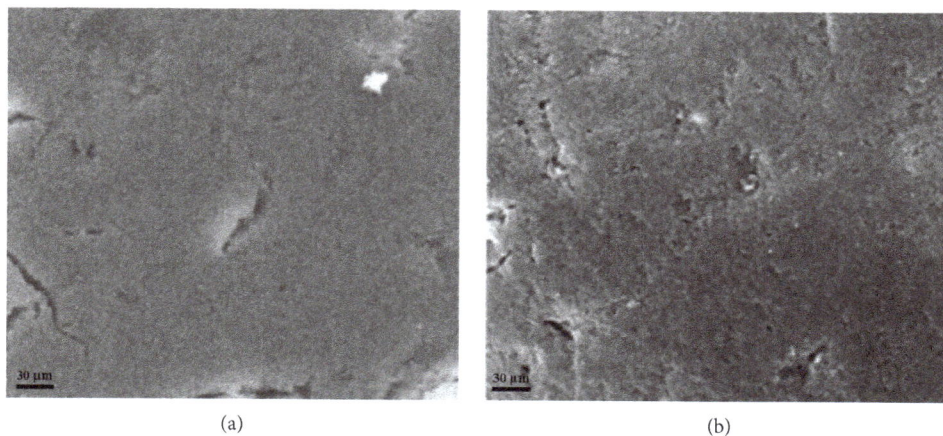

FIGURE 4: The scanning electron micrographs of (a) CMCPS resin (b) MIPs.

FIGURE 5: The transmission electron microscope image of MIPs.

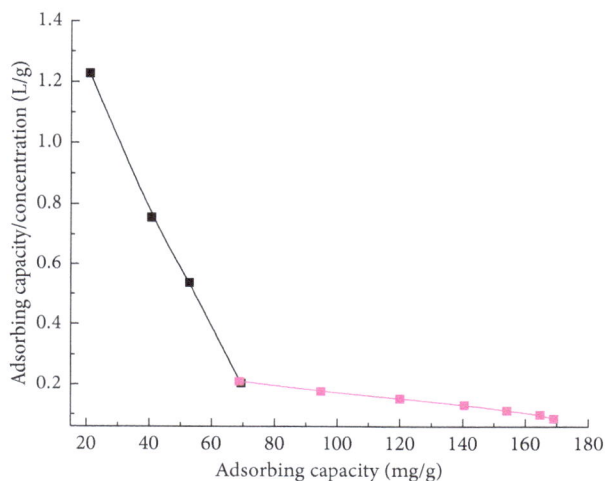

FIGURE 7: Scatchard's plot for imazethapyr MIPs.

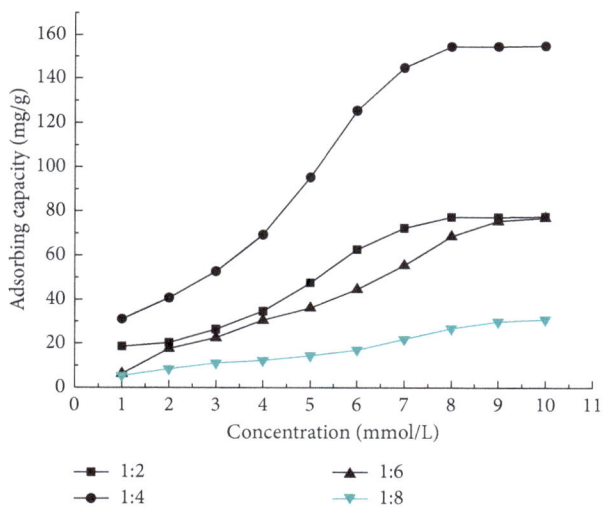

FIGURE 6: Effect of the ratio of template and monomer on adsorption.

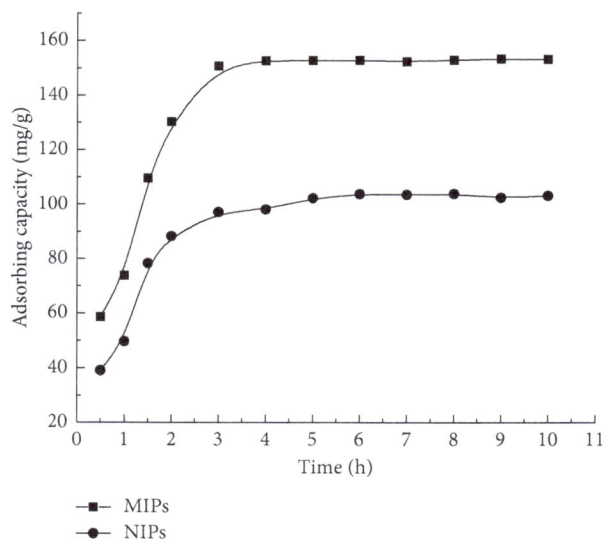

FIGURE 8: Dynamic adsorptions of MIPs and NIPs.

selectivity of the MIPs, analogues of imazethapyr, imazapyr, and carbendazim were selected. The experimental results were plotted in Figure 9, which showed MIPs a much higher selectivity to imazethapyr than NIPs. Selectivity coefficients for imazapyr imprinted material and

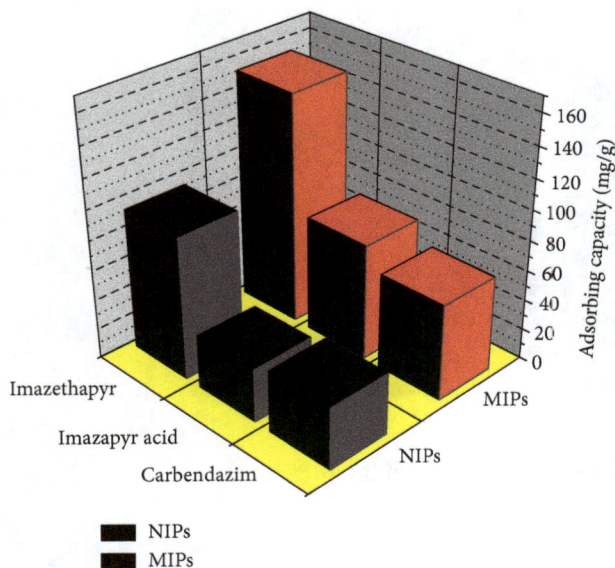

FIGURE 9: Selective adsorptions for the imazethapyr, imazapyr, and carbendazim.

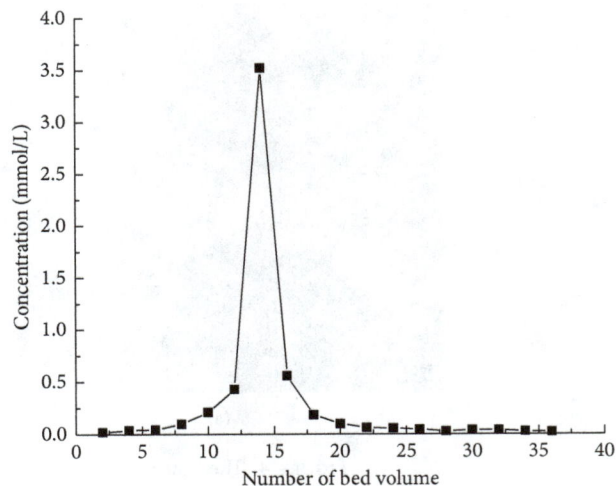

FIGURE 10: Elution curves of imazethapyr on MIPs.

TABLE 2: Recoveries of imazethapyr in soil samples.

Sample	Spiked conc. ($\mu g/g$)	Measured conc. ($\mu g/g$)	Repeatability (RSD%, $n = 3$)	Recovery (%)
	0.050	0.0455	6.5	91.1
Soil	0.500	0.487	3.7	97.5
	1.00	0.955	5.8	95.5

nonimprinted material to imazethapyr were 4.26 and 2.12, respectively. They were 3.17 and 1.59 for carbendazim, respectively. Although the structures of imazapyr and carbendazim are very similar with template imazethapyr, the MIPs could still specifically recognize imazethapyr based on the imprinted sites. The results further demonstrate the high selectivity of the MIPs.

3.6. Desorption Property. The dynamic desorption curve of the MIPs adsorbing imazethapyr in a saturated state is shown in Figure 10. This desorption curve was cuspidal without trail formation. The result showed that the desorption ratios in 22 BV and 24 BV reach 93.28% and 97.92%, respectively, which indicated that combined imazethapyr on the MIPs was easy to be desorbed, namely, the MIPs had an excellent eluting property. It was convenient to reuse MIPs.

3.7. Validation of the MIPs-SPE Procedure. To evaluate the suitability of the proposed sample preparation technology based on the combination of MIPs and SPE techniques, a method for the analysis of imazethapyr in soil via HPLC was developed. Regression equations, correlation coefficient (r), linearity range, limits of detection (LOD), and limit of quantification (LOQ) were investigated.

Matrix-matched calibration standard was applied to eliminate matrix interference effects for the more accurate results. Good linearity ($R > 0.9995.$) was obtained in the range of 0.10–5.0 $\mu g/mL$ of imazethapyr. The LOD at a signal-to-noise ratio of 3 and LOQ at a signal-to-noise ratio of 10 was 15.0 and 50.0 ng/g, respectively. Then, the recovery tests of MIPs were performed at three spiked levels (0.050, 0.500, and 1.00 $\mu g/g$) to evaluate the accuracy of the method. As shown in Table 2, the recoveries of imazethapyr in the

spiked samples were in the range of 91.1–97.5%, with relative standard deviation (RSD) in the range of 3.7%–6.5%, indicating that the MIPs can be used for the selective enrichment of imazethapyr in soil. The results fulfilled the demands for analysis. Then, recoveries showed no significant decrease after the MIPs-SPE was used for more than 10 times, indicating its good stability, reusability, and potential for practical analysis.

Taking into account the difficulties of this residue determination and the strong matrix effects that affect this type of analysis, performing a validation in a given soil by different SPE adsorbents demonstrated the method's applicability. To this aim, the use of MIPs, NIPs, and C_{18} is the best way to give realistic information on the method's performance. In this study, the prepared 500 mg MIPs, NIPs, and C_{18} were applied as the SPE adsorbent for the analysis of imazethapyr in the same soil from Yinchuan (Ningxia, China). The chromatograms spiked with 0.050 $\mu g/g$ of imazethapyr after treating with the MIPs-SPE, NIPs-SPE, and C_{18}-SPE columns are shown in Figures 11(a)–11(c) at the same HPLC test conditions. In terms of the separation of chromatogram peaks, the consequence of using MIPs was better than that of NIPs and C_{18}. As shown in Figure 11(a), the imazethapyr showed a better extraction effect and a higher chromatographic peak than Figure 11(b), which demonstrated its highly selective recognition of MIPs. However, no target compound was detected but a lot of matrix interference peaks after C_{18} were extracted (Figure 11(c)), which further indicated that MIPs could enhance the enrichment efficiency and selectivity of the extraction and analysis.

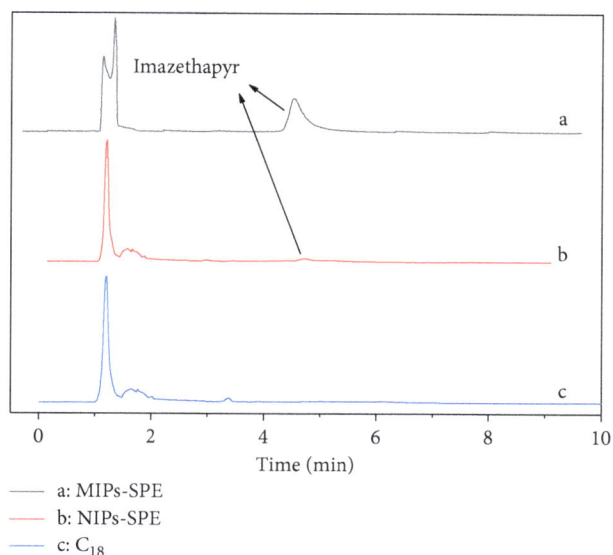

FIGURE 11: The chromatograms of soil samples with spiked imazethapyr ($0.05\,\mu g/g$) for MIPs-SPE cartridge (a), NIPs-SPE cartridge (b), and C_{18}-SPE cartridge (c).

3.8. Analysis Conditions of HPLC. The flow rate was $1.0\,mL/min$, and the detection wavelength was set at $258\,nm$ (the maximal absorbance wavelength of imazethapyr). All the separation and detection were performed on the Diamonsil C_{18} column ($150 \times 4.6\,mm$). The mobile phase was acetonitrile-0.5% acetic acid glacial ($40:60$, V/V).

4. Conclusions

In this work, we prepared MIPs using SI-ATRP with imazethapyr as the template and investigated the recognition properties of the MIPs. MIPs possessed excellent recognition ability and combining selectivity for imazethapyr. However, the combining capacity of MIPs for the contrast substances, imazapyr and carbendazim, whose structures were closely similar to imazethapyr, was low. The experimental results revealed that this material could be successfully used as absorbents of SPE coupled with HPLC to separate, enrich, and detect the trace imazethapyr in soil samples. Furthermore, recoveries showed no significant decrease after the MIPs-SPE was used for more than 10 times, which showed that MIPs had good stability and reusability. This study demonstrated that the MIPs were ideal materials in concentration and purification of trace imazethapyrin complex materials and also provides a model for qualitative and quantitative analysis of imazethapyrin in other complex matrices samples. In addition, the method in this study has provided important reference for further monitoring studies to improve soil safety.

Conflicts of Interest

The authors declare that there are no conflicts of interest.

Authors' Contributions

Yanqiang Zhou, Yinhui Yang, and Meihua Ma contributed equally to this work.

Acknowledgments

This work was financially supported by the National Natural Science Foundation of China (Nos. 21565001 and 31271868), Key Project of North Minzu University (No. 2015KJ30) and Graduate Innovation Project of North Minzu University (No. YCX18077).

References

[1] P. Perucci and L. Scarponi, "Effects of the herbicide imazethapyr on soil microbial biomass and various soil enzyme activities," *Biology and Fertility of Soils*, vol. 17, no. 3, pp. 237–240, 1994.

[2] W. Zhang, E. P. Webster, K. J. Pellerin, and D. C. Blouin, "Weed control programs in drill-seeded imidazolinone-resistant rice (Oryza sativa)," *Weed Technology*, vol. 20, no. 4, pp. 956–960, 2006.

[3] H. F. Qian, H. J. Hu, Y. Y. Mao et al., "Enantioselective phytotoxicity of the herbicide imazethapyr in rice," *Chemosphere*, vol. 76, no. 7, pp. 885–892, 2009.

[4] H. R. Nodeh, W. A. W. Ibrahim, M. A. Kamboh, and M. M. Sanagi, "New magnetic graphene-based inorganic-organic sol-gel hybrid nanocomposite for simultaneous analysis of polar and non-polar organophosphorus pesticides from water samples using solid-phase extraction," *Chemosphere*, vol. 166, pp. 21–30, 2017.

[5] C. C. Sun, S. Chen, Y. Y. Jin et al., "Effects of the herbicide imazethapyr on photosynthesis in PGR5- and NDH-deficient arabidopsis thaliana at the biochemical, transcriptomic, and proteomic levels," *Journal of Agricultural and Food Chemistry*, vol. 64, no. 22, pp. 4497–4504, 2016.

[6] H. F. Qian, X. Han, Q. N. Zhang, Z. Q. Sun, L. W. Sun, and Z. W. Fu, "Imazethapyr enantioselectively affects chlorophyll synthesis and photosynthesis in arabidopsis thaliana," *Journal of Agricultural and Food Chemistry*, vol. 61, no. 6, pp. 1172–1178, 2013.

[7] M. Ramezani, N. Simpson, D. Oliver, R. Kookana, G. Gill, and C. Preston, "Improved extraction and clean-up of imidazolinone herbicides from soil solutions using different solid-phase sorbents," *Journal of Chromatography A*, vol. 1216, no. 26, pp. 5092–5100, 2009.

[8] J. L. D. O. Arias, C. Rombaldi, S. S. Caldas, and E. G. Primel, "Alternative sorbents for the dispersive solid-phase extraction step in quick, easy, cheap, effective, rugged and safe method for extraction of pesticides from rice paddy soils with determination by liquid chromatography tandem mass spectrometry," *Journal of Chromatography A*, vol. 1360, pp. 66–75, 2014.

[9] P. Kaczyński, B. Łozowicka, M. Jankowska, and I. Hrynko, "Rapid determination of acid herbicides in soil by liquid chromatography with tandem mass spectrometric detection based on dispersive solid phase extraction," *Talanta*, vol. 152, pp. 127–136, 2016.

[10] L. Wang, F. Qiu, J. Li, and J. Pan, "Rapid and selective determination of melamine from eggs and milk based on magnetic surface molecularly imprinting technology," *Analytical Methods*, vol. 9, no. 48, pp. 6839–6848, 2017.

[11] M. Pei, X. Y. Zhu, and X. J. Huang, "Mixed functional monomers-based monolithic adsorbent for the effective extraction of sulfonylurea herbicides in water and soil samples," *Journal of Chromatography A*, vol. 1531, pp. 13–21, 2018.

[12] L. S. Sun, D. Y. Kong, W. D. Gu et al., "Determination of glyphosate in soil/sludge by high performance liquid chromatography," *Journal of Chromatography A*, vol. 1502, pp. 8–13, 2017.

[13] M. Kemmerich, G. Bernardi, M. B. Adaime, R. Zanella, and O. D. Prestes, "A simple and efficient method for imidazolinone herbicides determination in soil by ultra-high performance liquid chromatography-tandem mass spectrometry," *Journal of Chromatography A*, vol. 1412, pp. 82–89, 2015.

[14] W. X. Li, J. Mao, X. F. Dai et al., "Residue determination of triclopyr and aminopyralid in pastures and soil by gas chromatography-electron capture detector: dissipation pattern under open field conditions," *Ecotoxicology and Environmental Safety*, vol. 155, pp. 17–25, 2018.

[15] Y. Chen, T. Feng, G. Li, and Y. Hu, "Molecularly imprinted polymer as a novel solid-phase microextraction coating for the selective enrichment of trace imidazolinones in rice, peanut, and soil," *Journal of Separation Science*, vol. 38, pp. 301–308, 2015.

[16] T. Y. Li, L. S. Fan, Y. F. Wang et al., "Molecularly imprinted membrane electrospray ionization for direct sample analyses," *Analytical Chemistry*, vol. 89, no. 3, pp. 1453–1458, 2017.

[17] L. X. Chen, X. Y. Wang, W. H. Lu, X. Q. Wu, and J. H. Li, "Molecular imprinting: perspectives and applications," *Chemical Society Reviews*, vol. 45, no. 8, pp. 2137–2211, 2016.

[18] C. G. Xie, B. H. Liu, Z. Y. Wang, D. M. Gao, G. J. Guan, and Z. P. Zhang, "Molecular imprinting at walls of silica nanotubes for TNT recognition," *Analytical Chemistry*, vol. 80, no. 2, pp. 437–443, 2008.

[19] G. H. Yao, R. P. Liang, C. F. Huang, Y. Wang, and J. D. Qiu, "Surface plasmon resonance sensor based on magnetic molecularly imprinted polymers amplification for pesticide recognition," *Analytical Chemistry*, vol. 85, no. 24, pp. 11944–11951, 2013.

[20] S. Ansari, "Application of magnetic molecularly imprinted polymer as a versatile and highly selective tool in food and environmental analysis: recent developments and trends," *TrAC Trends in Analytical Chemistry*, vol. 90, pp. 89–106, 2017.

[21] G. D. Middeleer, P. Dubruel, and S. D. Saeger, "Molecularly imprinted polymers immobilized on 3D printed scaffolds as novel solid phase extraction sorbent for metergoline," *Analytica Chimica Acta*, vol. 986, pp. 57–70, 2017.

[22] A. Speltini, S. Andrea, F. Maraschi, M. Sturini, and A. Profumo, "Newest applications of molecularly imprinted polymers for extraction of contaminants from environmental and food matrices: a review," *Analytica Chimica Acta*, vol. 974, pp. 1–26, 2017.

[23] Diaz-Bao, M. R. Barreiro, J. M Miranda, A. Cepeda, and P. Regal, "Fast HPLC-MS/MS method for determining penicillin antibiotics in infant formulas using molecularly imprinted solid-phase extraction," *Journal of Analytical Methods in Chemistry*, vol. 2015, Article ID 959675, 8 pages, 2015.

[24] A. S. Yazdi and N. Razavi, "Application of molecularly-imprinted polymers in solid-phase microextraction techniques," *TrAC Trends in Analytical Chemistry*, vol. 73, pp. 81–90, 2015.

[25] Y. L. Niu, C. N. Liu, J. Yang et al., "Preparation of tetracycline surface molecularly imprinted material for the selective recognition of tetracycline in milk," *Food Analytical Methods*, vol. 9, no. 8, pp. 2342–2351, 2016.

[26] M. Rutkowska, J. P. Wasylka, C. Morrison, P. P. Wieczorek, J. Namiesnik, and M. Marc, "Application of molecularly imprinted polymers in analytical chiral separations and analysis," *TrAC Trends in Analytical Chemistry*, vol. 102, pp. 91–102, 2018.

[27] J. D. Dai, J. M. Pan, L. C. Xu et al., "Preparation of molecularly imprinted nanoparticles with superparamagnetic susceptibility through atom transfer radical emulsion polymerization for the selective recognition of tetracycline from aqueous medium," *Journal of Hazardous Materials*, vol. 205-206, no. 1, pp. 179–188, 2012.

[28] F. N. Andrade, C. E. D. Nazario, A. D. S. Neto, and F. M. Lanças, "Development of an on-line molecularly imprinted solid phase extraction by liquid chromatography-mass spectrometry for triazine analysis in corn samples," *Analytical Methods*, vol. 8, no. 5, pp. 1181–1186, 2016.

[29] J. Wang, Y. X. Sang, W. H. Liu, and X. H. Wang, "The development of a biomimetic enzyme-linked immunosorbent assay based on the molecular imprinting technique for the detection of enrofloxacin in animal-based food," *Analytical Methods*, vol. 9, no. 47, pp. 6682–6688, 2017.

[30] V. Coessens, T. Pintauer, and K. Matyjaszewski, "Functional polymers by atom transfer radical polymerization," *Progress in Polymer Science*, vol. 26, no. 3, pp. 337–377, 2001.

[31] Y. Li, Y. F. Wu, L. Yuan, and S. Q. Liu, "Application of atom transfer radical polymerization in biosensing," *Chinese Journal of Analytical Chemistry*, vol. 40, no. 12, pp. 1797–1802, 2012.

[32] P. Krys and K. Matyjaszewski, "Kinetics of atom transfer radical polymerization," *European Polymer Journal*, vol. 89, pp. 482–523, 2017.

[33] J. D. M. Morales, R. J. S. Leij, A. Carranz, J. A. Pojman, F. D. Monte, and G. L. Bárcenas, "Free-radical polymerizations of and in deep eutectic solvents: green synthesis of functional materials," *Progress in Polymer Science*, vol. 78, pp. 139–153, 2018.

[34] J. Tan, Z T. Jiang, R. Li, and X. P. Yan, "Molecularly-imprinted monoliths for sample treatment and separation," *TrAC Trends in Analytical Chemistry*, vol. 39, no. 39, pp. 207–217, 2012.

[35] J. B. Dong, J. B. Wu, J. Yang et al., "Preparation of high-capacity IDA chelating resin and its adsorption properties," *Chemical Journal of Chinese Universities*, vol. 34, no. 3, pp. 714–719, 2013.

20

Towards a Dual Lateral Flow Nanobiosensor for Simultaneous Detection of Virus Genotype-Specific PCR Products

Dimitra K. Toubanaki ⑩ and Evdokia Karagouni

Laboratory of Cellular Immunology, Department of Microbiology, Hellenic Pasteur Institute, 127 Vas. Sofias Ave., 11521 Athens, Greece

Correspondence should be addressed to Dimitra K. Toubanaki; dtouban@pasteur.gr

Academic Editor: Chih-Ching Huang

Nervous necrosis virus (nodavirus) has been responsible for mass mortalities in aquaculture industry worldwide, with great economic and environmental impact. A rapid low-cost test to identify nodavirus genotype could have important benefits for vaccine and diagnostic applications in small- and medium-scale laboratories in both academia and fish farming industry. A dual lateral flow biosensor for simultaneous detection of the most prevalent nodavirus genotypes (RGNNV and SJNNV) was developed and optimized. The dual biosensor consisted of two antibody-based test zones, indicative of each genotype, and a control zone. The positive signals were visualized by gold nanoparticles functionalized with anti-biotin antibody, and the detection was completed within 20 min. Optimization studies included antibody type and amount determination for test zone construction, gold nanoparticle conjugate type selection for high signal generation, and detection assay parameter determination. Following optimization, the biosensor was evaluated with healthy and RGNNV-nodavirus-infected fish samples. The proposed assay's cost was estimated to be less than 3 €, including the required reagents and biosensor. This work presents important steps towards making a dual lateral flow biosensor for nodavirus genotyping; further evaluation with clinical samples is needed before the test is appropriate for diagnostic kit development.

1. Introduction

Diseases of viral etiology have been wreaking havoc in the aquaculture industry, which is considered of strategic importance for Greek, European, and worldwide economies. Viral diseases often wipe out entire stocks within days of onset of infection with major economic and environmental costs [1, 2]. One such disease is viral nervous necrosis (VNN), also known as vacuolating encephalopathy and retinopathy (VER) or encephalomyelitis. VNN causes high mortalities in larvae and juveniles of 120 farmed and wild marine fish species, in geographically diverse areas including Europe, Australia, North America, and many parts of Asia. In many cases, the mortality rates may reach 100% within one week after infection, and even after recovering from the disease, the surviving fish are inclined to perform poorly [3–5].

Nervous necrosis virus (NNV), also known as nodavirus, has been recognized as the causative agent of VNN. Fish nodavirus belongs to the Nodaviridae family and the *Betanodavirus* genus. The virus is round shaped and non-enveloped, 23–25 nm in diameter with icosahedral structure. The virus genome is bisegmented; that is, it is formed by two positive-sense RNA molecules which are single-stranded (RNA1 and RNA2), while it does not contain a poly(A) sequence at the 3′ end [6]. The RNA1 sequence encodes an RNA-dependent RNA polymerase (RdRp) [7] while RNA2 encodes the capsid or coat protein [6]. A subgenomic transcript of RNA1, called RNA3, encodes two other non-structural proteins (B1 and B2) [5]. Phylogenetic analysis of *Betanodaviruses* indicated the presence of four distinct clusters of isolates: SJNNV (Striped Jack), TPNNV (Tiger Puffer), BFNNV (Barfin Flounder), and RGNNV (Red-Spotted Grouper) [8].

Nodaviruses belonging to different genotypes have different host ranges [9], and a particular viral strain can infect specific fish species at different geographical locations [10]. Diverse optimal in vitro growth temperatures have been associated with different nodavirus genotypes [11], a fact that seems to correspond with different in vivo pathogenicities. Thus, host specificity can be directly related to the viral phenotype and/or genotype [12–14]. As suggested, specific nodavirus genotypes have particular host ranges with distinct geographic distributions, revealing the virus' ability to adapt to different water temperatures [14, 15]. As recorded in epidemiological studies, the RGNNV genotype can be found in various warm-water fish species, especially groupers and sea bass, having the widest geographic distribution. The BFNNV genotype can be detected in cold-water marine fish species, while the TPNNV genotype has been found in a few fish species [5]. Even though it was believed that the SJNNV genotype could infect only Japanese fish species, it was recently detected in South Europe aquaculture sites [16]. More specifically, nodavirus strains isolated from the Atlantic coast of South Europe or the Mediterranean basin were found to belong to both SJNNV and RGNNV genotypes. Moreover, the simultaneous occurrence of those genotypes in a single animal has been found by phylogenetic analysis, indicating either reassortment or dual viral infection of the fish [13, 17–19].

Analysis of nodavirus genetic variation would vastly benefit the rational development of effective vaccines and diagnostic reagents. Molecular methods such as polymerase chain reaction (PCR) are extensively used for nodavirus detection [20–23], yet they cannot distinguish the different genotypes, which is vital for a complete strategy to eliminate the nodavirus from aquacultures effectively. The golden standard for *Betanodavirus* genotype evaluation is sequencing [24, 25]. However, its routine use for genotype screening is difficult since the method requires specialized and expensive instrumentation and software, while it is time-consuming as well. Genotyping techniques as restriction fragment length polymorphism (RFLP) [26] and a combination of RT-PCR and blot hybridization [17] have also been proposed for discrimination of nodavirus RGGNV and SJNNV genotypes. Our research group has recently developed a tetra-primer allele-specific PCR-based methodology, for detection of the RGNNV and SJNNV genotypes, in a rapid, specific, and sensitive format [27, 28]. Tetra-primer PCR is an allele-specific PCR methodology which relies on the amplification of the genotype-specific products simultaneously in a single PCR run, using four primers: a pair of outer (external) primers and two internal primers that are genotype-specific [29]. The tetra-primer PCR method is promising for nodavirus genotyping by medium-size research laboratories and fish farms; however, the amplicons detection by agarose gel electrophoresis or real-time PCR instrumentation are limiting its application in the field with a low-cost format.

Lateral flow paper biosensors (LFB) provide a tool, which is ideal for sensitive, reproducible, and accurate detection of PCR products, in a rapid way, implanted successfully in research laboratory setups. LFBs are prefabricated paper strips containing dry reagents that are activated by applying a sample-containing solution. They are designed for disposable single use and for applications where an on/off signal is sufficient [30, 31]. Lateral flow biosensors have been used as the detection method for analytes including DNA, mRNA, miRNA, proteins, biological agents, and chemical contaminants [32]. Our research group has developed a lateral flow biosensor for nodavirus amplification product detection enabling rapid and accurate positive virus sample visualization [33].

To further facilitate nodavirus genotyping with a promising technique, the aim of the present study was the development and optimization of a dry-reagent lateral flow biosensor for simultaneous visual detection of two different nodavirus genotypes, namely RGNNV and SJNNV. The dry-reagent biosensor was prepared by selecting the proper antibodies and optimizing their deposited amounts. Next, gold nanoparticles, which serve as signal reporters, were modified by conjugation with anti-biotin antibody. In a proof-of-principle test, viral samples were prepared by extracting RNA from healthy and infected fish samples and subjected to tetra-primer PCR for simultaneous amplification of SJNNV and RGNNV genotypes. Application of PCR products on functional dual lateral flow biosensor allowed detection of the genotype of the present virus by naked eye (visual). Knowledge of the correct nodavirus genotype is a valuable tool allowing more effective diagnosis and treatment of disease pathologies.

2. Experimental Section

2.1. Oligonucleotides. Synthesis of the oligonucleotides used in the present study was performed by Eurofins Genomics AT (Vienna, Austria). The primers were designed with respect to the publicly available RGNNV and SJNNV genotype sequences (GenBank accession numbers: Y08700.1 and NC_003449.1, resp.), as described before [17, 27]. The degenerate primers UpExtNdv (5′-ACACCTGA(A/G)GA(G/C)AC(T/C) ACCGCTCC(C/A)AT-3′) and DpExtNdv (5′-C(C/G)CCA(A/T)CTGTGAA(T/C)GTCTTGTT(A/G)AAGT(C/T) (A/G)TCCC-3′) were utilized as external primers (upstream and downstream, resp.). The upstream internal primer UpInSJNdv (5′-GATTTCGTTC-CATTCTCTTG-3′) was indicative for the SJNNV genotype and labelled with the hapten digoxigenin (Dig) at the 5′ end. The downstream internal primer DpInRGNdv (5′-GATTTCGTTCCATTCTCTTG-3′) was specific for the RGNNV genotype and labelled with the hapten fluorescein (Fluor) at the 5′ end. The SJNNV-specific probe (Probe_SJNdv: 5′-AGTGTCTCCAGCTTTCTTCC-3′) and the RGNNV-specific probe (Probe_RGNdv: 5′-CCACAA-ATGATTTCAAGTCC-3′) were both 5′ biotinylated (B). Oligonucleotides B-dA$_{20}$ (5′-biotin-AAAAAAAAAAAA-AAAAAAAA-3′), B-dC$_{20}$ (5′-biotin-CCCCCCCCCCCCC CCCCCCC-3′), dig-dT$_{20}$ (5′-digoxigenin-TTTTTTTTTT-TTTTTTTTTT-3′), and fluor-dG$_{20}$ (5′-fluorescein-GGGG-GGGGGGGGGGGGG GGG-3′) were used as reference oligonucleotides.

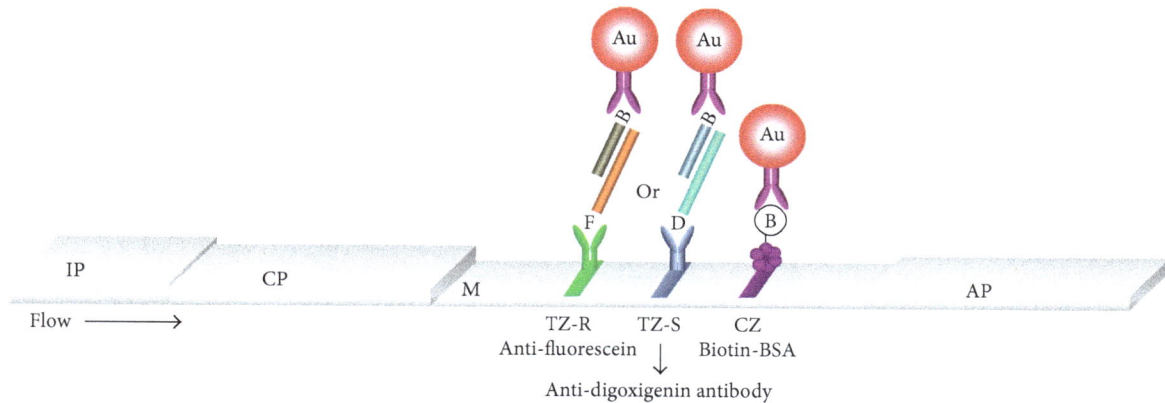

FIGURE 1: Principle of the dual nanoparticle-based lateral flow biosensor for simultaneous detection of nodavirus SJNNV and RGNNV genotypes. Two test zones (anti-fluorescein antibody (TZ-R) and anti-digoxigenin antibody (TZ-S)) and a control zone (biotinylated BSA (CZ)) have been deposited on the diagnostic membrane. The sample, containing the respective genotype of the target analyte (tetra-primer PCR product with fluorescein (F) for RGNNV genotype or digoxigenin (D) for SJNNV genotype), is hybridized with a biotinylated genotype-specific oligonucleotide probe and applied on the conjugation pad, where functionalized gold nanoparticles (Au) with anti-biotin antibody have already been added. Following that, the biosensor is immersed in the developing buffer, the sample and the nanoparticles are immobilized on the appropriate test zone, and the positive signal is visible by the naked eye, as a red zone. The excess nanoparticles bind to the control zone of the biosensor. The image shows a side view of the lateral flow biosensor. IP: immersion pad; CP: conjugation pad; M: diagnostic membrane; AP: absorbent pad. The assay components are not in scale.

2.2. Reference Plasmids.

Two reference plasmids, specific for each genotype (GenScript, Piscataway, NJ, USA), were used as targets for tetra-primer PCR optimization studies. The sequences of the pRGNNV and pSJNNV are described in detail in [27]. A partial sequence of RGNNV coat protein gene (295 bp) and a part of SJNNV coat protein gene (301 bp) were cloned in pUC57 by EcoRV, based on the respective reference sequences.

2.3. Assay Principle.

The principle of the dual lateral flow biosensor is illustrated schematically in Figure 1. The genotype-specific PCR for RGNNV- and SJNNV-specific amplification products has been described in detail in [27]. Briefly, total RNA isolated from fish samples was subjected to reverse transcription reaction and a single PCR with two sets of primers (tetra-primer PCR) was performed with the produced cDNA. Tetra-primer PCR consisted of phase I, where the external primer set amplify a segment that spans the highly variable genomic region of interest, and phase II, where a lower annealing temperature is applied and the inner primers (genotype-specific primers) anneal to opposite strands. The inner primers pair off with the external primers to guide a bidirectional amplification that uses the long PCR product as a template and generates short genotype-specific fragments, although amplification of the long product continues to some degree. The inner primers were designed with a digoxigenin or a fluorescein moiety at their 5′ end; thus, the short products were labelled with digoxigenin for the SJNNV genotype or fluorescein for the RGNNV genotype. The amplified DNA hybridized in solution with the genotype-specific probes SJNNV and RGNNV, which were labelled at their 5′ end with biotin, comprising a segment complementary to their respective target. The mixture was applied to the conjugate pad of the

biosensor, which was then immersed into the developing solution. The solution migrated along the LFB by capillary action and rehydrated the anti-biotin-conjugated gold nanoparticles. The hybrids were captured from immobilized anti-digoxigenin or anti-fluorescein at the respective test zone (TZ-S/TZ-R) of the biosensor and interacted with the biotinylated probes. As a result, there was accumulation of gold nanoparticles and generation of a characteristic red line at the proper test zone of the biosensor. The excess nanoparticles were captured from immobilized biotinylated BSA at the control zone of the LFB, hence generating a red line that confirmed the proper function of the biosensor. The biosensor detects only the short, genotype-specific PCR products and not the long ones. The latter hybridizes to both probes but is not captured at the test zones since it lacks a labelled end (the outer primers are unmodified). The genotype was assigned by the results of the LFB. The presence of an anti-fluorescein red zone (TZ-R) and absence of an anti-digoxigenin red zone indicated the RGNNV genotype. The SJNNV genotype was characterized by a red zone of anti-digoxigenin (TZ-S). Theoretically, presence of both genotypes in a sample would result in red zones for both immobilized antibodies.

2.4. Preparation of Antibody-Conjugated Gold Nanoparticles.

Gold nanoparticle (Au NP) functionalization with anti-biotin antibody was performed following the previously described protocol [34]. Briefly, 1 mL of gold nanoparticles solution (30 nm) (Sigma-Aldrich, Steinhem, Germany) was adjusted to pH 9 with addition of the appropriate amount (~25 μL) of borax solution (200 mM) (AppliChem, Darmstadt, Germany). Anti-biotin antibody (4 μg; Sigma-Aldrich, Steinhem, Germany) was diluted in 200 μL of borax solution (20 mM) and was mixed with the Au NP solution, with gradual addition by stirring (4 times × 50 μL). Following incubation at room temperature for 45 min, 100 μL of 10% bovine serum albumin (BSA-AppliChem, Darmstadt, Germany) diluted in borax solution (20 mM) were added, and the final mixture was incubated at room temperature (10 min). The excess of reagents was removed by centrifugation (4500 ×g, 1 hr). The resulting pellet was redispersed, and the wash solution (1 mL 1% BSA in a 2 mM borax solution) was added. The supernatant was discarded after centrifugation (4,500 ×g, 10 min). Finally, the red pellet was redispersed in 100 μL of an aqueous storage solution (0.1% BSA and 0.1% NaN$_3$ in 2 mM borax).

The preparation of anti-BSA gold nanoparticles, as described in [35], was as follows: 1 mL aliquots of Au NP solutions (5 and 10 nm, Sigma-Aldrich, Steinhem, Germany) were adjusted to pH 9 with addition of 200 mM borax. A solution containing 4 μg of anti-BSA antibody (Sigma-Aldrich, Steinhem, Germany) diluted in borax (20 mM) was added to each gold solution followed by stirring, and then the mixtures were incubated at room temperature for 45 min. Human serum albumin (HSA 10%, Sigma-Aldrich, Steinhem, Germany) in 20 mM borax was added in the mixture, which was incubated at room temperature (10 min). The excess reagents were removed by centrifugation (3500 ×g for 60 min), and the pellet was redispersed in 1 mL wash solution (1% HSA in 2 mM borax). After centrifugation (3500 ×g, 10 min), the supernatant was discarded, and the red pellet was redispersed in 100 μL of storage solution (0.1% HAS, 0.1% NaN$_3$, 2 mM borax). All incubations were carried out in the dark.

The antibody-gold nanoparticle conjugates were stored at 4°C.

2.5. Preparation of the Dual Lateral Flow Biosensor.

The dual dry reagent lateral flow biosensors (4 × 60 mm) were prepared as described before in [34]. The biosensors' parts are positioned on a plastic adhesive backing as follows: A nitrocellulose diagnostic membrane (M: HF240MC100, 25 mm in length; Millipore, Billerica, MA, USA) was placed on a laminated card by the manufacturer. A glass fiber conjugate pad (CP: GFCP000800, 8 mm; Millipore, Billerica, MA, USA) is added below the membrane, a cellulose immersion pad (IP: CFSPOO1700, 17 mm; Millipore, Billerica, MA, USA) is positioned below the conjugate pad, and a cellulose absorbent pad (AP: same as the immersion pad) is placed just above the membrane. Each pad is overlapping its adjacent pads (~2 mm) to make certain that the solution will migrate through the biosensor. The construction of the two test zones and the control zone was done utilizing the TLC applicator, Linomat 5, and the WinCats software (Camag, Muttenz, Switzerland). The zones were formed by loading anti-fluorescein antibody (TZ-R: polyclonal anti-fluorescein antibody; monoclonal anti-fluorescein antibody), anti-digoxigenin antibody (TZ-S: Roche Diagnostics, Mannheim, Germany), and biotinylated BSA (bBSA: Thermo Fisher Scientific Inc., Rockford, IL, USA) on the membrane and were located at 10, 15, and 20 mm distance from the edge of the membrane, respectively. In details, for the TZ-R zone, a solution consisting of 500 mg/L anti-fluorescein antibody, 50 mL/L methanol, and 20 g/L sucrose in freshly prepared

100 mM NaHCO$_3$ buffer (pH 8.5) was loaded at a density of 500 ng per LFB. For TZ-S zone, a solution containing 500 mg/L anti-digoxigenin antibody, 50 mL/L methanol, and 20 g/L sucrose in 100 mM NaHCO$_3$ buffer (pH 8.5) was loaded at a density of 500 ng per 4 mm membrane. Finally, for the control zone, a solution consisting of 4 g/L bBSA, 50 mL/L methanol, and 20 g/L sucrose in PBS (PBS: 0.14 M NaCl, 2.7 mM KCl, 10 mM sodium phosphate, and 1.7 mM potassium phosphate, pH 7.4) was loaded at a density of 1.6 μg per 4 mm membrane. The membrane was dried in an oven for 1 h at 80°C, and the sensors were assembled as described above. All biosensors were cut (4 mm width) utilizing a Guillotine cutter and stored dry at room temperature.

2.6. Fish Samples, RNA Extraction, and cDNA Preparation.

All samples used in the present study were European sea bass (Dicentrarchus labrax). One fish which was infected with nodavirus was collected from a sea-cage fish farm in Epidavros (Saronikos Gulf). Healthy fishes were reared (8 months) in experimental facilities of the Hellenic Centre for Marine Research (Athens, Greece), and used as negative controls. Fish retinas were isolated using aseptic techniques, transferred in sterile tubes, and stored at −80°C until use.

The RNeasy Mini kit (Qiagen, Hilden, Germany) was used for total RNA extraction, according to the manufacturer's instructions. Measurements of the absorbance at 260 nm (A_{260}) with a Nanodrop 1000 spectrophotometer (Thermo Fisher Scientific, Delaware, USA) confirmed that the isolated RNA was pure while it also extrapolated its concentration.

The purified total RNA was reverse transcribed (RT) with Superscript II reverse transcriptase (Invitrogen, Carlsbad, CA). The RT reaction consisted of 0.5 mM dNTPs (dNTPs: dATP, dTTP, dCTP, dGTP; HT Biotechnology, Cambridge, UK), 2.5 μM dT$_{20}$ oligonucleotide, RNase-free H$_2$O and extracted total RNA (100 ng). The reaction mixture was incubated at 65°C (5 min), quickly chilled on ice (0°C, 1 min), followed by addition of the first-strand buffer (1x), dithiothreitol (0.1 M), and RNase OUT RNase inhibitor (40 U; Invitrogen, Carlsbad, CA) and incubation at 42°C (2 min). The enzyme SSII RT (200 units) was added, and the final mixture was incubated at 42°C (50 min). The enzyme was deactivated by heating the mixture at 70°C (15 min), and the produced cDNA was stored at −20°C.

2.7. Tetra-Primer PCR for Nodavirus Genotyping.

The tetra-primer PCR amplification was performed with GoTaq Flexi DNA polymerase (0.625 units; Promega, WI, USA) in GeneAmp PCR System 9700 cycler (Applied Biosystems, NY, USA). The reaction mixtures contained 1 × GoTaq Flexi Buffer, 200 μM of each dNTP, 0.75 mM MgCl$_2$, 2 μL of cDNA or 5 ng of reference plasmid, 0.25 μM of each of UpExtNdv and DpExtNdv primers, and 1 μM of each of Dig-UpInSJNdv and Fluor-DpInRGNdv primers, in 25 μL final volume. The reactions' cycling conditions were incubation at 95°C (5 min), followed by the first phase of tetra-primer PCR (10 cycles of 94°C (15 s), 60°C (30 s), 72°C (30 s)), and the second phase of amplification (30 cycles of 94°C (15 s), 50°C

(30 s), 72°C (30 s)). After completion of the cycles, the mixture was incubated at 72°C (7 min) and cooled to 4°C. The absence of contamination was confirmed by addition of negative controls (containing water instead of cDNA) in each PCR series.

2.8. Dual Lateral Flow Biosensor Detection Assay of Reference Oligonucleotides.

Four reference oligonucleotides were utilized as target sequences: oligonucleotides B-dA$_{20}$ and B-dC$_{20}$ were designed to contain a biotin molecule in their 5' end, in order to interact with Au NPs functionalized with anti-biotin antibody. The oligonucleotide dig-dT$_{20}$ was designed with a digoxigenin molecule in its 5' end in order to be immobilized by the anti-digoxigenin antibody (TZ-S zone) while the oligonucleotide fluor-dG$_{20}$ was designed with a fluorescein molecule in its 5' end to interact with immobilized anti-fluorescein antibody (TZ-R zone). Two target mixtures were prepared: Dig-mixture consisted of 1 pmol B-dA$_{20}$, 1 pmol dig-dT$_{20}$, and ddH$_2$O; Fluor-mixture consisted of 1 pmol B-dC$_{20}$, 1 pmol fluor-dG$_{20}$, and ddH$_2$O. The mixtures were denatured at 95°C, for 3 min, and left to hybridize at 37°C for 10 minutes. Five microlitres of each target mix was applied on the LFBs. Next to them, 10 μL of anti-biotin Au NPs was applied, and the LFBs were dipped in the developing solution (60 mL/L glycerol, 10 g/L SDS in PBS, and 10 mL/L Tween-20, pH 7.4). The signal was visible after 20 min. After completion of the assay, the LFBs were scanned with a desktop scanner (HP Scanjet G4050, HP, California, USA), and the band densities were quantified with ImageJ software [36].

2.9. Optimization of the Dual Lateral Flow Biosensor Preparation

2.9.1. Anti-Fluorescein Test Zone Construction with Monoclonal or Polyclonal Anti-Fluorescein Antibody.

For the TZ-R zone with monoclonal anti-fluorescein antibody, a solution consisting of 500 mg/L anti-fluorescein antibody (Millipore, Billerica, MA, USA), 50 mL/L methanol, and 20 g/L sucrose in freshly prepared 100 mM NaHCO$_3$ buffer (pH 8.5) was loaded at a density of 500 ng per LFB. For the TZ-R zone with polyclonal anti-fluorescein antibody, 500 mg/L anti-fluorescein antibody (Meridian, Memphis, TN, USA) were mixed with the abovementioned buffer and loaded at a density of 500 ng per LFB. The procedures were performed as described in Section 2.5.

2.9.2. Test Zone Construction with Various Amounts of Anti-Digoxigenin and Anti-Fluorescein Antibodies.

In order to perform the antibody amount optimization studies, two mixes were prepared for each amount; that is, for TZ-S zone construction, two solutions were prepared: Solution 1 contained 250 mg/L while solution 2 contained 500 mg/L of anti-digoxigenin antibody diluted in 50 mL/L methanol and 20 g/L sucrose in 100 mM NaHCO$_3$ buffer (pH 8.5). The TZ-R zone construction was tested with solution 3, consisting of 75 mg/L of polyclonal anti-fluorescein antibody or 500 mg/L of

the same antibody (solution 4) in 50 mL/L methanol, and 20 g/L sucrose in freshly prepared 100 mM NaHCO$_3$ buffer (pH 8.5).

2.10. Dual Lateral Flow Biosensor Detection Assay of Reference Oligonucleotides with Signal Enhancement with Gold Nanoparticle Conjugates.

The target mixtures were prepared as before (Section 2.8). Five microlitres of each target mix was added to the biosensors' conjugation pad, where 5 μL of Au NPs functionalized with anti-biotin was already placed. The LFBs were immersed into the developing solution (250 μL), and 5 min later, 5 μL of Au NPs functionalized with anti-BSA antibody was added to the conjugation pad. The biosensors were redipped into the developing solution, and the assay was completed within 20 min. Finally, the biosensors were scanned with a desktop scanner, and the band densities were quantified with ImageJ software.

2.11. Dual Lateral Flow Biosensor Detection Assay of Nodavirus Tetra-Primer PCR Products.

Detection of the nodavirus tetra-primer PCR products was performed as follows: aliquots of PCR solutions were mixed with 90 mM NaCl, 1 pmol of each biotin-labelled genotype-specific nodavirus probe and 1 × PCR buffer (final volume: 5 μL). The mixtures were incubated at 95°C (3 min), followed by hybridization (10 min, 25°C). The resulting mixtures were added to the conjugation pad, where 10 μL of Au NPs functionalized with anti-biotin antibodies was already placed. The biosensors' immersion pads were then dipped in tubes containing the developing solution (250 μL). The visual detection was completed within 20 min. Longer times did not affect the assay results. After completion of the assay, the LFBs were scanned with a desktop scanner, and the ImageJ software was utilized for band density quantification.

2.12. Optimization of the Dual Lateral Flow Biosensor Detection Assay.

The dual LFB assay for tetra-primer PCR RGNNV- and SJNNV-specific product detection was optimized by comparing the detection specificity and efficiency obtained, using (i) different amounts of oligonucleotide detection probes (0.5–4 pmol/1 pmol of target; i.e., 0.5/1/2, and 4 pmol of probes) and (ii) different annealing temperatures (i.e., 25, 37, and 42°C). The parameter that resulted in the highest amount of specific signal in the appropriate test zone and the smallest amount of nonspecific signal was chosen as the optimum condition in each case.

3. Results and Discussion

3.1. Optimization of the Dual Lateral Flow Biosensor Preparation.

The proposed dual biosensor format was developed by our research group and has been successfully exploited on pharmacogenetic studies for cytochrome c single nucleotide polymorphism genotyping, combined with oligonucleotide ligation reaction [34]. In that study, the biosensor was used for detection of the double-labelled single-stranded DNA products of a ligation reaction. In the present work, our aim was the detection of a hybridized

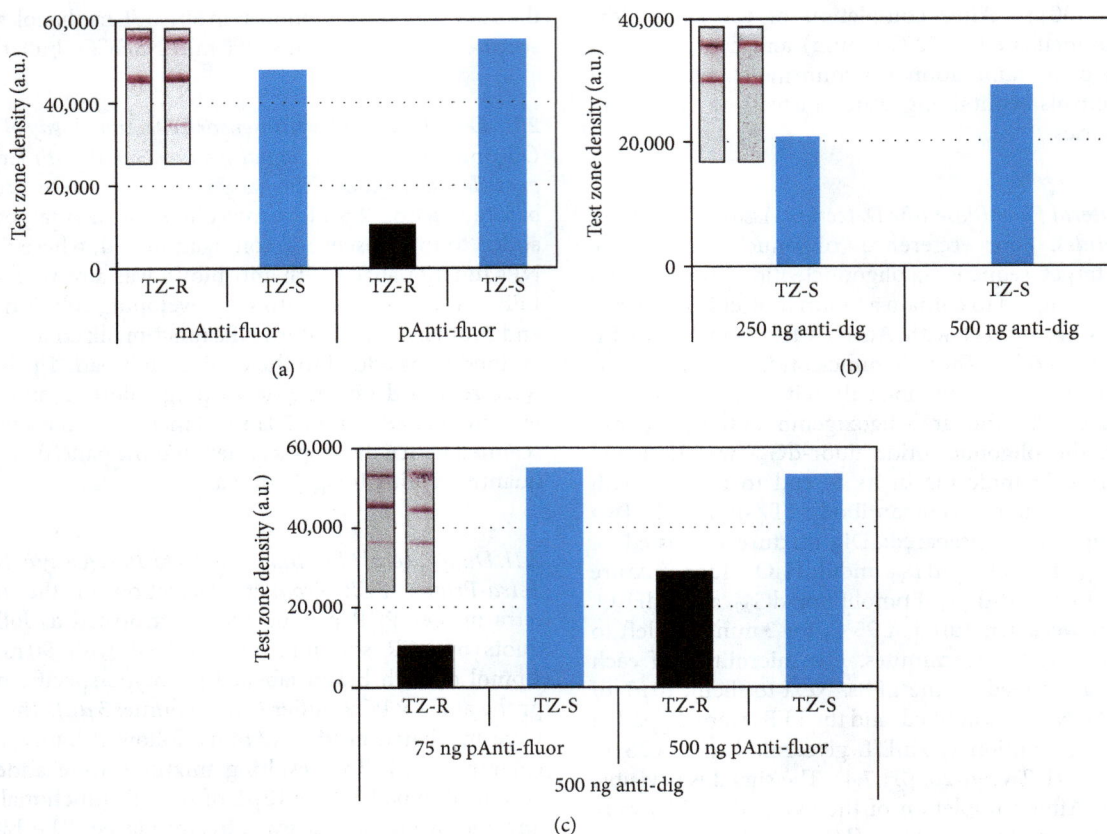

FIGURE 2: Dual lateral flow biosensor optimization studies. Representative lateral flow biosensors and signal intensity graphs for (a) effect of the use of monoclonal (mAnti-fluor) versus polyclonal (pAnti-fluor) anti-fluorescein antibody; (b) effect of the anti-digoxigenin (anti-dig) antibody amount for test zone construction; and (c) effect of the polyclonal anti-fluorescein antibody amount for test zone construction. All tests were performed with dig- and fluor-reference target mixtures. Signal is visualized with anti-biotin Au NPs. TZ-S: anti-digoxigenin zone; TZ-R: anti-fluorescein zone.

complex between a hapten-labelled PCR product and a biotin-labelled genotype-specific oligonucleotide probe. In an effort to increase the signal generation ability of the proposed biosensor, several optimization studies regarding the LFB construction were performed. The construction of the test zones is the most critical part of the developed assay since several parameters could affect the assays' specificity and sensitivity. The factors which were studied include the use of monoclonal versus the polyclonal antibody for the anti-fluorescein test zone construction and the deposited antibody amount for both test zones. Signal formation is affected by the gold nanoparticle accumulation on the biosensor zones; therefore, the application of a signal enhancement methodology [35] was also investigated. All optimization studies were performed with reference oligonucleotide mixtures as described in Section 2.8.

3.1.1. Monoclonal versus Polyclonal Anti-Fluorescein Antibody for Test Zone Construction.

The proposed dual LFB consisted of two test lines made by anti-digoxigenin (TZ-S) and anti-fluorescein antibodies (TZ-R) and a control zone which was made by biotinylated BSA, absorbed by the membrane. The signal visualization was realized by Au NPs conjugated with anti-biotin antibodies. The anti-digoxigenin

antibody performance for test zone construction was evaluated in the previously mentioned study [34], as well as other independent studies [37, 38]. Therefore, the same type of anti-digoxigenin antibody was utilized for the TZ-S construction in the present study. However, the use of the polyclonal fluorescein antibody for the construction of TZ-R zone was only evaluated in our previous study. In an attempt to increase the dual LFB specificity, a commercially available monoclonal anti-fluorescein antibody was tested in parallel with the previously used polyclonal antibody. As shown in Figure 2(a), the use of the monoclonal antibody did not result in any signal. Fluorescein has many isoforms, and possibly the fluorescein moiety of the utilized modified oligonucleotides could not interact with the monoclonal antibody. On the contrary, the polyclonal antibody resulted in satisfactory signal density, possibly because more antigenic epitopes were recognized in fluorescein and higher antigen amounts of antibody were immobilized; thus, more fluorescein hapten-labelled oligonucleotide was visualized.

3.1.2. Antibody Amount for Test Zone Construction Effect.

The immobilized anti-fluorescein and anti-digoxigenin antibody amounts were examined next. Two concentrations of each

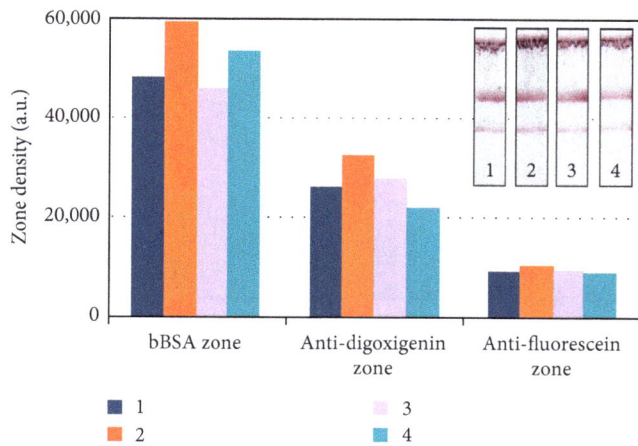

FIGURE 3: Effect of anti-biotin functionalized gold nanoparticle amount and signal enhancement with nanoparticle aggregates. Representative lateral flow biosensors and signal intensity graphs for Dig- and Fluor-reference target mixtures. Signal is visualized with (1) 30 nm gold nanoparticles functionalized with anti-biotin antibodies (5 μL); (2) 30 nm Au NPs functionalized with anti-biotin antibodies (10 μL); (3) 30 nm gold nanoparticles functionalized with anti-biotin antibodies (5 μL) and 10 nm anti-BSA conjugated Au NPs (5 μL); and (4) 30 nm gold nanoparticles functionalized with anti-biotin antibodies (5 μL) and 5 nm anti-BSA conjugated NPs (5 μL).

antibody were tested. The amount of anti-digoxigenin antibody on the TZ-S was initially studied (Figure 2(b)). Use of 500 ng of antibody per 4 mm biosensor resulted in the optimum signal compared with 250 ng of antibody (1.4-fold increase). These results are in accordance with [34]. The amount of anti-fluorescein antibody was subsequently tested. The used concentrations were 75 and 500 ng of antibody per 4 mm biosensor (Figure 2(c)). The optimum results were obtained with 500 ng of anti-fluorescein (2.9-fold increase). The use of 75 ng of the antibody resulted in a faint signal, in contrast with the results obtained in [34], possibly due to variations in the nitrocellulose membrane characteristics and additives between the two different providers.

3.1.3. Effect of the Au NP Anti-Biotin Conjugate Amount and Signal Enhancing NP Complex.

Recently, our research group developed a signal amplification methodology in one-step for nucleic acid detection lateral flow biosensors based on gold nanoparticles [35]. The "dual gold signal enhancement method" uses Au NPs functionalized with two different antibodies. The first type of Au NPs consists of nanoparticles functionalized with anti-biotin antibodies (30 nm), blocked with BSA, and the second conjugate contains anti-BSA antibody on Au NPs with different sizes (5 nm/10 nm). In that approach, the 30 nm anti-biotin conjugated Au NPs (which are used for signal generation) are forming complexes with anti-BSA conjugated Au NPs resulting in formation of large gold NP aggregates which enhance the LFB signal. Since the proposed biosensor was based on anti-biotin conjugated gold nanoparticles, the signal enhancement methodology was tested with that

format. The results are presented in Figure 3. Each particle combination was tested with both Dig- and Fluor-reference mixtures. Four conjugate types were evaluated: (1) 30 nm gold nanoparticles functionalized with anti-biotin antibodies (5 μL), which was, also, the basic quantity for detection; (2) 30 nm Au NPs functionalized with anti-biotin antibodies (10 μL) since the final volume of the gold conjugates of the dual format was 10 μL; (3) 30 nm gold nanoparticles functionalized with anti-biotin antibodies (5 μL) and 10 nm anti-BSA conjugated Au NPs (5 μL), and (4) 30 nm gold nanoparticles functionalized with anti-biotin antibodies (5 μL) and 5 nm anti-BSA conjugated NPs (5 μL). When the formulation (1) (5 μL of 30 nm anti-biotin conjugated Au NPs) was used as the standard or default condition, a signal increment of both test zones was observed following the addition of 10 nm gold anti-BSA conjugate. However, the control zone signal slightly decreased. In contrast, when 5 nm gold anti-BSA conjugate was added in 5 μL of 30 nm anti-biotin conjugates, test zone signal decreased, and control zone signal increased. Thus, even though the signal enhancement method increased the signal intensity with the use of 10 nm anti-BSA Au NPs, compared with Au NPs basic formulation, the use of the double quantity (10 μL) of 30 nm Au NPs functionalized with anti-biotin antibodies gave more intense signals in all test and control zones. In order to keep the dual lateral flow biosensor format as simple as possible, the use of a single Au NP conjugate was preferred, and the 30 nm gold nanoparticles functionalized with anti-biotin antibodies with 10 μL amount was used throughout the study.

3.2. Optimization of the Dual Lateral Flow Biosensor Detection Assay

3.2.1. Oligonucleotide Probe Amount Effect.

Optimization studies for assessment of the oligonucleotide probe impact in the hybridization reaction mixtures were performed. The oligonucleotide probes were tested in amounts of 0.5–4 pmol/1 pmol of target (Figures 4(a) and 4(b)). Both test zones resulted in optimum signals when 1 pmol of probe was used and decreased with higher amounts of probe. This observation is attributed to the fact that at high levels, the amount of biotin-labelled probe exceeds the binding capacity of the anti-biotin-functionalized nanoparticles, and the biotin-labelled probe which hybridizes to the specific target sequence competes with the unhybridized probe for binding to limited anti-biotin conjugated NPs. When higher amounts of biotinylated probes are used, the amount of nanoparticles that bind to the free probe is increasing. Even though these nanoparticles move along the LFB, they cannot be immobilized from the deposited antibodies in the test zones, and the red bands become fainter [39].

3.2.2. Hybridization Temperature Effect.

The hybridization temperature effect on target PCR product and specific oligonucleotide probe hybridization was tested with a temperature of 25–42°C. As observed in Figures 4(c) and 4(d), the density of both test zones is more intense at 25°C. In higher temperatures, the densities are decreasing slightly,

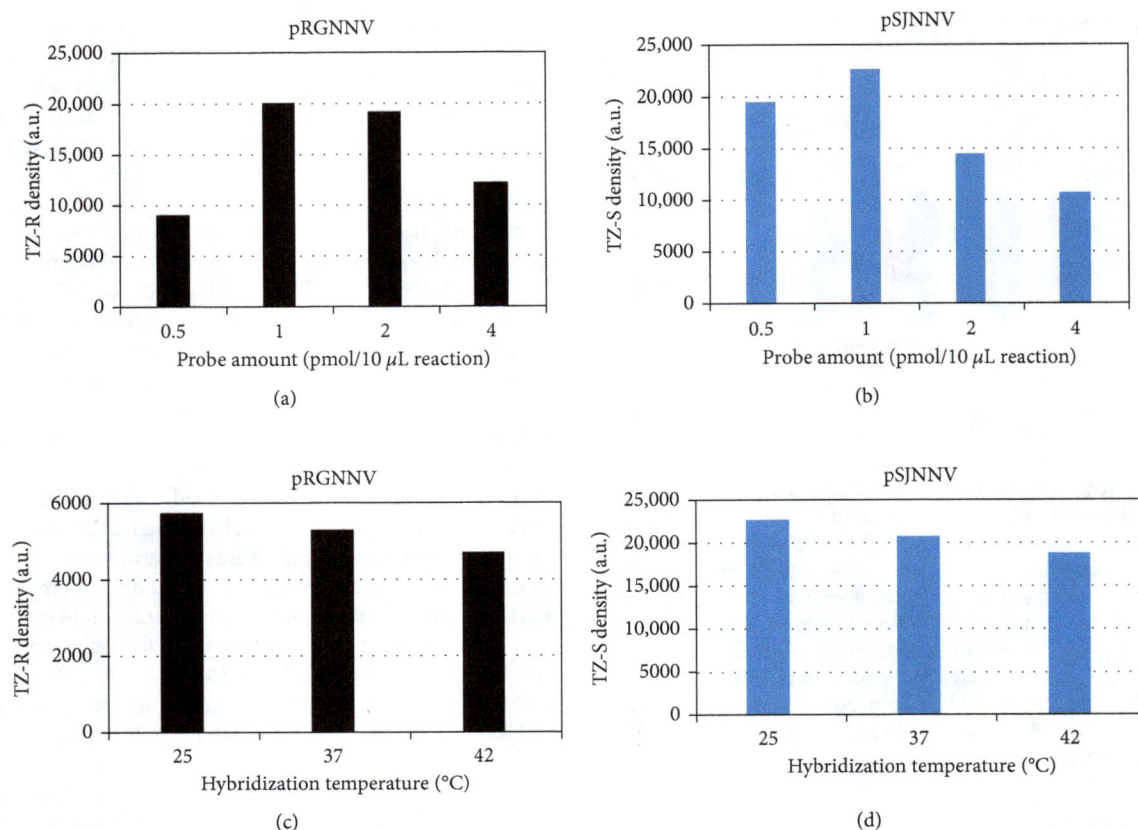

(a)

(b)

(c)

(d)

FIGURE 4: Dual lateral flow biosensor assay optimization studies. Representative lateral flow biosensors and signal intensity graphs for (a) oligonucleotide probe amount effect on plasmid pRGNNV amplification product hybridization mixture; (b) oligonucleotide probe amount effect on plasmid pSJNNV amplification product hybridization mixture; (c) hybridization temperature effect on the pRGNNV PCR product and specific oligonucleotide probe hybridization; and (d) hybridization temperature effect on the pSJNNV PCR product and specific oligonucleotide probe hybridization. The signal is visualized with anti-biotin Au NPs. TZ-S: anti-digoxigenin zone; TZ-R: anti-fluorescein zone.

due to inhibition of the hybridization. For that reason, the 25°C temperature was chosen for the optimum hybridization in all subsequent experiments.

3.3. Dual Lateral Flow Biosensor Assay Reproducibility. The

dual lateral flow assay reproducibility was assessed since it is one of the most important parameters for successful biosensor development. The proposed assay reproducibility was assessed with simultaneous application of the Dig- and Fluor-reference mixtures on the dual biosensors. Six biosensors which were prepared in different batches (i.e., LFBs 1 and 2: batch 1; LFBs 3 and 4: batch 2; and LFBs 5 and 6: batch 3) were tested with the reference mixtures, and the test and control zones intensities were measured. The results are presented in Figure 5. Analysis of the TZ-R test zone of the LFBs gave a coefficient of variation (CV) of 3.9%, while analysis of the TZ-S test zone resulted in 8.2% CV. The CV for the control zone was 9.2%, indicating excellent reproducibility.

3.4. Proof-of-Principle Nodavirus Amplification Product Detection with Dual Lateral Flow Biosensor. The dual lateral

flow biosensor was used to detect amplification products of both genotype-specific plasmids (pRGNNV and pSJNNV), one nodavirus infected *D. labrax* sample and one healthy (negative) *D. labrax* sample, as proof-of-principle for the proposed assay. All samples were subjected to tetra-primer PCR, and the PCR products were hybridized with the genotype-specific probes. Both probes were added in the hybridization mixture, and the resulting amplification product-probe complexes were applied on the biosensors. The present genotype was assigned by a single biosensor, and the results are shown in Figure 6. The presence of a single TZ-R zone for the pRGNNV product and a single TZ-S zone for the pSJNNV confirmed the specificity of the proposed dual LFB for each genotype. The noninfected sample did not show any signal in the test zones, correctly indicating the absence of nodavirus and further confirming the LFB specificity. The infected sample was positive with low signal intensity, and it was correctly classified as RGNNV. The genotype of the sample was previously determined by direct sequencing by an independent research group.

As mentioned in our previous studies [27, 28, 33] and independent research groups [40], there is a tremendous difficulty to obtain virus samples of various strains.

(a)

(b)

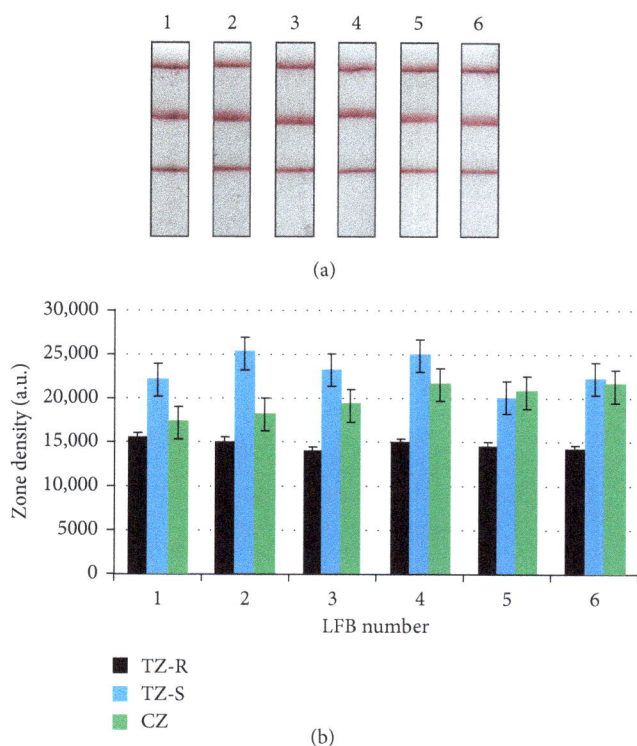

FIGURE 5: Reproducibility study. (a) Typical images of lateral flow biosensors with anti-biotin-gold nanoparticle conjugates after applying Dig- and Fluor-reference target mixtures as targets. (b) Optical density of the test and control zones. Graph of the intensities of the lateral flow biosensor test and control zones for reproducibility assessment (CV_{TZ-R}: 3.9%; CV_{TZ-s}: 8.2%; CV_{CZ}: 9.2%, $n = 6$). TZ-R: anti-fluorescein zone; TZ-S: anti-digoxigenin zone; CZ: control zone; CV: coefficient of variation.

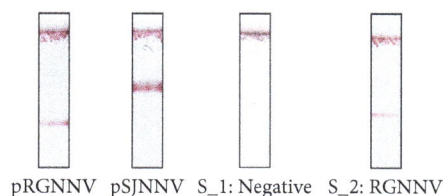

FIGURE 6: Visual detection of nodavirus genotype-specific products with dual lateral flow biosensors. Representative LFBs with amplification products of tetra-primer PCR performed with pRGNNV and pSJNNV reference plasmids and a healthy (S_1) and an infected (S_2) *D. labrax* sample.

The location of samples belonging to the SJNNV genotype was not feasible, and all samples previously analyzed by our research group belonged to the RGNNV genotype. Therefore, the present work was merely focused on the dual lateral flow biosensor optimization, contributing towards a fully developed nanobiosensor for nodavirus genotyping. Analysis of the plasmid tetra-primer PCR products, along with amplification products from one healthy and one nodavirus-infected sample confirmed the feasibility of the proposed biosensor. Studies for collection of a high number of fresh

samples from different geographical regions, in order to obtain both nodavirus genotypes, to fully validate the proposed methodology are in progress by our research group.

4. Conclusions

The proposed dual lateral flow biosensor constitutes a step forward to a robust, rapid, and accurate tool for fish virus genotype assessment with ease and low cost. The assay can be utilized as a potential detection system for virus genotyping by small- and medium-size research labs and the aquaculture industry, providing the means for effective vaccine and diagnostic development. The results demonstrate the optimization studies for a rapid single-step assay, which requires low amount of the analyzed sample and provides simultaneous amplification and genotyping of nodavirus DNA in a single, closed-tube methodology. The assay was optimized in terms of the biosensors' preparation and the detection assay parameters, demonstrating attractive characteristics with respect to specificity and reproducibility. The optimum goal for the proposed methodology is to replace the costly sequencing for virus genotyping, since such simple-to-use and low-cost methods are ideal for medium-scale laboratories.

The main advantage of the proposed method compared with previously used methods (i.e., gel electrophoresis and melting analysis) is that the dual biosensor minimizes the need for specialized and costly instrumentation and reagents. Therefore, it enables rapid and low-cost genotyping of nodavirus by visual detection of the RGNNV/SJNNV amplification product. Also, the tetra-primer PCR product is directly hybridized with genotype-specific probes without prior purification from the excess of primers and dNTPs, and the hybridization mixture is applied on the biosensors' conjugate pad, minimizing the possibility for contamination. Use of the genotype-specific probe and product detection by hybridization provides extra sequence confirmation, in contrast with electrophoresis that provides only the size of the amplification products. The visual detection of the genotype-specific product is completed in 20 min, and the overall assay can provide a samples' genotype in less than 4 hours. Finally, the lateral flow biosensor format minimizes the requirements for highly qualified personnel for performing the test and interpreting the results.

Conflicts of Interest

The authors declare that there are no conflicts of interest regarding the publication of this article. The founding sponsors had no role in the design of the study; in the collection, analyses, or interpretation of data; in the writing of the manuscript, and in the decision to publish the results.

Authors' Contributions

Dimitra K. Toubanaki conceived, designed, and performed the experiments; Dimitra K. Toubanaki and Evdokia Karagouni analyzed the data; Dimitra K. Toubanaki wrote the paper; Evdokia Karagouni proofread the manuscript. All authors have approved the present manuscript.

Acknowledgments

The research project was implemented within the framework of the Action "Supporting Postdoctoral Researchers" of the Operational Program "Education and Lifelong Learning" (Action's Beneficiary: General Secretariat for Research and Technology) and was cofinanced by the European Social Fund (ESF) and the Greek State.

References

[1] L. Barazi-Yeroulanos, *Regional Synthesis of the Mediterranean Marine Finfish Aquaculture Sector and Development of a Strategy for Marketing and Promotion of Mediterranean Aquaculture (MedAquaMarket)*, Studies and Reviews General Fisheries Commission for the Mediterranean, Rome, Italy, 2010.

[2] D. Pillai, J. R. Bonami, and J. Sri Widada, "Rapid detection of *Macrobrachium rosenbergii* nodavirus (MrNV) and extra small virus (XSV), the pathogenic agents of white tail disease of *Macrobrachium rosenbergii* (De Man), by loop-mediated isothermal amplification," *Journal of Fish Diseases*, vol. 29, no. 5, pp. 275–283, 2006.

[3] B. L. Munday, J. Kwang, and N. Moody, "*Betanodavirus* infections of teleost fish: a review," *Journal of Fish Diseases*, vol. 25, no. 3, pp. 127–142, 2002.

[4] Q. K. Doan, M. Vandeputte, B. Chatain, T. Morin, and F. Allal, "Viral encephalopathy and retinopathy in aquaculture: a review," *Journal of Fish Diseases*, vol. 40, no. 5, pp. 717–742, 2017.

[5] J. Z. Costa and K. D. Thompson, "Understanding the interaction between *Betanodavirus* and its host for the development of prophylactic measures for viral encephalopathy and retinopathy," *Fish and Shellfish Immunology*, vol. 53, pp. 35–49, 2016.

[6] K. Mori, T. Nakai, K. Muroga, M. Arimoto, K. Mushiake, and I. Furusawa, "Properties of a new virus belonging to nodaviridae found in larval striped jack (*Pseudocaranx dentex*) with nervous necrosis," *Virology*, vol. 187, no. 1, pp. 368–371, 1992.

[7] T. Nagai and T. Nishizawa, "Sequence of the non-structural protein gene encoded by RNA1 of striped jack nervous necrosis virus," *Journal of General Virology*, vol. 80, no. 11, pp. 3019–3022, 1999.

[8] T. Nishizawa, M. Furuhashi, T. Nagai, T. Nakai, and K. Muroga, "Genomic classification of fish nodaviruses by molecular phylogenetic analysis of the coat protein gene," *Applied and Enviromental Microbiology*, vol. 63, no. 4, pp. 1633–1636, 1997.

[9] T. Iwamoto, Y. Okinaka, K. Mise et al., "Identification of host-specificity determinants in *Betanodaviruses* by using reassortants between striped jack nervous necrosis virus and sevenband grouper nervous necrosis virus," *Journal of Virology*, vol. 78, no. 3, pp. 1256–1262, 2004.

[10] R. Thiery, C. Arnauld, and C. Delsert, "Two isolates of sea bass, *Dicentrarchus labrax* L., nervous necrosis virus with distinct genomes," *Journal of Fish Diseases*, vol. 22, pp. 201–207, 1999.

[11] T. Iwamoto, T. Nakai, K. Mori, M. Arimoto, and I. Furusawa, "Cloning of the fish cell line SSN-1 for piscine nodaviruses," *Diseases of Aquatic Organisms*, vol. 43, no. 2, pp. 81–89, 2000.

[12] Y. Ito, Y. Okinaka, K. Mori et al., "Variable region of *Betanodavirus* RNA2 is sufficient to determine host specificity," *Diseases of Aquatic Organisms*, vol. 79, no. 3, pp. 199–205, 2008.

[13] V. Panzarin, E. Cappellozza, M. Mancin et al., "In vitro study of the replication capacity of the RGNNV and the SJNNV *Betanodavirus* genotypes and their natural reassortants in response to temperature," *Veterinary Research*, vol. 45, no. 1, p. 56, 2014.

[14] N. Vendramin, A. Toffan, M. Mancin et al., "Comparative pathogenicity study of ten different *Betanodavirus* strains in experimentally infected European sea bass, *Dicentrarchus labrax* (L.)," *Journal of Fish Diseases*, vol. 37, no. 4, pp. 371–383, 2014.

[15] N. Chérif, R. Thiéry, J. Castric et al., "Viral encephalopathy and retinopathy of *Dicentrarchus labrax* and *Sparus aurata* farmed in Tunisia," *Veterinary Research Communications*, vol. 33, no. 4, pp. 345–353, 2009.

[16] J. M. Cutrín, C. P. Dopazo, R. Thiéry et al., "Emergence of pathogenic *Betanodaviruses* belonging to the SJNNV genogroup in farmed fish species from the Iberian Peninsula," *Journal of Fish Diseases*, vol. 30, no. 4, pp. 225–232, 2007.

[17] B. Lopez-Jimena, N. Cherif, E. Garcia-Rosado et al., "A combined RT-PCR and dot-blot hybridization method reveals the coexistence of SJNNV and RGNNV *Betanodavirus* genotypes in wild meagre (*Argyrosomus regius*)," *Journal of Applied Microbiology*, vol. 109, no. 4, pp. 1361–1369, 2010.

[18] G. P. Skliris, J. V. Krondiris, D. C. Sideris, A. P. Shinn, W. G. Starkey, and R. H. Richards, "Phylogenetic and antigenic characterization of new fish nodavirus isolates from Europe and Asia," *Virus Research*, vol. 75, no. 1, pp. 56–67, 2001.

[19] R. Thiéry, J. Cozien, C. de Boisséson, S. Kerbart-Boscher, and L. Névarez, "Genomic classification of new *Betanodavirus* isolates by phylogenetic analysis of the coat protein gene suggests a low host-fish species specificity," *Journal of General Virology*, vol. 85, no. 10, pp. 3079–3087, 2004.

[20] T. Nishizawa, K. Mori, T. Nakai, I. Furusawa, and K. Muroga, "Polymerase chain reaction (PCR) amplification of RNA of striped jack nervous necrosis virus (SJNNV)," *Diseases of Aquatic Organisms*, vol. 18, pp. 103–107, 1994.

[21] L. Bigarré, M. Baud, J. Cabon, K. Crenn, and J. Castric, "New PCR probes for detection and genotyping of piscine *Betanodaviruses*," *Journal of Fish Diseases*, vol. 33, no. 11, pp. 907–912, 2010.

[22] L. Dalla Valle, V. Toffolo, M. Lamprecht et al., "Development of a sensitive and quantitative diagnostic assay for fish nervous necrosis virus based on two-target real-time PCR," *Veterinary Microbiology*, vol. 110, no. 3-4, pp. 167–179, 2005.

[23] K. Hodneland, R. Garcia, J. A. Balbuena, C. Zarza, and B. Fouz, "Real-time RT-PCR detection of *Betanodavirus* in naturally and experimentally infected fish from Spain," *Journal of Fish Diseases*, vol. 34, no. 3, pp. 189–202, 2011.

[24] T. Nishizawa, K. Mori, M. Furuhashi, T. Nakai, I. Furusawa, and K. Muroga, "Comparison of the coat protein genes of five fish nodaviruses, the causative agents of viral nervous necrosis in marine fish," *Journal of General Virology*, vol. 76, no. 7, pp. 1563–1569, 1995.

[25] V. Panzarin, A. Fusaro, I. Monne et al., "Molecular epidemiology and evolutionary dynamics of *Betanodavirus* in southern Europe," *Infection, Genetics and Evolution*, vol. 12, no. 1, pp. 63–70, 2012.

[26] T. Iwamoto, K. Mori, M. Arimoto, and T. Nakai, "High permissivity of the fish cell line SSN-1 for piscine

nodaviruses," *Diseases of Aquatic Organisms*, vol. 39, no. 1, pp. 37–47, 1999.

[27] D. K. Toubanaki, M. Margaroni, and E. Karagouni, "Development of a novel allele-specific PCR method for rapid assessment of nervous necrosis virus genotypes," *Current Microbiology*, vol. 71, no. 5, pp. 529–539, 2015.

[28] D. K. Toubanaki and E. Karagouni, "Genotype-specific real-time PCR combined with high-resolution melting analysis for rapid identification of red-spotted grouper nervous necrosis virus," *Archives of Virology*, vol. 162, no. 8, pp. 2315–2328, 2017.

[29] S. Ye, S. Humphries, and F. Green, "Allele specific amplification by tetra-primer PCR," *Nucleic Acids Research*, vol. 20, no. 5, p. 1152, 1992.

[30] G. A. Posthuma-Trumpie, J. Korf, and A. van Amerongen, "Lateral flow (immuno)assay: its strengths, weaknesses, opportunities and threats. A literature survey," *Analytical and Bioanalytical Chemistry*, vol. 393, no. 2, pp. 569–582, 2009.

[31] C. Parolo and A. Merkoçi, "Paper-based nanobiosensors for diagnostics," *Chemical Society Reviews*, vol. 42, no. 2, pp. 450–457, 2013.

[32] X. Gao, L. P. Xu, S. F. Zhou, G. Liu, and X. Zhang, "Recent advances in nanoparticles-based lateral flow biosensors," *American Journal of Biomedical Sciences*, vol. 6, no. 1, pp. 41–57, 2014.

[33] D. K. Toubanaki, M. Margaroni, and E. Karagouni, "Nanoparticle-based lateral flow biosensor for visual detection of fish nervous necrosis virus amplification products," *Molecular and Cellular Probes*, vol. 29, no. 3, pp. 158–166, 2015.

[34] D. K. Toubanaki, T. K. Christopoulos, P. C. Ioannou, and C. S. Flordellis, "Identification of single-nucleotide polymorphisms by the oligonucleotide ligation reaction: a DNA biosensor for simultaneous visual detection of both alleles," *Analytical Chemistry*, vol. 81, no. 1, pp. 218–224, 2009.

[35] D. K. Toubanaki, M. Margaroni, and E. Karagouni, "Dual enhancement with a nanoparticle-based lateral flow biosensor for the determination of DNA," *Analytical Letters*, vol. 49, no. 7, pp. 1040–1055, 2016.

[36] M. D. Abramoff, P. J. Magalhaes, and S. J. Ram, "Image Processing with ImageJ," *Biophotonics International*, vol. 11, no. 7, pp. 36–42, 2004.

[37] J. K. Konstantou, P. C. Ioannou, and T. K. Christopoulos, "Dual-allele dipstick assay for genotyping single nucleotide polymorphisms by primer extension reaction," *European Journal of Human Genetics*, vol. 17, no. 1, pp. 105–111, 2009.

[38] J. K. Konstantou, A. C. Iliadi, P. C. Ioannou et al., "Visual screening for JAK2V617F mutation by a disposable dipstick," *Analytical and Bioanalytical Chemistry*, vol. 397, no. 5, pp. 1911–1916, 2010.

[39] K. Glynou, P. C. Ioannou, T. K. Christopoulos, and V. Syriopoulou, "Oligonucleotide-functionalized gold nanoparticles as probes in a dry-reagent strip biosensor for DNA analysis by hybridization," *Analytical Chemistry*, vol. 75, no. 16, pp. 4155–4160, 2003.

[40] Z. D. Su, C. Y. Shi, J. Huang et al., "Establishment and application of cross-priming isothermal amplification coupled with lateral flow dipstick (CPA-LFD) for rapid and specific detection of red-spotted grouper nervous necrosis virus," *Virology Journal*, vol. 12, no. 1, p. 149, 2015.

21

A Simple Separation Method of the Protein and Polystyrene Bead-Labeled Protein for Enhancing the Performance of Fluorescent Sensor

Hye Jin Kim,[1] Dong-Hoon Kang,[2] Seung-Hoon Yang,[3] Eunji Lee,[4] Taewon Ha,[5] Byung Chul Lee,[6] Youngbaek Kim,[5] Kyo Seon Hwang,[1] Hyun-Joon Shin,[2] and Jinsik Kim [4]

[1]Department of Clinical Pharmacology, Kyung Hee University, Seoul 02447, Republic of Korea
[2]Center for Bionics, Korea Institute of Science and Technology, Seoul 02792, Republic of Korea
[3]Systems Biotechnology Research Center, Korea Institute of Science and Technology, Gangneung 25451, Republic of Korea
[4]Department of Medical Biotechnology, Dongguk University, Seoul 04620, Republic of Korea
[5]Center for Nano-Photonics Convergence Technology, Korea Institute of Industrial Technology (KITECH), Gwangju 61012, Republic of Korea
[6]Center for BioMicrosystems, Korea Institute of Science and Technology (KIST), Seoul 02792, Republic of Korea

Correspondence should be addressed to Jinsik Kim; lookup2@dongguk.edu

Academic Editor: Subhankar Singha

Dielectrophoresis- (DEP-) based separation method between a protein, amyloid beta 42, and polystyrene (PS) beads in different microholes was demonstrated for enhancement of performance for bead-based fluorescent sensor. An intensity of $\nabla|E|^2$ was relative to a diameter of a microhole, and the diameters of two microholes for separation between the protein and PS beads were simulated to $3\,\mu m$ and $15\,\mu m$, respectively. The microholes were fabricated by microelectromechanical systems (MEMS). The separation between the protein and the PS beads was demonstrated by comparing the average intensity of fluorescence (AIF) by each molecule. Relative AIF was measured in various applying voltage and time conditions, and the conditions for allocating the PS beads into $15\,\mu m$ hole were optimized at 80 mV and 15 min, respectively. In the optimized condition, the relative AIF was observed approximately 4.908 ± 0.299. Finally, in $3\,\mu m$ and $15\,\mu m$ hole, the AIFs were approximately 3.143 and -1.346 by 2 nm of protein and about -2.515 and 4.211 by 30 nm of the PS beads, respectively. The results showed that 2 nm of the protein and 30 nm of PS beads were separated by DEP force in each microhole effectively, and that our method is applicable as a new method to verify an efficiency of the labeling for bead-based fluorescent sensor $\nabla|E|^2$.

1. Introduction

Labeling is one of the essential processes for analyzing and tracking the biomolecules and proceeds by conjugating the molecules with various materials such as isotope markers [1], photochromic compounds [2], and fluorescence polystyrene (PS) beads [3–5]. Especially, fluorescent PS beads have competitive price, high accessibility, and controllability so that various biomolecules such as protein [6], cell [3], and

deoxyribonucleic acid (DNA) [7] have been conjugated with fluorescent PS beads followed by being quantified and qualified [8–10]. Qin et al. separated the protein which conjugated with surface-modified fluorescent PS [11], and Fakih et al. multidetected the viral DNA using gold nanoparticle-coated fluorescent PS beads [12]. However, fluorescent PS beads are not conjugated with biomolecules perfectly; in other words, their labeling efficiency is under 100%, and consequently, not only biomolecules conjugated with fluorescent PS beads but also nonconjugated biomolecules existed in the analyte. The nonconjugated biomolecules decrease the accuracy and sensitivity in analyzing and tracking of the biomolecules; hence, a method is required for separating the biomolecules conjugated with PS beads and nonconjugated biomolecules, namely, the residue molecules. So, residue biomolecules after labeling need to be separated from the biomolecules, which are conjugated with labels, ideally. Although centrifugation approaches [13, 14] and fluidic-based approaches [15] are suitable for separating the residue molecules, these approaches are complex and require an additional process.

Dielectrophoresis (DEP), resulting from inhomogeneous electric fields, has been utilized for the specific manipulation of the particles, cells, and viruses as well as biomolecules such as DNA and even single protein, because of its simplicity, efficiency, and usability [16–18]. The intensity of the DEP force is dominated by the size of the molecules and the strength of the electric fields, which occurred between the electrodes, so that various molecules are affected by a different intensity of the force according to the size of the molecules and the structure of the electrode. Lapizco-Encinas et al. concentrated and separated the live and dead bacteria with insulator-based DEP (iDEP) [19]. But two types of bacteria were separated according to different types of DEP force, negative DEP and positive DEP, and Chen et al. suggested a simplified dielectrophoretic-based microfluid device for particle separation [20]. But these approaches were limited for observing the various molecules simultaneously or consisted of the complex structure.

Here, we suggest a simple method to separate the nonconjugated protein, namely, the residue protein, and the protein conjugated with PS beads with the DEP force, which is applicable for verifying the efficiency of labeling between protein and PS beads. The protein and PS beads were separated into two microholes with different diameters and formed on a single electrode according to the intensity of DEP force induced. The intensity of the DEP force increased when the diameter of the microhole was smaller, and thus, stronger repulsive force and attractive force occurred in the small microhole than the large one. Consequently, the smaller molecules, residue protein, were allocated into the small microhole, whereas the bigger molecules, PS beads, were expelled from that small microhole followed by allocating into the large microhole; it means that the protein and PS beads were separated. To verify the separation between protein and PS beads, approximately 2 nm of the protein, amyloid beta 42, and 30 nm fluorescent PS beads were used. The diameter of the two microholes for separating the protein and PS beads was optimized to 3 μm and 15 μm, respectively, by calculating the intensity of $\nabla|E|^2$ in each microhole with COMSOL simulation. Also, an applied voltage to induce the DEP force was optimized to 80 mV, which induced a difference in the $\nabla|E|^2$ force approximately 9.059-fold between two microholes. The microholes were fabricated by microelectromechanical systems (MEMS) technique, and separation between the protein and polystyrene beads was demonstrated by comparing a relative averaged intensity of fluorescence (AIF) by each protein and each PS bead.

2. Materials and Methods

2.1. Theory. The molecule present in an inhomogeneous electric field, E, is influenced by the DEP force, F_{DEP}, which is expressed as follows [21]:

$$F_{DEP} = 2\pi r^3 \varepsilon_m K(\omega) \cdot \nabla |\vec{E}|^2, \tag{1}$$

where r, ε_m, and $K(\omega)$ represent the radius of molecules, the effective permittivity of liquid, and the Clausius–Mossotti factor, respectively. The E and gradient of the electric field $\nabla|E|^2$ are described as follows:

$$\vec{E} = -\frac{\Delta V}{\Delta \vec{d}}, \tag{2}$$

$$\nabla|\vec{E}| = \frac{\partial^2 V}{\partial d_x^2}\hat{x} + \frac{\partial^2 V}{\partial d_y^2}\hat{y} + \frac{\partial^2 V}{\partial d_z^2}\hat{z}, \tag{3}$$

where V and d are the applying voltage and the distance between the electrodes, respectively. On the basis of (1) and (3), the intensity of the DEP force can be modified through (4) as follows:

$$F_{DEP} \propto \frac{r^3}{d^2}. \tag{4}$$

The intensity of $\nabla|E|^2$ is calculated with a finite-element model (FEM) in the AC/DC module of COMSOL Multiphysics software 5.2 (COMSOL Inc., USA).

2.2. Materials. Thirty nanometres of the carboxylate-modified polystyrene (PS) bead labeled with fluorescence (Sigma-Aldrich Inc., Korea) and 2 nm of the TAMRA-labeled beta-amyloid (1–42) protein (AnaSpec Inc., USA) were used to verify the separation of molecules, whose excitation/emission wavelength ($\lambda_{ex}/\lambda_{em}$) was ~470/505 nm and ~544/572 nm, respectively. The protein and PS beads were diluted with 1 mM PBS buffer (Corning Korea Co. Ltd., Korea) to create a 1 ng·mL^{-1} solution.

2.3. Fabrication of Microholes. Microholes were fabricated by a standard MEMS process. First, an insulation layer, 300 nm of SiO$_2$, and an electrode layer, 30 nm of tantalum (Ta) and 150 nm of platinum (Pt), were sequentially deposited on the 4-inch silicon (Si) wafer by thermal oxidation and sputtering, respectively. Next, an AZ GXR 601 photoresist (AZ Electronic Materials, Luxembourg) was coated by a spin coater (30 s, 3000 rpm) and exposed (3.8 s, 12 mW·cm^{-2}). Then, the hole patterns were etched by inductively coupled plasma etching (Oxford Instruments), and the photoresist was stripped by Microwave Plasma Asher (Plasma-Finish, Germany).

2.4. System Setup for Molecules Separation and Fluorescence Analysis. A DG4062 Series waveform generator (Rigol Technologies Inc., USA) (frequency range: up to 60 MHz; voltage range: up to 10 V), which applies a sinusoidal AC voltage for inducing the DEP force in the microhole on

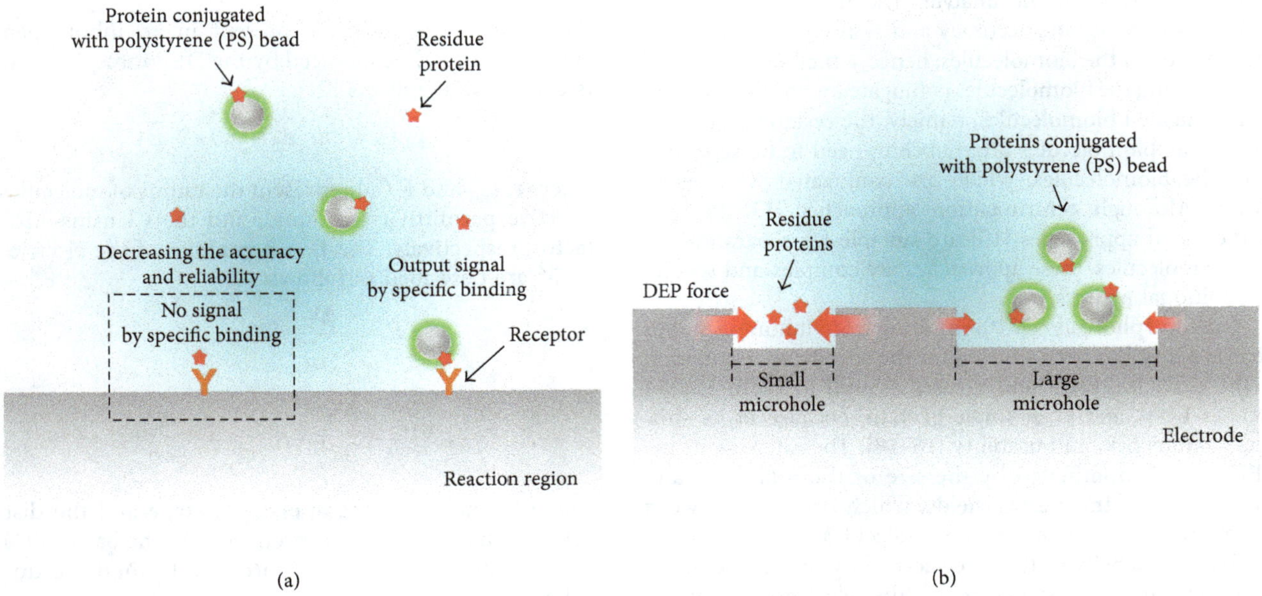

FIGURE 1: Schematic illustration of a simple separation method of the protein and protein conjugated with polystyrene (PS) beads. (a) Intensity of fluorescence by a specific binding of the protein conjugated with PS beads decreased due to a specific binding of the non-conjugated protein, expressed as residue protein. (b) Residue protein and protein conjugated with PS beads were separated by the dielectrophoresis (DEP) force in different microholes, respectively.

a single electrode, was used. The intensity of fluorescence was observed via an electron-multiplying charge-coupled device (ANDORiXonEM), an oil immersion 100x lens (Nikon Corp., Japan) (NA: 1.4), and an Eclipse Ti inverted microscope (Nikon Corp.) equipped with a halogen lamp and a 593 nm (bandwidth: 40 nm) filter and was analyzed by Image-Pro Plus 6.0 (Media Cybernetics Inc., USA). The average intensity of fluorescence (AIF) was calculated by values measured at five random positions in the microhole, and a value of a relative AIF was calculated by dividing the AIF values measured at each condition by the value in the reference condition.

3. Results and Discussion

Protein, nonconjugated with PS beads after labeling, does not emit the fluorescence but binds to the receptor specifically so that it impedes the specific binding between the ideally conjugated protein with PS beads and receptor, followed by decreasing the accuracy and reliability in the molecules' analyzing and tracking process (Figure 1(a)). Thus, the nonconjugated protein, namely, the residue protein, should be separated from the protein conjugated with PS beads. When alternating current (AC) voltage is applied to the electrode with microholes, the protein is allocated into each microhole according to the intensity of the DEP force that occurred in each microhole. The intensity of the DEP force is related to the diameter of the molecules and the distance between the electrodes, namely, the size of microholes, as described in (4), and consequently allocates different molecules into each microhole, respectively:

residue protein, smaller than the PS beads, is allocated into the small microhole, whereas the PS beads are placed in the large microhole—two molecules separate into small and large microholes, respectively (Figure 1(b)).

In order to separate the residue protein and conjugated protein with PS beads in each microhole, 4.5 kDa of amyloid beta, whose diameter was calculated to be approximately 2 nm, was used as a residue protein, and the conjugated protein with PS beads was simplified to just PS beads. The length and width of the electrode were fixed to 27 μm and 21 μm, respectively, and the diameter of the small microhole, d, and pitch between two microholes, p, were fixed to 3 μm. The intensity of the applied AC voltage and size of microholes were optimized via the COMSOL simulation (Figure 2(a)). Firstly, maximum intensity of $\nabla|E|^2$ in the small microhole was simulated according to the applied AC voltage (Figure 2(b)). Black line and scatter showed the maximum intensity of $\nabla|E|^2$ that occurred in the small microhole, and red line and scatter indicated the size of protein, which was allocated into the small microhole, depending on the intensity of the applied AC voltage. The intensity of $\nabla|E|^2$ increased parabolically and size of the protein decreased accordingly. The results signified that approximately 30 mV voltage, which resulted in $\nabla|E|^2$ with intensity approximately 2.310×10^{13} V^2·m^{-3}, was required to place 2 nm of the protein in the 3 μm hole. Also, the maximum intensity of $\nabla|E|^2$ that occurred in the other microhole was simulated according to the size of the other microhole at the condition that applied 30 mV AC voltage (Figure 2(c)). The diameter of the other microhole was expressed as a ratio to the diameter for the 3 μm hole, and the

(a)

(b)

(c)

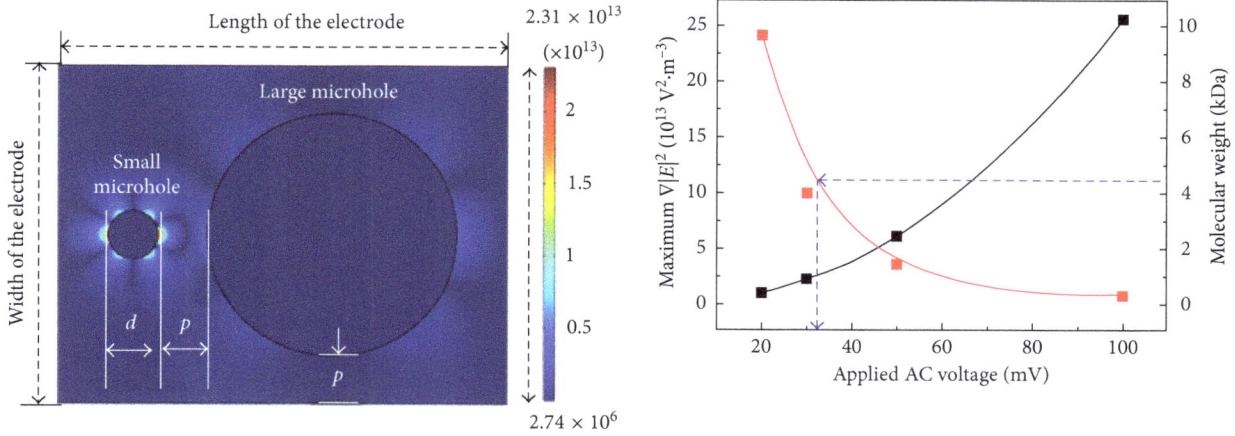

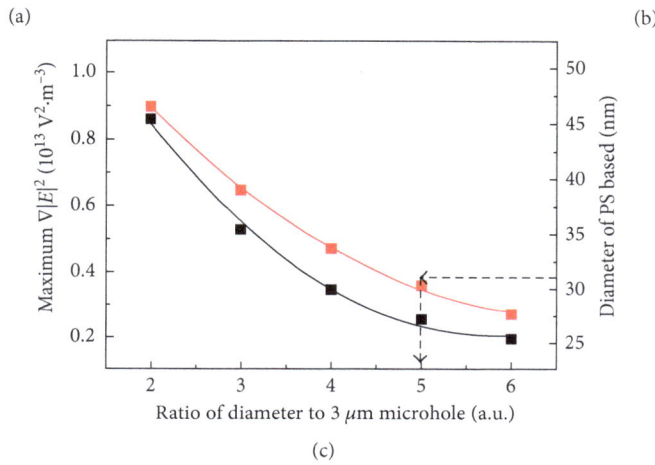

FIGURE 2: Simulation for separating the molecules by DEP force in the two microholes. (a) Distribution of $\nabla|E|^2$ at the top view of the electrode. The diameter of the small microhole and pitch between two microholes were represented as "d" and "p," respectively. (b) According to the applied voltage, maximum intensity of $\nabla|E|^2$ occurring in the 3 μm hole increased, whereas molecular weight of the protein, allocated into the microhole, decreased. (c) Maximum intensity of $\nabla|E|^2$ in the other microhole and the diameter of PS beads, allocated into the hole, decreased according to the increase of the diameter.

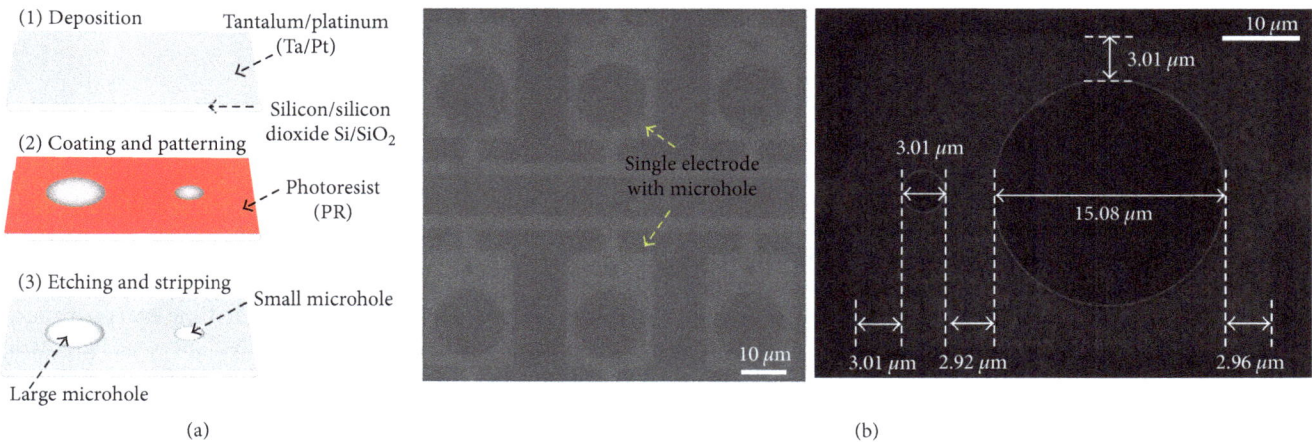

(a)

(b)

FIGURE 3: Fabrication of the single electrode consisting of two different-sized microholes by MEMS technology. (a) Schematic illustration of the fabrication process of the microholes. (b) Microscopic image of the two different microholes.

intensity of $\nabla|E|^2$ decreased according to the increase in the ratio of the diameter. In the 6 μm microhole (the ratio was 2), intensity of $\nabla|E|^2$ was about 0.861×10^{13} V^2·m^{-3} and it is too

strong to allocate 30 nm of the PS beads into the hole. $\nabla|E|^2$ was approximately 0.528×10^{13} V^2·m^{-3}, 0.347×10^{13} V^2·m^{-3}, 0.255×10^{13} V^2·m^{-3}, and 0.194×10^{13} V^2·m^{-3} in each value of

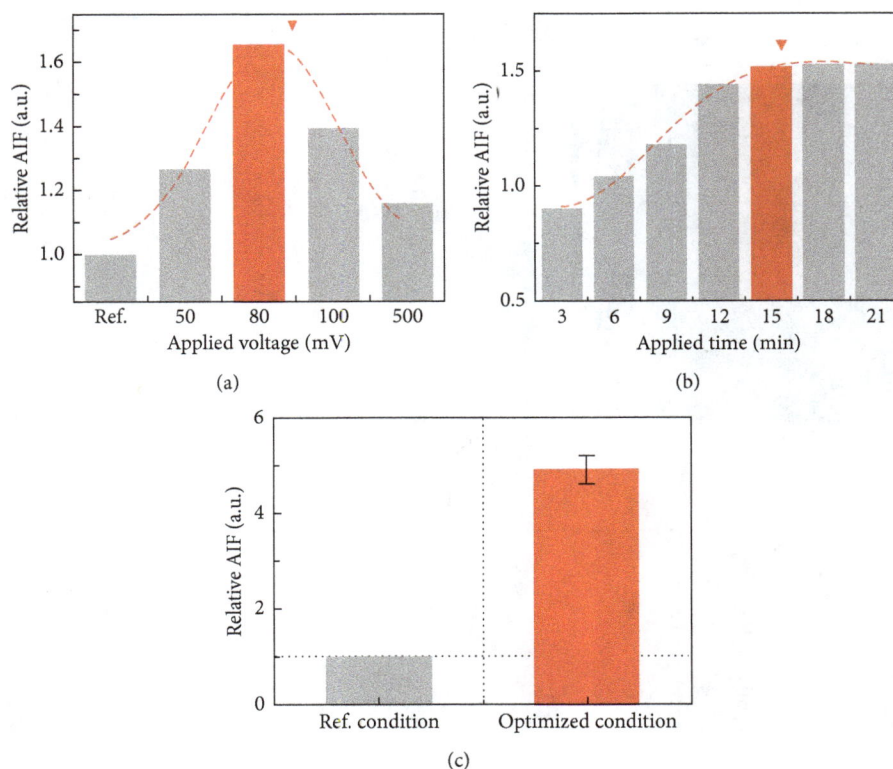

FIGURE 4: Optimization of the DEP condition by measuring the average intensity of fluorescence (AIF) of the PS beads in the 15 μm hole. Relative AIF was verified according to (a) the applied voltage and (b) the applied time of AC voltage. (c) Relative AIF by the PS beads in the 15 μm hole was compared in each reference and optimized DEP condition.

the ratio, and an optimized size of the microhole for placing 30 nm of the PS beads into the hole was verified to be 15 μm. Thus, the two microholes for separating the residue protein and protein conjugated with PS beads were optimized to 3 μm and 15 μm, respectively, whose difference in the intensity was approximately 9.059-fold.

Two microholes in the electrode were produced via a standard MEMS process on a 4-inch silicon (Si) wafer. The fabrication process consisted of 3 steps (Figure 3(a)), and details of the process are described in Material and Methods. The diameter of the fabricated small and large microholes was approximately 3 μm and 15 μm, respectively, and pitch between two microholes was about 3 μm (Figure 3(b)).

Various conditions of DEP, intensity of the DEP force and applied time, were optimized by measuring the relative AIF resulting from placing the PS beads in the 15 μm hole. In order to optimize the intensity of the DEP force, the applied frequency required for the DEP force to occur was fixed at 50 MHz. Firstly, relative AIF was measured in various applying voltage conditions ranging from 0 V (ref.) to 500 mV, and consequently, it was maximized at 80 mV (Figure 4(a)). The values were approximately 1, 1.265, 1.655, 1.396, and 1.1604 in ref., 50 mV, 80 mV, 100 mV, and 500 mV, respectively. The results indicated that the PS beads were most effectively placed in the microhole by the DEP force induced by the applied voltage 80 mV; consequently, the intensity of the applied voltage was settled to 80 mV. Also, in order to optimize the applied time condition of DEP force, the

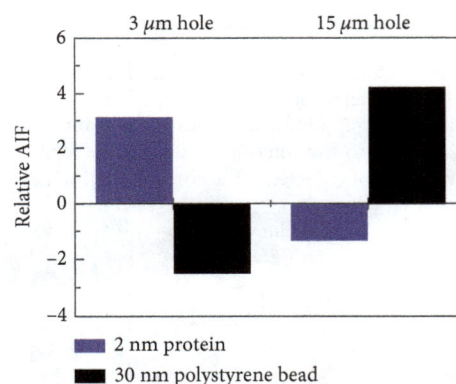

FIGURE 5: Relative AIF by 2 nm of the protein and 30 nm of the PS beads in 3 μm and 15 μm holes, respectively.

relative AIF was measured according to time every 3 minutes up to 21 minutes (Figure 4(b)). The AIF increased gradually depending on the time up to 15 min and was saturated afterward: each value of the AIF was approximately 0.907, 1.048, 1.184, 1.450, 1.526, 1.535, and 1.536 according to the applied time. Hence, the applied voltage and time were optimized to 80 mV and 15 minutes, respectively, and the relative AIF was observed to be approximately 4.908 ± 0.299 in the optimized condition (Figure 4(c)). The result demonstrated that the PS beads were allocated into the 15 μm hole effectively.

Finally, based on these results, a separation of the protein and PS beads in $3\,\mu m$ and $15\,\mu m$ holes, respectively, was demonstrated (Figure (5)). It was also confirmed by comparing the AIF by each molecule in the two microholes at the previous optimized condition. The relative AIF by 2 nm of the protein in the $3\,\mu m$ hole was a positive value, but the value by 30 nm of the PS beads was negative, and the values were approximately 3.143 and −1.346, respectively, whereas in the $15\,\mu m$ hole, the relative AIFs by the protein and the PS beads showed an opposite sign compared with the previous values, and the values were approximately −2.515 and 4.211, respectively. The negative value of the AIF indicated that the molecules were moving far away owing to the strong DEP force in the microhole, and the positive value of the AIF signified that the molecules were attracted and trapped into the microhole by the DEP force. Thus, the results signified that the DEP force allocated 2 nm of the protein and 30 nm of the PS beads into $3\,\mu m$ and $15\,\mu m$ holes, respectively. The results demonstrated that 2 nm of the protein and 30 nm of the PS beads were separated by DEP force in each microhole, effectively.

4. Conclusions

In this paper, a simple method for separation between 2 nm of the protein and PS beads into different microholes, respectively, by the DEP force was demonstrated. In order to separate two molecules, the diameter of the two microholes was simulated and the intensity of the DEP force induced in the microholes was calculated via simulation. The optimized diameter of the two microholes was $3\,\mu m$ and $15\,\mu m$, and a difference in the DEP force between two microholes was approximately 9.059-fold. The condition of the DEP force to separate two molecules was optimized experimentally: intensity of the AC voltage was 80 mV and the applied time was 15 minutes. The molecules which were separated by the DEP force in each microhole were verified by measuring the relative AIF by each molecule. In $3\,\mu m$ and $15\,\mu m$ holes, the AIFs were approximately 3.143 and −1.346 by 2 nm of the protein and about −2.515 and 4.211 by 30 nm of the PS beads, respectively. Consequently, the results demonstrated that 2 nm of the protein and 30 nm of the PS beads were separated by DEP force in each microhole, effectively. Our method has high expandability in separation of various-sized molecules, and furthermore, it is applicable for verification of the labeling efficiency.

Conflicts of Interest

The authors declare that there are no conflicts of interest regarding the publication of this paper.

Acknowledgments

This work was mainly supported by the Dongguk University Research Fund of 2017. Taewon Ha and Youngbaek Kim are grateful for financial support from the Korea Institute of Industrial Technology (Project no. EO170047).

References

[1] X. Chen, L. M. Smith, and E. M. Bradbury, "Site-specific mass tagging with stable isotopes in proteins for accurate and efficient protein identification," *Analytical Chemistry*, vol. 72, no. 6, pp. 1134–1143, 2000.

[2] M. Lummer, F. Humpert, M. Wiedenlübbert, M. Sauer, M. Schüttpelz, and D. Staiger, "A new set of reversibly photoswitchable fluorescent proteins for use in transgenic plants," *Molecular Plant*, vol. 6, no. 5, pp. 1518–1530, 2013.

[3] H. Sahoo, "Fluorescent labeling techniques in biomolecules: a flashback," *RSC Advances*, vol. 2, no. 18, pp. 7017–7029, 2012.

[4] D. Jung, K. Min, J. Jung, W. Jang, and Y. Kwon, "Chemical biology-based approaches on fluorescent labeling of proteins in live cells," *Molecular BioSystems*, vol. 9, no. 5, pp. 862–872, 2013.

[5] M. M. Bonar and J. C. Tilton, "High sensitivity detection and sorting of infectious human immunodeficiency virus (HIV-1) particles by flow virometry," *Virology*, vol. 505, pp. 80–90, 2017.

[6] C. Obermaier, A. Griebel, and R. Westermeier, "Principles of protein labeling techniques," *Proteomic Profiling: Methods and Protocols*, vol. 1295, pp. 153–165, 2015.

[7] T. T. Weil, R. M. Parton, and I. Davis, "Making the message clear: visualizing mRNA localization," *Trends in Cell Biology*, vol. 20, no. 7, pp. 380–390, 2010.

[8] J. Lu, G. Getz, E. A. Miska et al., "MicroRNA expression profiles classify human cancers," *Nature*, vol. 435, pp. 834–838, 2005.

[9] B. S. Edwards, T. Oprea, E. R. Prossnitz, and L. A. Sklar, "Flow cytometry for high-throughput, high-content screening," *Current Opinion in Chemical Biology*, vol. 8, no. 4, pp. 392–398, 2004.

[10] X. H. Gao and S. M. Nie, "Quantum dot-encoded mesoporous beads with high brightness and uniformity: rapid readout using flow cytometry," *Analytical Chemistry*, vol. 76, no. 8, pp. 2406–2410, 2004.

[11] L. Qin, X. W. He, W. Zhang, W. Y. Li, and Y. K. Zhang, "Surface-modified polystyrene beads as photografting imprinted polymer matrix for chromatographic separation of proteins," *Journal of Chromatography A*, vol. 1216, no. 5, pp. 807–814, 2009.

[12] H. H. Fakih, M. M. Itani, and P. Karam, "Gold nanoparticles-coated polystyrene beads for the multiplex detection of viral DNA," *Sensors and Actuators B: Chemical*, vol. 250, pp. 446–452, 2017.

[13] E. Fernández-Vizarra, M. J. López-Pérez, and J. A. Enriquez, "Isolation of biogenetically competent mitochondria from mammalian tissues and cultured cells," *Methods*, vol. 26, no. 4, pp. 292–297, 2002.

[14] U. Michelsen and J. von Hagen, "Isolation of subcellular organelles and structures," in *Methods in Enzymology*, pp. 305—328, Academic Press, Cambridge, MA, USA, 2009.

[15] A. A. S. Bhagat, H. Bow, H. W. Hou, S. J. Tan, J. Han, and C. T. Lim, "Microfluidics for cell separation," *Medical and Biological Engineering and Computing*, vol. 48, pp. 999–1014, 2010.

[16] S. Paracha and C. Hestekin, "Field amplified sample stacking of amyloid beta (1-42) oligomers using capillary electrophoresis," *Biomicrofluidics*, vol. 10, no. 3, article 033105, 2016.

[17] T. D. Mai, F. Qukacine, and M. Taverna, "Multiple capillary isotachophoresis with repetitive hydrodynamic injections for performance improvement of the electromigration pre-concentration," *Journal of Chromatography A*, vol. 1453, pp. 116–123, 2016.

[18] L Zheng, J. P. Brody, and P. J. Burke, "Electronic manipulation of DNA, proteins, and nanoparticles for potential circuit assembly," *Biosensors and Bioelectronics*, vol. 20, no. 3, pp. 606–619, 2004.

[19] B. H. Lapizco-Encinas, B. A. Simmons, E. B. Cummings, and Y. Fintschenko, "Dielectrophoretic concentration and separation of live and dead bacteria in an array of insulators," *Analytical Chemistry*, vol. 76, no. 16, pp. 1571–1579, 2004.

[20] X. Chen, Y. Ren, W. Liu et al., "A simplified microfluidic device for particle separation with two consecutive steps: induced charge electro-osmotic prefocusing and dielectrophoretic separation," *Analytical Chemistry*, vol. 89, no. 17, pp. 9583–9592, 2017.

[21] A. Ramos, H. Morgan, N. G. Green, and A. Castellanos, "Ac electrokinetics: a review of forces in microelectrode structures," *Journal of Physics D: Applied Physic*, vol. 31, no. 18, p. 2338, 1998.

22

In Situ Miniaturised Solid Phase Extraction (m-SPE) for Organic Pollutants in Seawater Samples

B. Abaroa-Pérez,[1] G. Sánchez-Almeida,[2] J. J. Hernández-Brito,[1] and D. Vega-Moreno ⓘ[2]

[1]Plataforma Oceánica de Canarias (PLOCAN), Las Palmas, Spain
[2]Chemistry Department, Universidad de Las Palmas de G.C (ULPGC), Las Palmas, Spain

Correspondence should be addressed to D. Vega-Moreno; daura.vega@ulpgc.es

Academic Editor: Bengi Uslu

Solid phase extraction (SPE) is a consolidated technique for determining pollutants in seawater samples. The current tendency is to miniaturise systems that extract and determine pollutants in the environment, reducing the use of organic solvents, while maintaining the quality in the extraction and preconcentration. On the other hand, there is a need to develop new extraction systems that can be fitted to in situ continual monitoring buoys, especially for the marine environment. This work has developed a first model of a low-pressure micro-SPE (m-SPE) for persistent organic pollutants (POPs) that can be simply applied to in situ monitoring in the marine environment. This system reduces the volumes of sample and solvents required in the laboratory in comparison with conventional SPE. In the future, it could be used in automated or robotic systems in marine technologies such as marine gliders and oceanographic buoys. This system has been optimised and validated to determine polycyclic aromatic hydrocarbons (PAH) in seawater samples, but it could also be applied to other kinds of persistent organic pollutants (POPs) and emerging pollutants.

1. Introduction

Interest in controlling and monitoring different kinds of organic pollutants in marine environments has grown [1–4], due to the harm they can do to the marine environment and human health [5]. One example of these are polycyclic aromatic hydrocarbons (PAHs) that are considered as priority pollutants by the European Union (EU) and the Environmental Protection Agency (EPA) because they are carcinogenic and they can genetically mutate [3, 5–8] and what is more, these compounds could activate oxidative stress of DNA, hence damaging metabolic activation and the generation of reactive kinds of oxygen [9, 10] making the extraction, preconcentration, and determination of these compounds in the environment very important [11, 12]. PAHs are ubiquitous pollutants in the environment, with special importance in seawater [4, 11, 13, 14], sediments [15], plankton, and filtering organisms [5, 10, 16].

The concentration of PAHs in seawater is normally in the range of 0.05 to 0.25 $\mu g \cdot L^{-1}$ [6, 17], due to their low solubility in water [1, 5, 18, 19]. A high concentration generally indicates PAH pollution of recent anthropogenic origin [6].

Over time, these compounds tend to accumulate in solid matrixes like sediment and marine plastic, with a strong tendency to bioaccumulate [16]. That is why new analytical methods are required that allow them to be monitored in situ while maintaining current levels of sensitivity and selectivity [6, 11, 20, 21].

The most commonly used techniques for determining PAHs are gas chromatography with mass spectrometry (GC-MS) [1, 2, 4] and high-pressure liquid chromatography with ultraviolet-visible detector or diode array ultra-violet-visible detector and fluorescence detector [11, 22–25]. In order to enhance the sensitivity and selectivity of the analyses, a first stage of extraction, purification, and preconcentration is required [5].

There are several preconcentration techniques for organic pollutants in liquid matrixes, such as liquid-liquid extraction (LLE), supercritical fluid extraction (SFE), and solid phase extraction (SPE) [5, 26, 27].

Solid phase extraction (SPE) is currently a highly consolidated technique for extracting pollutants from liquid samples [11, 28]. SPE gives high recoveries with a low consumption of organic solvents and high preconcentrations if volumes of

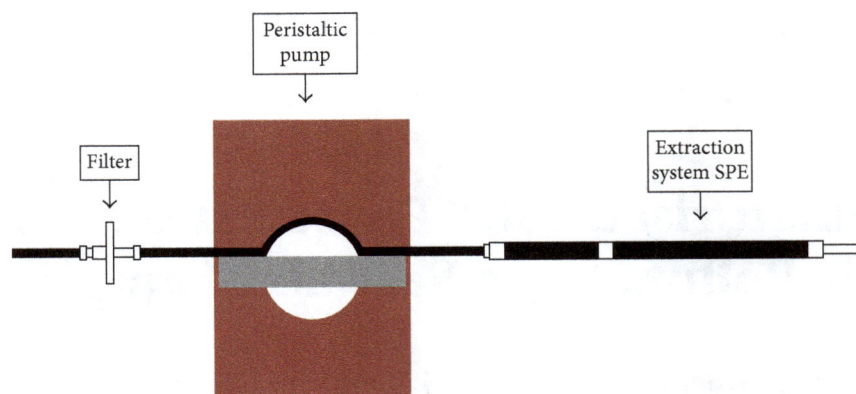

FIGURE 1: Miniaturised solid phase extraction system (m-SPE).

water of around 1 litre are filtered. SPE has been widely used for extracting hydrocarbons and other persistent pollutants from seawater and other kinds of marine samples [1, 3, 4, 12, 29]. However, these laboratory studies require large volumes of seawater that are processed in laboratories, rather than directly at the place where the sampling is done, entailing the transport and storage of the samples, which makes the sampling operation more difficult [14]. This explains the special interest in miniaturising the extraction and its application in situ, to facilitate enormously the logistics of sampling, and also opening up the possibility of future automation.

There are other systems that have attempted to miniaturise extraction, like solid phase microextraction (SPME) [5, 30–32] or fabric phase sorptive extraction (FPSE) [33], which combine extraction and preconcentration in a single step [6, 17, 34]. The disadvantage of SPME is that it is not a very robust or reproducible system, and it is very difficult to handle in situ in the marine environment. FPSE is a new technique that has yet to be tested on marine samples.

There are very few studies that consider monitoring the effects of a polycyclic aromatic hydrocarbon spill; most have been done with biomarkers, without any analytical quantification [16]. Portable systems are required to facilitate the task and reduce time and material, which allow a study to be conducted in situ.

The m-SPE presented here has the advantage of its simplicity, low cost, and ease of installation in the place the sampling is to be conducted (in situ). This method has been validated for extracting and preconcentrating PAHs in seawater.

2. Materials and Methods

2.1. Developing an m-SPE System. The miniaturised solid phase extraction system in question is shown in Figure 1 included an ISMATEC peristaltic pump, model: ISM 846 (60 rpm, dimensions $125 \times 88 \times 135$ mm), with SKALAR connectors, model: 3091 with a theoretical flow rate of $0.14 \, \text{ml·s}^{-1}$. Behind the sample reservoir, there is a fibre-glass Whatman GF/C filter (porosity of $1.2 \, \mu m$) to eliminate possible solids from the seawater that could interfere with the analysis. The peristaltic pump pumps water up to the miniaturised SPE cartridge, which consists of a TYGON tube (inert, SC0359)

with a diameter of 4.8 mm filled with the appropriate solid SPE sorbent for each analysis.

2.2. Chemical Reagents. The PAHs studied were fluoranthene, chrysene, benzo(b)fluoranthene, dibenzo(a,h)anthracene (Sigma-Aldrich®), and benzo(a)pyrene (Supelco®). The initial individual standard was dissolved in HPLC-grade acetonitrile (Panreac®).

A mixture of the 6 PAHs was prepared at a concentration of $10 \, \text{mg·L}^{-1}$ in methanol to study the recovery rates (LiChrosolv® Reag. Ph Eur Methanol gradient grade for liquid chromatography, Merck®). The seawater samples to be analysed are enriched with this mixture to validate the m-SPE system. In this case, 1 L of prefiltered seawater (Whatman GF/C glass-fibre filter, porosity of $1.2 \, \mu m$) was contaminated to eliminate any possible solutes that could interfere in the analysis. The concentration of seawater used to optimise the system was $0.2 \, \mu g·L^{-1}$.

2.3. Solid Phase Extraction Procedure (m-SPE). Miniaturising the SPE system is based on manufacturing sorbent cartridges that can be coupled to a peristaltic pump (avoiding the traditional vacuum pump). These cartridges were prepared with 0.3 g of Envi-18 (Supelco) silica gel, placed inside a tube with an internal diameter of 4.8 mm and 6 cm long. There is an IDEX 5 mm ISM560 joint at each end and a piece of polyethylene frit (Supelco) with a porosity of $20 \, \mu m$ on the inside of each joint.

Samples of one litre of seawater with $0.2 \, \mu g·L^{-1}$ of each of the six PAHs analysed were used to optimise extraction. After sampling, 50 mL of Milli-Q water is added through the system and it is left to dry, hence minimising the amount of water present before extraction. The presence of water can trigger a lower recovery and low reproducibility [28]. Finally, the pollutants are disorbed with methanol, and the first mL of extract is collected for analysis by HPLC with fluorescence detector.

2.4. PAH Analysis by High-Pressure Liquid Chromatography (HPLC) with Fluorescence Detector. The analysis of the samples was conducted in Varian® 230, fitted with a ProStar

3012 binary pump, which requires up to three entry lines of solvent and a ProStar Varian 410 self-sampler. The analytes of interest are put through a ProStar 363 fluorescence detector. The valve in the column is a 500-LC, with a Microsorb–MV 100-5 C18 ODS 150 × 4.6 mm × 1/4″ column.

The chromatographic column was kept at 30°C throughout the HPLC process to prevent variability due to environmental conditions. Consideration was given to the excitation and emission wavelengths of each of the PAHs to be analysed [35], and the range of an excitation length of 260 nm and an emission length of 440 nm was determined for the fluorescence detector, which enables the entire spectrum of the different PAHs to be seen.

The work was done on a gradient with a mobile phase A, methanol : water, in a proportion of 80 : 20, and a mobile phase B of 100% methanol. The method was applied on a gradient, lasting 18 minutes. It starts with 100% A, and then progressively increases the proportion of B until this reaches 100% B for 14 minutes. The last 4 minutes are to reestablish the initial conditions, ending with 100% A after 18 minutes.

3. Results and Discussion

3.1. Optimising the Miniaturised Solid Phase Extraction System (m-SPE).

In order to study the best SPE sorbent, extractions were made under different conditions for a reference sample (seawater with 0.2 μg·L^{-1} for the 5 PAHs studied). The main conditions studied were the kind of sorbent used and how much of it, in grams.

3.1.1. Comparison of Different Solid Sorbents.

The different kinds of cartridges to be used in the SPE are classified in accordance with the analytes of interest [17]. In this work, the right adsorbent for studying PAHs in seawater was assessed. Different brands and models of SPE cartridges were used for the filling. The cartridges used were the Supelco Envi-18, Thermo® scientific Hypersep SCX, and the Interchim® Upti-Clean.

The same procedure was used with each kind of cartridge. It was run three times with 1 litre of prefiltered seawater contaminated with 0.2 μg·L^{-1}. The pollutants are disorbed with 1 mL of methanol, and then they are analysed in the HPLC. Two different amounts were used to determine the best sorbent, 0.4 g in the first, and then the two best results were studied with 0.3 g for each kind of sorbent.

Thermo scientific Hypersep SCX is the sorbent that presents the lowest recovery percentage with 0.4 g of sorbent (Figure 2), which is why it was eliminated from the next study, using a smaller amount of sorbent. Figure 3 shows that the recovery percentage for the Supelco Envi-18 sorbent shows better results than the Interchim Upti-Clean.

3.1.2. Comparison of Different Amounts of Sorbent.

Once the kind of sorbent to be used was optimised, the results for the different doses of sorbent (in grams) were analysed. The three different doses of Envi-18 silica gel in the extraction cartridge were compared: 0.3, 0.2, and 0.1 grams. One litre of

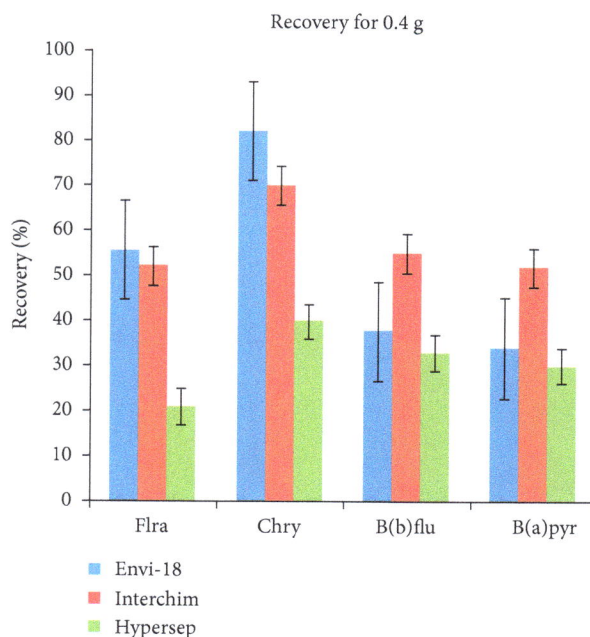

Figure 2: Comparison of different sorbents using a dosage of 0.4 grams: fluoranthene (Flra), chrysene (Chry), benzo(b)fluoranthene (B(b)flu), and benzo(a)pyrene (B(a)pyr).

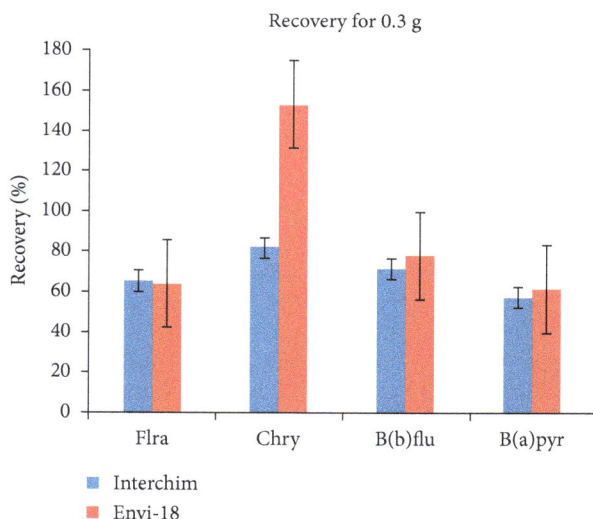

Figure 3: Comparison of the best two sorbents using a dosage of 0.3 grams: fluoranthene (Flra), chrysene (Chry), benzo(b)fluoranthene (B(b)flu), and benzo(a)pyrene (B(a)pyr).

prefiltered seawater was contaminated with 0.2 μg·L^{-1}, and it was run through the extraction system with each of the dosages. The results are shown in Figure 4, showing that the concentrations are higher in the m-SPE cartridge with the highest dose of silica gel.

3.2. Analytical Reproducibility and Application.

The analytical method proposed was assessed under the optimum conditions mentioned above, giving a relative standard

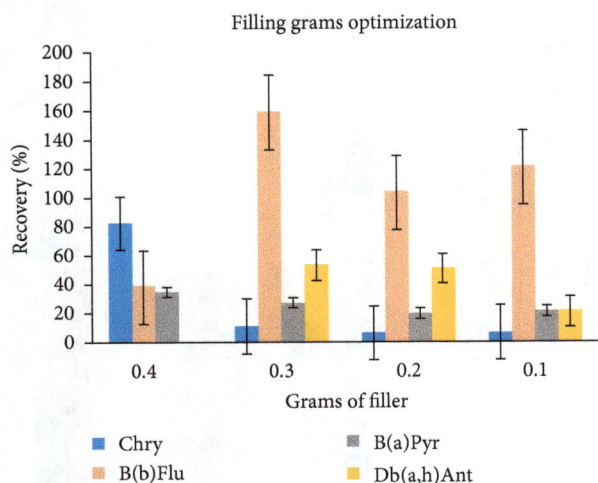

FIGURE 4: Optimising the amount of sorbent (in grams): chrysene (Chry), benzo(b)fluoranthene (B(b)Flu), benzo(a)pyrene (B(a)pyr), and dibenzo(a,h)anthracene (Db(a,h)ant).

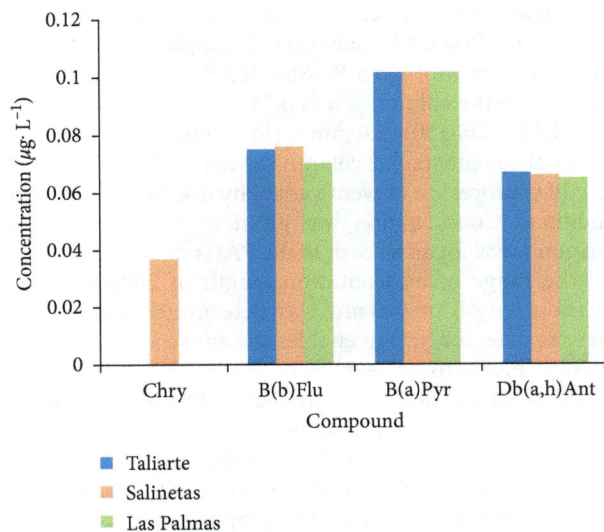

FIGURE 5: Results of applying the method to real, uncontaminated samples at different points of Gran Canaria, Canary Islands, Spain.

TABLE 1: Results of the relative standard deviation (% RSD), limit of detection (LOD), and limit of quantification (LOQ).

Compound	Abbreviation	% RSD	LOD (ng·L^{-1})	LOQ (ng·L^{-1})
Chrysene	Chry	4.21	0.22	0.72
Benzo(b) fluorantene	B(b)Flu	9.57	0.20	0.67
Benzo(a) pyrene	B(a)Pyr	10.27	0.30	1.00
Dibenzo(a,h) anthracene	Db(a,h)ant	9.88	0.02	0.06

deviation (RSD) for an extraction of a sample of 1 litre at 0.2 μg·L^{-1} of the mixture of PAHs.

The RSDs (%) obtained are shown in Table 1 and are around 4.21 and 10.27% for each PAH analysed. The limit of detection study gave very low results, as did the limit of quantification.

The applicability of the method was assessed using real samples in situ (without spike) collected from different places on the island of Gran Canaria, Canary Islands, Spain: two points in the east of the island, (Port of Taliarte and Port of Salinetas) and one further north, where the largest port of the island is located (Port of Las Palmas de Gran Canaria). In each case, 1 litre of surface seawater was collected per sample, with three samples taken at each point.

The volume taken for each sample was measured by collection time based on its flow rate, giving a total of 119 minutes per sample. The results of the real samples analysed are shown in Figure 5. The PAHs with the highest molecular weight are more hydrophobic than the PAHs with low molecular weight [36–38] (the chrysene is the most soluble, and dibenzo(a,h)anthracene is the least soluble), and this is reflected in the concentrations obtained, as they always have a lower presence in water.

The concentrations in the different areas of Gran Canaria varied significantly in several cases below the limit of detection. These figures do not exceed the limits permitted by the legislation in effect [39].

4. Conclusions

The m-SPE in this study was developed to extract PAHs for seawater samples and presented a robust PAH extraction capacity, even for very low concentrations in liquid samples, thus, guaranteeing that the method is able to detect and quantify concentrations below the limits set by law. This means that it is a reliable method for assessing concentrations of directive 2008/105/CE [39].

The reproducibility of the method could be improved by using pollutant separation with gas chromatography with mass spectroscopy (GC-MS) that offers improvements against high-pressure liquid chromatography (HPLC) [2, 9, 21, 40], showing greater sensitivity in the analysis of these pollutants.

These results show the feasibility of the in situ extraction process using miniaturised, solid phase extraction. The methodology developed in this study is simple, fast, easy, and allows for in situ sampling. It is also a sustainable methodology because the use of organic solvents is minimal. It represents the first step towards automating extraction in ports and coastal areas such that monitoring can be conducted more frequently without the need for frequent sampling.

The potential of this system is that it can be fitted to submarine vehicles and oceanic buoys, allowing for continual, efficient, and low-cost monitoring of the quality of the ocean.

Conflicts of Interest

The authors declare that they have no conflicts of interest.

Acknowledgments

This work is based on the MACMAR project (miniaturised embedded analytical chemistry in submarine robotics), C2016/90. This study was funded by the University of Las Palmas de Gran Canaria (ULPGC15_06). B. Abaroa-Pérez is the beneficiary of a research grant INNOVA Canarias 2020 grant from the Fundación Universitaria de Las Palmas, from the 2016 call for proposals. The authors would like to thank the Oceanic Platform of the Canary Islands (Plataforma Oceánica de Canarias, PLOCAN) for their magnificent support and sharing their research facilities with us.

References

[1] S. Frenna, A. Mazzola, S. Orecchio, and N. Tuzzolino, "Comparison of different methods for extraction of polycyclic aromatic hydrocarbons (PAHs) from Sicilian (Italy) coastal area sediments," *Environmental Monitoring and Assessment*, vol. 185, no. 7, pp. 5551–5562, 2013.

[2] Z. Vecra, A. B. Šková, J. Sklenská, and P. Mikuška, "A large volume injection procedure for GC-MS determination of PAHs and PCBs," *Chromatographia*, vol. 61, no. 3-4, pp. 197–200, 2005.

[3] I. Urbe and J. Ruana, "Application of solid-phase extraction discs with a glass fiber matrix to fast determination of polycyclic aromatic hydrocarbons in water," *Journal of Chromatography A*, vol. 778, no. 1-2, pp. 337–345, 1997.

[4] A. Filipkowska, L. Lubecki, and G. Kowalewska, "Polycyclic aromatic hydrocarbon analysis in different matrices of the marine environment," *Analytica Chimica Acta*, vol. 547, no. 2, pp. 243–254, 2005.

[5] J. J. S. Rodríguez and C. P. Sanz, "Fluorescence techniques for the determination of polycyclic aromatic hydrocarbons in marine environment: an overview," *Analusis*, vol. 28, no. 8, pp. 710–717, 2000.

[6] V. Khalili-Fard, K. Ghanemi, Y. Nikpour, and M. Fallah-Mehrjardi, "Application of sulfur microparticles for solid-phase extraction of polycyclic aromatic hydrocarbons from sea water and wastewater samples," *Analytica Chimica Acta*, vol. 714, pp. 89–97, 2012.

[7] E. Martinez, M. Gros, S. Lacorte, and D. Barceló, "Simplified procedures for the analysis of polycyclic aromatic hydrocarbons in water, sediments and mussels," *Journal of Chromatography A*, vol. 1047, no. 2, pp. 181–188, 2004.

[8] Y. Wan, X. Jin, J. Hu, and F. Jin, "Trophic dilution of polycyclic aromatic hydrocarbons (PAHs) in a marine food web from Bohai Bay, North China," *Environmental Science & Technology*, vol. 41, no. 9, pp. 3109–3114, 2007.

[9] J. Ma, R. Xiao, J. Li, J. Yu, Y. Zhang, and L. Chen, "Determination of 16 polycyclic aromatic hydrocarbons in environmental water samples by solid-phase extraction using multi-walled carbon nanotubes as adsorbent coupled with gas chromatography-mass spectrometry," *Journal of Chromatography A*, vol. 1217, no. 34, pp. 5462–5469, 2010.

[10] A. Valavanidis, T. Vlachogianni, S. Triantafillaki, M. Dassenakis, F. Androutsos, and M. Scoullos, "Polycyclic aromatic hydrocarbons in surface seawater and in indigenous mussels (*Mytilus galloprovincialis*) from coastal areas of the Saronikos Gulf (Greece)," *Estuarine, Coastal and Shelf Science*, vol. 79, no. 4, pp. 733–739, 2008.

[11] M. H. Habibi and M. R. Hadjmohammadi, "Determination of some polycyclic aromatic hydrocarbons in the Caspian seawater by HPLC following preconcentration with solid-phase extraction," *Iranian Journal of Chemistry and Chemical Engineering*, vol. 27, no. 4, pp. 91–96, 2008.

[12] S. A. Wise, L. C. Sander, and W. E. May, "Determination of polycyclic aromatic hydrocarbons by liquid chromatography," *Journal of Chromatography A*, vol. 642, no. 1-2, pp. 329–349, 1993.

[13] X. Song, J. Li, S. Xu et al., "Determination of 16 polycyclic aromatic hydrocarbons in seawater using molecularly imprinted solid-phase extraction coupled with gas chromatography-mass spectrometry," *Talanta*, vol. 99, pp. 75–82, 2012.

[14] C. E. Ramirez, C. Wang, and P. R. Gardinali, "Fully automated trace level determination of parent and alkylated PAHs in environmental waters by online SPE-LC-APPI-MS/MS," *Analytical and Bioanalytical Chemistry*, vol. 406, no. 1, pp. 329–344, 2014.

[15] V. Pino, J. H. Ayala, A. M. Afonso, and V. González, "Determination of polycyclic aromatic hydrocarbons in marine sediments by high-performance liquid chromatography after microwave-assisted extraction with micellar media," *Journal of Chromatography A*, vol. 869, no. 1-2, pp. 515–522, 2000.

[16] S. Sanz and J. Antonio, *Evaluación y seguimiento del Contenido en hidrocarburos aromáticos policíclicos (PAHS) en mejillón silvestre de la costa de galicia y cantábrico, antes y después del vertido del B/T Prestige*, vol. 323, Universidade da Coruña, Coruña, Spain, 2009, http://ruc.udc.es/dspace/bitstream/handle/2183/5670/SorianoSanz_JoseAntonio_TD_2009.pdf.

[17] T. M. Hii, C. Basheer, and H. K. Lee, "Commercial polymeric fiber as sorbent for solid-phase microextraction combined with high-performance liquid chromatography for the determination of polycyclic aromatic hydrocarbons in water," *Journal of Chromatography A*, vol. 1216, no. 44, pp. 7520–7526, 2009.

[18] L. Downes and Javier, *Métodos analíticos de extracción y test de bioaccesibilidad de hidrocarburos policíclicos aromáticos (PAHs) en sedimentos de las Islas Baleares*, 2015, http://dspace.uib.es/xmlui/handle/11201/1134.

[19] M. Karimi, F. Aboufazeli, H. R. L. Z. Zhad, O. Sadeghi, and E. Najafi, "Determination of polycyclic aromatic hydrocarbons in Persian gulf and Caspian sea: gold nanoparticles fiber for a head space solid phase micro extraction," *Bulletin of Environmental Contamination and Toxicology*, vol. 90, no. 3, pp. 291–295, 2013.

[20] E. R. Brouwer, A. N. J. Hermans, H. Lingeman, and U. A. T. Brinkman, "Determination of polycyclic aromatic hydrocarbons in surface water by column liquid chromatography with fluorescence detection, using on-line micelle-mediated sample preparation," *Journal of Chromatography A*, vol. 669, no. 1-2, pp. 45–57, 1994.

[21] V. Pino, J. H. Ayala, A. M. Afonso, and V. González, "Determination of polycyclic aromatic hydrocarbons in seawater by high-performance liquid chromatography with fluorescence detection following micelle-mediated preconcentration," *Journal of Chromatography*, vol. 949, no. 1-2, pp. 291–299, 2002.

[22] C. Miège, M. Bouzige, S. Nicol, J. Dugay, V. Pichon, and M. C. Hennion, "Selective immunoclean-up followed by liquid or gas chromatography for the monitoring of polycyclic aromatic hydrocarbons in urban waste water and sewage sludges used for soil amendment," *Journal of Chromatography A*, vol. 859, no. 1, pp. 29–39, 1999.

[23] Y. He and H. Lee, "Trace analysis by combined use of off-line solid-phase extraction, on-column sample focusing and U-shape flow cell in capillary liquid chromatography," *Journal of Chromatography A*, vol. 808, no. 1-2, pp. 79–86, 1998.

[24] A. Barranco, R. Alonso-Salces, A. Bakkali et al., "Solid-phase clean-up in the liquid chromatographic determination of polycyclic aromatic hydrocarbons in edible oils," *Journal of Chromatography A*, vol. 988, no. 1, pp. 33–40, 2003.

[25] J. L. Bernal, M. J. Nozal, L. Toribio et al., "Determination of polycyclic aromatic hydrocarbons in waters by use of supercritical fluid chromatography coupled on-line to solid-phase extraction with disks," *Journal of Chromatography A*, vol. 778, no. 1-2, pp. 321–328, 1997.

[26] G. M. Titato and F. M. Lanças, "Comparison between different extraction (LLE and SPE) and determination (HPLC and capillary-LC) techniques in the analysis of selected PAHs in water samples," *Journal of Liquid Chromatography & Related Technologies*, vol. 28, no. 19, pp. 3045–3056, 2005.

[27] G. Purcaro, S. Moret, and L. S. Conte, "Rapid SPE-HPLC determination of the 16 European priority polycyclic aromatic hydrocarbons in olive oils," *Journal of Separation Science*, vol. 31, no. 22, pp. 3936–3944, 2008.

[28] R. Marcé and F. Borrull, "Solid-phase extraction of polycyclic aromatic compounds," *Journal of Chromatography A*, vol. 885, no. 1-2, pp. 273–290, 2000.

[29] J. L. Santana-romero, M. Valdés-callado, and S. O. L. Lima, *Determinación de hidrocarburos aromáticos policíclicos ligeros en aguas superficiales de los ríos Almendares y Luyanó en La Habana*, vol. 43, 2012.

[30] J. L. Benedé, A. Chisvert, D. L. Giokas, and A. Salvador, "Determination of hydrophobic organic compounds in aqueous media," *Journal of Chromatography A*, vol. 1362, pp. 25–33, 2014.

[31] É. A. Souza-Silva, E. Gionfriddo, and J. Pawliszyn, "A critical review of the state of the art of solid-phase microextraction of complex matrices II. Food analysis," *TrAC Trends in Analytical Chemistry*, vol. 71, pp. 236–248, 2015.

[32] S. Montesdeoca-Esponda, Z. Sosa-Ferrera, and J. J. Santana-Rodríguez, "On-line solid-phase extraction coupled to ultra-performance liquid chromatography with tandem mass spectrometry detection for the determination of benzotriazole UV stabilizers in coastal marine and wastewater samples," *Analytical and Bioanalytical Chemistry*, vol. 403, no. 3, pp. 867–876, 2012.

[33] Y. Han, L. Ren, K. Xu et al., "Supercritical fluid extraction with carbon nanotubes as a solid collection trap for the analysis of polycyclic aromatic hydrocarbons and their derivatives," *Journal of Chromatography A*, vol. 1395, pp. 1–6, 2015.

[34] D. V. Moreno, Z. S. Ferrera, and J. J. S. Rodríguez, "SPME and SPE comparative study for coupling with microwave-assisted micellar extraction in the analysis of organochlorine pesticides residues in seaweed samples," *Microchemical Journal*, vol. 87, no. 2, pp. 139–146, 2007.

[35] Y. Li, S. Yoshida, Y. Chondo et al., "On-line concentration and fluorescence determination HPLC for polycyclic aromatic hydrocarbons in seawater samples and its application to Japan Sea," *Chemical and Pharmaceutical Bulletin*, vol. 60, no. 4, pp. 531–535, 2012.

[36] J. C. Antunes, J. G. L. Frias, A. C. Micaelo, and P. Sobral, "Resin pellets from beaches of the Portuguese coast and adsorbed persistent organic pollutants," *Estuarine, Coastal and Shelf Science*, vol. 130, pp. 62–69, 2013.

[37] C. E. Nika, E. Yiantzi, and E. Psillakis, "Plastic pellets sorptive extraction: low-cost, rapid and efficient extraction of polycyclic aromatic hydrocarbons from environmental waters," *Analytica Chimica Acta*, vol. 922, pp. 30–36, 2016.

[38] T. S. Bianchi, S. Findlay, and R. Dawson, "Organic matter sources in the water column and sediments of the Hudson River: use of total pigments as tracers," *Estuarine, Coastal and Shelf Science*, vol. 36, no. 4, pp. 359–376, 1993.

[39] U. Europea, "Relativa a las normas de calidad ambiental en el ámbito de la política de aguas," *Official Journal of the European Union*, vol. 348, pp. 84–97, 2008.

[40] K. Liapis and E. Bempelou, "Determination of polycyclic aromatic hydrocarbons in water by GC/MS/MS," *Hellenic Plant Protection Journal*, vol. 64, pp. 99–105, 2008.

23

Electrochemical Solvent Cointercalation into Graphite in Propylene Carbonate-Based Electrolytes: A Chronopotentiometric Characterization

Hee-Youb Song and Soon-Ki Jeongⓘ

Department of Chemical Engineering, Soonchunhyang University, Asan, Chungnam 336-745, Republic of Korea

Correspondence should be addressed to Soon-Ki Jeong; hamin611@sch.ac.kr

Academic Editor: Larisa Lvova

Interfacial reactions strongly influence the performance of lithium-ion batteries, with the main interfacial reaction between graphite and propylene carbonate- (PC-) based electrolytes corresponding to solvent cointercalation. Herein, the redox reactions of solvated lithium ions occurring at the graphite interface in 1 M·LiClO$_4$/PC were probed by chronopotentiometry, in situ atomic force microscopy (AFM), and in situ Raman spectroscopy. The obtained results revealed that high coulombic efficiency (97.5%) can be achieved at high current density, additionally showing the strong influence of charge capacity on the above redox reactions. Moreover, AFM imaging indicated the occurrence of solvent cointercalation during the first reduction, as reflected by the presence of hills and blisters on the basal plane of highly oriented pyrolytic graphite subjected to the above process.

1. Introduction

Graphite is extensively used for the fabrication of lithium-ion batteries (LIBs) [1–4] due to being capable of electrochemically intercalating lithium ions into graphene layers and thus enabling electrical energy storage during charging. However, the above intercalation reaction is strongly affected by the nature of the utilized electrolytes, which comprise lithium ions, counter anions, and organic solvents [5]. In electrolyte solutions, the interaction between lithium ions and organic solvents leads to the solvation of the former [6–8], which is undesired and should be inhibited due to solvated lithium ions being intercalated into graphite at more positive potentials than nonsolvated ones [9–12]. In addition, the continuous cointercalation of coordinated solvent leads to graphite exfoliation.

Despite its remarkable ionic conductivity at low temperature (melting point = −49°C), propylene carbonate (PC), widely used as a main organic solvent in primary lithium batteries, has not been applied to LIBs due to undergoing ceaseless cointercalation into graphite during the first

charging [13–16]. PC cointercalation is well known to occur at ~1 V versus Li$^+$/Li, hindering the lowering of the electrode potential to values corresponding to lithium intercalation (0.25–0.0 V versus Li$^+$/Li) and thus inducing graphite exfoliation prior to lithium intercalation during the first charging. Thus, solvent cointercalation significantly degrades the performance of LIBs and should be suppressed to allow reversible intercalation/deintercalation of lithium ions at graphite-negative electrodes.

The effective suppression of solvent cointercalation at graphite-negative electrode requires a deep understanding of numerous factors of influence. Commonly, the irreversible reduction of solvated lithium ions at ~1 V is the main interfacial reaction in PC-based electrolyte solutions, indicating the importance of understanding the redox behavior of PC-solvated lithium ions. Herein, we probed the above behavior by chronopotentiometry, characterizing the potential change upon application of negative/positive currents and investigated the effects of current density on the redox reactions of solvated lithium ions. Moreover, in situ atomic force microscopy (AFM) and in situ Raman spectroscopy

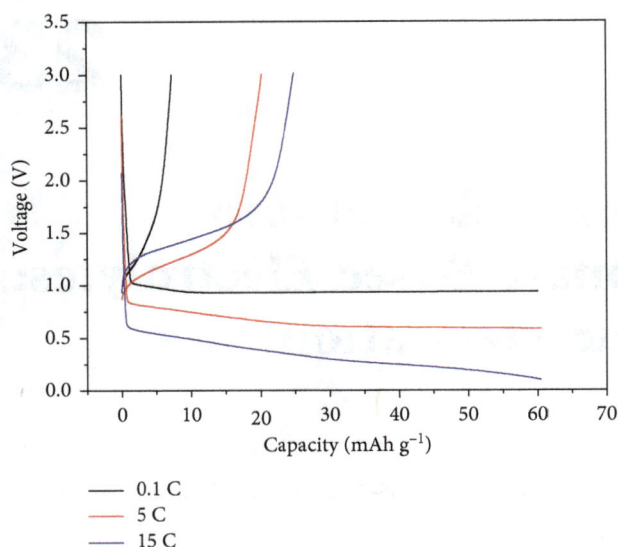

FIGURE 1: Voltage profiles of NG-7 at various C-rates in $1\,M\cdot LiClO_4/PC$, with the cell charged to $60\,mAh\cdot g^{-1}$ at each cycle.

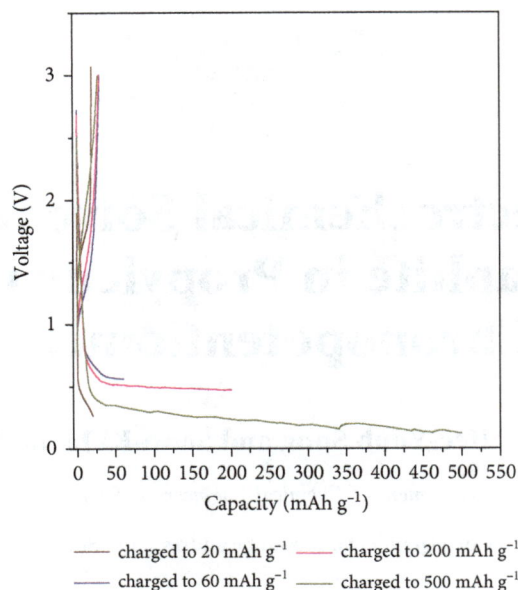

FIGURE 2: Voltage profiles of NG-7 at various charge capacities and a C-rate of 5 C in $1\,M\cdot LiClO_4/PC$.

were used to clarify certain aspects of solvent cointercalation during the first reduction reaction.

2. Materials and Methods

2.1. Preparation of Electrode Materials and Electrolyte Solution.
Natural graphite powder (NG-7, Kansai Coke, and Chemicals Co.) was used as an active electrode material for chronopotentiometry. A composite working electrode was prepared by coating copper foil (Nilaco Co.) with a 9 : 1 (w/w) mixture of NG-7 and poly(vinylidene difluoride) and drying it at 80°C in a vacuum oven (Yamato Scientific Co., DNE401) for 12 h. Highly oriented pyrolytic graphite (HOPG; Advanced Ceramics, ZYH grade, mosaic spread = 3.5 ± 1.5°) was used as a model electrode for in situ AFM imaging, which was carried out for freshly cleaved HOPG surfaces. Lithium foil (Honjo Metal Co.) was used as reference and counter electrodes in all electrochemical measurements, and a 1 M solution of $LiClO_4$ in PC (Kishida Chemical Co., battery grade) was used as an electrolyte.

2.2. Chronopotentiometry.
Chronopotentiometric measurements were performed using a battery test system (Hokuto Denko, HJ101SM6), with 2032 coin cells tested at various C-rates ($1\,C = 372\,mA\cdot g^{-1}$) to understand the effect of current density on the redox behavior of solvated lithium ions.

2.3. In Situ AFM.
The basal plane of HOPG was imaged in contact mode using a pyramidal silicon nitride tip (OLYMPUS Co., OMCL-TR800PSA) in $1\,M\cdot LiClO_4/PC$. AFM images ($5\,\mu m \times 5\,\mu m$) were automatically captured at each potential during cyclic voltammetry (CV) scans between 3.0 and 0.0 V (scan rate = $2\,mV\cdot s^{-1}$) using an AFM imaging system (Molecular Imaging, PicoSPM) equipped with a potentiostat (Molecular Imaging, PicoStat). All AFM characterizations

were performed at room temperature in an argon-filled glove box (Miwa, MDB-1B + MM3-P60S, dew point < −70°C).

2.4. In Situ Raman Spectroscopy.
An electrochemical (quartz) cell for in situ Raman spectroscopy was assembled in an argon-filled glove box, sealed, and removed from the glove box into ambient atmosphere. The 514.5 nm line of an argon-ion laser was scattered on HOPG during the first reduction by applying a constant current of 1 C. Raman spectra were collected using a triple monochromator (Jobin-Yvon, T64000) equipped with a multichannel charge-coupled device detector.

3. Results and Discussion

3.1. Effect of Current Density on Solvent Cointercalation.
Figure 1 shows voltage profiles recorded at various C-rates in $1\,M\cdot LiClO_4/PC$. During these measurements, the graphite-negative electrode was charged to $60\,mAh\cdot g^{-1}$ and instantly discharged without any rest time, providing insights into the redox behavior of solvated lithium ions above the lithium intercalation potential, with a different discharge capacity observed in each cycle reflecting the oxidation of solvated lithium ions. In general, solvent cointercalation (corresponding to the reduction of solvated lithium ions) takes place at ~1 V in PC-based electrolytes [13–16]. However, in this case, the solvent cointercalation potential decreased, and the discharge capacity increased as the current density increased from 0.1 to 5 or 15 C, indicating the occurrence of reversible redox reactions of solvated lithium ions at graphite and additionally showing that the oxidation of solvated lithium ions can be controlled by choosing an appropriate current density.

For further experiments, we selected a C-rate of 5 C and investigated the discharge capacity of fabricated cells to

(a)

(b)

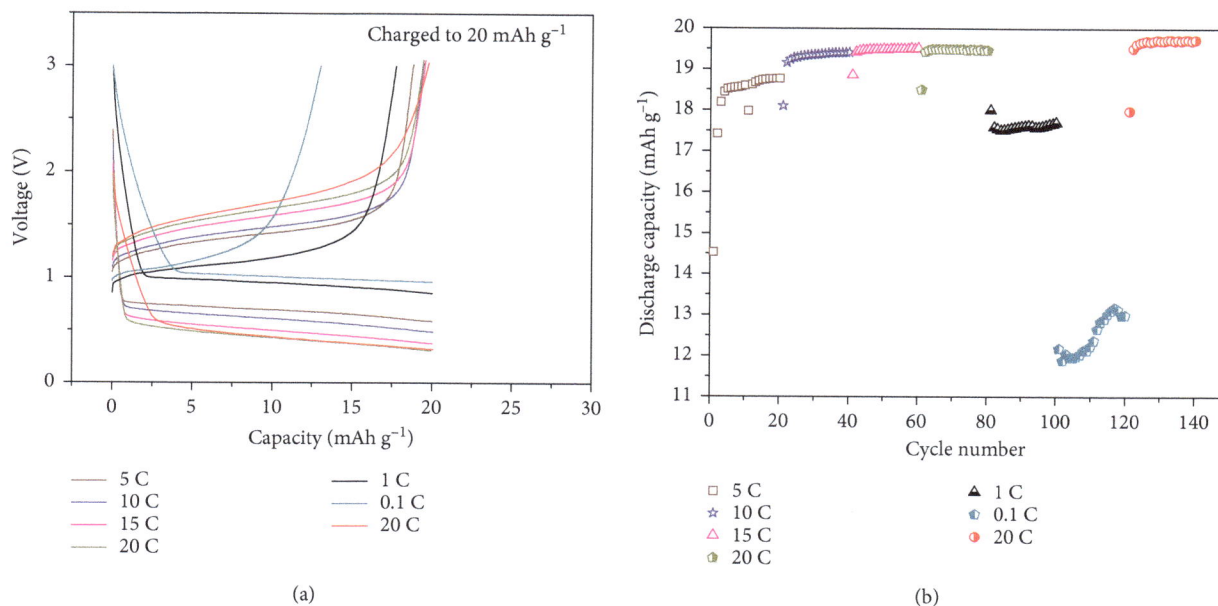

FIGURE 3: (a) Voltage profiles and (b) cycling performance of NG-7 at various C-rates in $1 M \cdot LiClO_4/PC$, with the cell charged to $20 \, mAh \cdot g^{-1}$.

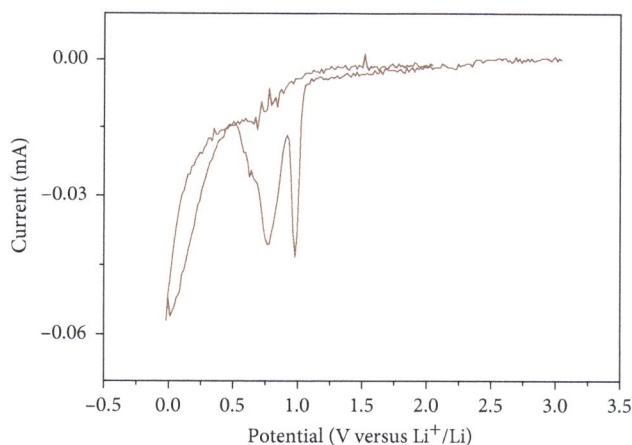

FIGURE 4: Cyclic voltammogram of HOPG during the first cycle in $1 M \cdot LiClO_4/PC$ recorded at a scan rate of $2 \, mV \cdot s^{-1}$.

understand the effect of charge capacity on the redox reactions of solvated lithium ions. Notably, no significant increase of discharge capacity ($\sim 20 \, mAh \cdot g^{-1}$) was observed as the charge capacity increased from 20 to $500 \, mAh \cdot g^{-1}$ (Figure 2). Moreover, coulombic efficiency increased with decreasing charge capacity, indicating that the redox reactions of solvated lithium ions could be controlled by varying both current density and charge capacity.

Figure 3(a) shows charge and discharge capacities obtained for a cell charged to $20 \, mAh \cdot g^{-1}$ above the lithium intercalation potential at various current densities. The corresponding coulombic efficiency was estimated as 65% at 0.1 C, increasing to ≥97.5% at 10, 15, and 20 C and thus indicating that most solvated lithium ions cointercalated into graphite were oxidized at the above high current densities

during discharge. Furthermore, the discharge capacity was sustained after 140 cycles (Figure 3(b)). Thus, the redox reactions of solvated lithium ions at graphite in PC-based electrolytes were demonstrated to be dependent on current density and charge capacity.

3.2. In Situ AFM and Raman Investigation of HOPG during the First Reduction.

Morphological changes of the HOPG basal plane were investigated by in situ AFM to obtain further insights into solvent cointercalation in PC-based electrolytes. Figure 4 shows a cyclic voltammogram of HOPG in $1 M \cdot LiClO_4/PC$, revealing the presence of three reduction peaks at potentials below 1.0 V during the first cycle. Prior to reduction, HOPG exhibited a flat surface comprising basal and edge planes (Figure 5(a)), with significant morphological changes observed after reduction; that is, hill-like structures were detected on the HOPG basal plane at ~ 1 V (Figure 5(b)). Similarly, Inaba et al. observed related structures on the surface of graphite after cointercalation of solvated lithium ions, ascribing the formation of blisters to the decomposition of solvated lithium ions within graphite [11, 13]. Accordingly, the reduction peak at ~ 1 V was attributed to solvent cointercalation into graphite, with blistering at decreased cell potentials (Figure 5(c)) explained as mentioned above. In addition, the height of the HOPG basal plane increased in a potential range of 0.49–0.0 V (Figure 5(d)).

We also employed in situ Raman spectroscopy to clarify the nature of the intercalate formed during the first reduction of graphite in constant-current mode in a PC-based electrolyte. In this case, the electrode potential did not drop to the lithium intercalation potential while the graphite electrode was charged to $20 \, mAh \cdot g^{-1}$ in the same manner as shown in Figure 3(a). However, both E_{2g2} (interior) and E_{2g2}(boundary) bands at 1583 and 1597 cm^{-1}, respectively,

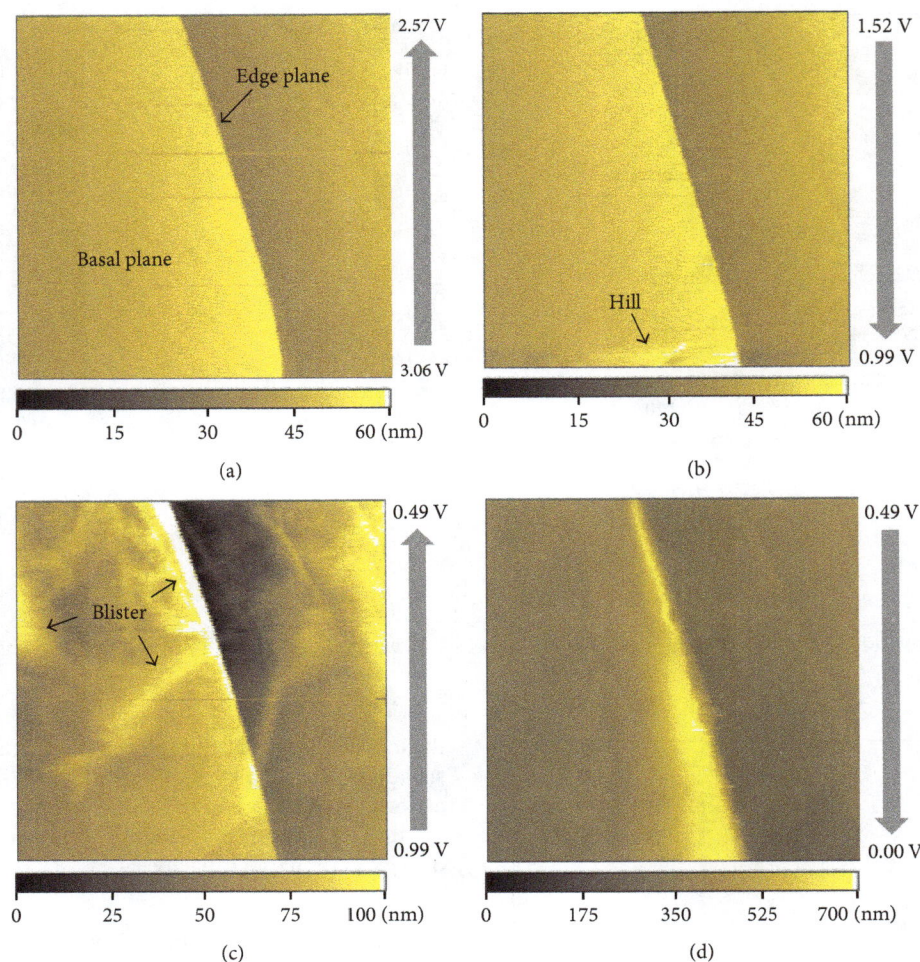

FIGURE 5: In situ AFM images of the HOPG basal plane ($5\,\mu m \times 5\,\mu m$) obtained at (a) 3.06–2.57, (b) 1.52–0.99, (c) 0.99–0.49, and (d) 0.49–0.0 V during the first reduction in $1\,M \cdot LiClO_4/PC$.

were observed during the first reduction reaction, indicating the absence and presence of intercalates within graphite layers, respectively (Figure 6). Moreover, the first reduction increased the intensity of the latter band and decreased that of the former, indicating that continuous cointercalation of PC-solvated lithium ions into graphite occurred at a potential higher than that of lithium intercalation. Based on the insights provided by in situ AFM and in situ Raman analysis, the results of chronopotentiometric characterization were attributed to the redox reactions of solvated lithium ions in the PC-based electrolyte.

4. Conclusions

Herein, we investigated the redox reactions of solvated lithium ions at graphite in $1\,M \cdot LiClO_4/PC$ by chronopotentiometry, in situ AFM, and in situ Raman spectroscopy, revealing that these reactions were reversible at high current density and thus highlighting the key role of the latter parameter. Moreover, the discharge capacity related to the oxidation of solvated lithium ions increased with increasing current density, and improved coulombic

efficiency was observed at high current densities of 10, 15, and 20 C in the case of charging to $20\,mAh \cdot g^{-1}$. AFM imaging revealed the appearance of hills and blisters on the otherwise smooth electrode surface during the first reduction, indicating the occurrence of solvent cointercalation and decomposition at graphite. Thus, the redox reactions of solvated lithium ions were shown to be the main interfacial reactions taking place at the graphite-negative electrode, with their control being possible by appropriate variation of current density and charge capacity.

Conflicts of Interest

The authors declare that they have no conflicts of interest.

Acknowledgments

This research was supported by the Basic Science Research Program through the National Research Foundation of Korea (NRF) funded by the Ministry of Education (No. NRF-2017R1A2B4010544). This work was supported by the Soonchunhyang University Research Fund.

$E_{2g2}(i)$ $E_{2g2}(b)$

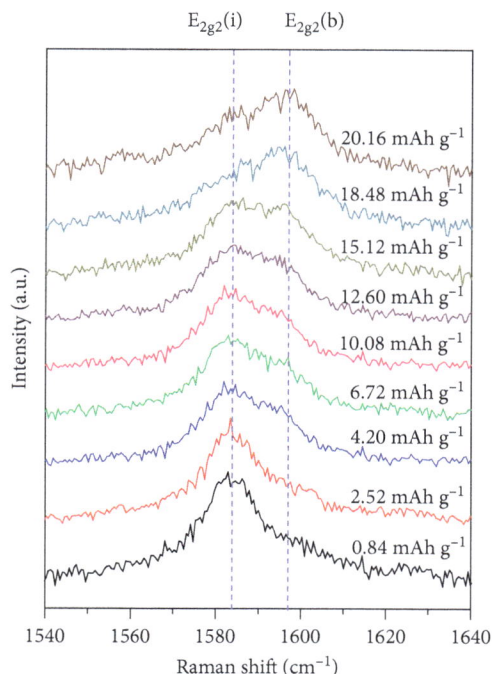

FIGURE 6: In situ Raman spectra of HOPG recorded during the first reduction in constant-current mode (1 C) in $1\,M\cdot LiClO_4/PC$. The electrochemical cell was charged to ~20 mAh·g^{-1}.

References

[1] M. S. Dresselhaus and G. Dresselhaus, "Intercalation compounds of graphite," *Advances in Physics*, vol. 30, no. 2, pp. 139–326, 1981.

[2] Z. Ogumi and M. Inaba, "Electrochemical lithium intercalation within carbonaceous materials: intercalation processes, surface film formation, and lithium diffusion," *Bulletin of the Chemical Society of Japan*, vol. 71, no. 3, pp. 521–534, 1998.

[3] M. Inaba, H. Yoshida, Z. Ogumi, T. Abe, Y. Mizutani, and M. Asano, "In situ Raman study on electrochemical Li intercalation into graphite," *Journal of the Electrochemical Society*, vol. 142, no. 1, pp. 20–26, 1995.

[4] D. Aurbach and Y. Ein-Eli, "The study of Li-graphite intercalation processes in several electrolyte systems using in situ X-ray diffraction," *Journal of the Electrochemical Society*, vol. 142, no. 6, pp. 1746–1752, 1995.

[5] K. Xu, "Nonaqueous liquid electrolytes for lithium-based rechargeable batteries," *Chemical Reviews*, vol. 104, no. 10, pp. 4303–4418, 2004.

[6] S.-K. Jeong, M. Inaba, Y. Iriyama, T. Abe, and Z. Ogumi, "Electrochemical intercalation of lithium ion within graphite from propylene carbonate solutions," *Electrochemical and Solid-State Letters*, vol. 6, no. 1, pp. A13–A15, 2003.

[7] Y. Yamada, Y. Koyama, T. Abe, and Z. Ogumi, "Correlation between charge-discharge behavior of graphite and solvation structure of the lithium ion in propylene carbonate-containing electrolytes," *The Journal of Physical Chemistry C*, vol. 13, no. 20, pp. 8948–8953, 2009.

[8] X. Bogle, R. Vazquez, S. Greenbaum, A. W. Cresce, and K. Xu, "Understanding Li$^+$-solvent interaction in nonaqueous carbonate electrolytes with ^{17}O NMR," *The Journal of Physical Chemistry Letters*, vol. 4, no. 10, pp. 1664–1668, 2013.

[9] J. O. Besenhard, M. Winter, J. Yang, and W. Biberacher, "Filming mechanism of lithium-carbon anodes in organic and inorganic electrolytes," *Journal of Power Sources*, vol. 54, no. 2, pp. 228–231, 1995.

[10] T. Abe, N. Kawabata, Y. Mizutani, M. Inaba, and Z. Ogumi, "Correlation between cointercalation of solvents and electrochemical intercalation of lithium into graphite in propylene carbonate solution," *Journal of The Electrochemical Society*, vol. 150, no. 3, pp. A257–A261, 2003.

[11] M. Inaba, Z. Siroma, A. Funabiki, and Z. Ogumi, "Electrochemical scanning tunneling microscopy observation of highly oriented pyrolytic graphite surface reactions in an ethylene carbonate-based electrolyte solution," *Langmuir*, vol. 12, no. 6, pp. 1535–1540, 1996.

[12] H.-Y. Song, T. Fukutsuka, K. Miyazaki, and T. Abe, "Suppression of co-intercalation reaction of propylene carbonate and lithium ion into graphite negative electrode by addition of diglyme," *Journal of the Electrochemical Society*, vol. 163, no. 7, pp. A1265–A1269, 2016.

[13] M. Inaba, Z. Siroma, Y. Kawatate, A. Funabiki, and Z. Ogumi, "Electrochemical scanning tunneling microscopy analysis of the surface reactions on graphite basal plane in ethylene carbonate-based solvents and propylene carbonate," *Journal of Power Sources*, vol. 68, no. 2, pp. 221–226, 1997.

[14] D. Aurbach, B. Markovsky, I. Weissman, E. Levi, and Y. Ein-Eli, "On the correlation between surface chemistry and performance of graphite negative electrodes for li ion batteries," *Electrochimica Acta*, vol. 45, no. 1-2, pp. 67–86, 1999.

[15] D. Aurbach, M. Koltypin, and H. Teller, "In situ AFM imaging surface phenomena on composite graphite electrodes during lithium insertion," *Langmuir*, vol. 18, no. 23, pp. 9000–9009, 2002.

[16] K. Xu, "Whether EC and PC differ in interphasial chemistry on graphite anode and how," *Journal of The Electrochemical Society*, vol. 156, no. 9, pp. A751–A755, 2009.

24

Sensitive Fluorescent Sensor for Recognition of HIV-1 dsDNA by using Glucose Oxidase and Triplex DNA

Yubin Li ⓘ,[1] Sheng Liu,[1] and Liansheng Ling ⓘ [2]

[1]College of Chemistry and Environment, Guangdong Ocean University, Zhanjiang 524088, China
[2]School of Chemistry, Sun Yat-Sen University, Guangzhou 510275, China

Correspondence should be addressed to Yubin Li; 2007liyubin@163.com and Liansheng Ling; cesllsh@mail.sysu.edu.cn

Academic Editor: Christophe A. Marquette

A sensitive fluorescent sensor for sequence-specific recognition of double-stranded DNA (dsDNA) was developed on the surface of silver-coated glass slide (SCGS). Oligonucleotide-1 (Oligo-1) was designed to assemble on the surface of SCGS and act as capture DNA, and oligonucleotide-2 (Oligo-2) was designed as signal DNA. Upon addition of target HIV-1 dsDNA (Oligo-3•Oligo-4), signal DNA could bind on the surface of silver-coated glass because of the formation of C•GoC in parallel triplex DNA structure. Biotin-labeled glucose oxidase (biotin-GOx) could bind to signal DNA through the specific interaction of biotin-streptavidin, thereby GOx was attached to the surface of SCGS, which was dependent on the concentration of target HIV-1 dsDNA. GOx could catalyze the oxidation of glucose and yield H_2O_2, and the HPPA can be oxidized into a fluorescent product in the presence of HRP. Therefore, the concentration of target HIV-1 dsDNA could be estimated with fluorescence intensity. Under the optimum conditions, the fluorescence intensity was proportional to the concentration of target HIV-1 dsDNA over the range of 10 pM to 1000 pM, the detection limit was 3 pM. Moreover, the sensor had good sequence selectivity and practicability and might be applied for the diagnosis of HIV disease in the future.

1. Introduction

Double-stranded structure, reported in 1953 [1], is the natural state of DNA. Meanwhile, double-stranded DNA (dsDNA) detection is of particular importance in gene therapy, diagnosis, and monitoring fatal infections caused by viruses and diseases that are associated with genetic alterations [2–8]. Routine protocols for sequence-specific recognition of dsDNA, for instance, are performed by using zinc finger DNA-binding proteins [9, 10], polyamides [11, 12], and triplex-forming oligonucleotides [13–17]. Triplex-forming oligonucleotide-based methods commonly require the protonation of cytosine for the formation of C•GoC ("•" denotes the Watson–Crick bond, and "o" denotes the Hoogsteen bond) in parallel triplex DNA structure. This has the performance of binding the major groove of dsDNA and exerting high sequence specificity, which have been extensively used for the analysis of dsDNA [18].

Enzymes are highly applicable in biosensors as recognition and signaling elements for the detection of specific molecules due to the features such as high sensitivity and good selectivity [19]. Given these favorable characteristics, glucose oxidase (GOx) not only can catalyze the oxidation of glucose but also is one of the cheapest and most stable redox enzymes. Additionally, on one hand, GOx has been used in constructing electrochemical [20–23], fluorescence [24–26], and colorimetric [27, 28] sensors for glucose. On the other hand, GOx is conjugated for recognition of biomolecules and acted as an amplifying label, which is successfully applied to establish sensors for proteins [29, 30] and DNA [31].

Herein, we explore the possibility to develop a fluorescent sensor for sensitive and sequence-specific detection of target dsDNA. The sequence of target dsDNA is from site 7960 to site 7991 of longest homopurine-homopyrimidine duplex strand in the human immunodeficiency virus 1 (HIV-1) dsDNA gene [32, 33]. The enzyme immunoassay (EIA) test has been authorized by Food and Drug Administration to recognize HIV by measuring humans' antibody response. However, there is a 25-day infectious window period for HIV EIAs [34]. The proposed method is expected to shorten this period. Besides, there are a lot of sensitive methods to detect HIV-1 RNA [35, 36], yet HIV-1 RNA may be cleaved during the process of sample preparation, which limits their application in real samples. HIV-1 RNA is integrated into the host gene forming double-stranded DNA in 3 days. Thus, the identification of HIV-1 dsDNA may play a significant role here. PCR-based tests for the detection of HIV-1 DNA are sensitive and specific, but their application in resource-limited areas is hindered due to the time consumingness of nucleic acid purification and the requirement of skilled processing and costly reagents and equipment [37]. Therefore, it is urgent to develop a new method to replace the PCR-based tests. This protocol takes advantage of the amplification property of GOx, the capture DNA is assembled on the surface of SCGS, and signal DNA is designed to conjugate with GOx. Upon addition of target HIV-1 dsDNA, GOx could be immobilized on the surface of SCGS. Thereby, the concentration of target HIV-1 dsDNA controlled the number of bound GOx, which could be detected with the fluorescence of oxidized HPPA. These methods have been applied to develop sensors for sequence-specific recognition of dsDNA based upon triplex formation [13–17]. Compared with the above methods, the proposed method is simpler, convenient, and time-saving.

2. Experimental

2.1. Materials, Chemicals, and Instrumentation.
Tri-(2-carboxyethyl) phosphine (TCEP) and bovine serum albumin (BSA) were purchased from Sigma-Aldrich (USA). Amicon filtration device and EZ-link sulfo-NHS-biotinylation kit were purchased by Thermo Fisher Scientific Inc. (USA). Horseradish peroxidase (HRP), streptavidin (SA), glucose oxidase (GOx), 3-(p-hydroxyphenyl)-propanoic acid (HPPA), DNase I, ammonia (25%), silver nitrate, and glucose were purchased from Sinopharm Chemical Reagent Co., Ltd. (Beijing, China). All oligonucleotides (Table 1) were synthesized and purified by Sangon Bioengineering Technology and Services Co., Ltd. (Shanghai, China). They were dissolved in PBS buffer. All chemicals were of analytical reagent grade. All solutions were prepared with ultrapure water.

PBS buffer (pH 6.0; 100 mM Na_2HPO_4, 100 mM NaH_2PO_4, and 100 mM $NaNO_3$) and Tris-HAc buffer (10 mM, pH 7.5; 2.5 mM $Mg(NO_3)_2$ and 0.5 mM $Ca(NO_3)_2$) were prepared for research.

The pH values of the solutions were measured by using a pHS-3E digital pH meter (Shanghai Leici Instrument Plant, China). The implementation of fluorescence measurements was with the help of RF-5301PC spectrofluorometer (Shimadzu, Japan), and the slit width was 5.0 nm. In addition, the excitation spectrum was set at 320 nm, while the emission spectra were collected from 380 nm to 450 nm. The circular dichroism spectra (CD spectra) were measured by using J-810 circular dichroism spectrum (Shimadzu, Japan), the range of emission wavelength were from 200 nm to 300 nm, the scanning speed was 100 nm·min^{-1}, the response time was 1 second, and its bandwidth was 1.71 nm. In addition, for the continuous scanning mode, the spectral scanning number was 3.

TABLE 1: Sequences of oligonucleotides.

Probes	Name	Sequence
Oligo-1	Capture DNA	5′-SH-(T)$_{12}$ CTT CCT TAT CTT CTT C-3′
Oligo-2	Signal DNA	5′-TTC CAC CTC TCT CTC T (T)$_{12}$-biotin-3′
Oligo-3	Target dsDNA	5′-TCT CTC TCT CCA CCT TCT TCT TCT ATT CCT TC-3′
Oligo-4		5′-GAA GGA ATA GAA GAA GAA GGT GGA GAG AGA GA-3′
Oligo-5	M-1 dsDNA	5′-TCT CTC TCT ACA CCT TCT TCT TCT ATT CCT TC-3′
Oligo-6		5′-GAA GGA ATA GAA GAA GAA GGT GTA GAG AGA GA-3′
Oligo-7	M-2 dsDNA	5′-TCT CTC TCT ACA CCT TCT TCT TCC ATT CCT TC-3′
Oligo-8		5′-GAA GGA ATG GAA GAA GAA GGT GTA GAG AGA GA-3′
Oligo-9	M-3 dsDNA	5′-TCT CCC TCT ACA CCT TCT TCT TCT ATT CCT TC-3′
Oligo-10		5′-GAA GGA ATA GAA GAA GAA GGT GTA GAG GGA GA-3′

2.2. Preparation of SCGS.
According to Li et al. [38], the SCGSs were prepared by the traditional silver mirror reaction.

2.3. Preparation of Biotin-Modified Glucose Oxidase.
Biotin can be connected to glucose oxidase (GOx) by the cross-linking agent sulfo-NHS-biotin, and the biotin-labeled glucose oxidase (biotin-GOx) was prepared as in [38].

2.4. Immobilization of Oligo-1 on the SCGS.
According to the reported methods, sulfhydryl capture DNA (SH-Oligo-1) can be immobilized on the SCGS [39, 40]. The SCGS was placed in PBS buffer which contains 1.4 µM Oligo-1. In order to eliminate free Oligo-1, the modified silver-coated glass was washed twice by using the same buffer.

2.5. Fabrication of the Sensor.
Firstly, the capture DNA-modified SCGS was submerged into different concentrations of target HIV-1 dsDNA (Oligo-3•Oligo-4) solution for 1 hour. Then, the slide was immersed into 50 nM signal DNA solution for another 1 hour. After washing twice by using PBS buffer, the slide was immersed in 3% BSA solution for 20 minutes to block possible remaining active sites. Then, the slide was dripped into 800 ng·mL^{-1} streptavidin solution and 50 µg·mL^{-1} biotin-GOx solution, respectively, for 10 minutes. The procedures mentioned above were carried out in PBS buffer at room temperature. After washing twice by using PBS buffer, the slide was immersed into Tris-HAc buffer containing 50 U DNase I for 1 hour at 37°C, and the slide was taken out. Then, 50 mM glucose was injected into the mixture for 2 hours at 37°C to yield H_2O_2. After that, 200 µM HPPA and 20 ng·mL^{-1} HRP were added into the abovementioned solution at the same time in darkness for

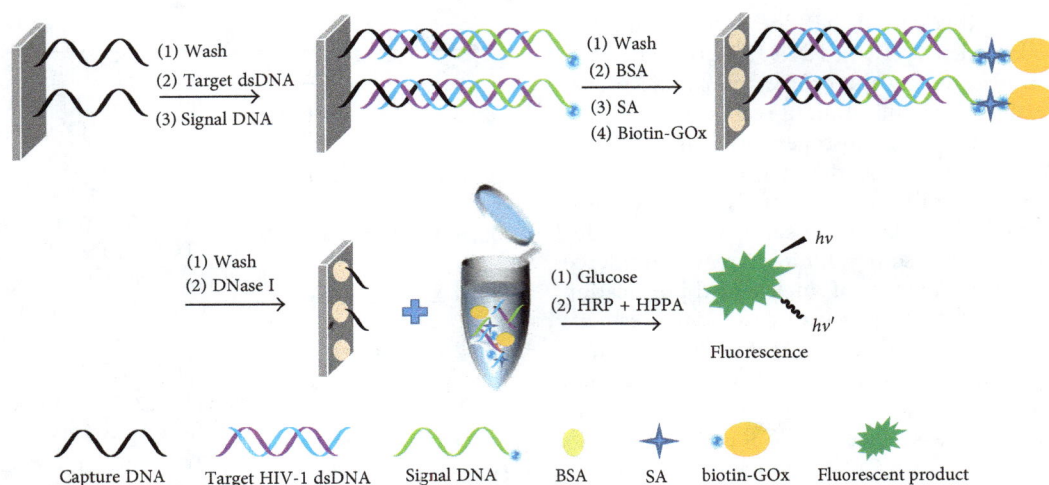

FIGURE 1: The scheme for fluorescent sensor for sequence-specific recognition of target dsDNA by using capture DNA and amplification by using glucose oxidase.

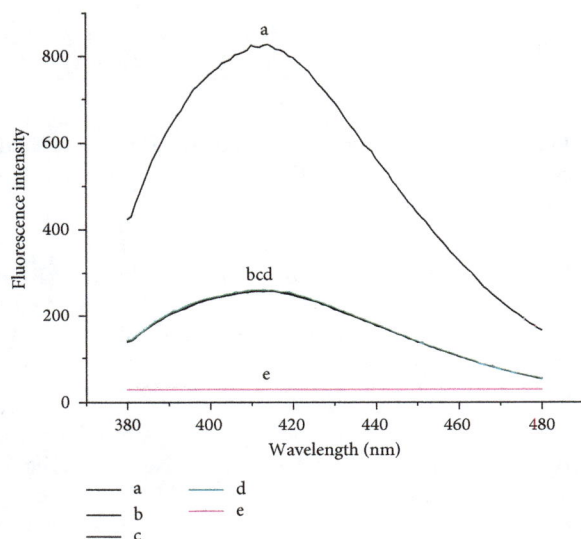

FIGURE 2: Fluorescence spectra of different mixtures under the same process. (a) Capture DNA + signal DNA + biotin-GOx + target HIV-1 dsDNA; (b) capture DNA + signal DNA + biotin-GOx; (c) target HIV-1 dsDNA + signal DNA + biotin-GOx; (d) capture DNA + target HIV-1 dsDNA + biotin-GOx; (e) capture DNA + target HIV-1 dsDNA + signal DNA.

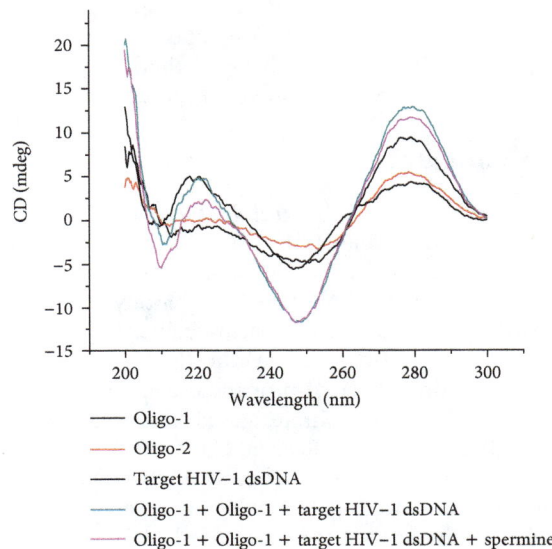

FIGURE 3: Circular dichroism (CD) spectroscopy of the sensor.

30 minutes at 37°C. Eventually, fluorescent spectra of the oxidative products of HPPA were recorded by using a RF-5301PC spectrofluorometer [41].

2.6. Preparation of Human Sera Samples.

Human sera was obtained from Guangdong Ocean University Campus Hospital (Zhanjiang, China), before the test with calf thymus DNA interference and recovery experiment in human sera, and each sample was dealt with the Amicon filtration device 10,000 to remove small molecules. And the calf thymus DNA was digested into smaller fragments before using.

3. Results and Discussion

3.1. Design of the Sensor. The scheme of the sensor is depicted in Figure 1. Capture DNA is for the purpose of aggregating on the surface of SCGS by use of Ag–S bond. Signal DNA can bind on the surface of SCGS due to the addition of target HIV-1 dsDNA. GOx can also bind to signal DNA by streptavidin-biotin bond after addition of streptavidin and biotin-GOx. And the concentration of GOx immobilizing on the surface of SCGS is dependent on that of target HIV-1 dsDNA. In order to avoid the nonspecific adsorption of streptavidin and GOx, DNase I is used to cleave the DNA strand from the surface of SCGS, and the bound glucose oxidase is transferred into the buffer. Thus, the concentration of target dsDNA is transduced into the concentration of H_2O_2 which is the oxidative product of glucose in the presence of GOx. Then, HPPA can be oxidized into the

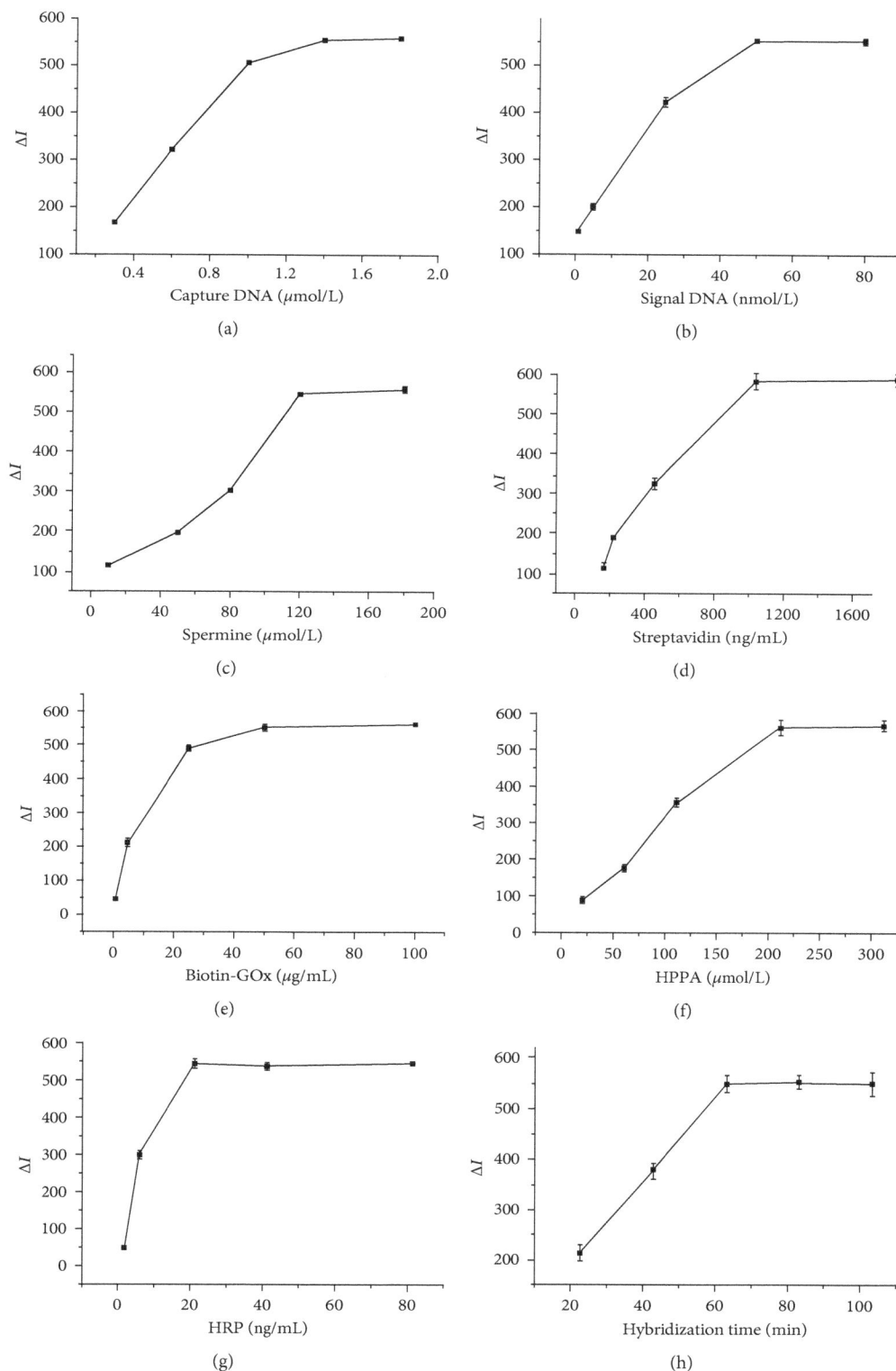

FIGURE 4: (a) Effects of capture DNA concentrations; (b) effects of signal DNA concentrations; (c) effects of spermine concentrations; (d) effects of streptavidin concentrations; (e) effects of biotin-GOx concentrations; (f) effects of HPPA concentrations; (g) effects of HRP concentrations; (h) effects of time of hybridization. When each of these factors is optimized, the other factors maintain optimal conditions (1.4 μM capture DNA, 50 nM signal DNA, 120 μM spermine, 800 ng·mL^{-1} SA, 50 μg·mL^{-1} biotin-GOx, 200 μM HPPA, 20 ng·mL^{-1} HRP, and 60 min of time of hybridization).

fluorescent product by H_2O_2 under the catalysis of HRP. Finally, the concentration of target HIV-1 dsDNA was estimated with the fluorescence intensity of oxidized HPPA.

3.2. Fluorescence Spectrum. To examine the feasibility of the sensor, the fluorescence signal is illustrated in Figure 2. It was demonstrated that there was no fluorescence without biotin-GOx (curve e). Moreover, the fluorescence intensities of the mixture were weak in the absence of target HIV-1 dsDNA (curve b), capture DNA (curve c), or signal DNA (curve d). However, the intensity increases were enhanced dramatically with the addition of 1000 pM target HIV-1 dsDNA (curve a).

3.3. Circular Dichroism (CD) Spectroscopy of the Sensor. Circular dichroism (CD) spectroscopy is an effective tool and has been extensively used in the study of DNA structure. Figure 3 demonstrates that the CD spectroscopy of capture DNA (Oligo-1) was almost the same as that of signal DNA (Oligo-2). This was the classic spectroscopy of single-stranded DNA which had a weak positive Cotton effect peak around 275 nm and a weak negative Cotton effect peak at 249 nm. There was a strong peak at 218 nm, which indicated the helicity of dsDNA [42]. When target HIV-1 dsDNA was added into Oligo-1 and Oligo-2, the negative peak of 210 nm increased apparently, which was the marker of the triplex DNA [43]. Upon addition of spermine, the negative peak of 210 nm increased obviously. This phenomenon explained that spermine can increase the stability of triplex DNA. Overall, these results were consistent with that of the absorption spectra.

3.4. Optimization of the Experimental Conditions. To gain the optimal results, the following factors that affected the method performance were optimized, including the concentration of capture DNA, signal DNA, spermine, streptavidin, biotin-GOx, HPPA, HRP, and hybridization time. The optimal conditions were selected by obtaining the maximum change of fluorescence intensity (ΔI). ΔI was defined as $I_{target} - I_{blank}$, where I_{target} represents the fluorescence intensity of the mixture that contains target HIV-1 dsDNA and I_{blank} denotes the fluorescence intensity in the absence of target HIV-1 dsDNA. A good detection performance was obtained when 1.4 μM capture DNA, 50 nM signal DNA, 120 μM spermine, 800 ng·mL^{-1} SA, 50 μg·mL^{-1} biotin-GOx, 200 μM HPPA, and 20 ng·mL^{-1} HRP and 60 min of time of hybridization were used (Figure 4). The temperature can affect the formation of dsDNA; the effect of temperature is shown in Figure S1. When each of these factors is optimized, the other factors maintain optimal conditions.

3.5. Calibration Curve and Absorption Spectra for Target dsDNA. The relationship between the fluorescence intensity and the concentration of target HIV-1 dsDNA was studied under the optimum conditions. As shown in Figure 5, the fluorescence intensity increased with the concentration of target dsDNA over the range of 10 pM to 1000 pM, with

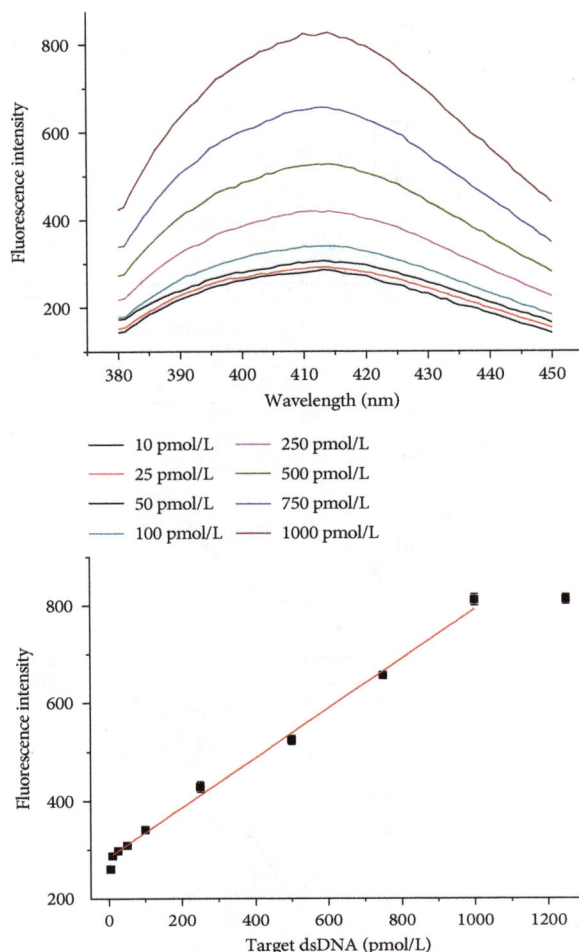

FIGURE 5: Calibration curve and absorption spectra for various concentrations of target dsDNA.

a linear regression equation of $I = 0.507\ C + 265.54$ (C: pM, $r = 0.996$, where C denotes the concentration of target HIV-1 dsDNA and I represents the fluorescence intensity) and a detection limit of 3 pM, which was obtained from the equation of $DL = 3\delta/slope$. The comparison with other methods is listed in Table 2.

3.6. Sequence Selectivity of the Sensor. For the purpose of exploring the sequence selectivity of the sensor, the sequence selectivity was investigated by means of replacing the target HIV-1 dsDNA with other three complementary dsDNAs, respectively. As shown in Figure 6, the intensity for target HIV-1 dsDNA was about 810, while that for single-base mismatched strand M-1 (Oligo-5•Oligo-6) and the two-base mismatched strands M-2 and M-3 (Oligo-7•Oligo-8 and Oligo-9•Oligo-10) were about 400, 320, and 315, respectively. These results illustrated that the proposed sensor had good sequence selectivity for target HIV-1 dsDNA.

3.7. Detection of Target HIV-1 dsDNA in Human Serum. To investigate the effect of other genes on the proposed strategy, test with calf thymus DNA interference aimed to recognize

TABLE 2: Comparison of our method with other methods for the determination of dsDNA.

Method	Analytical range	LOD	Application to samples	Reference
Polyamide microarrays method	1 nM–6 μM	1 nM	No	[12]
Strand-displacement amplification method	1 pM–250 pM	0.4 pM	No	[13]
Dynamic light-scattering method	59 pM–4061 pM	59 pM	No	[15]
Molecular beacon method	0.75 nM–50 nM	0.69 nM	No	[17]
Nicking enzyme amplification method	100 pM–200 nM	66 pM	Yes	[33]
Glucose oxidase amplification method	10 pM–1000 pM	3 pM	Yes	This work

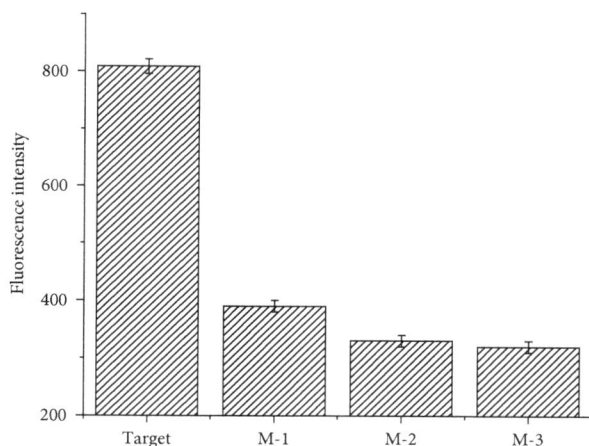

FIGURE 6: Sequence selectivity of the sensor. The concentration of target dsDNA, M-1, M-2, and M-3 was 1000 pM. Every point is the mean of three measurements. The error bar was the standard deviation.

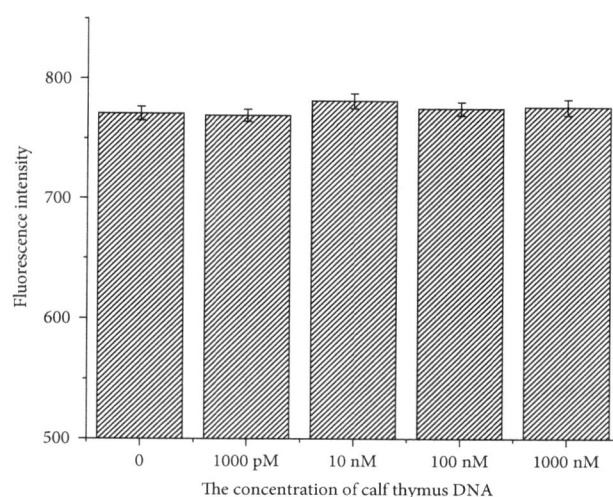

FIGURE 7: Effect of calf thymus DNA on the fluorescence of 1.0×10^{-9} M HIV-1 dsDNA.

TABLE 3: Recoveries of HIV-1 dsDNA from the spiked human serum samples with 1000 nM digested calf thymus DNA fragments.

Serum samples	Added HIV dsDNA (M)	Founded HIV dsDNA (M)	Recovery (%)	Relative standard deviation (%)
1	5.0×10^{-11}	5.16×10^{-11}	103.2	7.3
2	1.0×10^{-10}	1.13×10^{-10}	113.0	6.8
3	5.0×10^{-10}	4.65×10^{-10}	93.0	6.6
4	1.0×10^{-9}	0.98×10^{-8}	98.0	4.2

The values shown here are the average values from three measurements.

target dsDNA in the large amount of other dsDNA. As shown in Figure 7, there was little significant change in the fluorescence intensity with the addition of the concentration of calf thymus DNA increasing from 1000 pM to 1000 nM by using the proposed strategy. These results indicated that the proposed strategy could overcome the effect of other similar homopurine homopyrimidine sequences in the genome completely, and the effect of other similar homopurine homopyrimidine sequences in the genome could be ignored in the proposed method.

To assess the analytical application of the sensor in clinical specimens, the method was used to detect target HIV-1 dsDNA in the human serum. Since no HIV-1 dsDNA was found from the healthy volunteers' human serum, addition and recovery experiment was executed to evaluate the application of the sensor, and 1000 nM digested calf thymus DNA fragments were added into the serum as control DNA.

As demonstrated in Table 3, 5.0×10^{-11}–5.0×10^{-9} M of target HIV-1 dsDNA was added into each human serum, the recovery ranged from 93.0% to 103.2%, and the relative standard deviation values were in the range of 4.2%–7.3%, which meant that the application of the proposed sensor in real sample was possible.

4. Conclusions

In conclusion, through the amplifying property of GOx, a sensitive fluorescent sensor for sequence-specific recognition of target dsDNA was established. The target dsDNA which was selected from site 7960 to site 7991 of the HIV-1 dsDNA gene was designed as target HIV-1 dsDNA. Capture DNA was immobilized on the surface of SCGS by Ag–S bond, and with the help of biotin-SA, signal DNA can be conjugated with SA-biotin-GOx. Upon addition of target HIV-1 dsDNA, the GOx could be immobilized on the surface of silver-coated glass because of the formation of C•GoC in parallel triplex DNA structure. Thus, the concentration of target dsDNA controlled the number of bound GOx, which could be detected with the fluorescence of oxidized HPPA. Under the optimum conditions, the fluorescence intensity was proportional to the concentration of target HIV-1 dsDNA over the range from 10 pM to 1000 pM, with a detection limit of 3 pM. In addition, the sensor is target specific and is practicable. This is the first report of GOx amplification for sequence-specific recognition of dsDNA, and this assay might open a new avenue for applying in the diagnosis of HIV disease in the future.

Conflicts of Interest

The authors declare no conflicts of interest in publication of this research.

Acknowledgments

This work was supported by the Natural Science Foundation of Guangdong Ocean University (no. C16401) and Doctoral Initiating Project of Guangdong Ocean University Foundation for Natural Sciences (no. R17013).

References

[1] J. Watson and F. Crick, "Molecular structure of nucleic acids: a structure for deoxyribose nucleic acid," *Nature*, vol. 171, no. 4356, pp. 737–738, 1953.

[2] V. Avettand-Fenol, M. L. Chaix, S. Blanche et al., "LTR real-time PCR for HIV-1 DNA quantitation in blood cells for early diagnosis in infants born to seropositive mothers treated in HAART area (ANRS CO 01)," *Journal of Medical Virology*, vol. 81, no. 2, pp. 217–223, 2009.

[3] R. Dylla-Spears, J. E. Townsend, L. L. Sohn, L. Jen-Jacobson, and S. J. Muller, "Fluorescent marker for direct detection of specific dsDNA sequences," *Analytical Chemistry*, vol. 81, no. 24, pp. 10049–10054, 2009.

[4] Q. J. Gong, Y. D. Wang, and H. Y. Yang, "A sensitive impedimetric DNA biosensor for the determination of the HIV gene based on graphene-Nafion composite film," *Biosensors and Bioelectronics*, vol. 89, pp. 565–569, 2017.

[5] P. Zhang, C. S. Zhang, and B. Su, "Micropatterned paper devices using amine-terminated polydiacetylene vesicles as colorimetric probes for enhanced detection of double-stranded DNA," *Sensors and Actuators B: Chemical*, vol. 236, pp. 27–36, 2016.

[6] F. McKenzie, K. Faulds, and D. Graham, "LNA functionalized gold nanoparticles as probes for double stranded DNA through triplex formation," *Chemical Communications*, no. 20, pp. 2367–2369, 2008.

[7] K. M. Vasquez, L. Narayanan, and P. M. Glazer, "Specific mutations induced by triplex-forming oligonucleotides in mice," *Science*, vol. 290, no. 5491, pp. 530–533, 2000.

[8] V. C. Rucker, S. Foister, C. Melander, and P. B. Dervan, "Sequence specific fluorescence detection of double strand DNA," *Journal of the American Chemical Society*, vol. 125, no. 5, pp. 1195–1202, 2003.

[9] D. J. Segal and C. F. Barbas, "Custom DNA-binding proteins come of age: polydactyl zinc-finger proteins," *Current Opinion in Biotechnology*, vol. 12, no. 6, pp. 632–637, 2001.

[10] T. L. Roberts, A. Idris, J. A. Dunn et al., "HIN-200 proteins regulate caspase activation in response to foreign cytoplasmic DNA," *Science*, vol. 323, no. 5917, pp. 1057–1060, 2009.

[11] P. B. Dervan, "Molecular recognition of DNA by small molecules," *Bioorganic & Medicinal Chemistry*, vol. 9, no. 9, pp. 2215–2235, 2001.

[12] I. Singh, C. Wendeln, A. W. Clark, J. M. Cooper, B. J. Ravoo, and G. A. Burley, "Sequence-selective detection of double-stranded DNA sequences using pyrrole–imidazole polyamide microarrays," *Journal of the American Chemical Society*, vol. 135, no. 9, pp. 3449–3457, 2013.

[13] Y. B. Li, R. M. Li, L. Zou, M. J. Zhang, and L. S. Ling, "Fluorometric determination of Simian virus 40 based on strand displacement amplification and triplex DNA using a molecular beacon probe with a guanine-rich fragment of the stem region," *Microchimica Acta*, vol. 184, no. 2, pp. 557–562, 2017.

[14] S. P. Sau, P. Kumar, B. A. Anderson et al., "Optimized DNA-targeting using triplex forming C5-alkynyl functionalized LNA," *Chemical Communications*, vol. 44, pp. 6756–6758, 2009.

[15] X. M. Miao, C. Xiong, W. W. Wang, L. S. Ling, and X. T. Shuai, "Dynamic-light-scattering-based sequence-specific recognition of double-stranded DNA with oligonucleotide-functionalized gold nanoparticles," *Chemistry - A European Journal*, vol. 17, no. 40, pp. 11230–11236, 2011.

[16] Y. B. Li, X. M. Miao, and L. S. Ling, "Triplex DNA: a new platform for polymerase chain reaction–based biosensor," *Scientific Reports*, vol. 5, no. 1, pp. 1–8, 2015.

[17] Z. Y. Xiao, X. T. Guo, and L. S. Ling, "Sequence-specific recognition of double-stranded DNA with molecular beacon with the aid of Ag^+ under neutral pH environment," *Chemical Communications*, vol. 49, no. 34, pp. 3573–3575, 2013.

[18] T. Ihara, Y. Sato, H. Shimada, and A. Jyo, "Metalloregulation of triple helix formation by control of the loop conformation," *Nucleosides, Nucleotides and Nucleic Acids*, vol. 27, no. 9, pp. 1084–1096, 2008.

[19] E. Bakker, "Electrochemical sensors," *Analytical Chemistry*, vol. 76, no. 12, pp. 3285–3598, 2004.

[20] M. Wooten, S. Karra, M. Zhang, and W. Gorski, "On the direct electron transfer, sensing, and enzyme activity in the glucose oxidase/carbon nanotubes system," *Analytical Chemistry*, vol. 86, no. 1, pp. 752–757, 2014.

[21] H. Ju, C. M. Koo, and J. Kim, "Decoration of glassy carbon surfaces with dendrimer-encapsulated nanoparticles with a view to constructing bifunctional nanostructures," *Chemical Communications*, vol. 47, no. 45, pp. 12322–12324, 2011.

[22] E. C. Rama, A. Costa-García, and M. T. Fernández-Abedul, "Pin-based electrochemical glucose sensor with multiplexing possibilities," *Biosensors and Bioelectronics*, vol. 88, pp. 34–40, 2017.

[23] M. Rasmussen, R. West, J. Burgess, and I. D. Lee, "Bifunctional trehalose anode incorporating two covalently linked enzymes acting in series," *Analytical Chemistry*, vol. 83, no. 19, pp. 7408–7411, 2011.

[24] Y. Yi, J. Deng, Y. Zhang, H. Li, and S. Yao, "Label-free Si quantum dots as photoluminescence probes for glucose detection," *ChemComm*, vol. 49, no. 6, pp. 612–614, 2013.

[25] L. Li, F. Gao, J. Ye et al., "FRET-based biofriendly apo-GOx-modified gold nanoprobe for specific and sensitive glucose sensing and cellular imaging," *Analytical Chemistry*, vol. 85, no. 20, pp. 9721–9727, 2013.

[26] J. W. Liu, Y. Luo, Y. M. Wang, L. Y. Duan, J. H. Jiang, and R. Q. Yu, "Graphitic carbon nitride nanosheets-based ratiometric fluorescent probe for highly sensitive detection of H_2O_2 and glucose," *ACS Applied Materials & Interfaces*, vol. 8, no. 49, pp. 33439–33445, 2016.

[27] M. Ornatska, E. Sharpe, and D. S. Andreescu, "Paper bioassay based on ceria nanoparticles as colorimetric probes," *Analytical Chemistry*, vol. 83, no. 11, pp. 4273–4280, 2011.

[28] G. Darabdhara, B. Sharma, M. R. Das, R. Boukherroub, and S. Szunerits, "Cu-Ag bimetallic nanoparticles on reduced graphene oxide nanosheets as peroxidase mimic for glucose and ascorbic acid detection," *Sensors & Actuators B Chemical*, vol. 238, pp. 842–851, 2017.

[29] A. Singh, S. Park, and H. Yang, "Glucose-oxidase label-based redox cycling for an incubation period-free electrochemical immunosensor," *Analytical Chemistry*, vol. 85, no. 10, pp. 4863–4868, 2013.

[30] Y. B. Li and L. S. Ling, "Aptamer-based fluorescent solid-phase thrombin assay using a silver-coated glass substrate and signal amplification by glucose oxidase," *Microchimica Acta*, vol. 182, no. 9-10, pp. 1849–1854, 2015.

[31] J. Baur, C. Gondran, M. Holzinger, E. Defrancq, H. Perrot, and S. Cosnier, "Label-free femtomolar detection of target DNA by impedimetric DNA sensor based on poly(pyrrole-nitrilotriacetic acid) film," *Analytical Chemistry*, vol. 82, no. 3, pp. 1066–1072, 2010.

[32] U. Rathore, P. Saha, S. Kesavardhana et al., "Glycosylation of the core of the HIV-1 envelope subunit protein gp120 is not required for native trimer formation or viral infectivity," *Journal of Biological Chemistry*, vol. 292, no. 24, pp. 10197–10219, 2017.

[33] H. Zhu, M. Zhang, L. Zou, R. Li, and L. Ling, "Sequence specific recognition of HIV-1 dsDNA in the large amount of normal dsDNA based upon nicking enzyme signal amplification and triplex DNA," *Talanta*, vol. 173, pp. 9–13, 2017.

[34] S. M. Owen, "Testing for acute HIV infection," *Current Opinion in HIV and AIDS*, vol. 7, no. 2, pp. 125–130, 2012.

[35] C. Zhao, T. Hoppe, M. K. Setty et al., "Quantification of plasma HIV RNA using chemically engineered peptide nucleic acids," *Nature Communications*, vol. 5, p. 5079, 2014.

[36] S. Mercier-Delarue, M. Vray, J. C. Plantier et al., "Higher specificity of nucleic acid sequence-based amplification isothermal technology than of real-time PCR for quantification of HIV-1 RNA on dried blood spots," *Journal of Clinical Microbiology*, vol. 52, no. 1, pp. 52–56, 2014.

[37] T. Roberts, H. Bygrave, E. Fajardo, and N. Ford, "Challenges and opportunities for the implementation of virological testing in resource-limited settings," *Journal of the International AIDS Society*, vol. 15, no. 2, p. 17324, 2012.

[38] Y. B. Li, H. Zhang, H. Y. Zhu, and L. S. Ling, "A sensitive fluorescence method for sequence-specific recognition of single-stranded DNA by using glucose oxidase," *Analytical Methods*, vol. 7, no. 13, pp. 5436–5440, 2015.

[39] J. S. Lee, A. K. Lytton-Jean, S. J. Hurst, and C. A. Mirkin, "Silver nanoparticle–oligonucleotide conjugates based on DNA with triple cyclic disulfide moieties," *Nano Letters*, vol. 7, no. 7, pp. 2112–2115, 2007.

[40] K. Nemoto, T. Kubo, M. Nomachi et al., "Simple and effective 3D recognition of domoic acid using a molecularly imprinted polymer," *Journal of the American Chemical Society*, vol. 129, no. 44, pp. 13626–13632, 2007.

[41] X. Fan, H. Li, J. Zhao et al., "A novel label-free fluorescent sensor for the detection of potassium ion based on DNAzyme," *Talanta*, vol. 89, pp. 57–62, 2012.

[42] C. F. Jordan, L. S. Lerman, and J. H. Venable, "Structure and circular dichroism of DNA in concentrated polymer solutions," *Nature New Biology*, vol. 236, no. 64, pp. 67–70, 1972.

[43] M. Rosa, R. Dias, M. G. Miguel, and B. Lindman, "DNA–cationic surfactant interactions are different for double- and single-stranded DNA," *Biomacromolecules*, vol. 6, no. 4, pp. 2164–2171, 2005.

25

Systematic Evaluation of Chromatographic Parameters for Isoquinoline Alkaloids on XB-C18 Core-Shell Column using Different Mobile Phase Compositions

Ireneusz Sowa,[1] Sylwia Zielińska,[2] Jan Sawicki,[1] Anna Bogucka-Kocka,[3] Michał Staniak,[1] Ewa Bartusiak-Szcześniak,[1] Maja Podolska-Fajks,[1] Ryszard Kocjan,[1] and Magdalena Wójciak-Kosior ⓘD[1]

[1]Department of Analytical Chemistry, Medical University of Lublin, Chodźki 4a, 20-093 Lublin, Poland
[2]Department of Pharmaceutical Biology, Wroclaw Medical University, Borowska 211, 50-556 Wroclaw, Poland
[3]Department of Biology with Genetics, Medical University of Lublin, Chodźki 4a, 20-093 Lublin, Poland

Correspondence should be addressed to Magdalena Wójciak-Kosior; kosiorma@wp.pl

Academic Editor: Adam Voelkel

Chelidonium majus L. is a rich source of isoquinoline alkaloids with confirmed anti-inflammatory, choleretic, spasmolytic, antitumor, and antimicrobial activities. However, their chromatographic analysis is difficult because they may exist both in charged and uncharged forms and may result in the irregular peak shape and the decrease in chromatographic system efficacy. In the present work, the separation of main *C. majus* alkaloids was optimized using a new-generation XB-C18 endcapped core-shell column dedicated for analysis of alkaline compounds. The influence of organic modifier concentration, addition of salts, and pH of eluents on chromatographic parameters such as retention, resolution, chromatographic plate numbers, and peak asymmetry was investigated. The results were applied to elaborate the optimal chromatographic system for simultaneous quantification of seven alkaloids from the root, herb, and fruit of *C. majus*.

1. Introduction

Isoquinoline alkaloids such as coptisine, allocryptopine, protopine, berberine, chelidonine, sanguinarine, and chelerythrine are the main constituents of *Chelidonium majus* L. responsible for biological properties of the plant. They have analgesic, antispasmodic, antibacterial, antiviral, and antifungal activities. Moreover, they show cytotoxic and antiproliferative effects against various types of cancer cell lines [1–4]. Because of the broad spectrum of action, *C. majus* alkaloids are a subject of interest for pharmacology and toxicology; therefore, the effective analytical methods for their investigation are still being developed. Spectrophotometry [5, 6], capillary electrophoresis [7, 8], and thin-layer chromatography [9–12] have been used for this purpose; however, high-performance liquid chromatography is the most common for qualitative and quantitative analyses of *C. majus* [5, 13–17].

Due to the durability, silica-based RP-18 stationary phases are widely used for HPLC separation of plant extracts [18, 19]; however, RP chromatography of alkaloids is rather difficult because they may exist both as free bases and charged forms. Cationic forms strongly interact with residual silanol groups of the RP-type stationary phase and cause the occurrence of the dual retention mechanism (RP and ion-exchange retention mechanism) and result in the peak tailing, irreproducible retention, and poor system efficiency [20].

Due to the basic character of isoquinoline alkaloids, it would be preferable to conduct chromatographic separation at alkaline pH to avoid their ionization; however, silica-based adsorbents are unstable at this condition [21].

TABLE 1: The comparison of retention times and peak resolutions of investigated alkaloids in 20–25% of acetonitrile in water at different pH and ammonium acetate concentrations.

| | 20% ACN, pH = 3 | | | | 20% ACN, pH = 4 | | | | 25% ACN, pH = 3 | | | | 25% ACN, pH = 4 | | | |
| | 20 mM | | 10 mM | | 20 mM | | 10 mM | | 20 mM | | 10 mM | | 20 mM | | 10 mM | |
	t_R	R_S	t_R	R_S	t_R	R_S	t_R	R_S	t_R	R_S	t_R	R_S	t_R	R_S	t_R	R_S
Protopine	10.94	5.38	11.49	6.08	13.16	5.79	13.91	7.25	6.54	4.07	6.66	4.07	7.75	4.20	7.72	4.55
Allocryptopine	13.40	1.36	14.17	1.45	16.81	3.47	17.06	3.35	7.61	1.25	7.73	1.43	8.89	3.98	9.03	3.23
Chelidonine	14.07	4.11	14.90	3.88	18.79	2.68	19.29	3.05	7.97	2.56	8.17	2.60	10.07	1.21	10.07	1.29
Coptisine	16.02	13.65	16.84	14.26	20.35	14.18	21.03	14.95	8.73	8.72	8.92	9.41	10.5	9.57	10.55	9.08
Sanguinarine	24.58	5.62	26.39	5.45	33.71	4.59	37.27	4.07	11.76	5.65	12.11	5.70	14.8	5.38	14.78	5.11
Berberine	28.96	17.00	31.01	17.22	39.12	16.82	42.27	18.23	14.01	12.67	14.43	12.79	17.58	13.25	17.51	12.62
Chelerythrine	47.54	—	51.73	—	66.91	—	74.71	—	20.45	—	21.26	—	26.45	—	26.28	—

Different approaches may be used to eliminate these problems. Alkaline additives to mobile phases, mostly organic amines such as diethylamine, triethylamine, or dimethyloctylamine, are applied to suppress the ionization of the analyte and as silanol blockers [14, 15]. Addition of anionic ion-pairing reagents, for example, sodium dodecyl sulphate or salts (e.g., ammonium acetate, ammonium formate, and sodium phosphate) is also used to improve the chromatographic separation [5, 13, 17, 22]. On the other hand, the silanol-masking effect may be achieved by additional modification of the sorbent surface, for example, endcapping [23, 24].

XB-C18 sorbent is relatively new column filling with trimethylsilane endcapping and additional isobutyl chains. In the present work, an XB reversed-phase column was used to separate the isoquinoline alkaloids typically found in the *C. majus* extract. The influence of organic modifier concentration, addition of salts, and pH of eluents on chromatographic parameters such as retention, resolution, chromatographic plate numbers, and peak asymmetry was investigated. The results were applied to elaborate the optimal chromatographic system for simultaneous quantification of alkaloids from the root, herb, and fruit of *C. majus*.

2. Experimental

2.1. Chemicals and Reagents. Alkaloid standards such as protopine (Prot), allocryptopine (All), berberine (Berb), chelidonine (Che), chelerythrine (Chele), and sanguinarine (Sang) were purchased from Sigma (St. Louis, MO) and coptisine (Cop) from ChromaDex (USA). Ammonium acetate, ammonium formate, acetic acid, formic acid, HPLC-grade methanol (MeOH), and acetonitrile (ACN) were from Merck (Darmstadt, Germany). Water was deionized and purified by ULTRAPURE Millipore Direct-QVR 3UV-R (Merck, Darmstadt, Germany).

2.2. High-Performance Liquid Chromatography. Chromatography was carried out using a VWR Hitachi Chromaster 600 chromatograph (Merck, Darmstadt, Germany) with a spectrophotometric detector (DAD) and EZChrom Elite software (Merck).

The samples were analyzed on an XB-C18 reversed-phase core-shell column (Kinetex, Phenomenex, Aschaffenburg, Germany) (25 cm × 4.6 mm i.d., 5 μm particle size), at a temperature of 25°C and an eluent flow rate of 1 mL/min.

Chromatograms were recorded in the range of wavelength from 220 to 400 nm. The identity of compounds in plant extracts was confirmed by comparison of retention times and spectra with corresponding standards. Peak homogeneity was established comparing the spectrum recorded at the three peak sections upslope, apex, and downslope with the reference spectrum. Additionally, the chromatographic fractions eluted at the retention time characteristic for the investigated alkaloids were collected using a Foxy R1 fraction collector (Teledyne Isco, Lincoln, USA), and their identity was confirmed by direct injection mass spectrometry (micrOTOF-Q II, Bruker Daltonics, Bremen, Germany) using Compass DataAnalysis software version 4.1.

2.3. Sample Preparation. The extraction conditions were based on literature [13]. The root, leaf, and fruit of *C. majus* (1 g) were extracted in ultrasonic bath (3 × 15 min) with 10 mL of methanol acidified with 0.05 M HCl. Subsequently, the extracts were combined, evaporated to dryness, and dissolved in 20 mL of methanol.

3. Results and Discussion

Sufficient resolution between neighbouring peaks, symmetric peaks, and narrow peaks are the most important for the optimal chromatographic system. A stationary phase and a mobile phase have a crucial impact on these parameters. In our work, different variants of eluent compositions were tested for their suitability in HPLC of isoquinoline alkaloids on a new-generation XB-C18 endcapped core-shell column. The influence of the three variables: concentration of the organic modifier, salt, and pH on resolution (R_S), peak asymmetry (A_S), and system efficacy (N-theoretical plate numbers) for methanol/water and acetonitrile/water solvent systems was investigated.

3.1. Optimization of Chromatographic Condition. The chromatographic parameters were established in the range

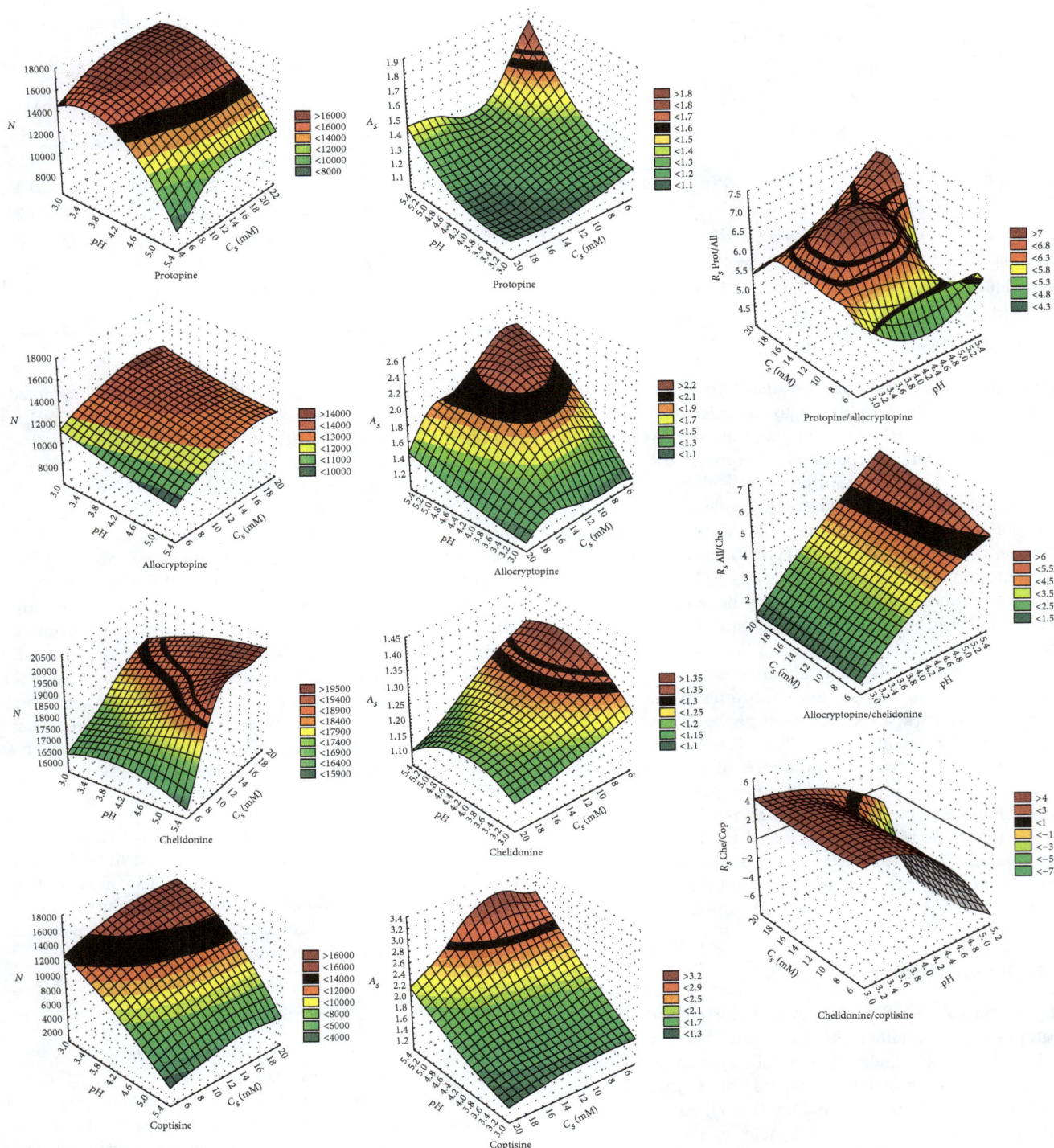

FIGURE 1: The relationship between theoretical plate numbers (N), peak asymmetry (A_s), resolution (R_s), and pH/ammonium acetate concentration.

of 20–40% of acetonitrile (ACN) and 30–40% of methanol (MeOH) in water. Acid (acetic or formic) to obtain appropriate pH (3–5.5) and salt (ammonium acetate or ammonium formate) at the concentration range 5–20 mM were added to tested eluents because blocking and suppressing the

ionization of residual silanol groups of the stationary phase were necessary to avoid peak splitting or broadening of basic compounds.

In ACN/water and MeOH/water systems, the amount of organic modifier strongly influence the alkaloid retention.

TABLE 2: The comparison of retention times and peak resolutions of investigated alkaloids in 35% of methanol in water at different pH and ammonium acetate concentrations.

| | pH = 3 | | | | pH = 4 | | | |
| | 20 mM | | 10 mM | | 20 mM | | 10 mM | |
	t_R	R_S	t_R	R_S	t_R	R_S	t_R	R_S
Protopine	8.05	0.63	8.34	0.77	10.67	1.12	10.51	1.10
Allocryptopine	8.47	0.96	8.92	0.76	11.56	2.25	11.23	2.36
Chelidonine	9.11	1.18	9.49	1.18	13.37	0.08	13.15	0.30
Coptisine	9.68	10.60	10.18	10.02	13.43	10.48	13.37	12.94
Berberine	15.88	2.57	17.23	2.75	23.82	7.93	23.9	14.47
Sanguinarine	17.73	13.12	19.47	13.26	36.85	6.21	48.31	5.71
Chelerythrine	29.85	—	33.4	—	50.16	—	63.02	—

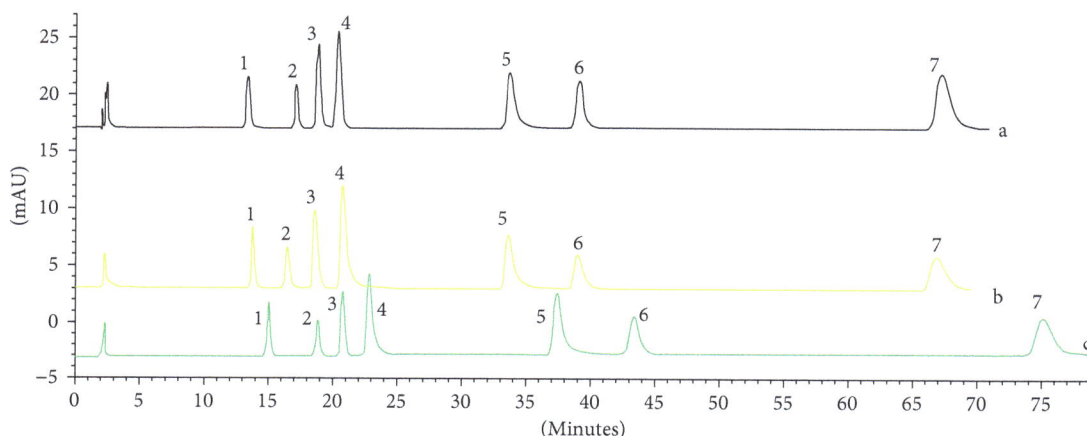

FIGURE 2: Exemplary chromatograms of the standard mixture obtained at (a) acetonitrile and 10 mM water solution of ammonium formate adjusted to pH 4 with formic acid (20 : 80, v/v), (b) acetonitrile and 20 mM water solution of ammonium acetate adjusted to pH 4 with acetic acid (20 : 80, v/v), and (c) acetonitrile and 10 mM water solution of ammonium acetate adjusted to pH 4 with acetic acid (20 : 80, v/v). (1) protopine, (2) allocryptopine, (3) chelidonine, (4) coptisine, (5) sanguinarine, (6) berberine, and (7) chelerythrine.

Taking into consideration the resolution, the total separation of investigated compounds was obtained for concentration of 20% ACN in the whole tested pH and salt concentration range. At 25% of ACN, the compound All/Che (at pH ≥ 4) or Che/Cop (at pH ≤ 4) partially coeluted. The exemplary R_s values are given in Table 1.

In higher concentrations, the majority of compounds were eluted below 10 min (k values for protopine, allocryptopine, and chelidonine were lower than 1), and the resolution was poor (Table S1). The concentration of ammonium acetate and pH also affected the alkaloid retention; at lower pH values and at higher salt amounts, retention times were shortened. Moreover, efficiency of the system (N), symmetry of peaks (A_s), and resolution (R_s) strongly depended on these variables. N, A_s, and R_s values versus pH and salt concentration are presented in Figure 1 and Figure S1.

As can be seen, at pH 5 and at concentration of salt 5 mM, theoretical plate number decreased and peak asymmetry increased significantly, and it resulted in the peak broadening and decreased peak resolution. In contrast to salt concentration, pH had a major impact on R_s. Resolution between Che/Cop, Cop/Sang, and Sang/Berb increased at

lower pH; in turn, for All/Che, the opposite effect occurred. Moreover, the change of elution order was observed for chelidonine and coptisine at pH 5.

Methanol showed lower elution strength, and the concentration in the range of 30–35% was required to obtain the elution of alkaloids at a reasonable time. Moreover, the order of elution strongly depended on pH and amount of organic modifier and salt (Table 2 and Table S2).

As can be seen, no composition of mobile phases provided sufficient separation within compounds with weaker retention such as protopine, allocryptopine, chelidonine, and coptisine, and all tested chromatographic parameters were worse for methanol/water than for acetonitrile/water eluents.

In further experiments, ammonium acetate/acetic acid in ACN/water eluents was replaced by ammonium formate/formic acid; however, it had a minor impact on chromatographic parameters. The R_s values and retention times did not differ significantly, and only a slight increase in system efficacy (narrower peaks) was observed. Exemplary chromatograms of the standard mixture obtained at various mobile phase compositions are shown in Figure 2.

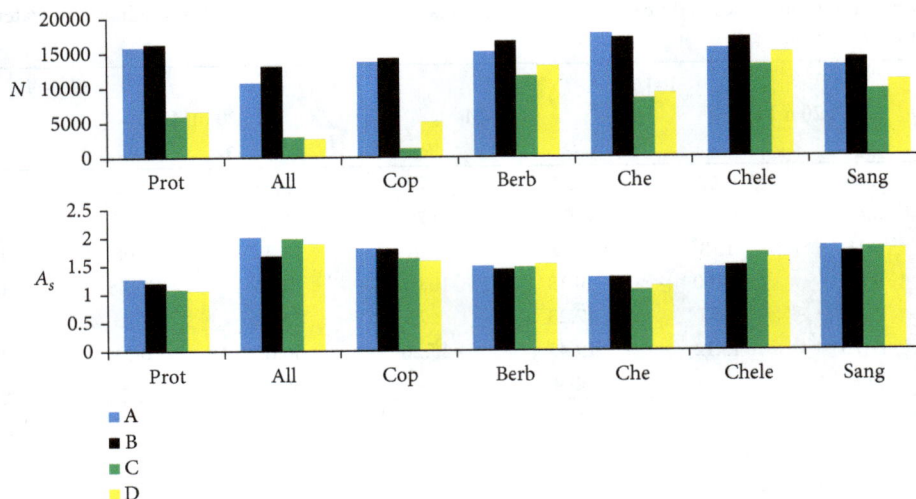

FIGURE 3: Comparison of theoretical plate numbers and peak asymmetry obtained on XB-C18 column for 20% of acetonitrile and 30% of methanol: (A) acetonitrile and 10 mM water solution of ammonium acetate adjusted to pH 4 with acetic acid; (B) acetonitrile and 10 mM water solution of ammonium formate adjusted to pH 4 with formic acid; (C) methanol and 10 mM water solution of ammonium acetate adjusted to pH 3 with acetic acid; (D) methanol and 10 mM water solution of ammonium acetate adjusted to pH 4 with acetic acid.

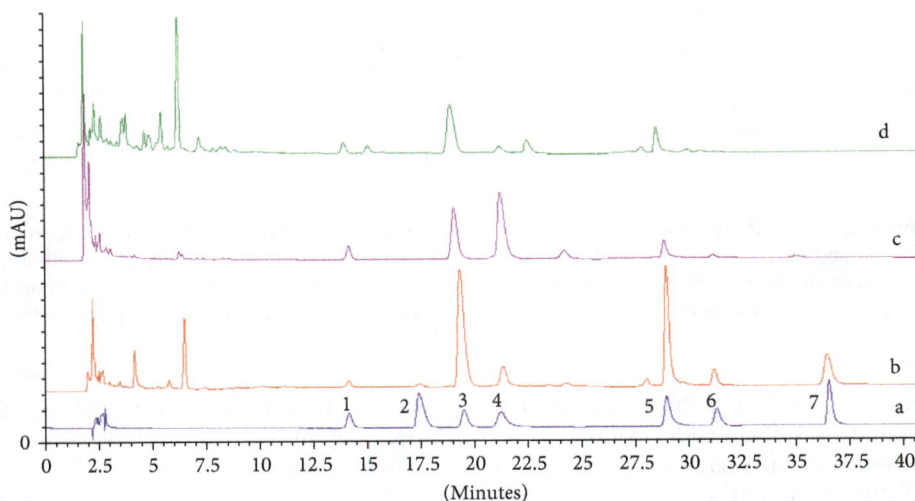

FIGURE 4: Example of an HPLC-DAD chromatogram of extracts from different parts of *C. majus*: (b) roots, (c) fruits, and (d) herbs and (a) standards of investigated alkaloids: (1) protopine, (2) allocryptopine, (3) chelidonine, (4) coptisine, (5) sanguinarine, (6) berberine, and (7) chelerythrine.

The comparison of N and A_s for various mobile phase compositions is presented in Figure 3.

Based on the obtained results, acetonitrile at concentration of 20% and water at pH 3-4 with addition of 10–20 mM ammonium acetate or ammonium formate were considered as optimal for isoquinoline alkaloid separation on an XB-C18 core-shell column.

A lot of chromatographic systems for RP separation of isoquinoline alkaloids in *C. majus* were described in literature; however, most of them were more complicated [14, 15, 22] or did not provide sufficient separation for quantitative analysis [25]. Due to additional modification of endcapped octadecyl silica by isobutyl chains in the XB-C18 stationary

phase, the interaction of the basic analyte with residual silanol decreased significantly. It allowed to conduct chromatographic separation using milder pH and lower amount of salt addition compared to eluents proposed in literature [13], and it is beneficial for the HPLC system.

3.2. Chromatographic Analysis of C. majus Extracts. Chromatography of the *C. majus* extract was carried out using the mobile phase consisting of ACN (solvent A) and 10 mM water solution of ammonium acetate adjusted to pH 4 with acetic acid (solvent B) (20 : 80, v/v). High R_s values between coptisine and sanguinarine allowed to use the simple

TABLE 3: The data used for identification of the investigated compounds.

	Retention time (min)	UV-Vis spectrum	m/z
Protopine	13.91		354.135
Allocryptopine	17.06		369.157
Chelidonine	19.29		353.126
Coptisine	20.86		320.092
Sanguinarine	28.25		332.092
Berberine	30.68		336.123
Chelerythrine	35.71		348.123

TABLE 4: The content of investigated alkaloids (mg/100 g ± SD) in different parts of *C. majus*.

	Root	Herb	Fruit
Protopine	15.7 ± 1.1	20.2 ± 1.6	22.4 ± 2.0
Allocryptopine	11.4 ± 0.8	—	—
Chelidonine	140.1 ± 10.4	65.2 ± 5.1	57.4 ± 5.1
Coptisine	50.1 ± 4.1	20.8 ± 1.5	247.2 ± 20.9
Sanguinarine	311.6 ± 22.4	29.8 ± 2.1	20.1 ± 1.9
Berberine	52.7 ± 3.9	0.8 ± 0.1	10.4 ± 0.9
Chelerythrine	100.3 ± 7.8	<LOQ	<LOQ

gradient program to shorten the total time of analysis. After 20 min, the elution strength of the mobile phase was increased to accelerate elution of strongly retained sanguinarine, berberine, and chelerythrine. The gradient program was as follows: A 20% and B 80% during 0–20 min, A 25% and B 75% during 20–27 min, and A 30% and B 70% during 27–40 min. The obtained chromatograms are presented in Figure 4.

The data used for identification of the investigated compounds are given in Table 3.

The results of quantitative determination of isoquinoline alkaloids in the root, leaf, and fruit of *C. majus* are given in Table 4, and validation parameters are summarized in Table S3.

Conflicts of Interest

The authors declare that there are no conflicts of interest regarding the publication of this paper.

References

[1] M. L. Colombo and E. Bosisio, "Pharmacological activities of *Chelidonium majus* L. (Papaveraceae)," *Pharmacological Research*, vol. 33, no. 2, pp. 127–134, 1996.

[2] M. Gilca, L. Gaman, E. Panait, I. Stoian, and V. Atanasiu, "*Chelidonium majus*–an integrative review: traditional knowledge versus modern findings," *Forschende Komplementärmedizin*, vol. 17, no. 5, pp. 241–248, 2010.

[3] R. Havelek, M. Seifrtova, K. Kralovec et al., "Comparative cytotoxicity of chelidonine and homochelidonine, the dimethoxy analogues isolated from *Chelidonium majus* L. (Papaveraceae), against human leukemic and lung carcinoma cells," *Phytomedicine*, vol. 23, no. 3, pp. 253–266, 2016.

[4] I. R. Capistrano, A. Wouters, F. Lardon, C. Gravekamp, S. Apers, and L. Pieters, "In vitro and in vivo investigations on the antitumour activity of *Chelidonium majus*," *Phytomedicine*, vol. 22, no. 14, pp. 1279–1287, 2015.

[5] K. Seidler-Łożykowska, B. Kędzia, J. Bocianowski et al., "Content of alkaloids and flavonoids in celandine (*Chelidonium majus* L.) herb at selected developmental phases," *Acta Scientiarum Polonorum Hortorum Cultus*, vol. 16, no. 3, pp. 161–172, 2017.

[6] M. Then, K. Szentmihályi, A. Sárközi, V. Illés, and E. Forgács, "Effect of sample handling on alkaloid and mineral content of aqueous extracts of greater celandine (*Chelidonium majus* L.)," *Journal of Chromatography A*, vol. 889, no. 1-2, pp. 69–74, 2000.

[7] Q. Zhou, Y. Liu, X. Wang, and X. Di, "Microwave-assisted extraction in combination with capillary electrophoresis for rapid determination of isoquinoline alkaloids in *Chelidonium majus* L.," *Talanta*, vol. 99, pp. 932–938, 2012.

[8] M. Kulp and O. Bragina, "Capillary electrophoretic study of the synergistic biological effects of alkaloids from *Chelidonium majus* L. in normal and cancer cells," *Analytical and Bioanalytical Chemistry*, vol. 405, no. 10, pp. 3391–3397, 2013.

[9] W. Jesionek, E. Fornal, B. Majer-Dziedzic, Á. M. Móricz, W. Nowicky, and I. M. Choma, "Investigation of the composition and antibacterial activity of Ukrain™ drug using liquid chromatography techniques," *Journal of Chromatography A*, vol. 1429, pp. 340–347, 2016.

[10] I. Malinowska, M. Studziński, H. Malinowski, and M. Gadzikowska, "Retention and separation changes of ternary and quaternary alkaloids from *Chelidonium majus* L. by TLC under the influence of external magnetic field," *Chromatographia*, vol. 80, no. 6, pp. 923–930, 2017.

[11] Á. Sárközi, G. Janicsák, L. Kursinszki, and Á. Kéry, "Alkaloid composition of *Chelidonium majus* L. studied by different chromatographic techniques," *Chromatographia*, vol. 63, no. S13, pp. 81–86, 2006.

[12] A. Bogucka-Kocka and D. Zalewski, "Qualitative and quantitative determination of main alkaloids of *Chelidonium majus* L. using thin-layer chromatographic-densitometric method," *Acta Chromatographica*, vol. 29, no. 3, pp. 385–397, 2017.

[13] L. Kursinszki, Á. Sárközi, Á. Kéry, and É. Szöke, "Improved RP-HPLC method for analysis of isoquinoline alkaloids in extracts of *Chelidonium majus*," *Chromatographia*, vol. 63, no. S13, pp. S131–S135, 2006.

[14] H. Wu and L. Du, "Ionic liquid-liquid phase microextraction for the sensitive determination of sanguinarine and chelerythrine in Chinese herbal medicines and human urine," *Journal of Liquid Chromatography & Related Technologies*, vol. 35, no. 12, pp. 1662–1675, 2012.

[15] J. Paulsen, M. Yahyazadeh, S. Hänsel, M. Kleinwächter, K. Ibrom, and D. Selmar, "13,14-dihydrocoptisine–the genuine alkaloid from *Chelidonium majus*," *Phytochemistry*, vol. 111, pp. 149–153, 2015.

[16] C. Grosso, F. Ferreres, A. Gil-Izquierdo et al., "Box–Behnken factorial design to obtain a phenolic-rich extract from the aerial parts of *Chelidonium majus* L.," *Talanta*, vol. 130, pp. 128–136, 2014.

[17] A. Borghini, D. Pietra, C. di Trapani, P. Madau, G. Lubinu, and A. M. Bianucci, "Data mining as a predictive model for *Chelidonium majus* extracts production," *Industrial Crops and Products*, vol. 64, pp. 25–32, 2015.

[18] X. Wei, H. Shen, L. Wang, Q. Meng, and W. Liu, "Analyses of total alkaloid extract of *Corydalis yanhusuo* by comprehensive RP × RP liquid chromatography with pH difference," *Journal of Analytical Methods in Chemistry*, vol. 2016 Article ID 9752735, 8 pages, p. 8, 2016.

[19] N. Yu, C. He, G. Awuti, C. Zeng, J. Xing, and W. Huang, "Simultaneous determination of six active compounds in Yixin Badiranjibuya granules, a traditional Chinese medicine, by RP-HPLC-UV method," *Journal of Analytical Methods in Chemistry*, Article ID 974039, vol. 2015, p. 9, 2015.

[20] F. Gritti and G. Guiochon, "Physical origin of peak tailing on C_{18}-bonded silica in reversed-phase liquid chromatography," *Journal of Chromatography A*, vol. 1028, no. 1, pp. 75–88, 2004.

[21] J. J. Kirkland, J. B. Adams Jr., M. A. van Straten, and H. A. Claessens, "Bidentate silane stationary phases for reversed-phase high-performance liquid chromatography," *Analytical Chemistry*, vol. 70, no. 20, pp. 4344–4352, 1998.

[22] N. A. Gañán, A. M. A. Dias, F. Bombaldi et al., "Alkaloids from *Chelidonium majus* L.: Fractionated supercritical CO_2 extraction with co-solvents," *Separation and Purification Technology*, vol. 165, pp. 199–207, 2016.

[23] F. Gritti and G. Guiochon, "Chromatographic estimate of the degree of surface heterogeneity of reversed-phase liquid chromatography packing materials. II-endcapped monomeric C18-bonded stationary phase," *Journal of Chromatography A*, vol. 1103, no. 1, pp. 57–68, 2006.

[24] F. Gritti and G. Guiochon, "Influence of the degree of coverage of C_{18}-bonded stationary phases on the mass transfer mechanism and its kinetics," *Journal of Chromatography A*, vol. 1128, no. 1-2, pp. 45–60, 2006.

[25] H. Prosen and B. Pendry, "Determination of shelf life of *Chelidonium majus, Sambucus nigra, Thymus vulgaris* and *Thymus serpyllum* herbal tinctures by various stability-indicating tests," *Phytochemistry Letters*, vol. 16, pp. 311–323, 2016.

Permissions

List of Contributors

Kang-Bong Lee
Green City Technology Institute, Korea Institute of Science and Technology, Hwarang-ro 14 gil 5, Seoul 02792, Republic of Korea

Kyung Min Kim
Green City Technology Institute, Korea Institute of Science and Technology, Hwarang-ro 14 gil 5, Seoul 02792, Republic of Korea
Department of Chemistry, Korea University, Anam-ro, Seongbuk-gu, Seoul 136-701, Republic of Korea

Yun-Sik Nam and Yeonhee Lee
Advanced Analysis Center, Korea Institute of Science and Technology, Hwarang-ro 14 gil 5, Seoul 02792, Republic of Korea

Yuna Jung
Department of Biomedical Science, Graduate School, Kyung Hee University, 26 Kyungheedae-Ro, Dongdaemun-Gu, Seoul 02447, Republic of Korea

Junyang Jung and Youngbuhm Huh
Department of Biomedical Science, Graduate School, Kyung Hee University, 26 Kyungheedae-Ro, Dongdaemun-Gu, Seoul 02447, Republic of Korea
Department of Anatomy and Neurobiology, College of Medicine, Kyung Hee University, 26 Kyungheedae-Ro, Dongdaemun-Gu, Seoul 02447, Republic of Korea

Dokyoung Kim
Department of Biomedical Science, Graduate School, Kyung Hee University, 26 Kyungheedae-Ro, Dongdaemun-Gu, Seoul 02447, Republic of Korea
Department of Anatomy and Neurobiology, College of Medicine, Kyung Hee University, 26 Kyungheedae-Ro, Dongdaemun-Gu, Seoul 02447, Republic of Korea
Center for Converging Humanities, Kyung Hee University, 26 Kyungheedae-Ro, Dongdaemun-Gu, Seoul 02447, Republic of Korea

Mateusz Kowalcze, Jan Wyrwa, Małgorzata Dziubaniuk and Małgorzata Jakubowska
AGH University of Science and Technology, Faculty of Materials Science and Ceramics, Mickiewicza 30, 30-059 Kraków, Poland

Rui Lv, Shuya Cui, Yangmei Zou and Li Zheng
Department of Chemistry and Chemical Engineering, Mianyang Teacher's College, Mianyang, Sichuan 621000, China

Yongjie Xu, Xiangrong Luo and Zhishun Lu
Department of Laboratory Medicine, Guizhou Provincial People's Hospital, College of Basic Medicine, Guizhou University, Guiyang 550002, Guizhou, China

Nana Geng and Mingsong Wu
Special Key Laboratory of Oral Diseases Research, Higher Education Institutions of Guizhou Province, Zunyi Medical University, Zunyi 563099, Guizhou, China

Maria Pilo, Roberta Farre, Elisabetta Masolo, Angelo Panzanelli, Gavino Sanna, Nina Senes, Ana Sobral and Nadia Spano
Department of Chemistry and Pharmacy, University of Sassari, 07100 Sassari, Italy

Joanna Izabela Lachowicz
Department of Chemical and Geological Sciences, University of Cagliari, Monserrato, 09042 Cagliari, Italy

Gabriela Islas, Jose A. Rodriguez, Irma Perez-Silva and Israel S. Ibarra
Área Académica de Química, Universidad Autónoma del Estado de Hidalgo, Carretera Pachuca-Tulancingo Km. 4.5, 42076 Pachuca, Hidalgo, Mexico

Jose M. Miranda
Departamento Química Analítica, Nutrición y Bromatología, Facultad de Veterinaria, Universidad de Santiago de Compostela, Pabell'on 4 planta bajo, Campus Universitario s/n, 27002 Lugo, Spain

Fadi Alakhras
Department of Chemistry, College of Science, Imam Abdulrahman Bin Faisal University, Dammam 31441, Saudi Arabia

Vesna Antunović
Faculty of Medicine, University of Banja Luka, 78000 Banja Luka, Bosnia and Herzegovina

Slavna Tešanović, Danica Perušković, Nikola Stevanović, Rada Baošić, Snežana Mandić and Aleksandar Lolić
Faculty of Chemistry, University of Belgrade, 11000 Belgrade, Serbia

Bing Wang, Jincui Gu, Boyi Chen and Chengfeng Xu
Key Laboratory of Advanced Textile Materials and Manufacturing Technology, Ministry of Education, Zhejiang Sci-Tech University, Hangzhou 310018, China

Hailing Zheng and Yang Zhou
Key Scientific Research Base of Textile Conservation, State Administration for Cultural Heritage, China National Silk Museum, Hangzhou 310002, China

Zhiqin Peng and Zhiwen Hu
Institute of Textile Conservation, Zhejiang Sci-Tech University, Hangzhou 310018, China

Hui Zhang, Huijie Jiang, Xiaojing Zhang, Shengqiang Tong and Jizhong Yan
College of Pharmaceutical Science, Zhejiang University of Technology, No. 18 Chaowang Road, Hangzhou 310014, China

Xiangping Liu, Xuemin Jing and Guoliang Li
College of Animal Science and Veterinary Medicine, Heilongjiang Bayi Agricultural University, Daqing 163319, China

Nan Zhan, Feng Guo, Shuai Zhu and Zhu Rao
The Key Laboratory of Eco-Geochemistry, Ministry of Nature Resources and National Research Center for Geoanalysis, Beijing 100037, China

Guoguang Wu
School of Chemical Engineering and Technology, China University of Mining and Technology, Xuzhou, Jiangsu 221116, China

Xiaohua Ma
School of Chemical Engineering and Technology, China University of Mining and Technology, Xuzhou, Jiangsu 221116, China
Henan Key Laboratory of Biomolecular Recognition and Sensing, College of Chemistry and Chemical Engineering, Shangqiu Normal University, Shangqiu, Henan 476000, China

Yuehua Zhao, Zibo Yuan, Yu Zhang, Ning Xia, Mengnan Yang and Lin Liu
Key Laboratory of New Optoelectronic Functional Materials (Henan Province), College of Chemistry and Chemical Engineering, Anyang Normal University, Anyang, Henan 455000, China

Qiao-Qiao Li, Yu-Xiu Yang, Zhi-Ning Xia and Feng-Qing Yang
School of Chemistry and Chemical Engineering, Chongqing University, Chongqing 401331, China

Jing-Wen Qv and Yuan-Jia Hu
State Key Laboratory of Quality Research in Chinese Medicine, Institute of Chinese Medical Sciences, University of Macau, Macau, China

Guang Hu
School of Pharmacy and Bioengineering, Chongqing University of Technology, Chongqing 400054, China

Keabetswe Masike and Ntakadzeni Madala
Department of Biochemistry, University of Johannesburg, Auckland Park 2006, South Africa

G. B. Slepchenko, T. M. Gindullina, M. A. Gavrilova and A. Zh. Auelbekova
National Research Tomsk Polytechnic University, Tomsk 634050, Russia

Çiğdem Arpa and Itır Aridaşir
Chemistry Department, Hacettepe University, Beytepe, 06800 Ankara, Turkey

Yanqiang Zhou, Meihua Ma, Zhian Sun, Shanshan Wu and Bolin Gong
College of Chemistry and Chemical Engineering, North Minzu University, Yinchuan 750021, China

Yinhui Yang
College of Chemistry and Chemical Engineering, Beijing Institute of Technology, Beijing 100081, China

Dimitra K. Toubanaki and Evdokia Karagouni
Laboratory of Cellular Immunology, Department of Microbiology, Hellenic Pasteur Institute, 127 Vas. Sofias Ave., 11521 Athens, Greece

Hye Jin Kim and Kyo Seon Hwang
Department of Clinical Pharmacology, Kyung Hee University, Seoul 02447, Republic of Korea

Dong-Hoon Kang and Hyun-Joon Shin
Center for Bionics, Korea Institute of Science and Technology, Seoul 02792, Republic of Korea

Seung-Hoon Yang
Systems Biotechnology Research Center, Korea Institute of Science and Technology, Gangneung 25451, Republic of Korea

Eunji Lee and Jinsik Kim
Department of Medical Biotechnology, Dongguk University, Seoul 04620, Republic of Korea

Taewon Ha and Youngbaek Kim
Center for Nano-Photonics Convergence Technology, Korea Institute of Industrial Technology (KITECH), Gwangju 61012, Republic of Korea

Byung Chul Lee
Center for BioMicrosystems, Korea Institute of Science and Technology (KIST), Seoul 02792, Republic of Korea

B. Abaroa-Pérez and J. J. Hernández-Brito
Plataforma Oce´anica de Canarias (PLOCAN), Las Palmas, Spain

G. Sánchez-Almeida and D. Vega-Moreno
Chemistry Department, Universidad de Las Palmas de G.C (ULPGC), Las Palmas, Spain

Hee-Youb Song and Soon-Ki Jeong
Department of Chemical Engineering, Soonchunhyang University, Asan, Chungnam 336-745, Republic of Korea

Yubin Li and Sheng Liu
College of Chemistry and Environment, Guangdong Ocean University, Zhanjiang 524088, China

Liansheng Ling
School of Chemistry, Sun Yat-Sen University, Guangzhou 510275, China

Ireneusz Sowa, Jan Sawicki, Michał Staniak, Ewa Bartusiak-Szcześniak, Maja Podolska-Fajks, Ryszard Kocjan and Magdalena Wójciak-Kosior
Department of Analytical Chemistry, Medical University of Lublin, Chod´zki 4a, 20-093 Lublin, Poland

Sylwia Zielińska
Department of Pharmaceutical Biology, Wroclaw Medical University, Borowska 211, 50-556 Wroclaw, Poland

Anna Bogucka-Kocka
Department of Biology with Genetics, Medical University of Lublin, Chod´zki 4a, 20-093 Lublin, Poland

Index

www.ingramcontent.com/pod-product-compliance
Lightning Source LLC
Chambersburg PA
CBHW070154240326
41458CB00126B/4594